PHASE STABILITY IN METALS AND ALLOYS

McGraw-Hill Series in Materials Science and Engineering

AZÁROFF *Introduction to Solids*
BARRETT AND MASSALSKI *Structure of Metals*
BRICK, GORDON, AND PHILLIPS *Structure and Properties of Alloys*
ELLIOT *Constitution of Binary Alloys, First Supplement*
PAUL AND WARSCHAUER *Solids under Pressure*
SHEWMON *Diffusion in Solids*
WERT AND THOMSON *Physics of Solids*

PHASE STABILITY IN METALS AND ALLOYS

Edited by

PETER S. RUDMAN
JOHN STRINGER
ROBERT I. JAFFEE

Metal Science Group, Battelle Memorial Institute
Columbus Laboratories

BATTELLE INSTITUTE
MATERIALS SCIENCE COLLOQUIA
Geneva and Villars, Switzerland,
March 7–12, 1966
Robert I. Jaffee, Chairman

McGRAW-HILL BOOK COMPANY
New York San Francisco Toronto London Sydney

PHASE STABILITY IN METALS AND ALLOYS

 Library of Congress Catalog Card Number 66-24468

54238

1234567890MP72106987

To PROFESSOR WILLIAM HUME-ROTHERY,

whose work has made the art of metallurgy a science

PARTICIPANTS

P. A. BECK *Department of Metallurgy, University of Illinois, Urbana, Illinois, U.S.A.*

A. L. BEMENT *Reactors Metals Research, Battelle Memorial Institute, Richland, Washington, U.S.A.*

A. P. BLANDIN *Laboratoire de Physiques des Solides associé au CNRS, Faculté des Sciences, Orsay, France*

WALTER BOLLMANN *Battelle Memorial Institute, Geneva-Carouge, Switzerland*

D. G. BRANDON *Battelle Memorial Institute, Geneva-Carouge, Switzerland*

LEO BREWER *Lawrence Radiation Laboratory, University of California, Berkeley, California, U.S.A.*

ROBERT COLLINS *Department of Mathematics, Royal College of Advanced Technology, Salford, U.K.*

H. M. DAVIS *Metallurgy and Ceramics Division, U.S. Army Research Office, Durham, N.C., U.S.A.*

ULRICH DEHLINGER *Institut für Theoretische und Angewante Physik, Techniche Hochschule, Stuttgart, Germany*

JOHN DROWART *Université Libre de Bruxelles, Brussels, Belgium*

POL DUWEZ *Department of Materials Science, California Institute of Technology, Pasadena, California, U.S.A.*

J. E. ENDERBY *Department of Physics, The University, Sheffield, U.K.*

GÉRARD FOURNET *Laboratoire Central des Industries Électriques, Fontenay-aux-Roses, France*

F. C. FRANK *H. H. Wills Physical Laboratory, Bristol University, Bristol, U.K.*

JACQUES FRIEDEL *Service de Physique des Solides, University of Paris, Orsay, France*

VOLKER HEINE *Cavendish Laboratory, University of Cambridge, Cambridge, U.K.*

MICHAEL HOCH *Department of Materials Science, University of Cincinnati, Cincinnati, Ohio, U.S.A.*

WILLIAM HUME-ROTHERY *Department of Metallurgy, University of Oxford, Oxford, U.K.*

R. I. JAFFEE *Metal Science Group, Battelle Memorial Institute, Columbus, Ohio, U.S.A.*

LARRY KAUFMAN *ManLabs, Inc., Cambridge, Massachusetts, U.S.A.*

OSWALD KUBASCHEWSKI *Metallurgy Division, National Physical Laboratory, Teddington, U.K.*

FRITZ LAVES *Institute für Kristallographie und Petrographie, Eidgenössische Technische Hochschule, Zurich, Switzerland*

W. M. LOMER *Solid State Physics Division, Atomic Energy Research Establishment, Harwell, U.K.*

T. B. MASSALSKI, *Mellon Institute, Pittsburgh, Pennsylvania, U.S.A.*

R. B. MCLELLAN *Department of Mechanical Engineering, Rice University, Houston, Texas, U.S.A.*

A. D. MCQUILLAN *Department of Physical Metallurgy, University of Birmingham, Birmingham, U.K.*

J. L. MEIJERING *Inorganic Chemistry Laboratory, Technical University, Delft, Netherlands*

M. V. NEVITT *Metallurgy Division, Argonne National Laboratory, Argonne, Illinois, U.S.A.*

HANS NOWOTNY *Institut für Physikalische Chemie, Vienna University, Vienna, Austria*

R. A. ORIANI *Fundamental Research Laboratory, U.S. Steel Research Center, Monroeville, Pennsylvania, U.S.A.*

W. B. PEARSON *National Research Council, Ottawa, Ontario, Canada*

A. J. PERRY *Battelle Memorial Institute, Geneva-Carouge, Switzerland*

R. M. PICK *Centre d'Études Nucleaires, Saclay, France*

P. S. RUDMAN *Metal Science Group, Battelle Memorial Institute, Columbus, Ohio, U.S.A.*

E. I. SALKOVITZ *Department of Metallurgy, University of Pittsburgh, Pittsburgh, Pennsylvania, U.S.A.*

ALFRED SEEGER *Max Planck Institut für Metallforschung, Stuttgart, West Germany*

JOHN STRINGER *Metal Science Group, Battelle Memorial Institute, Columbus, Ohio, U.S.A.*

W. E. WALLACE *Department of Chemistry, University of Pittsburgh, Pittsburgh, Pennsylvania, U.S.A.*

CLARENCE ZENER *College of Science, Texas A and M University, College Station, Texas, U.S.A.*

PREFACE

Herein are the proceedings of the First Battelle Materials Science Colloquium, "Phase Stability in Metals and Alloys," held in Geneva and Villars, Switzerland, March 7 to 12, 1966.

The problem considered was why under a given set of conditions—temperature, pressure, composition—a given structure or phase occurs. By design, a variety of viewpoints were presented: electron-band theory, chemical bond theory, lattice dynamics, and the geometry of space filling. The participants represented a wide range of disciplines: metallurgists, physicists, chemists, crystallographers, thermodynamicists and mathematicians.

The first day's meeting was devoted to introductory reviews by five distinguished scientists who have contributed importantly for many years and have greatly influenced thought in this field.

Four subsequent days were devoted consecutively to:

1. Pure metals and substitutional solutions
2. Intermetallic phases
3. Interstitial phases and solutions
4. Liquids and defect solids

In general, in a morning session of three hours, six formal papers were presented. As time allowed, there was some limited discussion following such presentations, and in these proceedings these discussions are appended to the respective papers. After an afternoon of relaxation the participants reconvened for a three-hour Agenda Discussion on the subject of the day. The Agenda Discussions were tape-recorded, and a more-or-less literal transcription was made by the Agenda Discussion Secretaries and presented to the Agenda Discussion Chairman for that day's topic. These transcriptions the respective Agenda Discussion Chairmen have condensed, rearranged, organized, added commentary to, and interpreted, each at his own discretion. These so-edited Agenda Discussions are included in these proceedings following the formal papers of the topic of the day. A final three-hour Summary Agenda Discussion handled in the same manner is the final paper in these proceedings.

At the end of the meeting, an informal dinner was held at which Professor Hume-Rothery was invited to speak about his professional career. This autobiographical sketch proved to be such an interesting summary of the development of the subject of the colloquium that it is presented on page xi.

The Battelle Materials Science Colloquia, of which this was the first, was made possible through the support of Battelle Institute and the continued encouragement of its president, Dr. B. D. Thomas and its director of research in the physical sciences, Dr. F. J. Milford.

The Battelle Geneva members of the organizing committee were Dr. Walter Bollmann and Dr. David G. Brandon. Dr. Hugo Thiemann, director of the Battelle Geneva Laboratories generously provided all needed services. The assistance of our charming and efficient secretaries, Misses Monika Freund, Brigitte Straub, and Lily van Notten is gratefully acknowledged. Finally, we wish to thank Mr. R. Bosson, our able projectionist.

P. S. Rudman
John Stringer
Robert I. Jaffee

AUTOBIOGRAPHICAL SKETCH OF WILLIAM HUME-ROTHERY

I was trained as a chemist, and then studied metallurgy under Sir Harold Carpenter at the Royal School of Mines with the object of introducing the subject at Oxford. As a chemist, I first looked for valency relations in alloys, and began work on very electropositive metals such as calcium, magnesium, and sodium in the hope of finding well-defined valency compounds. It was somewhere about 1924 that I realized the importance of the fact that intermetallic compounds were conductors of electricity. This suggested that all the electrons were not bound into stable groups and, hence, that the ordinary valency relations would not be followed.

In seeking for a solution to the problem, I was attracted by the electron-lattice theory of Lindemann which was in vogue at the time. This led to the concept of electron-atom ratios, and it was in this way that the idea of electron concentration as a controlling factor came into being. The electron-lattice theory is now admitted to be an incorrect approach, and it remains an example of the way in which a theory may be of value, even though it turns out to be quite wrong.

I was able to show that the β-brass phases corresponded to a constant electron-atom ratio of 1.5. This was the first step toward a theory of alloy structures, and for this I was responsible. The idea was then taken up with enthusiasm by Westgren and Bradley; and it is they who are entitled to the credit for its application to the γ-brass phases. Others then took up the work, and in this way a theory of copper, silver, and gold alloys was gradually developed.

Meanwhile I had returned to Oxford where I carried out research on a very small scale with one or two students. We began work on alloys of copper and silver with zinc, gallium, indium, germanium, and tin. Most people thought us quite mad, working with outlandish metals like gallium and germanium, but it was this work which led to the discovery of the atomic size factor principle, and of the existence of a constant electron concentration at the α-solid solution boundaries in copper and silver alloys.

At this time I was fortunate in having a few outstanding research workers from Oxford. First of these was Dr. P. W. Reynolds who introduced the electrochemical factor to the β and β' phases in copper, silver, and gold alloys. He wanted to continue metallurgical work, but the metallurgical industry would not pay him enough, and when I last heard of him he was Director of Research of the ICI Agricultural Division at Billingham—which shows what metallurgy loses when it doesn't pay enough.

Another was Geoffrey Raynor who added greatly to both theory and experiment and is now Professor at Birmingham University.

The Second World War then intervened, and we did research work on various problems that were sent us. We did our best to help, but I cannot say that we were really thrilled by the work. After the war, we returned to constitutional research and went on to iron alloys and alloys of transition metals, and were pleasantly surprised to find that electron-concentration principles turned up there, some of which have been described at this colloquium.

Metallurgy at Oxford became an independent subject in 1956, and we moved into a new laboratory in 1957—all very nice, but all a size too small.

You will see, therefore, that in one way and another I have had an interesting and a happy life. I owe a great debt to many friends and colleagues, and in honoring me tonight you are paying tribute to many others who have helped the work along.

William Hume-Rothery

Villars, Switzerland
March 12, 1966

CONTENTS

Part Three INTERMETALLIC PHASES

Part Four INTERSTITIAL PHASES AND SOLUTIONS

Part Five LIQUIDS AND DEFECT SOLIDS

Part One

INTRODUCTORY LECTURES

FACTORS AFFECTING THE STABILITY OF METALLIC PHASES

William Hume-Rothery

Department of Metallurgy, University of Oxford, Oxford, England

THE METALLIC ATOMS

In order to discuss the stabilities of metallic structures we may divide the elements into the four classes shown in the periodic table (Fig. 1).

Class I. Elements whose outer electrons are in *s* or *p* states and which favor *sp* hybrid bonding. In the alkaline earths the hybridization is of the *sd* type, but the *s* function is predominant. In the B subgroups the $(ns)^2$ grouping tends to become an inert pair, and *p* bonding may result.

Class II. The transition metals in which *sd* or *spd* bonding is favored, the proportion of the *d* function being a maximum somewhere in the middle of each transition series.

Class III. The lanthanide and actinide elements in which the 4*f* and 5*f* subgroups are filling up. At the beginning of each series, the *f* electron clouds may extend far enough to take part in bonding, but on proceeding along the series, the *f* electron clouds contract in size and fall in energy,

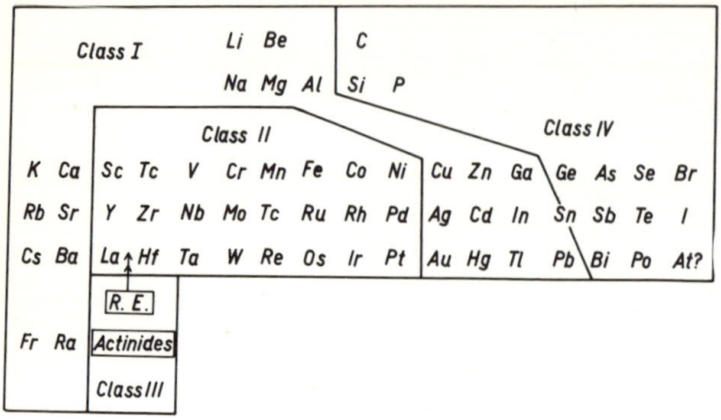

Fig. 1 The periodic table of the elements.

and no longer affect bonding, except that they may introduce magnetic energy terms.

Class IV. Elements which tend to complete their octets of electrons by covalent or ionic bonding.

These are the four main types of element, although there is some overlapping from one class to another—for example Sn belongs to both class I and class IV. For many purposes of alloy chemistry, Cu, Ag, and Au are properly regarded as univalent metals of class I although in the crystals of these elements there is additional cohesion which may be the result of a van der Waals attraction between the outermost electrons of the ions.

THE CONDENSED PHASES OF THE ELEMENTS

We may now consider the stabilities of the condensed phases which result when the atoms of a given element unite together. The boiling point represents the maximum temperature at which a condensed phase is stable at atmospheric pressure. We may use the term *solid ratio* to denote the ratio of the melting point to the boiling point in degrees Kelvin; it represents the fraction of the total condensed range occupied by the solid.

Figure 2 shows the solid and liquid ranges for the elements of classes I and IV. At the beginning of each period, the stabilities of the condensed phases increase on passing through groups IA → IIA → IIIA → IVA, and through groups IIB → IIIB → IVB. But the solid ratios vary greatly. For the alkalies the solid ratio is about 1:3, and in the alkaline earths it increases to between 1:2 and 2:3. On passing to Al, it decreases to about 1:3, and then increases on passing to Si. This curious alternation in the values of the solid ratios is also found in the B subgroups on passing through Zn → Ga → Ge, and from Cd → In, but is not continued on passing to Sn, whose low melting point is not accompanied by a low boiling point.

The very low solid ratios for Ga and In are remarkable and are not understood.

In Fig. 2 the behavior of *B* and *C* is remarkable in that, although the condensed range is exceedingly high, the liquid phase almost vanishes at atmospheric pressure. The figure shows that, taken as a whole, the *s* or *p* or *sp* bonding of the class I and IV elements favors the liquid state as much as or slightly more than the solid state. Except for group II, and for *B* and *C*, there are few elements in Fig. 1 where solid ratios are greater than 1:2.

Figure 3 shows the corresponding solid ratios for the class II transition metals and the adjacent alkaline earths. The boiling points of the transition metals are much higher than those of the class I elements. The solid ratios are much more regular, and are usually greater than 1:2, suggesting that *sd* or *spd* bonding favors the solid state as much as the liquid. Figure 3 emphasizes what is now generally recognized, namely that there is a difference between the sequence of properties in the first and in the later transition series.

BONDING PROCESSES

We may now ask what conclusions we can draw from the above facts about the nature of the bonding forces in our solid and liquid metals.

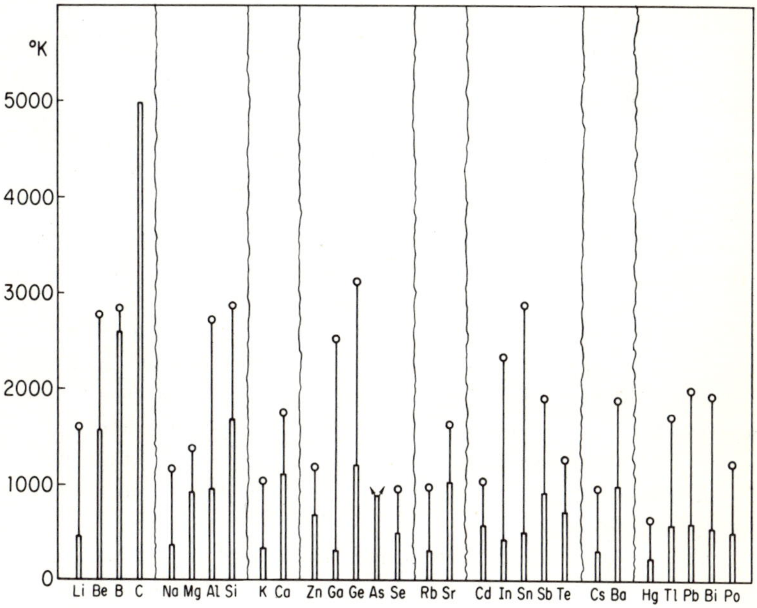

Fig. 2 Solid and liquid ranges for elements of classes I and IV.

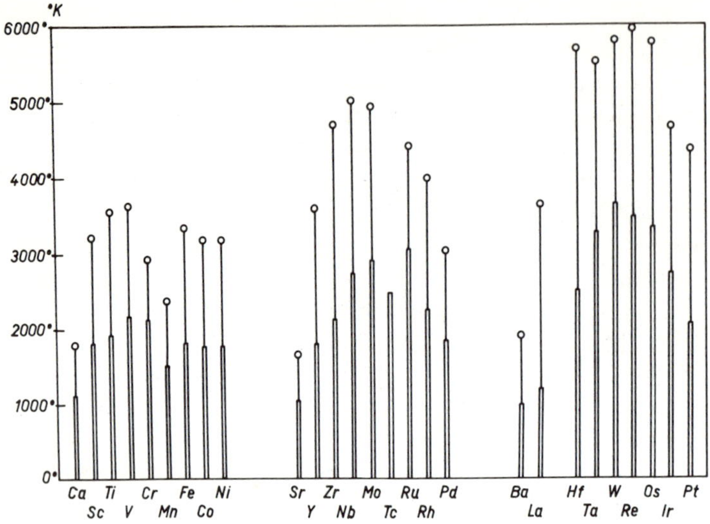

Fig. 3 Solid and liquid ranges for transition metals.

In a general way we shall expect that, on going down the periodic table, the increasing nuclear charge, partly neutralized by screening effects, will increase the attraction between the bonding electrons and the nuclei. This increase will be interrupted abruptly each time the principal quantum number n is increased. The directional characteristics of a given type of bond are also weakened by increasing n. This weakening effect is shown by the general weakening of metallic bonding on passing through Li $\rightarrow$ Na $\rightarrow$ K $\rightarrow$ Rb $\rightarrow$ Cs, and of covalent directional bonding on passing through C $\rightarrow$ Si $\rightarrow$ Ge $\rightarrow$ Sn (gray).

For a given type of bonding, we shall expect the cohesion to increase with the number of bonding electrons per atom. This effect is shown clearly at the beginning of each period, where the increasing number of bonding electrons reinforces the effect of increasing atomic number. In the long periods this increase stops somewhere in the middle of the transition series. Pauling takes this to indicate that electrons begin to enter nonbonding atomic orbitals, whereas the supporters of the collective electron theories say that the repulsive part of the wave function is becoming more marked. Recent work suggests that these two views are not so much in contradiction as might be supposed. For Fe, the calculations of Wood show that the electrons of lowest energy in the d bands have electron clouds that are more diffuse than in the free atoms, and may therefore be more likely to bond, whereas those of higher energy have electron clouds less diffuse than in the free atoms, and therefore are less likely to bond. There is thus a suggestion of a division into diffuse bonding and less diffuse nonbonding electron clouds.

The slight weakening in cohesion on passing through K → Rb → Cs is the result of increasing quantum numbers of the bonding electrons. In contrast to this, there is a continual increase in the strength of cohesion on passing from the elements of groups VA → VIIIB in the first transition series to the corresponding elements in the second and the third transition series. This is in spite of the fact that the interatomic distances increase on passing through V → Nb, Cr → Mo, . . . , and remain almost unaltered on passing to the third long period. The suggestion is clearly that, for a given group, the number of bonding electrons per atom increases on passing from the first to the second to the third transition series; this is in agreement with the predominant valencies in the chemistry of these elements.

It may be said quite definitely that the data of Fig. 3, together with the variation in compressibility and interatomic distances and many facts of alloy chemistry, do not agree with any scheme that allots a constant valency to the elements from group VI to group VIIIC in all three long periods. The details of the Pauling valency schemes are, therefore, almost certainly incorrect, although the idea of atomic nonbonding *d* electrons may well be true.

THE STRUCTURES OF THE METALS

The simple concept of the building up of octets of electrons enables us to understand the $(8 - N)$ or valency structures of the elements of class IV. Electron theory can predict the body-centered cubic structures of the alkali metals, and the preference for cubic close packing shown by Cu. Otherwise we have little understanding of why the different metals assume the crystal structures that are found. As we shall explain later, in the alloys of the elements from Mo → Pd and W → Pt, the composition limits of the phase boundaries of the body-centered cubic, close-packed hexagonal, and face-centered cubic phases suggest strongly that these are related to characteristic electron-atom ratios, but there is no quantitative theory of this.

We can understand why, if an element exists in several modifications of which the body-centered cubic structure is one, this is always stable at the highest temperature. For the relatively loose bcc packing results in high amplitudes of vibration in certain directions, and the resulting contribution to the $-TS$ term of the free energy becomes increasingly important as the temperature rises.

Recent work on the effect of pressure on crystal structure has confirmed the expectation that some bcc metals would become close-packed under high pressure. The A_3 transformation of iron is lowered by pressure.

A striking result of this work has been to show that Si, Ge, and gray Sn transform to the white Sn structure under pressure, and this structure

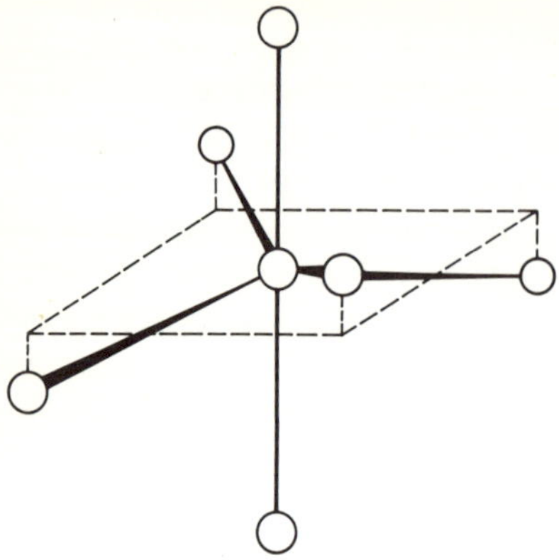

Fig. 4 The structure of white Sn.

also results from the compression of InSb, GaSb, and AlSb. The white Sn structure thus appears to be of fundamental importance and is not understood. It is shown in Fig. 4, and if this arrangement of atoms is stretched vertically and compressed horizontally, it becomes the structure of gray Sn (diamond type) in which the interatomic distances are smaller but the atomic volume is larger than for white Sn. Several types of hybrid *dsp* bonds would give the (4 + 2) coordination, and with four electrons in six orbitals, the metallic properties would be accounted for but there is no real understanding of the structure.

FACTORS AFFECTING THE STABILITY OF PHASES

We may now consider what factors determine whether two kinds of atom unite together to form phases that are stable either thermally or in a range of composition. We may postulate that the effects of the following factors are important.

1. Difference between the electronegativities, $\Delta\chi$, of the two metals. This is the most important factor, and increasing $\Delta\chi$ increases the tendency for two atoms to unite in either liquid or solid phases.

2. A tendency for atoms of elements near the ends of the short periods and B subgroups to complete their octets of electrons, and a similar but less marked tendency to fill the *d* shell in the later transition elements.

3. Size-factor effects, i.e., effects related to the atomic diameters of the elements, which may be defined as the closest distances of approach of the atoms in the crystals of the elements. This crude definition correlates better with the facts than do more sophisticated atomic diameters or atomic volumes.

4. A tendency for definite crystal structures to occur at characteristic numbers of electrons per unit cell which, if all atomic sites are occupied, is equivalent to saying that the structures occur at characteristic electron-atom ratios or electron concentrations.

5. Orbital-type restrictions. Structures whose hybrid bonding orbitals involve a very high proportion of *d* function may not mix with class I or class IV elements. Similarly, atoms giving rise to almost pure *p* bonding may not enter structures involving other types. A high value of $\Delta\chi$ may overcome these restrictions.

SUBSTITUTIONAL TERMINAL SOLID SOLUTIONS

We may now examine how the above factors affect the stabilities of substitutional terminal solid solutions.

The original picture of a substitutional solid solution as a random assembly of atoms on a uniformly expanded or contracted lattice was known to be incorrect as long ago as 1930. It is now generally recognized that the different sizes of the atoms result in local distortions of the lattice, and that short-range order (in the sense of a tendency to avoid like neighbors), or alternatively local clustering, is usually present. These effects have been studied by refined X-ray methods (diffuse and low-angle scattering), and have also been investigated by resistivity and specific-heat techniques. The information gained is usually qualitative rather than quantitative, and there is much need for a more detailed knowledge of the exact atomic displacements in a solid solution.

It is necessary to define terms very carefully. Thus in a 27 a/o Zn α-brass, there is definite short-range order in slowly cooled alloys. There is no clustering of Zn atoms in the crystals themselves, but Zn atoms may be attracted to the dislocations, and it may be said that there is clustering in the dislocated regions, although not in the main mass of the crystals.

Can we now say what will affect the stability—both thermal, and as regards range of composition—of a solid solution? Here it is more easy to say what will render a solid solution unstable, rather than what will increase its stability. The well-known size-factor principle states that if the atomic diameters of two metals differ by more than 14 to 15%, substitutional solid solutions become restricted. It is a negative principle, and we cannot say that a solid solution will be formed when the size factor is favorable. The lattice distortion produced by an increasing size factor favors the less regular structure of the liquid solution, and so the liquidus and solidus curves are depressed by the size-factor effect.

Orbital-type restrictions often make solid solutions unstable. Thus Cu and Cr, and Ag and Mo, are almost immiscible in the solid state, and even give rise to immiscible liquids, in spite of entirely favorable size factors. Here it seems that the bonding in Cr and Mo involves such a high proportion of *d* function that the Cu and Ag atoms cannot give what is required. Ag and Pd are completely miscible in the solid state but, on going back to Rh, the proportion of *d* function becomes too high and immiscibility results. In the same way, Au is immiscible with Ir, but is miscible with Pt at high temperatures, although the equilibrium diagrams show clearly that Au and Pt mix less readily than Ag and Pd. If our interpretation is correct, the reason is that the valency of Pt is greater than that of Pd and, hence, the proportion of *d* function is greater and the Au atom finds it difficult to give what is required.

We also find that elements such as Tl and Pb whose crystal structures suggest incomplete ionization and p bonding are sometimes immiscible in the solid state and even in the liquid state with elements with which we should expect some miscibility.

But these apparently electronic restrictions can be overcome by a sufficiently high electronegativity difference. Thus Be is freely soluble in Fe, although there is here no possibility of a $2d$ state.

The effect of electronegativity difference on the stability of a solid solution is not simple. It has been suggested that a large value of $\Delta\chi$ means a restricted solid solution, but this is not true unless the large electronegativity difference results in the formation of a stable intermediate compound. If such a compound is formed, the usual free-energy curve-tangent principle shows how increasing stability of the compound results in a decrease in the solid solubility limit. But in other cases a wide solid solution may be found with a large value of $\Delta\chi$. In the system Au-Mg, $\Delta\chi$ is as high as 1.3, but as much as 25 a/o Mg dissolves in Au. This is smaller than the value 40 a/o Cd in Ag where $\Delta\chi = 0.2$, but it is still a very large value. We may, in fact, argue that, where no very stable intermediate compounds are formed, increasing $\Delta\chi$ will favor the formation of both liquid and solid solutions, since unlike atoms will tend to attract each other—this concept for liquid solution was used by B. W. Mott.

As regards the thermal stability of a solid solution, Fig. 5a shows the simplest kind of equilibrium diagram which is formed when everything is favorable. Here we may say that the thermal stability of A is raised by solution of B while that of B is lowered by A. We call this the *melting-point effect* or m.p. effect, and provided that A and B are of the same general type (e.g., if both are transition metals, the m.p. effect is always present if A and B have the same crystal structure, even though intermediate phases are formed. If A and B have different structures the position is more complicated, but we may still see the m.p. effect in a general way if A and B are not of the same type.

The solution of one metal in another introduces lattice distortion, and

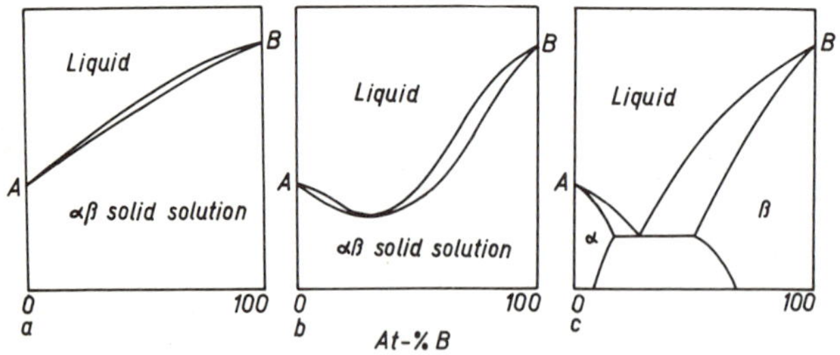

Fig. 5 Simple types of equilibrium diagrams. [*From J. Iron Steel Inst.* **203:**1182 (1965); *reproduced by courtesy of The Iron and Steel Institute.*]

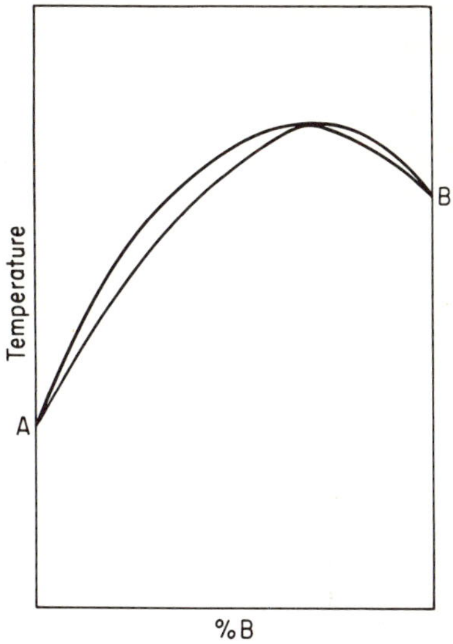

Fig. 6 An uncommon type of equilibrium diagram.

there may be electronic effects due to different valencies, types of bonding orbital, etc. These nearly always act so as to favor the less regular structure of the liquid, and may therefore reinforce or oppose the m.p. effect. With increasing magnitude of these disturbing factors, the diagram of Fig. 5*a* passes into those of Fig. 5*b* and *c*.

But diagrams of the type of Fig. 6 are very seldom found. The melting point of a solid solution is seldom greater than those of both constituent metals, and is usually below the straight line joining the melting points of the constituent metals. But there are occasional striking exceptions. Thus, the Cu-Al liquidus and solidus curves take the form of Fig. 7, and show an astonishingly small depression. This was thought to be due to the electropositive nature of Al, but this explanation is unconvincing because the effect is not found in the system Ag-Al or Au-Al where the electrochemical factor is greater.

The initial depression of liquidus and solidus curves can be expressed more quantitatively provided that the solid and liquid solutions are monatomic. The gradients of the liquidus and solidus curves then permit a calculation of ΔG, the free energy of transfer of solute from liquid to solid at infinite dilution. A positive sign of ΔG means that the liquidus and solidus are depressed, and a negative sign that they are raised.

The liquid stabilizing powers of the different elements are given by the ΔG values, and results for alloys of Fe with transition metals are shown in Fig. 8. In the first transition series Mn behaves anomalously, but for the remaining elements there is a clear minimum in the region of group VII for each transition series. The magnitude of ΔG is clearly determined by (1) the melting-point effect, and (2) the lattice-disturbing effects mentioned above. We may write

$$\Delta G = \text{m.p. effect} + \Delta F$$

where ΔF is the effect of lattice distortion, valency, etc. The m.p. effect can be estimated approximately, and in this way we can obtain ΔF; Fig. 9 shows the results obtained for alloys of Fe with transition metals.

Here the value for Zr is anomalous, but otherwise the ΔF values for the first transition series are systematically smaller than those for the second

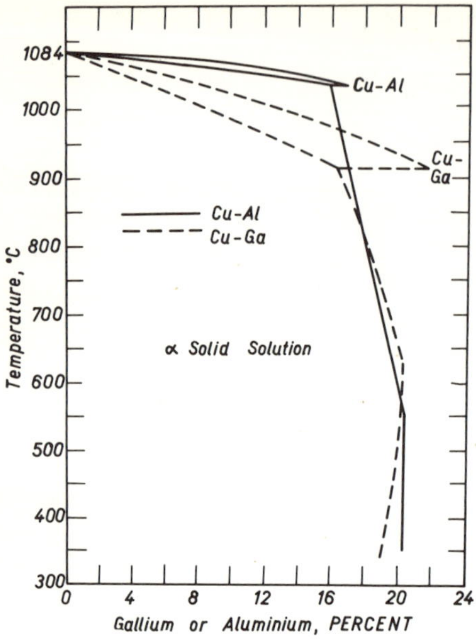

Fig. 7 Liquidus and solidus curves for Cu-rich Cu-Al and Cu-Ga alloys. (*From W. Hume-Rothery and G. V. Raynor, "The Structure of Metals and Alloys," The Institute of Metals, London,* 1962; *reproduced by courtesy of The Institute of Metals.*)

and third, as would be expected from the atomic diameters and electronic characteristics. A clear relation with the position in the periodic table is shown. In this way we separate the m.p. effect from the other disturbing factors and begin to understand the relative stabilities of liquid and solid phases. There is a clear group-number effect depending on the relative position of solvent and solute in the periodic table; a clear size-factor effect related to the atomic diameter; and a clear m.p. effect.

Similar effects are found for alloys of Cu with Zn, Ga . . ., and of Ag with Cd, In. . . . We may draw a line halfway between the liquidus and solidus, and call this the $\bar{x}$ line. If the equilibrium diagrams are drawn in terms of electron concentration, the $\bar{x}$ lines are roughly superposed. The departure from exact superposition is related to the lattice distortion at constant electron concentration, in the sense that increasing lattice distortion lowers the $\bar{x}$ line. It is the lattice distortion at constant electron concentration which is important—not that at constant atomic percentage of solute. The electronic effect was examined theoretically by Jones as long ago as 1937, but there has been an unfortunate failure to develop the subject further, although the qualitative principles are clear.

Apart from the size factor and the existence of a stable intermetallic compound, what other factors will limit the extent of a terminal solid solution? For the elements of class I, terminal solid solutions are more restricted as the valencies of solvent and solute are more different. Only for Cu alloys is there any real theory on these effects. Here H. Jones in 1937 gave calculations that predicted the markedly distorted Fermi surface in Cu, which was later established experimentally. He showed that the terminal solid solution of Zn, Ga, etc., in Cu would extend to an electron concentration of about 1.4, in agreement with the empirical rules advanced by the Oxford School.

It is well known that in Au alloys the α-solid solubility limits tend to occur at lower electron concentrations than in Cu and Ag alloys. This has led to speculations about valencies greater than 1 for Au, and to suggestions such as those of Cohen and Heine of the effect of solutes on the form of the Fermi surface. We have recently surveyed all the data, and find that there is a simple correlation between $\Delta\chi$, the electronegativity difference between the two metals, and the electron concentration at the α-solid solubility limit. High values of $\Delta\chi$ correlate with low values of the solubility limit in terms of electron concentration, and low values of $\Delta\chi$ with higher solubility limits. So far as generalization of the facts is concerned, there is no reason to look further than an electronegativity effect.

Recent work has shown that electron-concentration principles apply also to the solid solutions in the later face-centered cubic transition metals. Figure 10 shows the limits of solid solutions in Rh, Pd, Ir, and Pt in terms

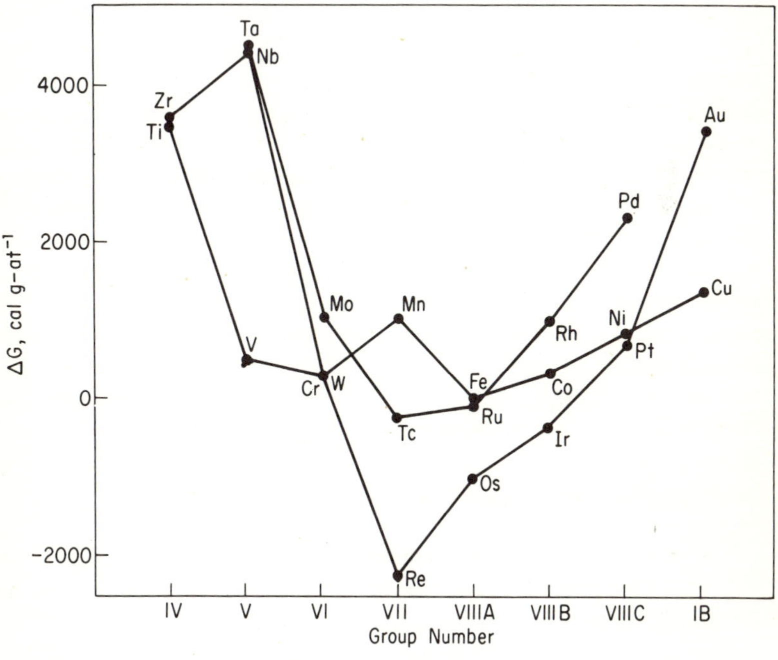

Fig. 8 Solute free-energy differences ΔG for the liquid-to-δ transformation of alloys of Fe with transition metals. [*From J. Iron Steel Inst.* **203:** 1186 (1965); *reproduced by courtesy of The Iron and Steel Institute.*]

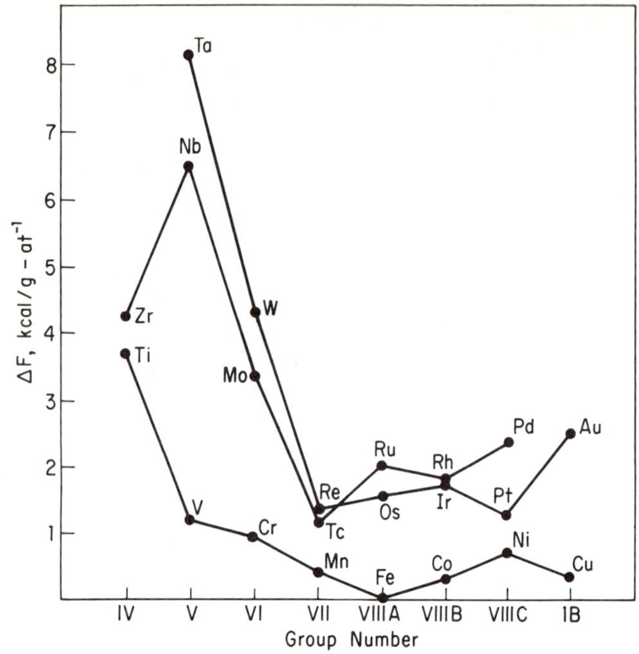

Fig. 9 ΔF values for the liquid-to-δ transformation in alloys of iron with transition metals. [*From J. Iron Steel Inst.* **203:** 1186 (1965); *reproduced by courtesy of The Iron and Steel Institute.*]

of average group number (AGN), which is the same as the electron concentration if all the electrons outside the rare-gas shell are counted. There are some exceptions, but the general tendency for the fcc solid solution to extend back to an AGN value of ~8.4 is very clear. A similar effect is found for solid solutions of V and Cr in fcc γ-Fe, Co, and Ni. The solid solutions of Rh and Ru in bcc Mo extend up to almost the same AGN value of 6.6.

In all this work there are the clearest signs of empirical principles, but the theoretical workers have not met the challenge.

INTERMEDIATE PHASES

We may now ask how our general stability factors will affect the intermediate phases in alloy systems, and we must again distinguish between thermal stability (i.e., melting point) and composition stability or range of composition over which a phase persists.

The effect of electronegativity difference is often very striking. Figure

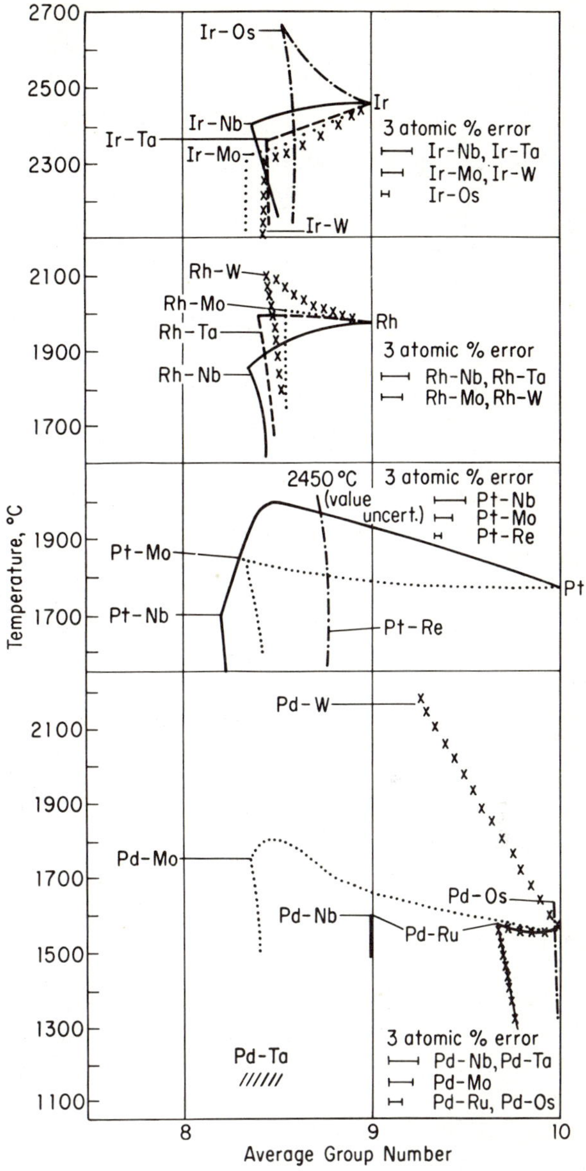

Fig. 10 Limits of terminal solid solutions in Rh, Pd, Ir, and Pt. The diagram is drawn in terms of average group numbers (AGN values).

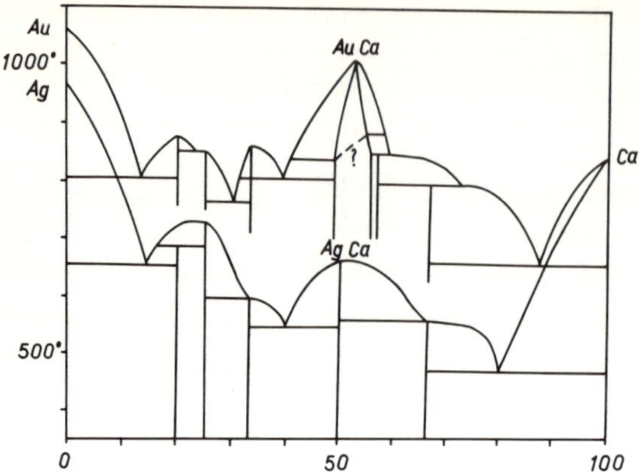

Fig. 11 The equilibrium diagrams of the systems Au-Ca and Ag-Ca.

11 compares the equilibrium diagrams of the systems Au-Ca ($\Delta\chi = 1.4$) and Ag-Ca ($\Delta\chi = 0.9$). Here, no fewer than five intermediate phases have the same compositions, and the uniformly higher melting points of the Au compounds are clear. Since the atomic diameters of Ag and Au are about the same, the differences in Fig. 11 are simple electronegativity effects.

We can obtain even more striking effects if an electronegativity difference is combined with some other factor such as the octet effect. Thus, in the system Mg-Sb, the melting point of the compound Mg_3Sb_2 is 1228°C and is much higher than that of either constituent. In this way it is quite easy to obtain phases whose melting points lie above the straight line joining those of the component metals, and, with atoms of comparable size, it is usually only by invoking the electronegativity effect that we can do this. In general, those electronegativity compounds do not show much variation in composition.

The Mg_3Sb_2 phase involves a combination of the electronegativity and the octet effect. Octets may be completed by electron transference (co-valency), or electron sharing, and we naturally expect electron transference to be favored by a high value of $\Delta\chi$. Pearson and his co-authors have shown how to express this quantitatively. In MgTe, the valency electron quantum numbers of the outer electrons are Mg(3) and Te(5), and the average quantum number, denoted by $\bar{n}$, is 4. Mooser and Pearson argue that the larger the value of $\bar{n}$, the weaker will be the directional character of covalent bonds, whereas increasing value of $\Delta\chi$ will favor ionic bonding. For the equiatomic *AB* compounds they then draw a diagram (Fig. 12) in which the coordinates are $\bar{n}$ and $\Delta\chi \times r_A/r_B$, where r_A and r_B are the radii of the anion and cation elements, respectively. In this way a remarkable generalization is found, and all the compounds with tetrahedral structures lie below the curve, and those with NaCl-type structures lie above. There is, as yet, no mathematical interpretation or theory of this.

The tetrahedral compounds are the simplest type of covalent compound in which an octet is completed by electron sharing. Pearson has shown that, in some cases, the stability of a phase can be regarded as due to the building up of an octet by the formation of a sublattice of one kind of atom within the crystal pattern. Thus, in CaSi, each Ca atom gives two electrons to the Si atom, so that the latter possesses six electrons, and then octets are completed by the formation of chains of Si atoms within the structure, each Si atom having two Si neighbors. In contrast to this, in $CaSi_2$, each Si atom can receive only one electron from the Ca atom and thus has five electrons. The structure is such that there are double layers of Si atoms, in which each has three Si neighbors, and the octets are completed by electron sharing.

If one kind of atom is sufficiently small for interstitial compounds to be formed, the electronegativity factor can be combined with other factors to give spectacular results. We can leave the metallic atoms as a normal metallic lattice with additional bonding to the interstitial atoms. In this way in compounds such as TiN and TiC, or WC or MoC we can obtain melting points as high as 3000 to 3500°C. The very stable phases of this kind are confined to the transition-metal compounds. The reason for this is probably that *dsp* hybrid bonds are much stronger than *sp* hybrids and that appropriate *dsp* hybrids exist. Thus, in TiC, the structure is of the NaCl type with each atom surrounded by six of the opposite kind. Here we may regard the very strong d^2sp^3 octahedral hybrids as bonding the Ti

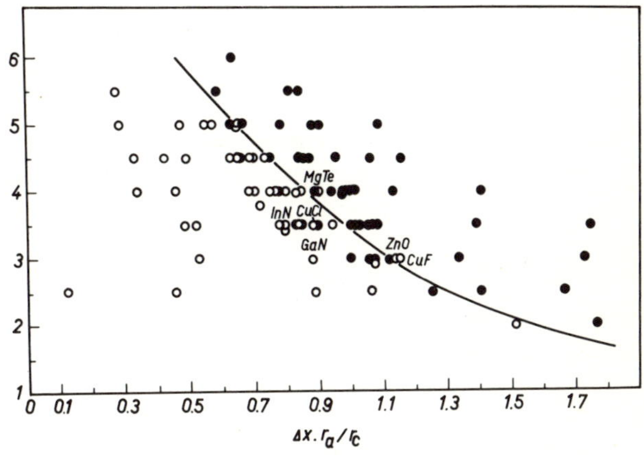

Fig. 12 The Mooser-Pearson diagram separating *AB* compounds into covalent and ionic types. [*From W. B. Pearson, J. Phys. Chem. Solids* **23**:103 (1962).]

atoms to its six C neighbors, which can themselves give rise to octahedral bonding by using the $sp^2(p)$ bonds suggested by Rundle, so that the C atoms complete their octets by means of ⅔ bonds. In addition, the electropositive nature of Ti relative to C or N means a further electronegativity effect.

Now what will happen if we deliberately reduce our electronegativity effect? For Cu, Ag, and Au alloys it seems that when $\Delta\chi < \sim 0.3$, typical electron-concentration equilibrium diagrams are formed provided that the size factors are favorable. In such cases the phases occur at their characteristic electron-atom ratios, and to a rough approximation the composition ranges are simply related to the electron-atom ratios, with lattice distortion at constant electron concentration as an additional complication.

The electron compounds of Cu, Ag, and Au are not of great thermal stability unless the electronegativity factor is superimposed. This may be seen by comparing the equilibrium diagrams of the systems (Fig. 13) Ag-Cd ($\Delta\chi = 0.2$), Au-Cd ($\Delta\chi = 0.7$), Ag-Mg ($\Delta\chi = 0.7$), and Au-Mg ($\Delta\chi = 1.2$) where increasing $\Delta\chi$ results in β phases with long-range order and increasing thermal stability. The equiatomic bcc or CsCl structure is clearly very attractive to a system with a high value of $\Delta\chi$, because it gives each atom eight neighbors of the opposite kind. So here, with a divalent solute, the 3:2 electron-atom ratio reinforces the electronegativity factor and we can understand the effect.

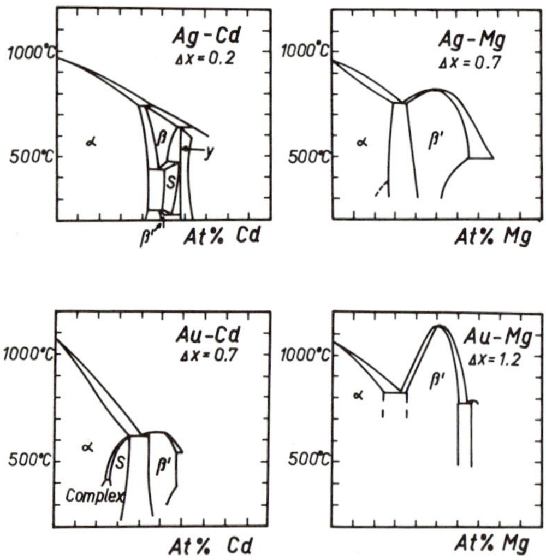

Fig. 13 Parts of the equilibrium diagrams of the systems Ag-Cd, Ag-Mg, Au-Cd, and Au-Mg. The differences between the electronegativities of the two metals are indicated.

Now what will happen if we go to a transition metal such as Ni? Here we find that in some cases, Ni acts as a zero-valent element, which implies that the *d*-shell effect is concerned. In such cases, it is with a trivalent solute that the equiatomic ratio corresponds to an electron concentration of 1.5. We can, thus, understand the great stability of the NiAl and CoAl ordered body-centered cubic structures (Fig. 14), since in these cases we have the cell electron-concentration effect and the *d*-shell effect working together, and supported by a slight electronegativity difference $\Delta\chi = 0.3$. If we pass to FeAl, we keep $\Delta\chi$ almost constant, but greatly reduce the *d*-shell effect and, although an FeAl superlattice exists, it is much less stable than NiAl or CoAl.

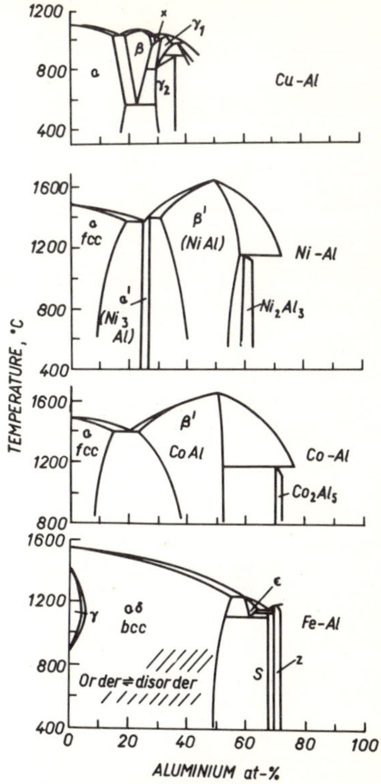

Fig. 14 Equilibrium diagrams of the systems Cu-Al, Ni-Al, Co-Al, and Fe-Al. (*From W. Hume-Rothery, "The Structures of Alloys of Iron," Pergamon, Oxford,* 1966; *reproduced by courtesy of Pergamon Press.*)

Electron compounds were first discovered for alloys of the B subgroup elements with Cu, Ag, and Au, and with the elements immediately preceding them in the periodic table (e.g., Pd, Co, Fe). Until recently they were thought to be confined to these alloys, but modern work has shown that analogous phases exist in binary transition-metal alloys. In the second and third long periods we have the sequence of crystal structures

bcc (groups VA, VIA) : cph (groups VII, VIIIA) : fcc (groups VIIIA, VIIIB)

Here it has been shown that cph phases can be formed by the alloying of a group VIA element with one of group VIIIB or VIIIC. Since the valencies are in dispute, it is better to describe these effects in terms of average group number, or AGN values, so that an equiatomic alloy, for example, Mo and Pd, has an AGN value of $(6 + 10)/2 = 8$. Figure 15 shows the phase boundaries of a number of these close-packed hexagonal phases, both terminal solid solutions in Ru, Re, and Os, and of intermediate phases. The tendency of the phase boundaries to lie at roughly constant AGN

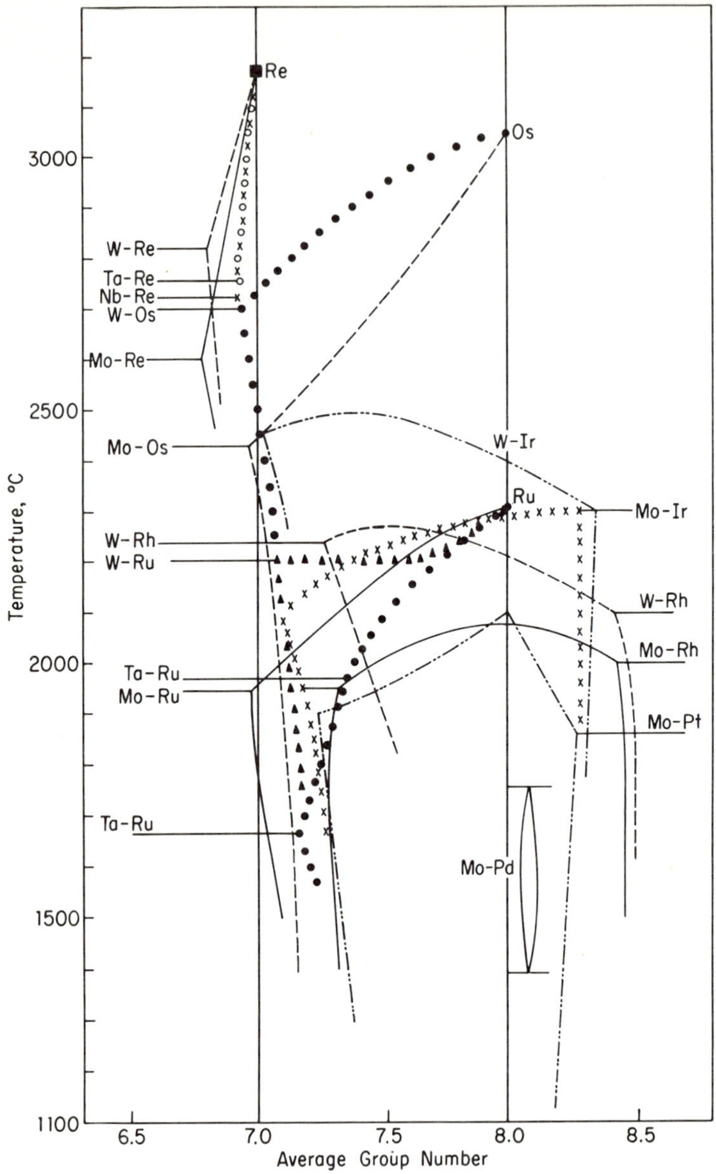

Fig. 15 Phase boundaries of some close-packed hexagonal phases in transition-metal alloys. [*From Acta Met.* **14:** 560 (1966); *reproduced by courtesy of Pergamon Press.*]

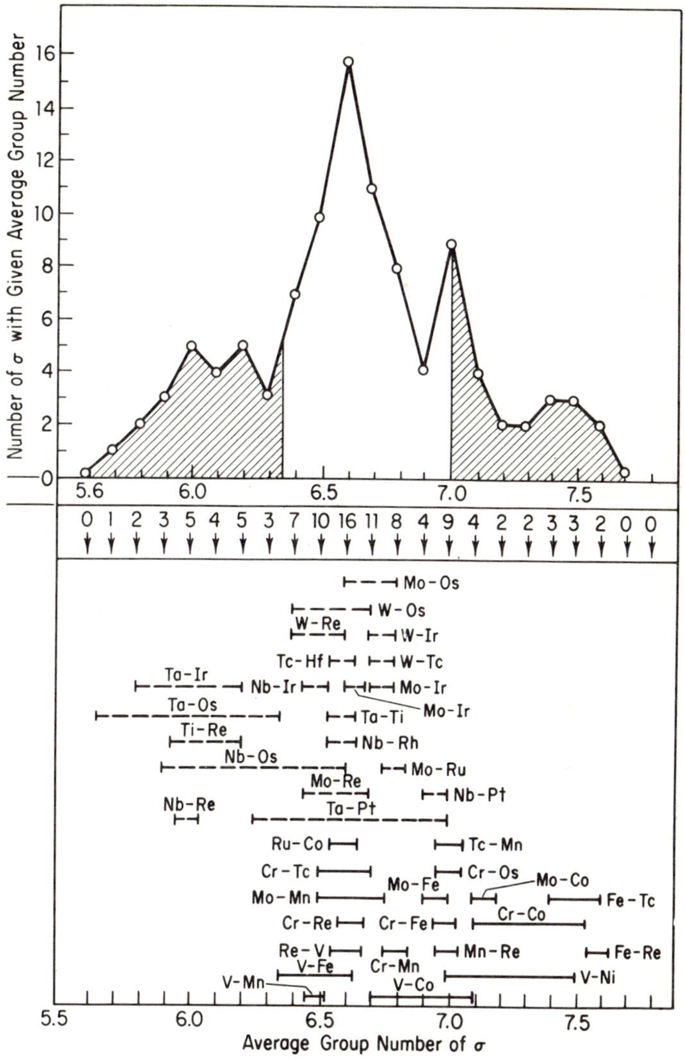

Fig. 16 Composition ranges of σ phases in terms of average group number. *(From W. Hume-Rothery, "The Structures of Alloys of Iron," Pergamon, Oxford,* 1966; *reproduced by courtesy of Institution of Metallurgists.)*

values is very striking, and these effects are quite similar to those found for Cu, Ag, and Au alloys.

Another example of a transition-metal phase whose composition is related to the AGN value is that of the well-known σ phases. Figure 16 shows the distribution of these phases as a function of the AGN value, and the tendency of them to occur predominantly in the range AGN = 6.4 to 6.8 is very striking.

The Laves phases present a particularly interesting and difficult problem. They were originally thought to be size-factor compounds with radius ratios near to that of the ideal value 1.225. Increasing knowledge has shown that they may be formed with radius ratios less than 1.05 and greater than 1.4, and several authors have suggested that it is no longer justifiable to regard them as size-factor compounds. In some cases the electron-cloud patterns of the atoms may be sufficiently deformable for the atoms to be regarded as being expanded or compressed so as to fit a structure based on the ideal framework. In other cases changes of valency may be assumed to account for changes in size, but it is sometimes difficult to see what is happening. It has long been known that in some cases the Laves phases are connected with electron concentration, and recently Beck has claimed that with transition metals Laves phases are not found with AGN values of 8 or more—there is a Co_2Ta (AGN $7\frac{2}{3}$) but no Co_2W (AGN 8). It has also been shown that Rh and Pd in the second transition series are less willing to form Laves phases than are Co and Ni, or Ir and Pt, in the first and third series. This agrees with the incompressible nature of the *d* electron clouds at the end of the second transition series.

With atoms of equal size, a close-packed arrangement can give a coordination number of not greater than 12. With atoms of unequal size, higher coordination numbers are possible. The very beautiful geometrical analysis of Frank and his collaborators has shown how the Laves phases, and other complex structures such as the σ phases, result from different ways of linking together coordination polyhedra of high coordination number. In this way we can understand the geometry of the structures, and we can say that they result in space being filled compactly. We may perhaps argue that, when the bonding is nondirectional, the atoms will tend to assemble together closely, and that these different complex structures simply represent the best possible ways of doing so. But it seems probable that there is some additional and not yet understood factor which determines the particular arrangement chosen. In the case of the σ phases, the connection with AGN or electron concentration is very clear. It seems that the packing cannot be achieved unless some electronic condition is satisfied which is not yet understood. In the case of the Laves phases there is no simple relation with electron concentration—Na_2K, Mg_2Ba, and Al_2La with electron concentrations of 1, 2, and 3, respectively, all have the Cu_2Mg structure. But there are clear suggestions that electron concentration is sometimes concerned. The position of the Laves phases is less clear than it appeared to be 20 years ago, and it is to be hoped that it will be more clear in 10 years time.

The stability of metallic phases is a fascinating and exciting problem. Experiment and empirical generalizations are ahead of quantitative theory, and we may perhaps hope that this conference will stimulate the theoretical physicists.

INFLUENCE OF ENTROPY IN PHASE STABILIZATION

Clarence Zener

College of Science, Texas A and M University,
College Station, Texas

INTRODUCTION

The stability of a phase is determined by two factors: the binding energy at absolute zero, and the entropy of the system, i.e., the uncertainties in the specification of the parameters of the system. Thus in Fig. 1, the α phase is stable at low temperatures because it has the higher binding energy, whereas the β phase becomes stable at high temperatures because it has the higher entropy. In my talk I shall consider only the second factor, namely entropy, particularly the influence of entropy in causing a change in phase with a rise in temperature.

Two major types of uncertainty contribute to the entropy of a phase: that coming from an uncertainty in the specification of the atomic parameters, and that coming from the uncertainty in the specification of the electronic parameters.

PHASE CHANGES INDUCED BY ATOMIC ENTROPY

Except at low temperatures the velocity distribution of the atoms is given by the Boltzmann formula, and is hence the same for all phases. The uncertainty in the precise atomic velocities thus does not contribute to the relative stability of the various possible phases.

The uncertainty in the position of the individual atoms may be resolved into two components. The one is the uncertainty in the equilibrium position of the atoms; the other is the uncertainty in the position of the atoms with respect to the equilibrium position about which they vibrate.

The uncertainties in the equilibrium positions of the individual atoms give rise to what is called order-disorder phase transformations. These are perhaps the most familiar type of phase transformations. The prototype is body-centered cubic β-brass (50 a/o Cu, 50 a/o Zn). The binding energy is greatest when each Cu atom is surrounded by only Zn atoms, and when each Zn atom is surrounded only by Cu atoms. The low-temperature phase is therefore completely ordered. At higher temperatures the entropy associated with disorder generally introduces disorder. As Bragg and Williams[1] first pointed out, as the temperature is gradually raised, disordering increases at a continually increasing rate, until finally a critical temperature is reached at which disorder occurs cataclysmically. This cataclysmic behavior is of course due to the fact that the ordering force is itself reduced by disordering. We have here the prototype of what is now commonly called a "cooperative phenomenon," a type of phenomenon that is now recognized as occurring in all branches of physics.

The original field-theory model of β-brass introduced by Bragg and Williams considered only long-range order. The specific heat of such a model became very high at the critical temperature and returned to its normal value just above this temperature. Experiments, however, indicated a peak specific heat at the critical temperature lower than that for only long-range order, and an anomalously high specific heat for some

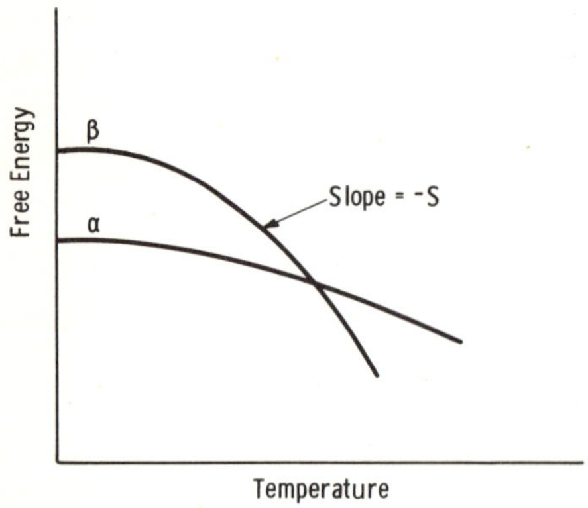

Fig. 1 Illustration of role of entropy in inducing phase transformations.

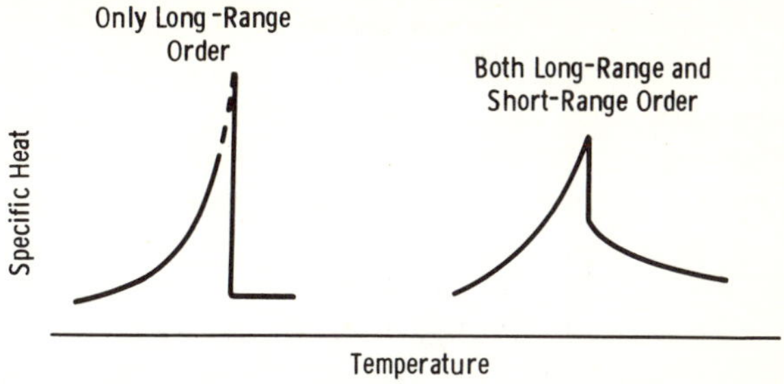

Fig. 2 Experimental manifestation of short-range order.

distance above the critical temperatures (Fig. 2). Attention was then focused by Bethe[2] upon a refinement of the original theory. As a consequence we now have a well-known theory of short-range order, again applicable to many areas in physics.

By short-range order one usually implies an isotropic short-range order, e.g., the probability that a given Cu atom will be surrounded by so many Zn atoms as nearest neighbors. Short-range order can also be anisotropic. Thus we may ask what is the preference that a Zn atom in brass has another Zn atom as nearest neighbor along a particular (111) axis. Anisotropy of ordering occurs when a tensile stress is applied along a particular crystallographic axis.[3,4] Whereas the ordering induced by a practical stress may be itself very small, such ordering may be readily detected by measurements of internal friction. Thus if the period of vibration of the stress is very long compared to the time of relaxation for ordering, stress and strain will be essentially in phase with each other. The same will be true if the period of vibration P is very short compared to the relaxation time t_1. Stress and strain will no longer be in phase when the period of vibration is comparable to the time of relaxation, as indicated in Fig. 3. A typical relaxation peak associated with anisotropic ordering is reproduced in Fig. 4. A study of these peaks has in fact developed into a useful metallurgical tool.[5] Thus a study of the shift in optimum P with temperature gives the dependence of the diffusion coefficient upon temperature. A study of the dependence of the height of the peak with temperature leads to a prediction that at a certain low temperature the lattice would become spontaneously ordered in an anisotropic manner, at least if an isotropic ordering did not take place first.

The most important ordering force in alloys resides in the size difference of the inner cores of the two constituent atoms. Since inner cores always strongly repel one another, the atoms will want to arrange themselves so that large core atoms have as nearest neighbors only small core atoms. Such an ideal arrangement may be attained in an alloy such as the bcc β-brass previously discussed. The Cu atoms are located on one of the

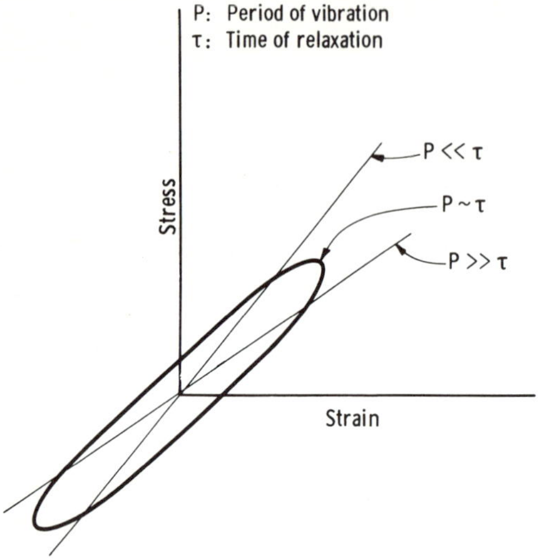

Fig. 3 Illustration of influence of an internal relaxation.

two interpenetrating simple cubic lattices, the Zn atoms on the other. In a 50-50 composition with an fcc lattice, such as CuAu, such an ideal arrangement cannot be attained. Thus the nearest neighbors of any one atom are nearest neighbors of one another. That arrangement in which the large-core Au atoms have the fewest number of nearest neighbors is an alternate stacking of (001) planes of Au atoms with (001) planes of Cu atoms. In such an arrangement each Au atom has only four Au atoms as nearest neighbors, but eight Cu atoms. The absence of Au-Cu core repulsions between adjacent (001) planes results in a contraction along the z axis. The resulting tetragonal structure has a c/a ratio less than unity. The observed entropy difference between the ordered fct and the disordered fcc AuCu is in complete agreement[6] with the theoretical value of $R \ln 2$. A list of other examples analogous to CuAu has been given by Hume-Rothery.[7]

We have now seen how an ordering of a face-centered lattice with a 50-50 composition results in a transformation from an fcc to fct lattice as the temperature is lowered. But if each crystal in a polycrystalline aggregate were to change from cubic to tetragonal symmetry, large-scale strain would ensue. Nature solves this transformation problem in an ingenious way. A cubic to a tetragonal transformation may be affected by the superposition of two shears, $(011)\langle 01\bar{1}\rangle$ and $(101)\langle 10\bar{1}\rangle$. The first shear may be accomplished without an overall change in dimensions of a crystal by means of successive layers of twins, as indicated in Fig. 5. The elastic stresses are confined to the immediate vicinity of the grain boundaries. The second shear takes place within each twin plate, each twin now transforming into a myriad of still finer twins. The overall result is that each originally cubic crystal now transforms into a double network of tetragonal twins. Such twin networks have been observed in many 50-50 alloys, including Cu-Au,[8,9] Co-Pt,[10,11] Fe-Pt.[12] Beautiful pictures of the double-twin system, as well as a detailed crystallographic analysis, have been given by Rowles, Barrett, and Guttman.[13]

A shear stress applied across a twin boundary tends to move the twin boundary in such a direction as will relieve the stress. 50-50 alloy systems which have undergone an fcc → fct transformation through a lowering of

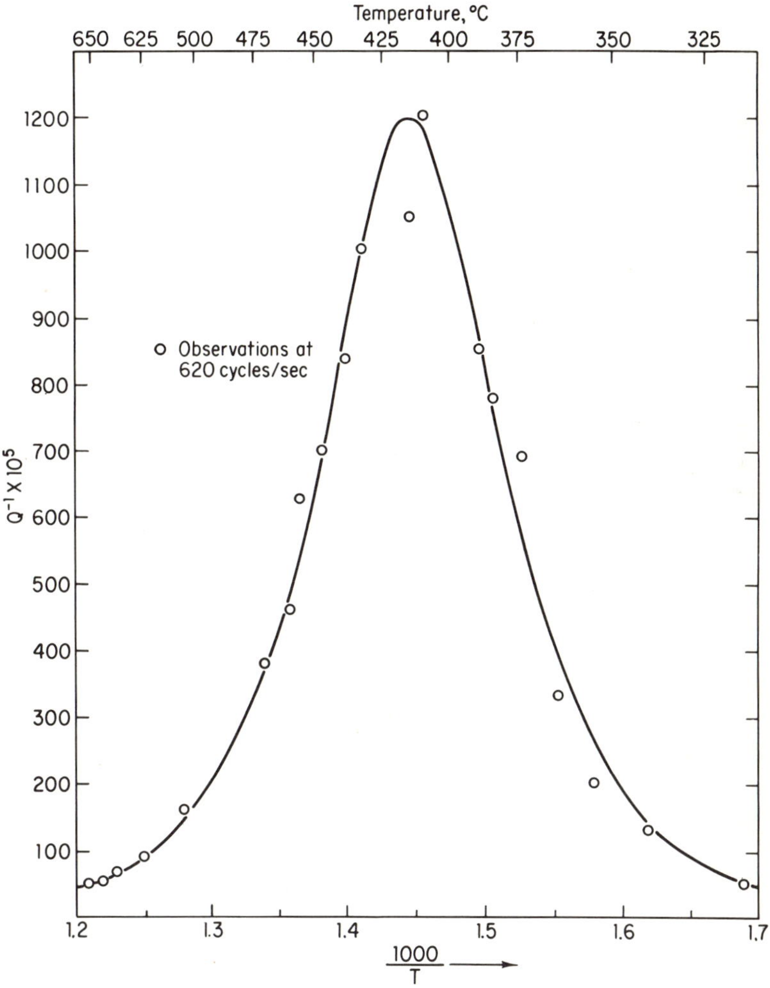

Fig. 4 Internal friction due to anisotropic local ordering in α-brass.

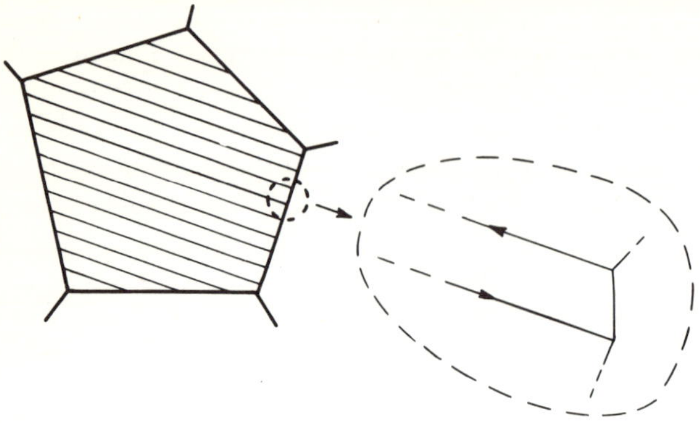

Fig. 5 Illustration of how nature solves the cubic → tetragonal transformation problem.

temperature may therefore be expected to show a large stress-strain hysteresis and related anelastic effects. Such effects in a system having undergone an fcc → fct transformation have been studied only in the Tl-In system. Here the anelastic effects are so high that Burkart and Read[14] have described the system as rubberlike.

Let us now pass to a consideration of the uncertainty in the location of an atom with respect to its equilibrium position, namely to a consideration of the entropy of vibration. As a crude generalization, we anticipate that the looser the packing, the greater will be this entropy of vibration. Thus we can expect close-packed structures at low temperatures, loosely packed structures at high temperatures (Fig. 6).

Apart from Fe, which will be considered later, whenever the bcc lattice is an allotropic modification, it is a high-temperature modification. This is true, for example, in Ca, Sr, Ti, Zr, Hf, Tl, Th, U, Np, and Pu. Because of this tendency for the bcc lattice to be a high-temperature allotropic form, Barrett suspected that those elements that are known only with the

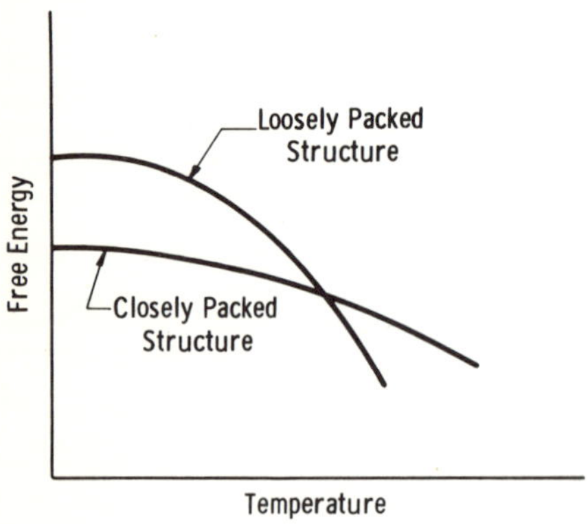

Fig. 6 Illustration of why loosely packed bcc structures are common at elevated temperatures.

bcc structure may in fact have a close-packed structure at low temperatures. He indeed found such close-packed modifications for Li[15,16] and for Na.[17,18]

We can be somewhat more specific in regard to the effect of vibrational entropy in the common β-phase type bcc structures with an electron-atom ratio of 3:2. Whereas it is commonly regarded that these structures are stabilized by the Brillouin zones, in truth we can only say that the Brillouin zones act as a stabilizing influence. Thus in no way is the β-phase bcc structure stable at room temperature unless stabilized by ordering. In the absence of ordering they simply flop into more stable structures upon a lowering of temperature. Metallurgists have a name for this kind of flopping: namely *martensitic transformation.*

We have here an example of a phase that is *mechanically* unstable at low temperatures, and yet is stabilized by entropy at high temperatures. As an idealized model of how a low-temperature mechanically unstable phase can be stabilized at high temperature by entropy, we consider the linear chain represented in Fig. 7. The atoms are acted upon by two types of forces. One arises from a fixed periodic potential. The other arises from springs which separate neighboring atoms. The springs are regarded as such that, acting alone, they would produce an interatomic spacing identical to the period of the fixed potential. At zero temperature the atoms are at rest. They will then all lie at the bottom of the potential valleys. As we raise the temperature, the atom will vibrate, and thereby acquire entropy. We then translate the atomic chain so that it lies on top of the mountains. The entropy of vibration will now be greater than when the atoms were in the valleys, for now the force constant due to the periodic potential opposes that due to the springs, resulting in a greater amplitude of vibration at a given temperature. At a sufficiently high temperature the free energy of the mountain-top (β) phase will be lower than that of the valley-bottom (α) phase.

We next inquire why the β-brass type structure is so commonly mechanically unstable at low temperatures. The lattice structure of β-brass

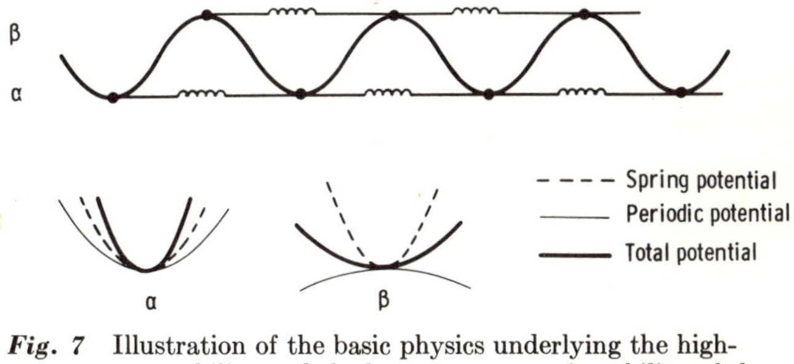

Fig. 7 Illustration of the basic physics underlying the high-temperature stability and the low-temperature instability of the bcc β-phase alloys.

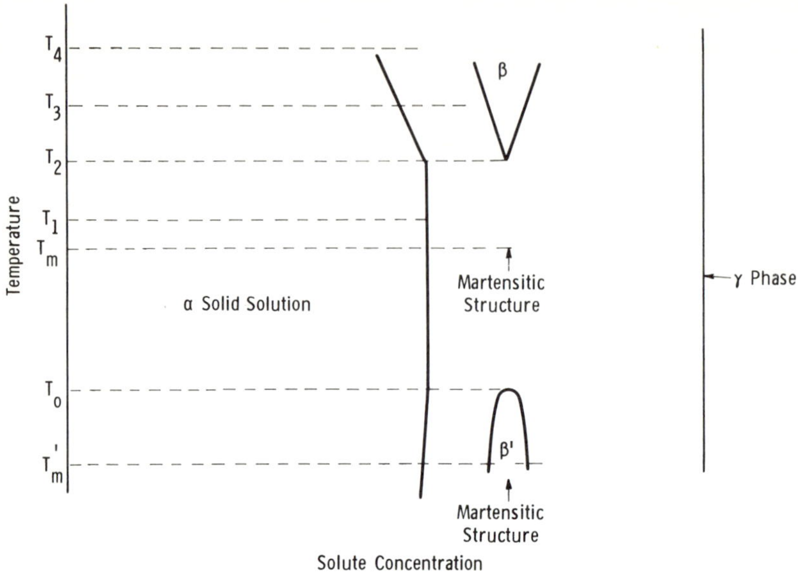

Fig. 8 A typical β-brass type of phase diagram.

is body-centered cubic (bcc). A bcc structure would be mechanically unstable if the energy of the conduction electrons were a function of volume only, not of shear strain, and if the repulsion interaction between the inner cores was short range. For then a pure shear across a (110) plane in a $\langle 1\bar{1}1 \rangle$ direction would leave the electronic energy unchanged, but would lower the interaction between the inner cores. The instability is perhaps most plainly seen by visualizing a bcc stacking of hard spheres, the whole ensemble of spheres being subjected to a hydrostatic compression. Such a system would be unstable with respect to a (110) $\langle \bar{1}10 \rangle$ shear. For with respect to a (110) $\langle 1\bar{1}0 \rangle$ motion, the unsheared cubic structure represents the structure of largest volume.

We are now in a position to understand the general features of metallic equilibrium diagrams containing β-type phases, such as is illustrated in Fig. 8. In Fig. 9 we show schematically the pertinent free-energy diagrams. Because of its anomalously large entropy of vibration, the disordered β phase has a free energy which with a rise in temperature drops rapidly with respect to the free energy of the other phases. Above T_2 the β-phase free-energy curve drops below the line which touches the free-energy curve of both the α phase and the γ phase. Above T_2 the β phase is therefore stable, but not below T_2. Below a certain temperature T_0, the bcc structure exists only in the ordered phase. If, as represented in Figs. 8 and 9, T_0 is below T_2, a temperature gap exists between the β and β' phases.

We denote by T_m the temperature at which the disordered β phase becomes mechanically unstable. If, as in the example of Fig. 7, T_m is below the temperature T_2, the mechanical instability is manifested as a martensitic transformation by a rapid quench. The quench must be sufficiently rapid to prevent the dissolution of the β phase in the more stable α and γ phases.

We denote by T'_m the temperature at which the ordered β' phase becomes mechanically unstable. Since ordering tends to stabilize the bcc lattice for a 50-50 composition, T'_m is below T_m. In the prototype example of CuZn, T'_m varies[19] from 150 to 250°K over the composition range for which β' exists. Kunze[20] has shown that the martensitic transformation does indeed proceed by an initial (110) $\langle 1\bar{1}0 \rangle$ shear as indicated above.

As indicated earlier, the driving force for ordering comes from a difference in size of the inner core. In the β structures of Cu, Ag, and Au with Zn, Cd, and Mg, the greatest difference in core radii occurs with AuMg, with $\Delta R = 0.71$ A. The ordered β phase of AuMg is so stable that it extends from 0°K to the melting temperature. This ordered phase is in fact so stable that the melting temperature of the ordered phase is above that of the two component metals. Next comes AuZn and AgMg, with $\Delta R = 0.63$ and 0.60 A. Here also the ordered phase extends from 0°K to the melting temperature. None of the other β phases are stable at very low temperatures.

Dr. Rudman has pointed out to the author that in those phases in which the bcc lattice is a high-temperature modification, the bcc phase is in fact

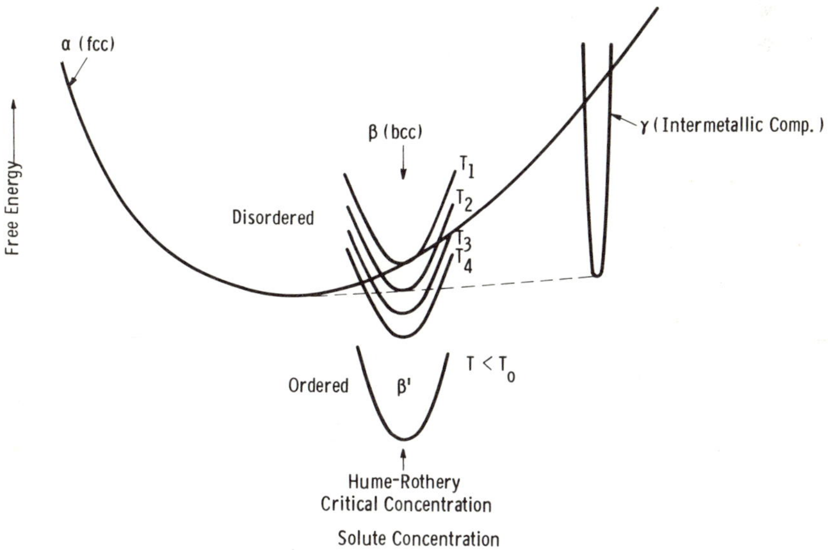

Fig. 9 Interpretation of origin of phase diagrams such as is illustrated in Fig. 8.

the denser phase.[35] We are therefore not justified in saying that La, Sr, . . . , Np, Pu transform to a bcc lattice at high temperatures because this lattice has a loose structure, and therefore the higher entropy. The only interpretation we can make is that here the closed inner s and p shells play the same repulsive role as do the inner closed d shells for β-brass. This interpretation predicts that when the high-temperature bcc lattice is rapidly quenched, it will transform by a martensite-like mechanism.

PHASE CHANGES INDUCED BY ELECTRON ENTROPY

Electron entropy may be separated into two essentially distinct components: that coming from a near degeneracy of states other than spin orientation, and that coming from an uncertainty in spin orientation.

Very little is known regarding the first component. The $\alpha \rightarrow \gamma$ transition in Fe indicates that the fcc γ-Fe has a higher entropy than the bbc α-Fe. Although this higher entropy had been interpreted[21] as indicating that the Debye temperature of the γ-Fe is lower than that for α-Fe, it has more recently been interpreted as confirming two electron states in γ-Fe. Apart from this one example, the author is not aware of any instances in which electronic transitions have been reported to be responsible for phase changes. In contrast, the entropy associated with the uncertainty in spin orientation is known to play an important role in many phase transformations.

As a first example[23] of the role electron spins play in phase transformations, we shall consider fcc Mn stabilized by small additions of alloying elements such as Cu. We have here a situation which, at least formally, parallels the example of ordering in Au-Cu alloys previously discussed. The electron spin of the Mn core may be regarded as having two possible orientations, namely up or down. Now two cores with parallel spins have an exchange repulsion. Hence upspin cores would like to have as nearest neighbors only downspin cores. But in a fcc lattice such a perfect ordering is impossible. The best that can be done is for alternating (001) planes to have upspins and downspins. Each atom will then have only four nearest neighbors of like spin, but eight nearest neighbors of opposite spin. The resulting structure is tetragonal with a c/a ratio less than unity, precisely as in the case of 50-50 Au-Cu alloys. At high-temperatures, spin entropy prevents spin ordering, and so one has the fcc structure. On quenching, the fcc lattice transforms into an fct lattice, with the double-twin structure discussed in the case of Au-Cu alloys. The anelastic behavior associated with this double-twin structure gives an internal friction peak below room temperature.[24] The spin ordering can of course be detected by neutron diffraction experiments, whereas in the Au-Cu alloys ordering is detected by X-ray diffraction. The association of the fct

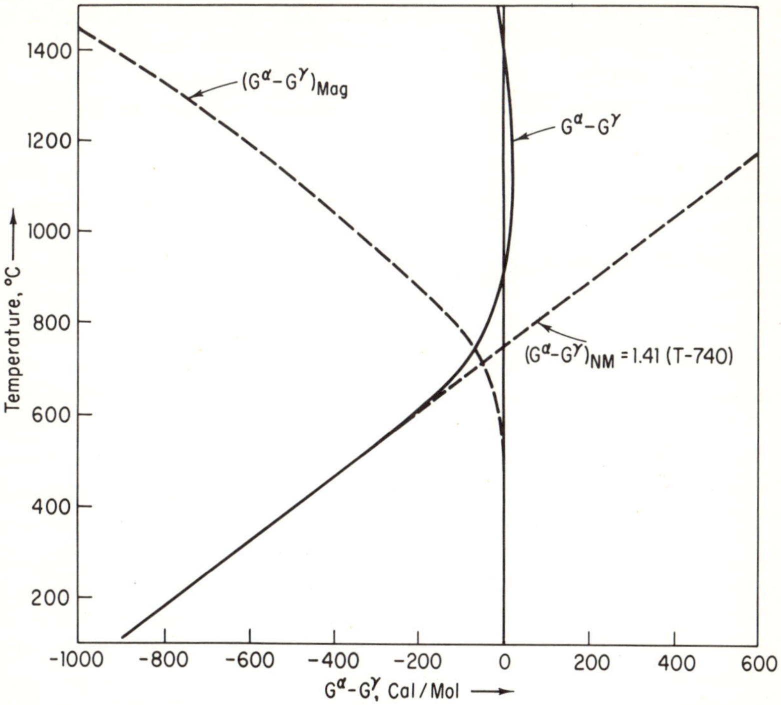

Fig. 10 Resolution of free-energy difference of α- and γ-Fe into a magnetic and a nonmagnetic component.

lattice with spin ordering disappears in a 15% Cu 85% Mn alloy with the fct $\rightleftharpoons$ fcc transition temperature[26] for the same alloy.

As a second example of the role electron spins play in phase transformations we shall consider the allotropic phases of pure Fe. The possibility that the electron spin may indeed play a dominating role in the phase changes of Fe and of Fe alloys may be seen by comparing the entropy rise associated with spin uncoupling with the entropy change in the $\alpha \rightarrow \gamma$ transition. The former is $R \ln 3 = 2.20$ entropy units per mole; the latter[27] is only 0.18.

The difference in free energy between the α and γ phases of pure Fe is shown in Fig. 10. Here the free-energy difference is resolved into two components.[28] The first component is just an extrapolation of the low-temperature data. It would be the sole component if the magnetic moments of the Fe atoms did not uncouple from one another in the α phase. The second component, $(G^\alpha\text{-}G^\gamma)_{\text{Mag}}$, comes from this magnetic uncoupling. We see that the spin uncoupling not only raises the α transition but also causes a reversal of this transformation back to bcc at still higher temperatures.

The Fe side of binary Fe alloys may be constructed from a knowledge of the G^α-G^γ function given in Fig. 10 and from a knowledge of the work $\Delta G_{\alpha\to\gamma}$ required to transfer one mole of solute atom from a pure α phase to a pure γ phase. If this work were temperature-independent, the binary diagrams could be directly constructed from Fig. 10. Thus an element with a positive ΔG^α would be a gamma-loop closer, and the phase diagram would be proportional to G^α-G^γ between the two transition temperatures for pure Fe, $T_{\alpha\to\gamma}$ and $T_{\gamma\to\alpha}$. Conversely, an element with a negative ΔG^α would be a gamma-loop opener. Its phase diagram would be proportional to G^α-G^γ below $T_{\alpha\to\gamma}$ and above $T_{\gamma\to\alpha}$. Such rough considerations enable us to at least interpret why some alloying elements are gamma-loop closers and others are gamma-loop openers.

Since the spin entropy plays such a dominant role in Fe transformations, we may expect the temperature dependence of $\Delta G_{\alpha\to\gamma}$ to be given, at least to a fair approximation, by the influence of short-range magnetic spin coupling. Such an influence may be represented by a vertical shear of the binary alloy diagram. In this way we may interpret[28] the drooping gamma loop of the Fe-Cr system. Conversely, theoretical arguments[29] have been presented which predict that all those gamma-loop closers without incomplete *d* shells will have an upward shear, thereby giving the gamma loop a tendency to point upward. Such a tendency is obvious in the Fe-Si system. By opening the γ loop in the Fe-Al system by the addition of Ni, Roesler demonstrated that the Fe-Al system also has a slight upward tilt.[30]

The original discussion[28] of the author about the effect of spin entropy upon the phase diagrams of binary Fe alloys was based on the thermodynamic data of Johannson.[31] More recent thermodynamic data[32] and analyses[33] leave the above general conclusions unaltered. In particular, Hillert[34] et al. have found by refined experimental techniques that the binary equilibrium diagrams are in better agreement with the simple theory[28] than the currently accepted diagrams.

REFERENCES

1. Bragg, W. L., and E. J. Williams: *Proc. Roy. Soc.* **A145**:699 (1934).
2. Bethe, H.: *Proc. Roy. Soc.* **A156**:552 (1935).
3. Zener, C.: *Trans. AIME* **152**:122 (1943).
4. Zener, C.: *Phys. Rev.* **71**:34 (1947).
5. For a review, see A. Nowick, *Progr. Metal Phys.* **4**:1–70 (1953).
6. Orr, R. L., J. Luciat-Labry, and R. Hultgren: *Acta Met.* **8**:431, 489 (1960).
7. Hume-Rothery, W.: "Electrons, Atoms, Metals and Alloys," 3d ed., Dover Publications, Inc., New York, 1963.
8. Haughton, J. L., and R. J. M. Payne: *J. Inst. Metals* **46**:457 (1931).
9. Harkei, D.: *Trans. ASM*, p. 210, 1943.
10. Beghart, E., and W. Köster: *Z. Metallk.* **32**:253 (1940).

11. Newkirk, J. B., A. H. Geisler, D. L. Martin, and R. Smoluchowski: *Trans. AIME* **188**:1249 (1950).
12. Lipson, H., D. Shoenberg, and G. V. Stupart: *J. Inst. Metals* **67**:333 (1941).
13. Bowles, J. S., C. S. Barrett, and L. Guttman: *Trans. AIME* **188**:1478 (1950).
14. Burkart, M. W., and T. A. Read: *Trans. AIME* **197**:1516 (1953).
15. Barrett, C. S., and O. R. Trautz: *Trans. AIME* **175**:579 (1948).
16. Barrett, C. S., and D. F. Clifton: *Trans. AIME* **188**:1329 (1950).
17. Barrett, C. S.: *J. Inst. Metals* **84**:43 (1955).
18. Barrett, C. S.: *Acta Cryst.* **9**:671 (1956).
19. Kunze, G.: *Z. Metallk.* **55**:277, 307, 405 (1964).
20. Kunze, G.: *Z. Metallk.* **53**:329, 392, 565 (1962).
21. Weiss, R. J., and K. J. Tauer: *Phys. Rev.* **102**:1490 (1956).
22. Kaufman, L., E. V. Clougherty, and R. J. Weiss: *Acta Met.* **11**:323 (1963).
23. Zener, C.: *Phys. Rev.* **81**:440 (1951).
24. Zener, C.: "Elasticity and Anelasticity of Metals," p. 161, University of Chicago Press, Chicago, 1948.
25. Meneghetti, D., and S. S. Sidhu: *Phys. Rev.* **105**:130 (1957).
26. Basinski, Z. S., and J. W. Christian: *J. Inst. Metals* **80**:659 (1951–1952).
27. Cleaves, H. E., and J. G. Thompson: "The Metal–Iron," p. 148, McGraw-Hill Book Company, New York, 1935.
28. Zener, C.: *Trans. AIME* **203**:619 (1955).
29. Roesler, U., H. Sato, and C. Zener: *Trans. ASM* **48A**:255 (1956).
30. Roesler, U.: *Trans. AIME* **206**:1285 (1956).
31. Johannson, H.: *Arch. Eisenhuettenw.* **11**:241 (1937).
32. Darken, L. S.: *Trans. AIME* **167**:546 (1946). Darken, L. S., and R. P. Smith: *Ind. Eng. Chem.* **43**:1815 (1951).
33. Oelsen, W.: *Stahl Eisen* **69**:468 (1949).
34. Hillert, M., T. Wada, and H. Wada: Communicated for publication.
35. Rudman, P.: *Trans. AIME* **233**:864 (1965).

VIEWPOINTS OF STABILITY OF METALLIC STRUCTURES*

Leo Brewer

Lawrence Radiation Laboratory and Department of Chemistry, University of California, Berkeley, California

INTRODUCTION

There are some serious problems in speaking to a heterogeneous group of this type. The difficulties may be best illustrated by an experience which we had at Berkeley. The Inorganic Materials Research Laboratory organized a symposium on metals at which one-third of the speakers were physicists, one-third were chemists, and one-third were metallurgists. As soon as the first few speakers had spoken, it became apparent that there existed problems of communication. Each of the three groups experienced difficulty in understanding the speakers of the other two groups. The vocabulary and concepts were quite unfamiliar, and the communication between the groups was poor. There were calls for interpreters. In recognition of this problem, the remaining speakers were urged to recognize the difficulties of communication and to present their ideas in terms that

* This research has been sponsored by the Atomic Energy Commission, whose support is gratefully acknowledged.

would be more generally understood by all. Once this problem had been recognized, the conference proceeded much more smoothly. However, it shortly became obvious to the members of each group that each of the other two groups had blind spots in their appraisal of metals. Each group had a limited view of the problem as a whole and often did not even recognize the existence of problems important to the other two groups. Here we clearly had an example of three blind men and an elephant, each describing the animal in terms of those portions that he could reach. In recognition of this previous experience, I hope that I will be able to express my ideas in terms that will be understandable to those who may not have had previous experience with the types of methods that I will be using. I hope that I may be able to point out new methods of approach to the understanding of metals that may not be familiar to all.

One of the difficulties in maintaining good communication between the various disciplines is that there are often decided differences in the general philosophies of approach to problems. The chemist, for example, makes full use of the well-established fundamental principles, but he is not content to restrict himself to areas of science which are so well developed that all properties of interest can be developed deductively from first principles. The chemist has developed a number of successful procedures that have allowed him to combine our present knowledge of the fundamental principles with various semiempirical approaches that have carried him far beyond the reach of purely deductive processes. For example, for most molecules it is still out of the question to hope for a detailed and accurate solution of Schrödinger's equation to describe their properties. The realistic chemist is not content to limit himself to the very few molecules that can be treated accurately by rigorous quantum-mechanical treatments.

A general description of the type of method used can be given in the following terms. We know that the interactions of the electrons with one another and with the nuclei are responsible for the properties of materials. These interactions are very complex, and we do not have the means of working them out in detail. However, it is recognized that certain properties are closely enough related that many of the complex interactions among the electrons affect them in a similar way. In some instances, one of these properties can be determined experimentally relatively easily, whereas for another property this may be very difficult. By recognizing that the influences of the electronic interactions upon the two properties have much in common and that the difference in the behavior of the two properties cancels out a great portion of the complex electronic interactions, one may often find relatively simple correlations between the variations of these two properties upon variation of some fundamental parameter, such as the number of bonding electrons or the type of electrons that are involved. That is the type of correlation that I wish to present.

In trying to develop such a correlation, it is important to have some physical picture or some type of model to serve as a guide. Chemists generally use one of two starting models for treating bonding. They may be termed the *molecular-orbital method* and the *valence-bond method*. Even though these two methods are not consistent with each other in their simplest forms, we do know that with suitable refinements both must yield the same results. However, it is not possible to use either method in full refinement, and it is found that for some properties one model is more useful than the other, whereas for other properties this is the reverse. Often the two approaches are most usefully combined to obtain a more complete understanding of chemical behavior.

For example, in my research work on the thermodynamic properties of high-temperature gases, it is important to be able to predict the ground and low-lying electronic states of diatomic molecules. The molecular-orbital approach is an extremely useful one for predicting the order of electronic states. Nevertheless, it is inadequate for meeting all our needs, and I have found it very useful to combine it with the valence-bond approach, where one starts with the electronic states of the separated atoms and considers the strength of bonding due to the various combinations of atomic electronic states to produce the various molecular states that are indicated by quantum-mechanical correlation rules. The molecular-orbital approach takes the nuclei in their final positions with the molecular orbitals already arranged in their proper order, as indicated by the theory, and it is a simple matter of then adding the available electrons to determine the resulting electronic state. The same ideas are used in metals. Whereas in a diatomic molecule a combination of a pair of atoms with s atomic orbitals produces a molecule with two sigma molecular orbitals, a combination of n atoms with s orbitals would produce a molecule or a crystal with n sigma orbitals, which would be spread out at different energies to produce the band that we talk about in band theory of metals. This model has been a very fruitful one for the understanding of metals. It is the common model used today, just as the molecular-orbital model is the common one used today for gaseous molecules when one is interested in the distribution of electronic states. However, just as it is useful in dealing with gaseous molecules to supplement the molecular-orbital theory with valence-bond theory which goes back to the separated atoms, it turns out likewise to be of value in understanding more of the detail of metallic interactions to consider the valence-bond approach.

In its simplest form the valence-bond theory requires that there be unpaired electrons in the separated atoms which can pair with each other to form an electron-pair bond between the atoms as the atoms are combined. Thus for Cl atoms with the electronic configuration s^2p^5 for the valence electrons of the ground state, only one unpaired electron is available for bonding because of the Pauli principle, and the diatomic molecule Cl_2 with one electron-pair bond is formed. The S atom with electronic configuration s^2p^4 can use two electrons for bonding and form rings with two electron-pair bonds per atom. Phosphorus (s^2p^3) forms a structure with puckered planes in which each P atom forms three electron-pair bonds. Silicon atoms with the configuration s^2p^2 for the ground state introduce a new consideration. The ground state has only two unpaired electrons, and only two bonds can be formed to each atom. However, there is an excited electronic configuration sp^3 which has four unpaired electrons and can therefore make four electron-pair bonds in the diamond structure. The promotion energy[1] of 95 kcal required to convert the ground state to the ^{5}S state with configuration sp^3 is more than offset by the energy of two additional bonds, and Si does form the four-coordinated diamond structure. The previously considered elements, P, S, and Cl, had no possibility of unpairing electrons by promotion without exciting to the next main shell, which would require a prohibitive promotion energy that could not be offset by additional bonding energy.

If one examines the actual processes taking place as Si atoms, for example, approach each other, it could actually be described in terms of gradually increasing interactions which would change the degree of configuration interaction and the relative energies of the different states to result, finally at close approach, in the stabilization of the sp^3 configuration. Rather than treat this complex process over the entire range of the approach of the atoms, it is much simpler to first prepare the atom in a valence state, that is, to put in sufficient energy to promote the Si atom from the ground electronic state to the appropriate excited electronic state which would have the maximum number of electrons available for bonding, and then to bring those properly prepared valence states of the atom together to produce the final electronic structure of the solid. From the laws of thermodynamics we know that the energy or enthalpy changes must be the same by either of the processes described. The two-step process by way of a valence state essentially breaks up a complex process into two much simpler processes.

The electrons that bind these nonmetals together are the same glue that binds the metals together, and the same procedure would be expected to be applicable to metals. For example, atomic Be, Mg, Zn, and Cd all have a ground electronic state in which there are no unpaired electrons. If the ^{1}S ground state were the only possibility, one would expect these elements to be very weakly bonded. They would be somewhat more strongly bonded than the rare gases because of the greater polarizability of the pair of s electrons, but nevertheless very weakly bonded. In actual practice, the bonding is much stronger than one would expect from consideration of only the ground electronic state. In terms of the valence-bond model, one would describe the bonding in terms of the two-step process by first promoting the ^1S state to a ^3P state in which there is one unpaired s electron and one unpaired p electron. Then the atoms in the ^3P valence state are brought together to produce a hexagonal close-packed structure in which there is an equal number of bonding s electrons and bonding p electrons.

For the transition metals the process is more complex in that there are s, p, and d electrons to be considered. This added complication has made it difficult to relate crystal structure to electronic structure. It has been long accepted since Hume-Rothery established the relationship[2] between numbers of electrons per atom and structures of many intermetallic compounds that there must be in many cases a strong correlation between the number of electrons per atom and the types of crystal structures, just as there was strong correlation between the structures of Cl, S, P, and Si and the number of electrons per atom.

Many years ago Niels Engel[3] proposed a similar correlation for all the metals: namely that the body-centered cubic structure is to be correlated with an electron configuration approximating the configuration $d^{n-1}s$; when the valence state corresponds to one s and one p electron ($d^{n-2}sp$), the hexagonal close-packed structure is expected; and when the valence state corresponds to one s and two p electrons, the face-centered-cubic structure is expected. The correlation was not well accepted when originally proposed, but this correlation and the application of it as Engel suggested provides a very simple way of predicting a great deal about the behavior of metals. In the almost two decades since Engel presented his ideas, there has been a tremendous expansion in the gathering of data, and it has turned out that these new data have very largely substantiated the predictions that come about through the application of the Engel theory. I have written three papers[4–6] in recent years which have reviewed the evidence, and they show that the correlation is indeed extraordinarily effective. Not only does it allow one to correctly predict the structures of most of the pure metals, but it also allows the correct prediction of most of the structures and composition ranges of the various intermetallic phases which arise in alloy systems.

Figure 1 will illustrate the application of the valence-bond model in terms of Engel's correlation. Here we see the relative promotion energies of the $d^{n-1}s$ and the $d^{n-2}sp$ electronic configurations for the transition metals of the second transition series, where n is the total number of valence electrons. There are other low-lying electronic con-

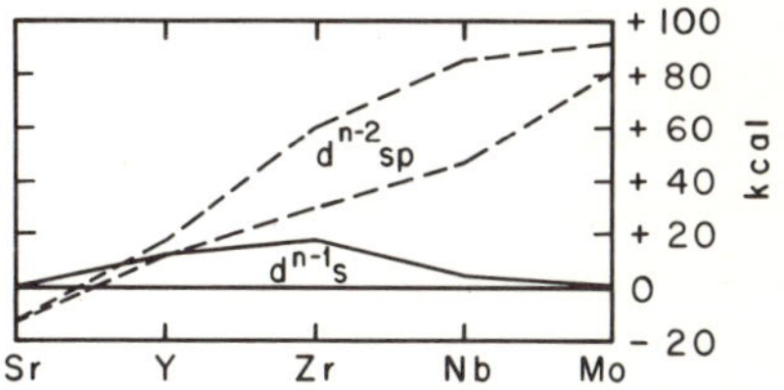

Fig. 1 Relative stabilities of $d^{n-1}s$ and $d^{n-2}sp$ electronic configurations for transition metals of the second transition series.

figurations for these elements for which not all the electrons are unpaired: for example, the $d^{n-2}s^2$ configuration. The valence model would require that we take only those configurations with all of the valence electrons unpaired for the left-hand side of the transition series where the promotion energies are small compared to the bonding energies.

Another way of putting this is that the resulting levels in the metal that correlate with the configurations involving paired s electrons, which must be localized on the atom and not used in bonding between the atoms, would be relatively high in energy and would not be filled; whereas levels of configurations involving all the electrons unpaired in the valence state, and therefore completely available for bonding in the metal, would lie much lower in energy and be the ones to be filled with electrons. Both the $d^{n-1}s$ and $d^{n-2}sp$ configurations in Fig. 1 have all the valence electrons unpaired and free for bonding. There are some differences in the bonding ability of d electrons compared to that of s and p electrons which will be discussed in detail shortly, but if we ignore these differences, we can still use the data shown in Fig. 1 to predict qualitatively the relative stability of the bcc and hcp structures for these elements. The energies are presented as bands or areas since there are a number of states corresponding to a given electronic configuration. The ranges of energies are those of the electronic states of highest multiplicity which correspond to the indicated electronic configurations. It can be seen that for the first three metals, Sr, Y, and Zr, the two configurations lie close enough together in energy so that one might expect that the bcc ($d^{n-1}s$) and hcp ($d^{n-2}sp$) structures would be close together in energy. This is found experimentally to be the case. All three of these elements have an hcp structure which converts to a bcc structure at higher temperatures. Since the bcc structure has a coordination number of 8 compared to a coordination number of 12 for the hcp structure, the bcc phase would be expected to have more vibration and a greater entropy, and therefore would be the one to become stabilized at the higher temperatures. For Nb and Mo it can be seen that the electronic configuration $d^{n-2}sp$ is so highly promoted compared to the configuration with only one s electron that one would predict that the hcp structure would be considerably unstable compared to the bcc structure. This again is in agreement with the experimental observations since only the bcc structure is known for Nb and Mo at all temperatures and pressures. Similar comparisons for the other transition metals indicate that the predictions that can be made with the spectroscopic data available are in agreement with the available experimental observations. To allow comparison between the predictions of the calculations to be presented below and the experimental observations, the crystal structures of the elements are presented in Table 1.

There are then two steps in the application of the methods that I wish

TABLE 1 ***Crystal Structures of the Elements***[a]

Li I	Be II	B * *											C IV >=[f]	N ≡[b]	O =[b]	F —[c]
Na I	Mg II	Al III											Si IV	P ↖[e]	S <[d]	Cl —
K I	Ca I III	Sc I II	Ti I II	V I	Cr I	Mn I III β χ	Fe I III I	Co III II	Ni III	Cu III	Zn II	Ga A11	Ge IV	As ↖	Se <	Br —
Rb I	Sr I II III	Y I II	Zr I II	Nb I	Mo I	Tc II	Ru II	Rh III	Pd III	Ag III	Cd II	In (III)[g]	Sn A5 IV	Sb ↖	Te <	I —
Cs I	Ba I	La I III II	Hf I II	Ta I	W I	Re II	Os II	Ir III	Pt III	Au III	Hg A10	Tl I II	Pb III	Bi ↖	Po	At
Fr	Ra	Ac														

Ce I III II	Pr I II	Nd I II	Pm	Sm I (II)[g]	Eu I	Gd I II	Tb I II	Dy I II	Ho I II	Er I II	Tm I II	Yb I III	Lu I II
Th I III	Pa *	U I σ (II)[g]	Np I **	Pu I * III ***	Am ? II	Cm ? II	Bk	Cf	Es	Fm	Md		Lw

[a] I: Body-centered cubic, A_2. II: Hexagonal close-packed, A_3. III: Cubic close-packed, A_1. IV: Diamond, A_4.

[b] Diatomic molecules with double or triple bonds.

[c] Diatomic molecule with a single electron-pair bond.

[d] Atoms that form two single bonds per atom to form rings or infinite chains.

[e] Three single bonds per atom, corresponding to a puckered planar structure.

[f] The graphite structure where one resonance form consists of two single and one double bond per atom.

[g] Parentheses indicate slight distortions.

Asterisks denote complex structure.

The structures are listed in order of temperature stability except for metastable diamond.

to describe. First we must examine[1] the available electronic configurations for the gaseous atoms and consider the promotion energies[1] required to obtain configurations which have their electrons completely available for bonding (for elements of the first six groups), and then compare the bonding energies of each of the configurations to determine which yields the lowest levels in the metal. If the valence state that contributes to the lowest levels in the metal is one that involves only one *s* electron, then the Engel correlation would predict a bcc structure. If the valence state contributing to the bulk of the low-lying levels in the metal is one involving *sp* electrons, one would expect the hcp structure. The theory is more difficult to apply to metals than to simple gaseous molecules where the structures resulting from various combinations of *s*, *p*, and *d* orbitals have been well worked out. For the metals the situation is complicated in that one must consider not only nearest neighbors but also more distant neighbors. For example, in comparing the hcp and the ccp structures, it is clear that one could not make any predictions about the relative stabilities of these two structures on the basis of any model that depended only upon nearest neighbor interactions. However, more complex calculations have been carried out by Altmann, Coulson, and Hume-Rothery[8] which show that characteristic structures are to be associated with specific electronic configurations.

One can now use the spectroscopic data in a more quantitative manner by using the experimental promotion energies required to achieve each of the two electronic configurations under consideration here, together with the experimental enthalpies of sublimation, to determine the bonding energies that result when the gaseous atoms in the corresponding valence state are condensed to the solid metal. The enthalpies of sublimation of the metals have been taken from published tabulations.[4,7] The promotion energies[1] for the $d^{n-2}sp$ electronic configuration are given in Table 2. Table 3 shows the corresponding promotion energies[1] from the ground electronic state of the gaseous atoms to the electronic configuration $d^{n-1}s$. The second numerical entry below each element is the promotion energy that would correspond to the configuration $d^{n-1.5}sp^{0.5}$, which would be the average of the promotion energies for the configuration with one *s* electron and the configuration with one *s* and one *p* electron. The second value is tabulated because the bcc structure is observed up to a total of one and a half electrons of the *s* or *p* type per atom, and thus the range of promotion energies from that for one *s* electron to that for one and a half *s*, *p* electrons must be considered for quantitative treatment of the data for bcc structure.

The calculation of stabilities of bcc structure based on the two configurations will be illustrated later and can be used to determine which configuration is the most appropriate. Generally one expects the bonding energies of *s* or *p* electrons to decrease with increasing internuclear distances as one goes down in the periodic table. One would expect the bonding energies to increase to the right in the periodic table because of increasing nuclear charge.

Figure 2 illustrates data for bonding energies of metals of the second transition series. The top curve is a curve of bonding energy due to an *s* or *p* electron going from Sr to Cd. The values for Sr and Cd can be fixed directly since they both have hcp structures corresponding to *sp* electronic configurations. The smooth curve drawn in between corresponds to the expected variation of bonding energy as nuclear charge is increased. For each of the elements in this series, the bonding energy was calculated from

TABLE 2 ***Promotion Energies of $d^{n-2}sp$ Valence State***

In kcal per mole

Ground-state configurations										
s^2	ds^2	d^2s^2	d^3s^2 except*	d^5s except*	d^5s^2	d^6s^2 except*	d^7s^2 except*	d^8s^2 except*	$d^{10}s$	$d^{10}s^2$
Mg										
63										
Ca	Sc	Ti	V	Cr	Mn	Fe	Co	Ni(d^8s^2)	Cu	Zn
43	45–46, 53	45–54, 72–80	47–54, 81–88	71–80	53	55–70	67–76	74–87	113–128	92–94
Sr	Y	Zr	Nb*(d^4s)	Mo	Tc	Ru*(d^7s)	Rh*(d^8s)	Pd*(d^{10})	Ag	Cd
41–43	43–49, 54	42–57, 66–74	48–61, 74–86	80–93	47–50	72–93	(95)–	<145–	161–185	86–91
Ba	La	Hf	Ta	W*(d^4s^2)	Re	Os	Ir	Pt*(d^9s)	Au	Hg
35–39	38–46, 50–52	51–79, 85–95	49–99	49–99	54–68	67–(90)	75–109	86–160	121–160	108–126
Ra	Ac									
37–48	39–68									

TABLE 3 ***Promotion Energies of $d^{n-1}s$ and $d^{n-1.5}sp^{0.5}$ Valence States***

In kcal per mole

Ground-state configurations									
s^2	ds^2	d^2s^2	d^3s^2 except*	d^5s except*	d^5s^2	d^6s^2 except*	d^7s^2 except*	d^8s^2 except*	$d^{10}s$
Mg									
137									
100									
Ca	Sc	Ti	V	Cr	Mn	Fe	Co	Ni	Cu
58	33, 51	19, 40	6–7	0	49–50	20–23, 50–51	10–15, 43–46	1–5	0
51	39–43, 48–52	32–60	26–48	36–40	51–52	38–47, 53–61	39–61	37–46	
Sr	Y	Zr	Nb*(d^4s)	Mo	Te	Ru*(d^7s)	Rh*(d^8s)	Pd*(d^{10})	Ag
52	31–33, 44	14–17, 31–32	0–3	0	7–12	0–9, 25–27	0–10, 26–31	19–29	0
47	37–49	28–53	24–45	40–47	27–31	36–60			
Ba	La	Hf	Ta	W*(d^4s^2)	Re	Os	Ir	Pt*(d^9s)	Au
26–27	8–12, 21–22	40–51, 59–64	28–38	8	34–49	15–37, 44–(51)	8–34, 37–48	0–29	0
30–33	23–37	46–79	39–69	32–47	44–58	41–52, 56–	42–78	43–	
Ra	Ac								
37–42	(15–20), 26–35								
38–45	26–51								

NOTE: The numerical entry below the symbol of an element gives values for the $d^{n-1}s$ valence state. Where a second entry is given, it is for the $d^{n-1.5}sp^{0.5}$ valence state.

the combination of sublimation enthalpies and promotion energies. The contribution of the *s* or *p* electrons was subtracted, the number of *s* and *p* electrons being determined by the structure. The resulting bonding energy was assigned to the remaining *d* electrons, and the lower curves in Fig. 2 correspond to the bonding energy per *d* electron available for bonding, (i.e., the remaining bonding energy divided by the number of unpaired *d* electrons).

The characteristic curve that is shown here is typical of the curves in Figs. 3 and 4 found for the other transition series, in that the bonding effectiveness of a *d* electron decreases as the number of *d* bonds increases, reaching a minimum at the maximum of five *d* bonds per atom, apparently owing to interference between the *d* bonds when there are too large a number. The *x*'s indicates the experimental points derived from hcp structures; the circles indicate the values derived from bcc structures. The bonding energy of an electron is largely independent of the particular structure, and one can draw a curve that can be used for predicting the bonding in yet unobserved phases of these metals with either the bcc or hcp structures.

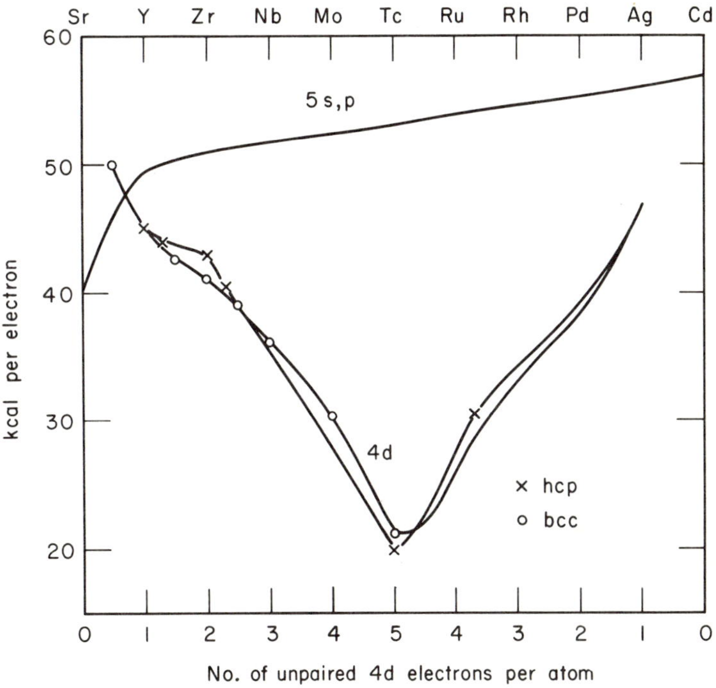

Fig. 2 Valence-state bonding enthalpy in kcal per mole per electron. Top curve: $E_{s,p}$, bonding enthalpy of 5*s* or 5*p* electrons, versus element. Bottom curve: E_d, bonding enthalpy of 4*d* electrons, versus number of unpaired *d* electrons.

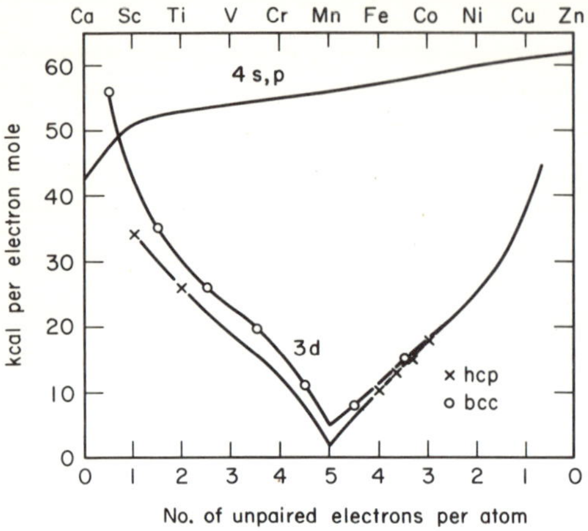

Fig. 3 Valence-state bonding enthalpy, in kcal per mole per electron. Top curve: $E_{s,p}$, bonding enthalpy of 4*s* or 4*p* electrons, versus element. Bottom curve: E_d, bonding enthalpy of 3*d* electrons, versus number of unpaired *d* electrons.

For example, for hcp Nb which has never been observed, one could predict the stability of such a phase by calculating the bonding energy that would result from the configuration d^3sp by multiplying by two the bonding energies given in Fig. 2 for an *s* or a *p* electron of Nb and multiplying by 3 the bonding energy for a *d* electron when there are a total of three *d* bonds per atom, as indicated by the ordinate at the bottom of Fig. 2. This bonding energy for three *d* electrons added to contribution of the two *sp* electrons yields the total bonding energy for the valence state d^3sp, condensing to produce hcp Nb. From this bonding energy must now be subtracted the promotion energy given in Table 2 for the configuration d^3sp, to yield an atomization energy of 158 kcal for hcp Nb. A similar procedure for the configuration d^4s yields an atomization energy of 173 kcal for the bcc structure which indicates that the hcp structure of Nb is quite unstable. A more detailed treatment to be presented below shows that a slightly less unstable hcp Nb is calculated using the configuration $d^{3.3}sp^{0.7}$.

This procedure has been carried out for all the transition metals, and Table 4 yields the bonding energies when the indicated valence states of the gaseous atoms are combined to form hcp solids.

Table 5 indicates the corresponding bonding energies for the formation of bcc structures. These can then be combined with the promotion energies given previously to calculate the heats of sublimation of either bcc or hcp structures for all the transition metals except Rh and Pd, where only rough estimates can be given for the hcp structure since the spectroscopic data for the $d^{n-2}sp$ configurations are incomplete or lacking for these two elements.

Before carrying out the calculations of the heats of sublimation given in Table 6, we must examine in more detail the process of bonding for transition metals. It might be useful at this point to translate from valence-bond terminology to molecular-orbital and electronic-band terminology. We have been speaking of the groups of electronic levels of a gaseous atom which correspond to a given electronic configuration. In some instances, such as shown for Y in Fig. 1, the energies of two configurations are so close that the electronic states are actually intermingled.

This is also shown in Table 7 where the ranges of energy of the sublevels of each electronic state of each electronic configuration of Y are tabulated. When we consider a diatomic molecule from the molecular-orbital point of view, we similarly find instances of molecular orbitals with similar energies such that the molecular electronic levels corresponding to each configuration are intermingled. Also, typically the electronic states corresponding to a given molecular orbital correlate to different electronic configurations of the atoms. When the gaseous atoms are condensed to the metal, the electronic configurations that participate in bonding are lowered in energy to produce a sequence of low-lying levels or bands that will be filled by electrons. When two electronic configurations are close in energy, there will be intermingling of levels and strong configuration interaction, and the bands cannot always be uniquely described in terms of an integral electronic configuration.

For example, in Table 5, the three valence electrons of Sc, Y, and La are designated as being 50% d, 33% s, and 17% p, or the $d^{1.5}sp^{0.5}$ configuration. This recognizes that the d^2s and dsp configurations are so close together that the lower levels of the electronic band, which will be filled, will be characterized as originating approximately equally from the two configurations. It was noted earlier that the bcc structure is expected not only for a metal with a $d^{n-1}s$ electronic configuration but that the bcc structure persists with increasing contribution from the $d^{n-2}sp$ configuration up to a limit of $d^{n-1.5}sp^{0.5}$. This is discussed in Ref. 5 in detail for the transition-metal alloy systems. It might be useful at this point to examine some alloys systems for the implications of a limiting electronic configuration.

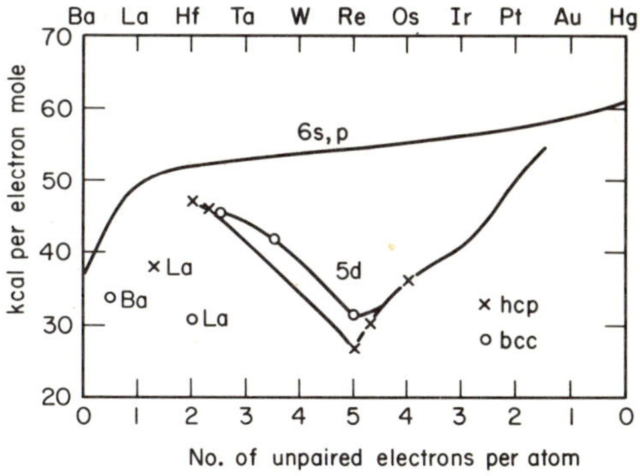

Fig. 4 Valence-state bonding enthalpy, in kcal per mole per electron. Top curve: $E_{s,p}$, bonding enthalpy of 6s or 6p electrons, versus element. Bottom curve: E:, bonding enthalpy of 5d electrons, versus number of unpaired d electrons.

TABLE 4 ***Bonding Energies of the $d^{n-2}sp$ Valence State in the Hexagonal Close-packed Structure***

Number of bonding electrons per atom										
2	3	4	5	6	7	6	5	4	2.4	2
In kcal per mole										
Mg										
98										
Ca	Sc	Ti	V	Cr	Mn	Fe	Co	Ni	Cu*	Zn
(84)	136	158	(165)	(158)	(121)	154	169	(171)	(135)	124
Sr	Y	Zr	Nb*	Mo*	Tc	Ru*	Rh*	Pd*	Ag*	Cd
80	144	188	(197)	(198)	205	205	(189)	(165)	(130)	114
Ba	La	Hf	Ta	W	Re	Os	Ir	Pt	Au*	Hg
(74)	142	198	(230)	(244)	241	255	(234)	(215)	(140)	(123)
In kcal per bonding electron										
Mg										
49										
Ca	Sc	Ti	V	Cr	Mn	Fe	Co	Ni	Cu*	Zn
(42)	45	40	(33)	(26)	(17)	26	34	(43)	(56)	62
Sr	Y	Zr	Nb*	Mo*	Tc	Ru*	Rh*	Pd*	Ag*	Cd
40	48	47	(39)	(33)	29	38	(43)	(49)	(54)	57
Ba	La	Hf	Ta	W	Re	Os	Ir	Pt	Au*	Hg
(37)	47	50	(46)	(41)	34	43	(47)	(54)	(58)	(62)

* Indicates that bonding energy is given for $d^{n-1.7}sp^{0.7}$ valence state. The number of bonding electrons per atom for Ru*, Rh*, and Pd* are, respectively, 5.4, 4.4, and 3.4. For Cu*, Ag*, and Au*, the $d^{9.3}sp^{0.7}$ valence state has 2.4 bonding electrons per atom.

TABLE 5 ***Bonding Energies of the $d^{n-1}s$ or $d^{n-1.5}sp^{0.5}$ Valence States in the Body-centered Cubic Structure***

Valence state and number of bonding electrons per atom											
1 s	2 $d^{0.5}sp^{0.5}$	3 $d^{1.5}sp^{0.5}$	4 $d^{2.5}sp^{0.5}$	5 $d^{3.5}sp^{0.5}$ except*	6 d^5s except*	6 $d^{5.5}sp^{0.5}$	5 $d^{6.5}sp^{0.5}$	4 $d^{7.5}sp^{0.5}$ except*	3 $d^{7.5}sp^{0.5}$ except*	2 d^9s	1 $d^{10}s$
In kcal per mole											
Na											
26											
K	Ca	Sc	Ti	V	Cr*($d^{4.5}sp^{0.5}$)	Mn	Fe	Co	Ni		Cu
21.5	92	129	144	149	131	119	138	(139)	(135)		(61)
Rb	Sr	Y	Zr	Nb*(d^4s)	Mo	Tc		Ru*(d^7s)	Rh*(d^8s)	Pd	Ag
19.5	85	138	174	173	157	(179)		(153)	(131)	(102)	(56)
Cs	Ba	La	Hf	Ta	W	Re	Os	Ir	Pt		Au
19	73	126	192	226	211	(228)	(218)	(197)	(167)		(59)
In kcal per bonding electron											
Na											
26											
K	Ca	Sc	Ti	V	Cr	Mn	Fe	Co	Ni		Cu
22	46	43	36	30	22	20	28	(35)	(45)		(61)
Rb	Sr	Y	Zr	Nb	Mo	Tc		Ru	Rh	Pd	Ag
20	43	46	44	35	26	(30)		(38)	(44)	(51)	(56)
Cs	Ba	La	Hf	Ta	W	Re	Os	Ir	Pt		Au
19	37	42	48	45	35	(38)	(44)	(49)	(56)		(59)

TABLE 6 $\Delta H°_{298}$ ***of Sublimation of BCC and HCP Phases of the Transition Metals***

In kcal per mole

	Ca	Sc	Ti	V	Cr	Mn	Fe	Co	Ni	Cu
BCC	41	90	112	123	95	68	100	(100)	(98)	(61)
HCP	(41)	91	113	(118)	(87)	(68)	(99)	102	(97)	(56)
	Sr	Y	Zr	Nb	Mo	Tc	Ru	Rh	Pd	Ag
BCC	38	101	146	173	157	(152)	(153)	(131)	(83)	(56)
HCP	39	102	146	(163)	(142)	158	155			
	Ba	La	Hf	Ta	W	Re	Os	Ir	Pt	Au
BCC	43	103	146	187	203	(184)	(177)	(155)	(124)	(59)
HCP	(41)	104	147	(181)	(189)	187	188	(159)	(129)	(55)

NOTE: Values in parentheses are for phases that are not stable at 1 atm pressure.

Figure 5 is a multidimensional phase diagram of the type given in Ref. 5. A horizontal line between Mo and Os, for example, represents a projection of the binary Mo-Os system along the temperature axis such that the composition range of each phase region represents the maximum extent at the optimum temperature. Thus the I phase region or bcc structure extends from Mo to 20 a/o Os, the A_{15} or Cr_3Si structure has a narrow composition range around Mo_3Os, the σ phase extends from 30 to 37 a/o Os, and the area labeled II corresponds to the hcp phase region from 48 to 100 a/o Os. A horizontal line drawn halfway between Re and Os would represent the sequence of phase regions resulting from adding Mo to an equimolal mixture of Re and Os. The Engel theory indicates that the diagram of this type will yield simple trends in solution behavior. The

TABLE 7 ***Relative Energies of States of Highest Multiplicity of the Electron Configurations of Y***

In kilocalories

d^2p	^4D, ^4S, (^4P)	103–107
dp^2	^4D, ^4F, ^4P	97–100
sp^2	^4P	66–67
d^2p	^4G, ^4F, ^4D	53–54, 59–61, 64–65
d^3	^4F, ^4P	53–54, 60–61
dsp	^4F, ^4D, ^4P	12–15, 16–18, 23
d^2s	^4F, ^4P	0–2, 12–13
s^2p	^2P	−1 to +2
ds^2	^2D	−31 to −30

elements on the right-hand side are the adjoining elements of the same transition series to minimize the effect of variation of atomic size and of internal pressure and thus isolate the effect of electron concentration. The ordinate on the right-hand side is, in effect, the number of electrons per atom varying from 6 for W to 10 for Pt. If a mixture of W, Re, Os, Ir, and Pt is taken and the average electron concentration computed, e.g., 7.5 valence electrons per atom, then the alloy behavior upon adding Mo to this mixture would be defined by the horizontal line from the 7.5 position halfway between Re and Os. Thus the Engel theory indicates the types of diagrams that will represent a maximum of information with a minimum of space, as Fig. 5 is in agreement with all the reliable data available. It will be noted that the regions labeled by I (bcc), σ (β-U), χ (α-Mn), II (hcp), and III (ccp) show the behavior of electron phases or electron compounds. That is, they are restricted to a limited range of electron concentrations or averaged electron-per-atom ratios. If one were to draw isoelectronic lines or lines of equal electron-per-atom concentrations, they would extend across the diagram in the manner of the boundaries of the electron phases listed above.

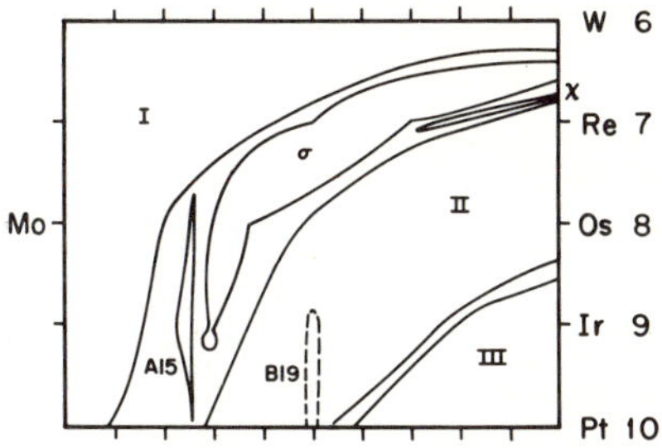

Fig. 5 Multicomponent phase diagram of Mo with third transition-series metals, W to Pt, projected along the temperature axis.

The implication of electron phases in terms of free-energy variation is that the relatively slow variation of free energy of the bcc phase as Os is added to Mo, for example, changes to a rapid rate of increase as the electron concentration approaches 6.5 electrons per atom. The limiting concentration will, of course, not be determined by the free energy of the bcc phase alone but also by the free energy of the saturating phase. Examination of all the available data and particularly the non-transition-metal diagrams as discussed by Brewer[5] indicates that the Hume-Rothery limit of 1.5 s, p electrons is a reasonable limit for the bcc structure. From the free-energy considerations, we recognize that the actual limits will vary somewhat depending upon the temperature, internal-pressure differences, and the nature of the saturating phase; and that these other factors must be considered as secondary corrections, but Fig. 5 shows that the limits of the bcc phase regions do approach closely to the $d^5sp^{1.5}$ limit. Generally the limits will fall short, with the difference increasing with increasing size difference and increasing internal pressure differences as one would expect. Reference 5 discusses in more detail the methods of establishing the electron-concentration limits for the various electron phases and the methods of dealing with factors such as size and internal-pressure differences which are clearly recognized as important factors in fixing the free energies in addition to electron concentrations.

Returning now to pure elements, the recognition of ranges of electron concentrations to be correlated with a given structure leaves some ambiguity as to the assignment of exact electronic configuration in some instances. We can see from the large difference in energy of the d^5s and d^4sp configurations of Mo, shown in Fig. 1 or in Tables 2 and 3, together with the bonding energies of d versus p electrons shown in Fig. 2, that all the low-lying levels of the electronic bands of metallic Mo will be derived almost exclusively from the d^5s configuration, and there is no question of what electronic configuration to assume in choosing the promotion energy from Table 5. However, we would not expect such simple behavior for Y, for example, for which the $d^{n-1}s$ and $d^{n-2}sp$ configurations are close in energy.

The relative contributions of the $d^{n-2}sp$ and $d^{n-1}s$ configurations to the bonding of the transition metals with five or fewer d electrons depends upon the relative bonding strengths of p and d electrons, $E_p - E_d$, as given in Figs. 2 to 4 compared to the relative promotion energies tabulated in Tables 2 and 3. Only a small portion of the curve of E_d versus n_d, the number of d electrons, is applicable for a given element, e.g., 0.5 electron per atom variation for the bcc structure, and 0.3 electron per atom variation for the hcp structure. If that small portion is approximated as a straight line, one can set up a simple equation for the energy as a function of n_d, the number of d electrons. Differentiation of this equation to solve for the condition of minimum energy yields the equation

$$n_d = \frac{E_{\text{prom}} - (E_p - E_d)}{S}$$

where E_{prom} is the energy needed to promote a d electron in $d^{n-1}s$ to a p electron in $d^{n-2}sp$, and S is the rate of change of E_d with decrease in number of d electrons of the slope of the plot of E_d versus n_d in Figs. 2 to 4. Of the quantities required to calculate n_d, E_{prom} is obtained directly from Tables 2 and 3. The promotion energies have a range of values corresponding to the different electronic states of a given configuration. The minimum promotion energies are used, for the reasons discussed in the appendix. The value of E_p is taken from Figs. 2 to 4 for the particular element in question. S is obtained from the slope of E_d versus n_d in the range of n_d values that are allowed by the Engel correlation. Thus all the quantities are uniquely fixed at the start of the calculation except E_d which will vary with n_d. If we use bcc Zr as an example, the first approximation of $E_d = 39$ kcal corresponding to a guess of $n_d = 2.5$ electrons yields $n_d = [28 - (51 - 39)]/5.8 = 2.75$. The second approximation uses $E_d = 38$

kcal, and n_d converges to 2.7 corresponding to the configuration $d^{2.7}sp^{0.3}$. In the appendix 1 where details of the arithmetic calculations are presented, atomizations energies of the $d^{2.5}sp^{0.5}$, $d^{2.7}sp^{0.3}$, and d^3s configurations are calculated to be 145, 146, and 146 kcal, respectively, illustrating the shallow nature of the minimum of energy in this example.

For transition elements with five or more d electrons, the promotion of a d to a p electron releases two electrons for bonding, and the energy varies as the sum of $E_p + E_d$ in contrast to the difference $E_p - E_d$ for elements with five or less d electrons. For these elements, much higher promotion energies can be offset, particularly to the far right of the transition series where the number of bonding d electrons is small and the sum $E_p + E_d$ is large. Differentiating the energy equation to obtain the minimum energy as above yields

$$n_d - 5 = \frac{E_{\text{prom}} - (E_p + E_d)}{S}$$

where n_d is the total number of d electrons and not the number of bonding d electrons shown in Figs. 2 to 4. For $n_d > 5$, the dumber of bonding d electrons will be $10 - n_d$.

The procedure outlined above and given in more detail in the appendix was used to characterize the optimum electronic configurations of the transition metals. In many instances, the energy varied clearly across the configuration range with the calculated minimum energy outside the range allowed by the Engel correlation, since the above equations do not incorporate any terms to produce an increase in free energy at the limits set by the Engel correlation. Configurations could clearly be assigned as the limiting configuration of lowest energy. For elements with the minimum energy at a configuration intermediate between the limiting configurations, the atomization energies were often almost identical for either limit. Since there was little effect upon the calculated energy, the entry in Tables 4 and 5 was chosen to minimize the number of starred entries corresponding to an element with a different electronic configuration from the other elements of the same group. In Figs. 2 to 4, values are plotted for both the limiting configurations when they are of closely the same stability.

With the optimum configurations clearly established we may proceed to the calculation of atomization energies of the various structures. For example, we previously calculated the atomization energies of bcc and hcp Nb on the basis of the d^4s and d^3sp configurations. The d^4s configuration is the optimum configuration for bcc Nb, and the atomization energy of 173 kcal is unchanged, but the optimum configuration of hcp Nb is the $d^{3.3}sp^{0.7}$ configuration which yields the valence-state bonding energy of 197 kcal listed in Table 4 and an atomization energy of 163 kcal after subtraction of promotion energy to indicate that hcp Nb is unstable by 10 kcal relative to bcc Nb. Table 6 presents the results for the other transition metals.

Table 6 in agreement with Table 1 shows the universal occurrence of a stable bcc structure for all the transition metals of the first six groups. It is of interest to comment on the primary factor responsible for the widespread occurrence of the bcc structure. It could be due to either a preference for d bonding over p bonding or to difference in promotion energies of the $d^{n-1}s$ and $d^{n-2}sp$. Examination of Figs. 2 to 4 and Tables 2 and 3 shows that except for Ca and Sr, the promotion energy alone is the primary factor. In fact, the greatly reduced bonding energy of the d electrons compared to the bonding energy of the p electrons when one or more d electrons are involved in bonding tends to stabilize the hcp structure, but in no instance is this stabilization sufficient to wipe out the advantage of the lower promotion energy of the $d^{n-1}s$ configuration and to make the bcc structure unstable at all temperatures.

On the right-hand side of the periodic table, the bonding energies of Fig. 3 yield the results of Table 6 that hcp Mn, Fe, and Co are close in energy to the bcc structures. It is difficult to predict the entropies of these phases because of the magnetic contributions. The higher atomization enthalpy of bcc Fe is sufficient to ensure its stability relative to hcp Fe at low pressures. However, bcc Co has a smaller atomization enthalpy than hcp Co by 2 kcal, which is too large to be overcome by any likely entropy difference. Mn is one of the unique elements like U and Pu for which several different electronic configurations yield closely the same stabilities. The values in Table 6 indicate that α-Mn and β-Mn have a substantial contribution from the d^5sp configuration but that there must be atoms with d^4sp^2 and d^6s configurations maintaining separate identities rather than averaging electron configurations with their neighbors.

The experimental spectral data clearly confirm that the bcc and hcp structures are to be correlated with $d^{n-2}s$ and $d^{n-2}sp$ configurations, and the theoretical calculations[8] are also in agreement with the Engel correlation. For the ccp (fcc) structure, the theoretical calculations[8] indicate correlation with $d^{n-3}p^3$ rather than with $d^{n-3}sp^2$ which Engel has suggested. Table 7 presents the spectral data for Y, for which all expected electronic states of high multiplicity of the possible configurations involving $4d$, $5s$, and $5p$ electrons are known except for the d^2p ^{4}P state and the states of the p^3 configuration. The comparison of dsp-ds^2, s^2p-sp^2, d^2s-d^2p, dsp-dp^2 differences indicates that the p^3 configuration would be expected well above 100 kcal. Examination of the spectral data for the other elements, for which data are less complete than Y, also indicates that the $d^{n-3}p^3$ configuration has too high a promotion energy to make a significant contribution to the metallic bonding. The theoretical calculations might be reconciled with the spectral data by using the $d^{n-1}p$ configuration since the calculations indicated that mixtures of d^n and $d^{n-3}p^3$ would yield the ccp structure. Except for La, which has an unusually low promotion energy of 50 kcal for the d^2p configuration, the promotion energies for the transition elements range from over 60 to 120 kcal. Accurate comparison with the $d^{n-3}sp^2$ configuration cannot be made since Y is the only transition element for which data on this configuration are known. This is due to the fact that transitions from the ground electronic state to states of this configuration are very weak because they involve multiple electron jumps and violate other spectral selection rules. However, as shown in Table 7 for Y, it is expected that the sp^2 configuration will not lie below the $d^{n-1}p$ configuration. However, only to the far left of the periodic table, where the d electron bonding is as strong as or stronger than the s, p bonding, will the $d^{n-1}p$ configuration have any advantage over the $d^{n-3}sp^2$ configuration. On the right-hand side of the periodic table, the $d^{n-3}sp^2$ configuration has a very large advantage in valence-state bonding energy over the $d^{n-1}p$, $d^{n-1}s$, and $d^{n-2}sp$ configurations since the $d^{n-3}sp^2$ configuration produces the greatest unpairing of electrons.

Thus it is possible that the $d^{n-1}p$ configuration might play a role in stabilizing fcc Ca, Sr, and La, but the bonding energies required do not seem reasonable. Because of the fact that all configurations that would be considered for the elements of the first six groups have all the valence electrons unpaired in the valence state, the possibility of significant contribution to the bonding by either $d^{n-1}p$ or $d^{n-3}sp^2$ for the other metals of the third through sixth groups can be excluded. For example, from the data of Fig. 2 and Table 7, one calculates that fcc Y is unstable by 20 kcal compared to hcp or bcc Y. It is difficult to reach a definite decision for La owing to uncertainty of spectral data for Sc and La. One method of extrapolation would estimate the sp^2 promotion energy for La to be much lower than Y, perhaps around 50 kcal, and another method would estimate about the same promotion energy for Y and La. Also the estimation of the bonding energy of La in the sp^2 valence state is difficult because the lanthanide contraction creates a discontinuity between La and Hf. Calculation of valence-state bonding energies for the metals on the right-hand side of the transition series indicates that the fcc structure with the sp^2 configuration or with the lower limit indicated by the Engel correlation of $d^{n-2.5}sp^{1.5}$ would be stable with respect to bcc and hcp structures with $d^{n-3}sp^2$ promotion energies of 110 to even 190 kcal. Examination of the data for the $d^{n-3}s^2p$ configuration and the trends across the transition series indicates that the promotion energies can be expected to fall in that range. When more spectral data become available for the transition metals, it should be possible to carry out calculations for the ccp structure as in Table 6.

APPENDIX *Calculation of Atomization Enthalpies*

The use of Figs. 2 to 4 and Tables 2 and 3 for calculating atomization enthalpies will be illustrated by the example of Zr. Although the difference between the energy and the enthalpy of atomization is within the uncertainty of these calculations, the values in Figs. 2 to 4 are designed to yield enthalpy values at 298°K. From Fig. 2 it is seen that the bonding enthalpy of the d^3s valence state is $3 \times 36 + 51$ or 159 kcal. The promotion energy of 14 kcal from Table 3 then yields an atomization of bcc Zr at 298°K of $159 - 14$ or 145 kcal to the ground state of the gaseous atom. For the configuration $d^{2.5}sp^{0.5}$, the bonding enthalpy is $2.5 \times 39 + 1.5 \times 51$ or 174 kcal, but the promotion energy is 28 kcal, yielding an atomization enthalpy of 146 kcal. Finally the configuration $d^{2.7}sp^{0.3}$, which is the configuration that yielded the most stable solid as discussed earlier, yields a valence-state bonding enthalpy of $2.7 \times 37.7 + 1.3 \times 51$ or 168 kcal. The promotion energy of 22 kcal then yields an atomization enthalpy of 146 kcal. When the promotion and bonding energies are so balanced that the optimum configuration is centered between the limiting configurations, the variation across the range is so small that calculations carried out to fractions of a kcal are necessary to show the variation accurately. Such an accuracy is not warranted for the present purposes.

In the above calculations, the promotion energy that is used should be in principle the average promotion energy of the range of electronic states that correspond to the molecular orbitals or band levels being filled. In the calculations upon which Figs. 2 to 4 and Tables 4 and 5 are based, the promotion energy was taken in every case as the lowest promotion energy of the bottom level of each group of levels. As long as the difference between the lowest level and the average level does not change abruptly from one element to the next, this is quite an adequate procedure, since the use of average promotion energies rather than the lowest would increase the promotion energies and the calculated bonding energies equally. Thus the curves of Fig. 2 to 4 would be uniformly raised. Since the observed points fall on smooth curves, there is direct evidence that there is no abrupt variation from one element to the next, and no appreciable error is introduced by the calculation procedure used here.

It is interesting to note not only that the curves of Figs. 2 to 4 are smooth but also that the bcc and hcp curves fall very close to each other. Differences in additional energy obtained by hybridization and other interactions could produce differences; it is a simplification to have the bonding energies so nearly independent of structure or configuration. This makes it easier to extrapolate the use of these curves, and common curves were assumed for both structures on the right-hand side of the periodic table where data for both structures are lacking. The closeness of the curves also provides a basis for using the curves to calculate values for fcc structures.

The electron bonding energies in Fig. 3 are plotted against the number of bonding *d* electrons. In view of the magnetism of the metals from Cr to Ni, it is clear that not all the unpaired electrons of the valence state are used in forming electron-pair bonds. In fact, the value of *S*, the rate of change of *d* bonding energy with the number of *d* electrons, is so large in Fig. 3 that one could calculate that Cr, for example, would have a higher bonding energy by not using one of its *d* electrons if the bonding of the remaining *d* electrons were based on the value higher up the curve corresponding to just the number of *d* electrons used in bonding. By using magnetic data, it would be of interest to calculate the plot of Fig. 3 on the basis of the actually bonding electrons. However, for the present purposes, Fig. 3 is adequate.

REFERENCES

1. Moore, C. E.: "Atomic Energy Levels," vols. 1–3, Government Printing Office, Washington, D. C., 1949, 1952, 1958.
2. Hume-Rothery, W., and G. V. Raynor: "Structures of Metals and Alloys," 4th ed., Institute of Metals, London, 1962.
3. Engel, N.: *Ingenioren,* N101 (1939), M1 (1940); Haandbogi Metallāre: Selskabet for Metalforskning, Copenhagen, 1945; Kemisk *Maandesblad* **30**(5):53; (6):75; (8):97; (9):105; (10):114 (1949); *Powder Met. Bull.* **7**:8 (1954); *ASM Trans. Quart.* **57**:619 (1964).
4. Brewer, L.: Thermodynamic Stability and Bond Character in Relation to Electronic Structure and Crystal Structure, pp. 221–235 in P. A. Beck (ed.), "Electronic Structure and Alloy Chemistry of the Transition Elements," Interscience Publishers, Inc., New York, 1963; Dover Publications, Inc., New York, 1965.
5. Brewer, L.: "Prediction of High Temperature Metallic Phase Diagrams," chap. 2, pp. 12–103, in V. F. Zackay (ed.), "High-strength Materials," John Wiley & Sons, Inc., New York, 1965.
6. Brewer, L.: A Most Striking Confirmation of the Engel Metallic Correlation, submitted to *Acta Met.*,1966.
7. Hultgren, R., R. L. Orr, P. D. Anderson, and K. K. Kelley: "Selected Values for the Thermodynamic Properties of Metals and Alloys," John Wiley & Sons, Inc., New York, 1963, and Supplement, 1965.
8. Altmann, S. L., C. A. Coulson, and W. Hume-Rothery, *Proc. Roy. Soc.* **240A**:145 (1957).

THE THERMODYNAMIC STABILITY OF METALLIC PHASES*

Oswald Kubaschewski

Metallurgy Division, National Physical Laboratory, Teddington, England

INTRODUCTION

To the thermochemist the *stability* of a phase means its free energy of formation from the component elements (in their standard states) relative to that of the *neighboring* phases or its free energy relative to that of other modifications of the same composition, at a given temperature and pressure, and to the thermochemist this is at present essentially an experimental problem.

The results are required firstly for the calculation of equilibria of technological significance and secondly as a basis for considerations of bond mechanism, that is, the fundamental causes of the thermodynamic properties. The present colloquium is concerned only with the second point. However, all attempts to date to derive the data by other independent means have failed or, at best, produced inaccurate results, owing to the involved nature of the underlying bond mechanisms. A number of models have been and are being devised and studied individually in order to recog-

* This work forms part of the research program of the National Physical Laboratory and is published with the permission of the Director.

nize their contribution to the stability of phases as a whole. One may call this *the physicist's approach.* Progress along these lines is naturally slow. *The chemist's approach* may be somewhat quicker but is less satisfactory because it is more empirical and, in the long run, less accurate. The chemist's method is to measure experimentally the thermochemical properties of a relatively large number of substances and to compare these among themselves and with other properties that appear to contribute to the thermodynamic stabilities. In the end, of course, the two approaches must lead to the same fundamental theory which, if it is to be uniform, will be quite involved.

Thermochemical data of reasonable accuracy that have been obtained to date mostly relate to alloys of relatively low melting point because the experimental difficulties are relatively small. This is, in a way, unfortunate because many of the low-melting-point metals belong to groups IIB to VB, to the so-called "metametals" and "semimetals," the bond mechanism of which appears to be even more complicated than that of the others.

Consequently, the next stage should be to explore the thermochemical properties of metals and alloys with higher melting points. The task is a formidable one, and it is appropriate at this stage to review briefly the present position of the experimental methods.

ASSESSMENT OF EXPERIMENTAL METHODS

Heat capacities and heats of transition are being obtained on a relatively large scale by the so-called "drop method." The method is quite useful for the investigation of pure metals and "pure" compounds where heat contents are required in order to convert the heat and entropy of a chemical reaction from one temperature to another. It is, in principle, not accurate enough for the evaluation of true heat capacities and even less so when the alloy undergoes a transformation during the drop involving some rearrangement of atoms. Some improvement of the method may be possible by using higher temperatures for the calorimeter into which the specimens are dropped.

However, for the measurement of true heat capacities and heats of transformation, the direct method, i.e., an adiabatic or similar calorimeter, is fundamentally superior. Accuracies that can be achieved above 1000°C with the available apparatus are of the order of a few calories per gram atom for a heat of transformation, and, with an appreciable financial effort, this could certainly be improved at least tenfold. Chemical interactions within the calorimeter will be the limiting factor.

Accurate determinations of heats of formation are a more difficult problem. Either difference methods or direct-reaction methods may be employed. In both cases the temperature of the calorimeter should be considerably higher than room temperature. *Liquid-metal calorimeters* are being used as difference methods, the liquid metal (e.g., Sn) acting as the solvent. Accuracies for the resulting heats of alloying of a few calories per gram atom have been claimed. However, this applies only where the solute metals and alloys are relatively noble and the heats of solution are of the same order as, or less than, the heat of formation. Undoubtedly, there is still room for further improvement of liquid-metal calorimetry, implying a wider application, before its limitations can be assessed.

Direct-reaction calorimetry appears to be more versatile, but higher temperatures of measurement than with the difference methods are required, and the accuracy consequently suffers. Dench[1] has described an adiabatic reaction calorimeter in which the heats of alloying of, for example, chromium alloys have been determined at 1400°C with an accuracy of ±50 cal/g atom to which the "chemical" errors have still to be added. It is believed that the limits, in terms of temperature, of adiabatic-reaction calorimetry for solids have been reached. The potentialities of heat-flow and isoperibol calorimetry at high temperatures remain to be explored.

Calorimetric measurements are also used for the determination of absolute entropies. Modern improvements have made it possible to measure C_p fairly near absolute zero. However, the uncertainty in the value of the zero entropy owing to frozen-in equilibria impairs the usefulness of the method. In general, the present author prefers a combination of calorimetric heats of formation or reaction with the results of free-energy measurements for the derivation of the entropies of reaction.

Two important experimental approaches to measuring free energies of alloying are known: dissociation-pressure methods and electromotive-force methods. The former rely on the determination of the difference between the logarithms of the dissociation pressure of the alloy and the vapor pressure of the volatile component(s). A number of methods are available of which Knudsen's effusion technique in its various forms is probably the most sensitive and is also the one most widely applied. After having made a comparative study with other vapor-pressure methods, we would confirm the superiority of the effusion technique, in particular when combined with a tracer method; but would consider its accuracy still too low for really good results at high temperatures. In terms of activity, its accuracy at, say, 1400°C is of the order of ±5% at best, but normally worse.

The mass spectrometer is coming to the fore as a means of measuring vapor pressures. It is too inaccurate for alloys but most useful in dealing with such types of evaporation where different, polyatomic, chemical species appear in the vapor.

Other vapor-pressure methods are applicable only to special cases, such as the Langmuir method for pure high-boiling-point elements or the transpiration method for low-boiling-point alloys.

The electromotive-force method has been and is considered the most reliable "free-energy method," provided a suitable electrolyte can be found and the component elements differ sufficiently in standard potentials. Considerable difficulties have been encountered with molten halides as the electrolytes when the temperature exceeded 500°C or so. Fluorides have occasionally been used, but a real advance was made when ionic-conducting oxide solutions, such as thoria-yttria, were introduced as suitable electrolytes. Accuracies for the resulting activities are difficult to estimate but are certainly better than those achieved by vapor-pressure measurements. The testing of the method is still in progress, and its limitations with respect to alloy thermodynamics cannot yet be assessed.

Here it is only possible to mention other methods, such as the evaluation of phase diagrams and distribution equilibria. They depend on very accurate work with regard to equilibration and chemical analysis and should not be used independently of calorimetric data.

As little use as possible should be made of the temperature coefficients of free-energy results for the derivation of heat and entropy data, in particular where equilibria with solid phases are involved. Such values derived from temperature coefficients are likely to be subject to errors which cannot be estimated and may be substantial.

Concluding this section, it should be added that the reliability of the available numerical information depends almost as much on the critical assessor as on the original experimentalist. No other field of physical metallurgy lends itself so much to checking and cross-checking of its data as chemical thermodynamics. The assessors, however, must be equally well acquainted with the fallacies of the experimental methods as with the fundamental principles of thermodynamics. Ten years ago, we tried to assess and tabulate the relevant data for alloys,[2] and meanwhile Hultgren and his colleagues[3] have expanded this tabulation to include the more recent publications.

ON TRANSFORMATIONS

In general, the thermochemist is inclined to take the relative stability of the pure metals for granted. The question, for instance, why the boiling points of Na and W, both body-centered cubic, differ so much is carefully avoided, and no satisfactory answer to this question has actually been offered from any source.

Zener, in his paper at this colloquium, indicates a possible reason why metals tend to have a close-packed structure at low temperatures and may convert into a bcc structure at higher temperatures: "The looser the packing, the higher the entropy of vibration." The free-energy differences involved can be quite small. Figure 1 shows the free-energy differences

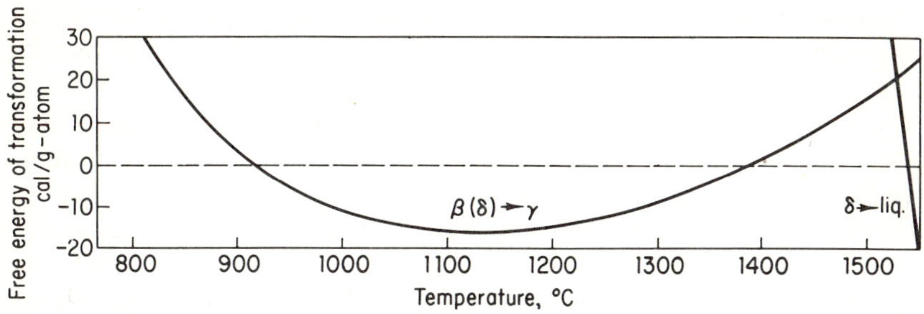

Fig. 1 Free energies of transformation and fusion of Fe.

between bcc and fcc Fe, as determined by Dench,[4] which in the γ range never exceeds 20 cal/g atom, only a little more than the energy assigned to the effects of strains, interfaces, or small amounts of impurities. Near the transformation points, the "driving force," i.e., the free-energy difference, is so small that one begins to wonder to what extent the effects observed by measurements of specific heats and other properties are related to true equilibria.

The situation is aggravated when more than one atomic species is present in quantity. The atomic rearrangement requires time, and the experimental procedures rarely seem to have ascertained whether true equilibrium had been attained. In the Au-Cu system, for instance, the thermochemical data[3] do not seem to allow a clear differentiation between the disordered and the various ordered phases.

In the case of melting phenomena, as previously described by Schneider and Heymer,[5] the situation is less obscure because the free-energy changes above and below the melting points are more pronounced. Even so, there is ample evidence, which cannot be recorded here, of a time factor which allows for structural changes that seem to affect the so-called premelting and postmelting phenomena.

The point the author has been trying to make here is that, except for qualitative considerations, much of the thought that has gone into explaining the phenomena near transitions in metals and alloys has been based on insecure experimental grounds. We shall return to this point later in the paper.

"REGULAR SOLUTIONS"

For estimating the entropy of concentrated metallic solutions, thermochemists often rely on the "regular solution model," introduced by J. Hildebrand, which implies that the excess entropy of mixing is zero at all concentrations: in other words, the excess free energy and the heat of mixing are identical.

Useful as this concept has been for approximate calculations, there is now sufficient experimental evidence to challenge its validity. Reasonably accurate heats and entropies of mixing are available for some 50 binary liquid systems and 8 solid systems that exhibit complete mutual solubility. Data for the integral quantities have been taken from the two compilations mentioned,[2,3] together with more recent critical evaluations by the present author. Most of the liquid systems are of relatively low melting point involving metals such as Bi, Sb, Pb, Sn, Tl, In, Hg, Cd, Zn, Al, Cu, Ag, Au, Mg, and Na. The solid systems referred to are Cr-Fe, Cr-Mo, Au-Ni, Au-Cu, Fe-Ni, Cu-Ni, Ag-Au and Mg-Cd.

In Fig. 2 is plotted $\Delta S^{xs}_{\max}/R$ versus $\Delta H_{\max}/\frac{1}{2}R(T_{e1} + T_{e2})$ where $\Delta S^{xs}_{\max}$ and $\Delta H_{\max}$ represent the integral maximum or minimum excess entropies and heats of mixing, respectively, per gram atom; and T_{e1} and T_{e2} are the boiling points of the component metals. Each cross or circle represents a binary system.

In Fig. 2 the regular solutions would be represented by a horizontal line at $\Delta S^{xs}_{\max}/R$ = zero. Instead, we find an inclined straight line passing through the ideal solution value. The accuracy with which the excess entropies are known is estimated at about ±0.2 cal/(deg) (g atom); this

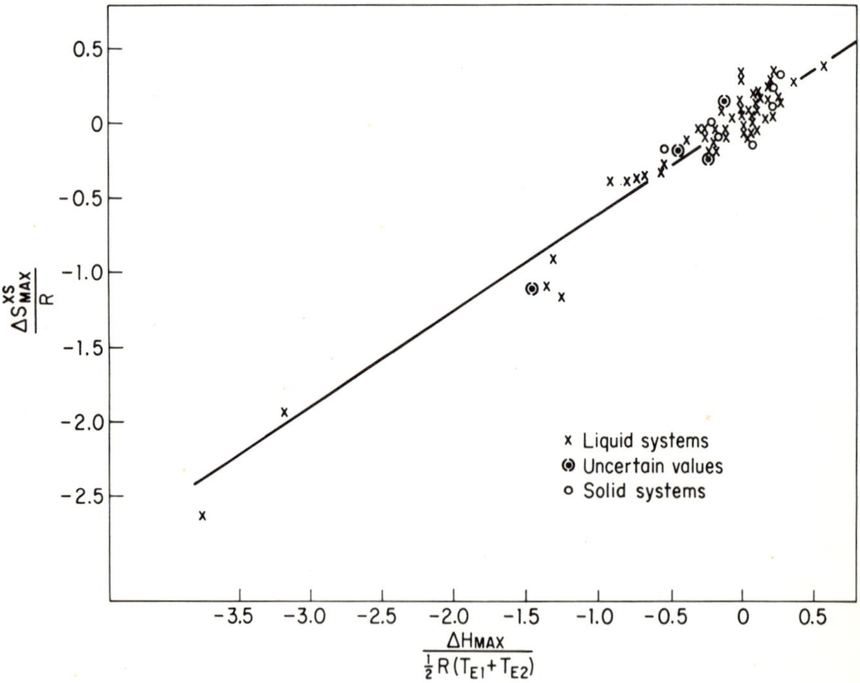

Fig. 2 Plot of maximum or minimum heats versus excess entropies of mixing of solid and liquid solutions.

corresponds to the scatter of the experimental values. The straight line may be represented by

$$\Delta S_{\max}^{xs} = \frac{0.64 \Delta H_{\max}}{\frac{1}{2}(T_{e1} + T_{e2})} \tag{1}$$

It should be pointed out that the systems considered in Fig. 2 are of apparently quite different bond mechanism (e.g., Na-Hg, Fe-Si, Ag-Au, Tl-Sb, Cr-Mo). Nevertheless, they all seem to fall approximately into a single pattern, indicating that the fundamental causes for the values of the excess functions are similar, and that interpretations of the one must also account for the other.

A similar relationship seems to exist between the excess volumes and the heats (or the excess entropies) of alloying. However, since the known volume changes on mixing are rather inaccurate for experimental reasons, the plot of $\Delta V_{\max}$ vs. $\Delta H_{\max}$ is not reproduced here.

Although the excess entropies of mixing of the solutions under consideration are generally known with an estimated accuracy of ± 0.2 cal/(deg)(g atom) and often better, it is usually difficult to locate the position of the maximum exactly in terms of concentration. All one can say is that the experimental data are consistent with the assumption that the maxima or minima in the ΔH and ΔS^{xs} curves occur at the same composition in a particular system. If this is correct, then it is possible to estimate maximum excess entropies from free-energy measurements alone when the heats of mixing are not known. It follows from Eq. (1) that the excess free energies are related to the excess entropies in the following manner,

$$\Delta S_{\max}^{xs} = \frac{\Delta G_{\max}^{xs}}{0.78(T_{e1} + T_{e2}) - T_{\text{obs}}} \tag{2}$$

where T_{obs} is the absolute temperature at which the free energy has been measured.

The experimental evidence also shows that symmetric ΔH or ΔS^{xs} vs. concentration curves are rare if they occur at all. This means that the representation of the heat of mixing by a parabola ($\Delta H = \text{const } N_1 N_2$), as is often done with "regular solutions," is an oversimplification not borne out by the experimental results. Attempts by the author to find some empirical relationship that would help in predicting the position of the maxima have so far failed.

The present author has also been unable to prove the validity of Eq. (1) by the application of fundamental models. Nevertheless, he uses Eq. (2) for estimates of entropies (and heats) from free-energy data, in preference to the use of the regular solution model, and the evidence has been presented here as a possible starting point for further considerations.

It may be added that an attempt has also been made to treat ordered intermetallic phases likewise, taking the excess entropy as identical with the measured entropy of formation. As expected, the scatter of the values in the plot of $\Delta S/R$ vs. $\Delta H/R(N_1T_{e1} + N_2T_{e2})$ is greater than in Fig. 2, owing obviously to the fact that complete order is rarely attained in such phases, possibly partly because of frozen-in equilibria. The corresponding curve thus attains the form of a "limiting relationship." More accurate data on phases in which equilibrium is truly assured will be needed.

BOND MECHANISMS IN ALLOYS

If one is to express phase stability quantitatively without reference to experimental free energies, one must start with models that describe the types of atomic bonds responsible for the observed enthalpies and entropies. Again it is easier to discuss changes in bonding on alloying than to derive the absolute values of bonding energy.

In limiting cases one recognizes three main types of bond: the polar, the covalent, and the metallic, of which the metallic bond is the least well understood. Once one moves away from the limiting cases, these and other bonds overlap to such an extent as to make the quantitative treatment very difficult. The concepts still remain rather vague, and different authors operate with different interpretations. The thermochemist cannot help feeling rather at a loss. He is inclined to revert to the phenomenological, and simply seek out independent physical properties of the metals and alloys to be compared with the observed thermochemical properties.

For long, Hume-Rothery has been pointing out three important factors responsible for the type of equilibrium diagram formed by two metals: the size factor, the electron concentration, and the electrochemical factor. The electrochemical factor corresponds to the polar bond, and the size factor is particularly important for the formation of metallic phases.

In 1958, the present author suggested[6] that the stability of intermetallic phases was essentially an atomic packing effect. Although this led to a cursory estimate of the heat of formation of intermetallic compounds from crystallographic data, the method has been accepted by only a few investigators, and the quantitative calculations involving a redefinition of the coordination number will therefore not be repeated here.

Since the present author continues, however, to use the concept for the estimation of the type of equilibrium diagram that may be formed by two metals, it has a direct bearing on phase stability and may be mentioned briefly. It is really an extension of Hume-Rothery's suggestion and, to some extent, of Bernal's earlier models. It should be confined to combinations of metals from groups IA to VIIIA, IB, and IIB except Hg.

It is concluded that the atoms in a stable intermetallic phase formed between these metals are always more closely packed than in the component metals, by virtue mainly of the difference in atomic sizes. For stability, it is essential that the larger atom should be the "stronger binder," i.e., have the higher heat of sublimation, because it is the larger atom, A, that will gain in coordination while the smaller atom, B, will lose. This implies that the bond energy, AB, is the mean of the bond energies AA and BB, in the first approximation.

Consequently, if the larger atom has the lower heat of sublimation, one would expect that the heats of formation are positive, and, above a certain size difference, liquid miscibility gaps will appear.

In order to test these assumptions, about 350 binary phase diagrams have been ordered in groups and entered into a diagram of $(r_A - r_B)/\frac{1}{2}(r_A + r_B)$ versus $(L_A - L_B)/\frac{1}{2}(L_A + L_B)$ where r_A and r_B are the atomic radii, and L_A and L_B the heats of sublimation, at 25°C, of the component metals. A mostly small correction term $(\epsilon_A - \epsilon_B)^2$ has been added to the heat-of-sublimation term in order to account for the "electrochemical factors," ϵ_A and ϵ_B being the electronegativities of the components.

The results are shown in the simplified representation of Fig. 3. The diagram exhibits certain areas indicating the preferences for the type of equilibrium diagram to be formed. Systems forming large ranges of disordered solutions have been included in the survey although these are not covered by the original concept.

The boundaries in Fig. 3 are not really as sharp as the lines in the figure suggest. There are "regions of transition" accounting for the much larger variety of types of equilibrium diagram, such as eutectic systems showing appreciable solid solubility, or more complicated systems showing substantial solid solubility, as well as a number of extensive intermetallic phase fields. A more detailed analysis of the original diagram is of considerable interest but would lead us too far here.

A sharp distinction between simple eutectic systems and those showing liquid miscibility gaps cannot be made because it depends on the melting points whether a miscibility gap appears in the phase diagram or not. Instead one ought to consider the heats of mixing which are, however, known only for a small proportion of the systems considered.

Of the 350 systems examined, less than 20 clearly do not fit into the present pattern (e.g., Cr-W, Mg-Ni, Ca-Na) but a number of the exceptions are systems involving Mn or U, the atomic diameters of which are difficult to assess. The so-called "Hume-Rothery systems" also do not agree well with the present pattern, possibly because of the relatively low coordination of pure Zn and Cd despite their hexagonal A_3 structure.

The rule represented by Fig. 3 has some similarity with the Hildebrand rule as extended by Mott,[7] but appears to agree better with the observations whether the original concept that led to its formulation is correct or not. Even so, Fig. 3 cannot be more than a crude, though useful, guide because too many simplifications have entered the basic considerations or, as C. Wagner once put it: "I am wary of attempts which consider atoms as constituents rather than ions and electrons."

There is much less difficulty in understanding the polar bond, but this colloquium is less concerned with strongly ionic compounds; and so it may be passed over.

Many authors do not differentiate distinctly between the metallic and covalent bonds. However, when considering the thermochemical properties of the alloys formed by the metametals and semimetals of groups IIB to VB, one wonders if such a distinction will not be necessary. The tendency of such combinations to exhibit positive heats and positive excess entropies of mixing is independent of the size factor and can qualitatively be explained by the partial destruction of interatomic bonds in the component metals on mixing.

Pauling assumed on the basis of quantum-mechanical calculations that the covalent-bond energy of a compound is the geometric mean of the

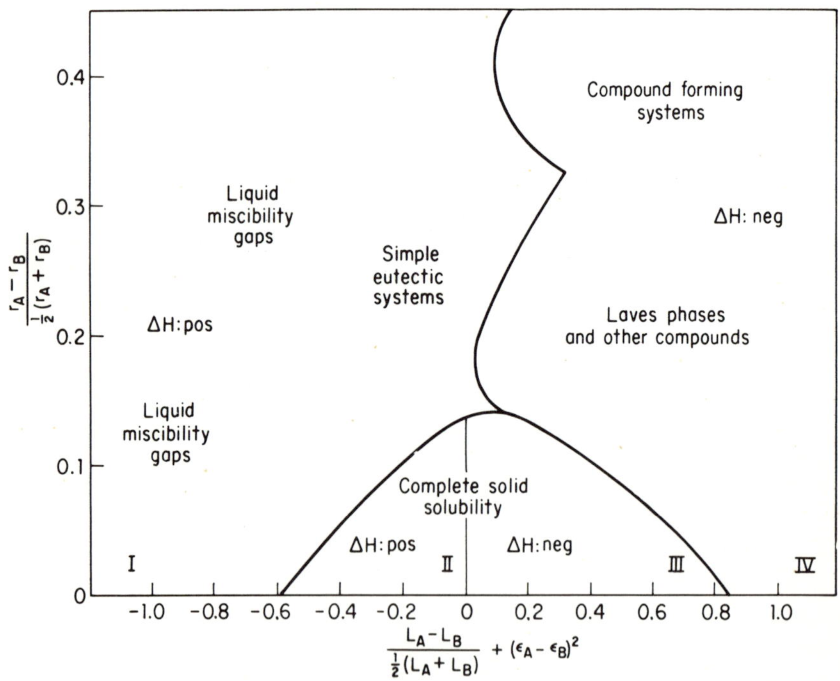

Fig. 3 Diagram indicating, in terms of atomic radius, heat of sublimation, and electronegativity of the component metals *A* and *B*, regions of preference for the formation of certain types of binary equilibrium diagrams.

TABLE 1 ***Maximum Integral Heats of Mixing in Binary Liquid Systems of Various Covalent Metals Corrected by an Electronegativity Term,*** $\Delta H_{\text{obs}} + 23{,}066\,(\epsilon_A - \epsilon_B)^2$

In cal per g atom

$\epsilon =$	1.2 Mg	1.5 Al	1.5 Zn	1.5 Cd	1.5 In	1.5 Tl	1.65 Hg	1.65 Pb	1.7 Sn	1.8 Bi	1.85 Sb
Sb			1970	2030	2030	2370		840	20	200	0
Bi	3480	3050	3200	2290	1630	980	660	255	255	0	
Sn	2300	1910	1730	1360	870	1095	450	390	0		
Pb	2350		2370	1200	750	160	145	0			
Ga			390	640							
Hg			620	−130	−30	170	0				
Tl			1430	560	135	0					
In		1220	780	350	0						
Cd	320		510	0							
Zn	650	615	0								
Al	(1000)	0									

bond energies of the component elements: $H_{AB} - \sqrt{H_{AA}H_{BB}} = 0$. If we apply this principle to the liquid mixtures of the metals in question, we would expect the heat of mixing to be more positive, the greater the difference in covalent bond strength in the component metals. In order to test this hypothesis, we may again revert to an empirical method of comparison, since we have no absolute measure for the covalent-bond energies in the present cases. In Table 1, the metals in question (to which Mg has been added) have been arranged in a certain order which corresponds approximately to the order of the dissociation energies of their diatomic gaseous molecules. The figures in the table represent, in calories per gram atom, the experimental maximum heats of mixing, ΔH_{obs}, for the respective binary mixtures to which the values of $23{,}066(\epsilon_A - \epsilon_B)^2$ have been added in order to eliminate the ionic contribution to the heat of mixing.

As the metals have been arranged in the table, one would expect the covalent heats of mixing to increase from the diagonal representing zero heats toward the upper left-hand corner, because in this direction the difference in the covalent-bond energies of the component metals is supposed to increase. Although there are exceptions, the values seem to support the view submitted above.

One could go one step further and try to derive relative covalent-bond energies for the liquid metals from the values in the table, but at this stage no useful purpose seems to be served by doing so, in particular as the relations are more complicated than might appear from the present argument. For example, when investigating the thermochemistry of the aluminides of the transition metals of the fourth row some time ago,[8] we found the heats of formation to be substantially more negative than either the electrochemical or the size factor could explain. We then compared the interatomic distances in the various aluminides which, it should be emphasized, show a great variety of crystallographic structure. It was found that in any one system the Me-Al distances (Me = metal) are nearly independent of composition but vary from system to system. The Al-Al distances do not vary much from system to system for a given composition, but do vary with composition, as can be seen from Fig. 4 where the average distances between Al atoms that touch, divided by 2.86 A, have been plotted; where the Al atoms do not touch, the next nearest distances have been taken.

The Δd_{Al} curve shows a minimum, that is, maximum contraction, at $N_{Al} = 0.7$ to 0.8. The relatively good agreement of the Δd_{Al} values to a single curve irrespective of the nature of the other metal, Me, is particularly striking in view of the variety of structures formed by the systems under consideration.

The shortening of an atomic distance may be taken to indicate a strength-

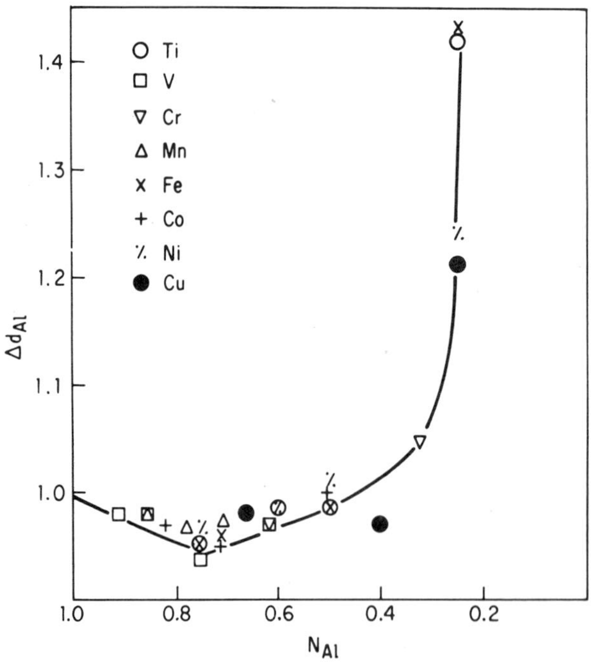

Fig. 4 Reduced shortest Al-Al distances plotted against composition for the aluminides of various transition metals.

ening of the bond, and a tendency was found for the maximum in the heat of formation to shift to compositions around 75 a/o Al unless the overall enthalpies of formation are strongly negative.

The present author is not in a position to account for the strengthening of Al bonds by the addition of the transition metals. Since neither the size factor nor the electrochemical factor accounts fully for the observed heats of formation, the changes in bond strength have provisionally been classified as of a covalent nature, and it follows that changes in covalent bonds can make a positive or negative contribution to the heat of formation.

FREE-ENERGY AND PHASE DIAGRAMS

In the preceding sections some observations have been recorded that occurred to the author and his colleagues while carrying out experimental measurements on the thermochemical properties of metallic solutions and intermetallic phases. The least, one hopes, these might accomplish is to provoke some thought. Further progress the present author will have to leave to those more qualified than he to deal with wave mechanics.

Let us now return to the actual measurements of the thermochemical properties of alloys to which we have contributed in the past. In Fig. 3 areas II and III represent the regions, in terms of atomic diameter and heat of sublimation, where two metals are expected to be completely soluble at some temperature in the solid state, provided the component metals have the same or nearly the same structure. In area II the heat of solution is expected to be positive; in area III negative. If the heat of mixing, or rather the excess free energy, is positive, the system must show a miscibility gap at some lower temperature, the mutual solubilities decreasing to nil at absolute zero. In cases where such a miscibility gap has not yet been detected, as for instance in the system Ni-Cu, the reason lies in the slow rate of decomposition. In the Cr-Mo system some alloys had to be annealed for over a year[9] in order to demonstrate clearly that decomposition occurred, the limits of the gap having been calculated from measured thermochemical data.

If the disordered solid solutions form with an absorption of heat and the system forms, in addition, one or more ordered phases, also with a positive ΔH, then these phases must "disappear" at some lower temperature. This type of system is exemplified by the Fe-Cr alloys. The thermochemical values of the disordered solid solutions and the σ phase were measured by various complementary methods.[9] The free-energy versus composition curves are shown in Fig. 5 for three temperatures. The disordered alloys have, of course, a higher entropy than the ordered ones, and the σ phase therefore becomes metastable at some higher temperature—in this case, above 815°C. Because of the different shapes of the enthalpy

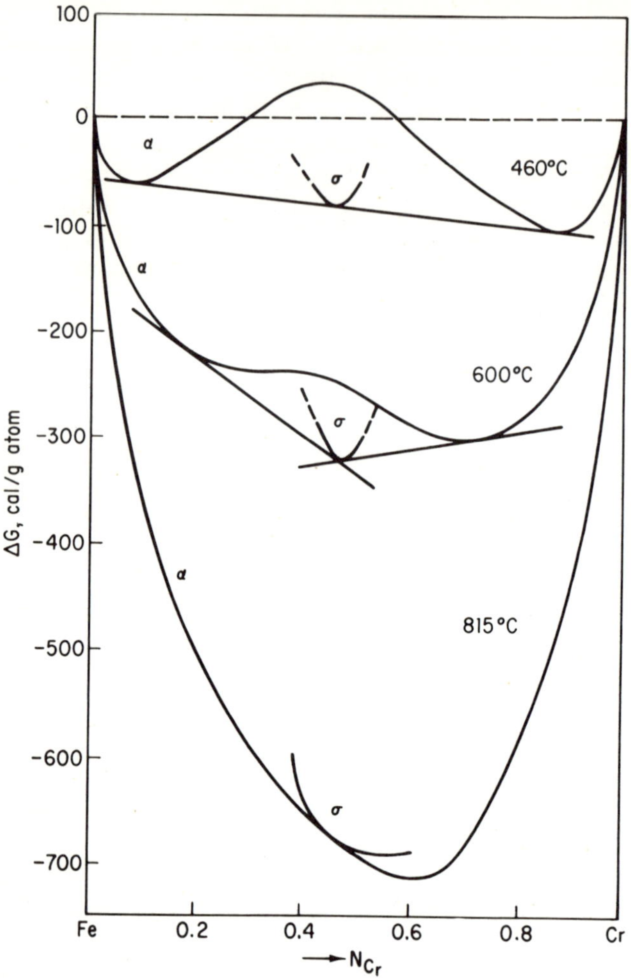

Fig. 5 Free energies of formation of the α and σ phases in the system Fe-Cr at three temperatures. [*From Kubaschewski and Chart, J. Inst. Metals* **93:**335 (1965), *reprinted by courtesy of The Institute of Metals.*]

curve (approximately a distorted parabola) and of the curve for the positional entropy [given by $-R(N_{\mathrm{Cr}} \ln N_{\mathrm{Cr}} + N_{\mathrm{Fe}} \ln N_{\mathrm{Fe}})$], the characteristic free-energy curves, as shown in the diagram, result at lower temperatures. While the inverted humps in the ΔG-N curve, though shifting, must persist at negative values with decreasing temperature to near absolute zero, the ΔG-N curve for σ must eventually disappear in the inverted ΔG trough, owing to its positive heat term—in the present case, at 460°C. The corresponding equilibrium diagram (Fig. 6) is expected to be typical for quite a large number of binary systems in region II (Fig. 3), for which diagrams only the upper part has as yet been established. Because of the long annealing times likely to be required to establish the lower limit of the intermediary phases, its knowledge will often be of little practical significance, but there will be a number of cases where the decomposition rates are within the lifetime of alloys used in plant or other technical appliances. The question of the stability of length standards is probably one that involves (apart from internal stresses) transformations proceeding slowly with time, whether these be ordering or separation of "impurities" over long periods of time. The change in length,[17] over the past 70 years, of the British yard made of a type of bronze may be a pertinent example, and one is surprised to learn that the standard meter at Paris made of Pt and Ir has "not" changed its length during the past 50 years or so.

In such cases, the best way of establishing the equilibrium diagram is to measure the thermochemical properties of all the phases at some convenient

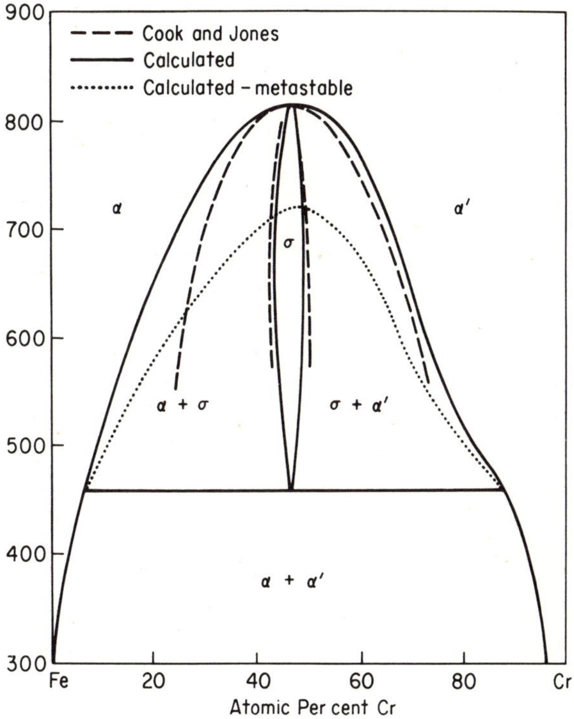

Fig. 6 Equilibrium diagram of the system Fe-Cr below 900°C. [*From Kubaschewski and Chart, J. Inst. Metals* **93:**335 (1965), *reprinted by courtesy of The Institute of Metals.*]

temperature, and then to calculate the phase boundaries at other temperatures from the isothermal free-energy curves.

Where the relative atomic sizes of the component metals are larger and the relative heats of sublimation smaller, the heats of mixing in the solid and liquid states become more positive. This results in more extensive solid miscibility gaps until they extend, hypothetically, above the melting points. Thus, eutectic systems with various degrees of solid solubility are formed. The strain energy (representing a positive heat term) is less in liquid mixtures than in the solid state, but the "packing effect" outlined on page 70 still pertains. Thus, for even higher differences in atomic diameter, we find systems exhibiting miscibility gaps in the liquid state, provided that the heat of sublimation of the larger atoms is much less than that of the smaller. Examples of phase-boundary calculations in such systems have been given in earlier papers.[9]

In region III of Fig. 3, one expects to find systems exhibiting large ranges of mutual solid solubility—in this case with negative heats of solution. If the third law is taken to imply that at absolute zero only completely ordered alloys exist in equilibrium, then any such system that shows complete mutual solid solubility should form ordered compositions at lower temperatures. In many systems, such as Ag-Au, these phases have not yet been found, again because of the slow rates of attainment of equilibrium. One may wonder whether anyone has yet examined, roentgenographically, Ag-Au alloys of ancient origin that have been "annealed" at room temperature for several thousand years. Possibly, the beginning of ordering might be detected in such alloys.

As an example from a system in which the disorder-order change has been found, the free-energy curves of the ordered and disordered phases in the system Cu-Au are shown in Fig. 7 at a temperature of 300°C. The points of contact of the tangents to the curves represent the phase boundaries at that temperature. There remains some considerable doubt whether the available thermodynamic data are really accurate enough to construct the whole equilibrium diagram from them. Measurements near the phase boundaries are likely to be affected by the slowness of equilibration, and the investigators often do not choose the best experimental methods nor allow sufficient time for attainment of equilibrium. Much of the thermochemical and thermal work on low-affinity systems is bedeviled by this human impatience. The accuracy of the accepted phase diagram of the Cu-Au system is therefore also subject to suspicion.

With high atomic ratios, the stability of the ordered phases is comparatively high owing to the "packing effect"—region IV in Fig. 3. Among the intermetallic structures, the Laves phases allow a particularly favorable packing with its optimum at a radius ratio of $r_A/r_B = 1.225$. Such phases may even exist, though with a lower stability, when the heat of sublimation of metal A is somewhat less than that of metal B. An example is the Laves phase KNa_2. At 0°C, this phase is in equilibrium with the liquid mixtures and with the solid solution in Na. Figure 8 shows the former equilibrium: ΔG_{liq} does not terminate at zero because the standard states are solid Na and K and the free energies of fusion thus enter the evaluation of the free energies of the liquid mixtures. The free energy of solution of Na in K has not been determined and has been neglected in drawing the tangent pertaining to this solution and the liquid.

The free energy of formation of KNa_2 is relatively small for the reason

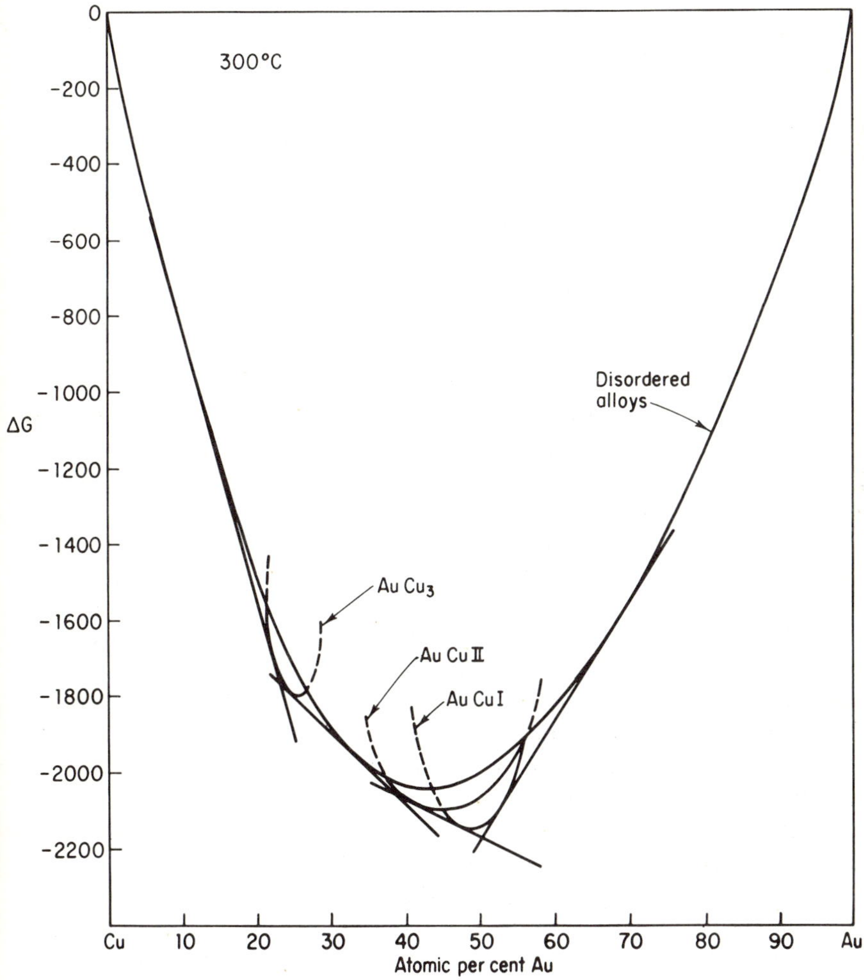

Fig. 7 Free energies of formation of the ordered and disordered phases in the system Cu-Au at 300°C.

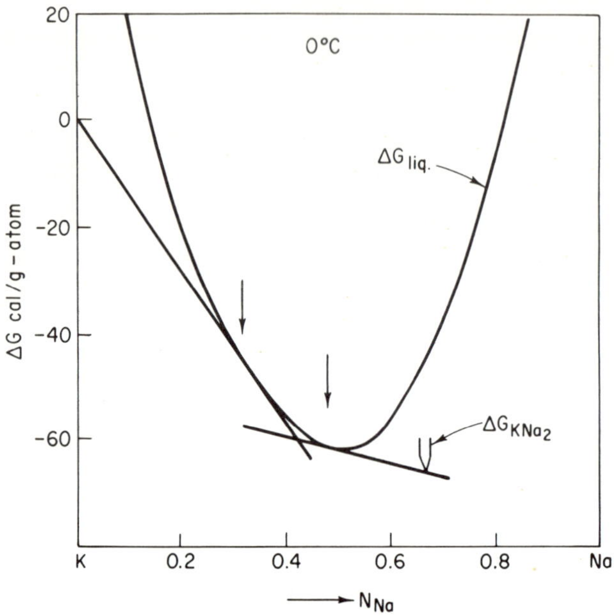

Fig. 8 Free energies of formation of the liquid and solid phases in the system K-Na.

mentioned. Most Laves phases are of higher stability. The free energy of formation of $TaCr_2$ for instance, has recently been measured[10] as -2200 cal/g atom, nearly independent of temperature, as compared with -65 cal/g atom for KNa_2 at 0°C.

The examples shown may suffice as a demonstration of how phase boundaries are obtained from thermochemical results. Quite a few more could be added. The recommended method of evaluating the phase boundaries is the graphical one, the so-called "common-tangent method" as demonstrated by Figs. 5, 7, and 8. In simple cases, mathematical expressions can be derived[9,11,12] and used for the calculation of the phase boundaries, but this requires simplifying assumptions concerning the thermodynamic data and their dependence on concentration with a resulting loss in accuracy. Otherwise the calculations become much too involved. However, the use of computers should be considered, in particular for the evaluation of multicomponent systems.

At present, one is too short of accurate thermochemical data for a large-scale application to phase-boundary determinations. On the other hand, the advantages of the method are impressive because it permits the establishment of phase boundaries in cases where equilibration is slow or very slow.

SOME NOTES ON RATE OF PHASE FORMATION

This point leads to the kinetic aspects of phase formation which are just as important as the thermodynamic aspects but much less tangible. The first difficulty arises when one has to account for the interplay between the effects of nucleation and growth where even the mode of nucleation may vary with temperature.[13] It is therefore not surprising that the application of diffusion data alone to precipitation rates has led to anomalous results.

It is difficult to see where the experience of the thermochemist can be of assistance, but since several implications concerning a connection between free energy and rate of formation or transition have been indicated in the foregoing discussions, the present author may be allowed to record a few observations made during our thermochemical work.

The primary condition for a martensitic transformation, a precipitation, or a dissolution to occur is, of course, that the system should attain a state of lower free energy in the process. The actual difference in free energy between the initial and final stages is probably a contributory factor to rate although the kineticists rarely seem to make quantitative use of free-energy data. However, the bulk data do not account for dislocations and other imperfections, nor for the role played by foreign nuclei.[13] Also, virtually no quantitative information is available on the free energy of transition precipitates.

It always strikes the present author how small the differences in free energy often are that provide the driving force—as may be seen from Figs. 1, 5, and 7, for instance. In these cases they are of the order of tens of calories per gram atom, and much less near the phase boundaries. It has been mentioned earlier that energies arising from mechanical strain, impurities, interfaces, and surfaces may amount to, say, 10 cal/g atom or more, so that the thermodynamic bulk free energies alone cannot represent the true energy conditions. Analysis of the latter is hampered by the low experimental accuracy with which the individual energy contributions can be measured.

In connection with the investigation of the thermodynamic properties of the σ-FeCr phase some observations were noted on the rate of the diffusionless transformation, $\alpha \rightarrow \sigma$. It appeared that the maximum rate of the transformation is not near the normal transformation temperature (815°C) but somewhere around 700°C. By determining the magnetic pull on various α alloys containing 43 to 50 a/o Cr after progressive times of annealing at 650°C, my associate, Mr. W. A. Dench, obtained some measure for the rate of transformation as a function of concentration. Between 43 and 47 a/o Cr, the rate was relatively fast (with an apparent maximum at 45%), but slowed down considerably above these compositions. Although it is difficult to explain the concentration dependence, the temperature dependence may be largely determined by the free energy. The maximum difference between the free energies of formation of σ and $\alpha + \alpha'$ (see Figs. 5 and 6) is found near 650°C. Although diffusion rates increase with temperature, $\Delta G_{\sigma \to \alpha}$ decreases, and, above 750°C, Cook and Jones[14] were unable to produce transformation, $\alpha \to \sigma$, within 7 days. It is concluded that, with martensitic transformations, free energy is a determining factor for rate. This is borne out by the many uncertainties in phase-boundary determinations as well as with thermochemical measurements near a transformation temperature.

The influence of the actual free-energy changes is probably less in reactions, such as precipitation and other forms of phase formation. Here, the extent of the structural changes involved appears to be a determining factor for nucleation and growth. Again, some observations made during our thermochemical studies may be invoked. We have mostly been concerned with the formation of substitutional alloys whereas most of the kinetic knowledge on precipitation and formation has been concerned with interstitial alloys.

An earlier calorimeter for the investigation of solid-solid reactions relied on the speed with which alloys would form from the metals a few hundred degrees below the melting points. The heats of formation measured, and thus the free energies of formation, were numerically rather high: between 5 and 15 kcal/g atom. It was found that high rates of formation could be expected where the heat derives mainly from an increase in bond strength rather than from a packing effect—e.g., aluminides of the transition metals.[8] The system Ni-Al-Ti, for instance, is dominated by three bcc phases, NiAl, $NiTiAl_2$, and TiAl. Little difficulty was encountered in achieving near-equilibrium when the temperature was allowed to rise, by self-heating after ignition, to 1000 to 1250°C where it remained for less than 1 minute. However, it was not possible under similar conditions to form, for example, Laves phases of similar melting point and heat of formation. Argent,[16] in particular, made a series of attempts, all unsuccessful, to synthesize Laves phases in such a calorimeter. It was necessary to use a calorimeter[1] operated at high temperature (1400°C) and to allow several hours for equilibration in order to study the thermodynamic properties of phases[10] such as $TaCr_2$.

CONCLUSIONS

The remarks made in the present address represent a thermochemist's point of view. His foremost task is the accurate measurement of the thermodynamic properties of metals and alloys, and will remain so for some time to come. He is also in the best position to assess the accuracy of the data used in any theory. New experimental methods will have to be developed and the available ones improved.

Meanwhile, the thermodynamic data determined experimentally may be used in conjunction with conventional work to establish metallurgical equilibrium diagrams more accurately in regions where equilibration is slow or otherwise hampered. Application of chemical thermodynamics in practice, though of a varied nature, is almost always phase-diagram work.

Inevitably, as a thermochemist progresses, he continues to ponder over the origin of the data he obtains. His approach is essentially empirical based on hypothetical models pertaining to bond mechanism. His lines of thought are less elaborate but more systematic than those of the metal physicist. Not all his contentions can be disregarded while a uniform theory of phase stability is being developed.

REFERENCES

1. Dench, W. A.: *Trans. Farady Soc.* **59**:1279 (1963).
2. Kubaschewski, O., and J. A. Catterall: "Thermochemical Data of Alloys," Pergamon Press, New York and London, 1956.
3. Hultgren, R., R. L. Orr, P. D. Anderson, and K. K. Kelley: "Selected Values for Thermodynamic Properties of Metals and Alloys," John Wiley & Sons, Inc., New York and London, 1963.
4. Dench, W. A., and O. Kubaschewski: *J. Iron Steel Inst.* **201**:140 (1963).
5. Schneider, A., and G. Heymer: Paper 4A, in "The Physical Chemistry of Metallic Solutions and Intermetallic Compounds," NPL Symposium 9, Her Majesty's Stationery Office, London, 1959.
6. Kubaschewski, O.: Paper 3C, in *ibid.*
7. Mott, B. W.: *Deut. Ges. Metallk. Hauptversamml. Essen*, 1965.
8. Kubaschewski, O., and G. Heymer: *Trans. Faraday Soc.* **56**:473 (1960).
9. Kubaschewski, O., and T. G. Chart: *J. Inst. Metals* **93**:329 (1965).
10. Feschotte, P., and O. Kubaschewski: *Trans. Faraday Soc.* **60**:1941 (1964).
11. Wagner, C.: *Acta Met.* **2**:242 (1954).
12. Meijering, J. L.: Paper 5A, in "The Physical Chemistry of Metallic Solutions and Intermetallic Compounds," *op. cit.*
13. McLean, D.: Conference on Some Topics in Metal Physics, p. 62, Institute of Metallurgists, 1966.
14. Cook, A. F., and F. W. Jones: *J. Iron Steel Inst.* **148**:217 (1943).
15. Kubaschewski, O.: *Trans. Faraday Soc.* **54**:814 (1958).
16. Argent, B. B.: Private communication, 1958.
17. Bigg, P. H., and P. Anderton: *Nature* **200**:730 (1963).

SPACE LIMITATIONS ON THE GEOMETRY OF THE CRYSTAL STRUCTURES OF METALS AND INTERMETALLIC COMPOUNDS

Fritz Laves

Institut für Kristallographie und Petrographie, Eidgenössische Technische Hochschule, Zurich, Switzerland

SPACE PRINCIPLE

Considering the fact that the atoms of different elements have different atomic structures, one could expect that each element would crystallize into a structure type of its own. As is well known, this is not the case. There are a few structure types that are highly preferred by the metallic elements. These types are the close-packed ones with coordination number (CN) 12, either the cubic Cu type or the hexagonal Mg type, and the body-centered cubic W type with CN 8. Among these structure types with coordination numbers 12 and 8 those with CN 12 are by far preferred. Considering the huge number of *possible* structure types with different coordination numbers, it is amazing that *one*, CN 12, is so preferred. This cannot be a matter of chance. As mathematicians have proved that the densest packings of equally sized spheres have CN 12, it becomes apparent that the geometrical properties of our three-dimensional space may be important factors for the realization of certain crystal structures, and as a consequence of this for the stabilization of certain phases. In the case of the elements these space properties limit the CN to a maximum of 12. An element that would "like" to have, for example, a CN 18 just cannot realize its "desire," and theoretical physicists have to put up with such laws of space.

The frequency of CN 12 suggests a tendency of the metallic elements to fill space as well as possible. If we assume that the atoms behave in crystals as though they were contacting spheres, a space-filling value can be defined as

ζ = volume of number of contacting spheres per unit cell divided by volume of unit cell

In the case of CN = 12, $\zeta = 0.74$. In the case of CN = 8, ζ is 0.68. The fact that many more elements crystallize with CN 12 ($\zeta = 0.74$) than with CN 8 ($\zeta = 0.68$) thus indicates that there is a tendency to fill space as well as possible. This tendency will be called the *space principle.*

SYMMETRY PRINCIPLE

On the other hand, it has to be mentioned that geometrically sphere packings exist with coordination numbers between 8 and 12, e.g., the γ-Pu structure with CN 10. Such packings consist of closest-packed sphere layers piled up in such a way that each sphere has six neighbors in its own layer and two times two neighbors in the adjacent layers. As the ζ value of such packings equals 0.71, the question arises: Why is such a structure so rarely realized compared with the frequency of the W structure type with CN 8? This fact appears to contradict the validity of the space principle.

To take account of this fact one may consider the *symmetry* differences between the W structure type and the sphere packings with CN 10. Whereas the W structure type is cubic, sphere packings with CN 10 are tetragonal or have an even lower degree of symmetry. Thus, a tendency to form arrangements of high symmetry is observable. This tendency will be called the *symmetry principle.*

As our three-dimensional space does not permit crystals to have a higher symmetry than cubic, another limiting factor on the geometry of crystal structures becomes apparent.

In addition to the symmetry principle, another spatial point of view may contribute to the stability of the W type. According to Frank and Kasper,[1] there is another way of defining coordination numbers which differs from the "classical" way (see below). According to Frank and Kasper's definition the W type would have CN 14, suggesting another kind of good space filling.

CONNECTION PRINCIPLE

In addition to the two principles mentioned before, a third one may now be discussed which is trivial as far as metallic element structures are concerned but appears to be important as far as intermetallic compounds are concerned: the *connection principle.*

To realize the meaning of this principle, it is at first necessary to define the concept of *connection.* To simplify the discussion, we confine ourselves to structures of two-component compounds A_mB_n. Connect each atom A and each atom B with all the other atoms present in the structure by straight lines. Let the shortest line between $A(B)$ atoms be $d_A(d_B)$, and the shortest line between A and B atoms be d_{AB}. Those $A(B)$ atoms that are connected by $d_A(d_B)$ form a *homogeneous* connection; those A and B atoms that are connected by d_{AB} form a *heterogeneous* connection. Such connections can be 0-, 1-, 2-, 3-dimensionally infinite and may form island, chain, net, and lattice connections correspondingly.

Considering these definitions, it can be frequently observed that metallic compounds tend to form *homogeneous* connections with relatively small $d_A(d_B)$ values compared with the d_{AB} values; i.e., the quotient

$$Q = \frac{d_A + d_B}{2d_{AB}}$$

tends to become small. Table 1 shows some examples. In this table

TABLE 1 ***Q Values of Pairs of Some Simple Structure Types That Have the Same Coordination Numbers, CN***

Structure	CN	$Q = \frac{d_A + d_B}{2d_{AB}}$
NaCl	8 and 8	1.41
NiAs	8 and 8	1.19
CaF_2	8 and 4	1.39
$CuAl_2$	8 and 4	1.06

Note that the more metallic structure types have smaller Q.

TABLE 2 ***Lattice Constants and d Values of*** **NiX** ***Compounds***

Compound	Structure type	Lattice constants a	c	c/a	d_{Ni}	d_X	d_{NiX}
NiO	NaCl	4.17			2.95	2.95	2.09
NiS	NiAs	3.42	5.30	1.55	2.65	3.42 3.30	2.38
NiSe	NiAs	3.66	5.33	1.46	2.67	3.66 3.40	2.50
NiTe	NiAs	3.95	5.36	1.36	2.68	3.95 3.52	2.64

Note (1) the difference of d_{Ni} in the oxide and in the other compounds, and (2) the rather constant d_{Ni} value compared to the d_X values when X = S, Se, Te.

structure types are compared which are not different as far as the radius ratio of the components is concerned. Table 2 gives some special d values of NiAs-type compounds taken from an earlier paper,[2] and Table 3 gives some c/a values from a later paper[3] in which the structure determination of Ni_2In and Cu_2In was presented and discussed in view of what is here called the connection principle. Both tables demonstrate strongly the tendency of Ni to form *homogeneous* chain connections.

In addition to the tendency to form homogeneous connections resulting in low Q values, the *connection principle* may include a tendency to form homogeneous connections of a high dimension number.

TABLE 3 ***Axial Ratios c/a and Atomic Percent*** **Ni** ***in*** **NiAs*****-type Compounds***

			Ni-S 1.55 ~50%
Ni-Ga 1.25 ~60%	Ni-Ge 1.28–1.30 67–57%	Ni-As 1.39 ~50%	Ni-Se 1.46 ~50%
Ni-In 1.23 67–61%	Ni-Sn 1.27 64–55%	Ni-Sb 1.31 60–50%	Ni-Te 1.35 50–33%
		Ni-Bi 1.31 ~55%	

Note the systematic changes as a function of the position of the electronegative partner in the periodic table.

A frequent structure of the elements, the fcc structure, corresponds best to the three principles mentioned above: highest possible space filling, highest possible symmetry, and highest possible connection dimension.

DISCUSSION OF SOME METALLIC COMPOUNDS

Before starting this discussion, I would like to state what I shall consider to be a compound. I shall call *compounds* such phases that are not continuously connected with one of their elementary components by solid solution.

Proceeding now to a discussion of the geometrical features of structure types, we shall see that from a statistical point of view the same three principles play roles that have been outlined above.

TABLE 4 ***Structure Types of Two-component Compounds Having 25 Representatives or More***

Structure type	*Number of representatives*
$MgCu_2$	158
CsCl	111
$AuCu_3$	108
$MgZn_2$	76
$CaZn_5$	58
β-U (or σ)	49
Mg	45
γ-Brass	45
W	40
Cr_3Si	34
CuAu	29
AlB_2	29
$CuAl_2$	25
	807

For several reasons we shall confine the discussion to two-component compounds of such components that have their position in the periodic table on the left side of the Zintl line (defined as lying between the rows Si, Ge, Sn, Pb and Al, Ga, In, Tl).

Consulting the index of Schubert's 1964 book[4] which is today the best available compilation of two-component compounds, we can count the number of structure types known today and the number of their representatives known today. If we do so, we find that there are about 200 structure types known with about 1,300 representatives.

Table 4 lists those structure types that have 25 representatives or more.

It is interesting to note that about 800 of the 1,300 structures known today crystallize into only 13 types. These figures clearly suggest an influence of space geometry on structure formation and, as a consequence of it, an influence of space geometry on the question of which stoichiometric compositions of metallic compounds are favored.

To explore the geometrical meaning of this statistical material further, Table 5 shows another arrangement of these 13 structure types by grouping them according to their CN, disregarding symmetry.

As in the case of the elements, we observe that the close-packed structures ($AuCu_3$, AuCu, Mg) are more frequent than those with CN 8 (W, CsCl). However, even more frequent are the MgX_2-type structures* ($MgCu_2$, $MgZn_2$). What may be the reason for such predominance of the MgX_2-type representatives?

TABLE 5 ***Structure Types Grouped According to Coordination Number***

$MgCu_2$	158	234
$MgZn_2$	76	
$AuCu_3$	108	182
Mg	45	
AuCu	29	
CsCl	111	151
W	40	
$CaZn_5$		58
β-U (σ)		49
γ-Brass		45
Cr_3Si		34
AlB_2		29
$CuAl_2$		25
		807

As can be seen from Fig. 1 which shows the $MgCu_2$ type as the cubic example of the MgX_2-type structures, these structure types follow very well the three geometrical principles mentioned above:

1. Good space filling (high CN)
2. High symmetry (cubic or hexagonal)
3. Good connection properties

As to the "good space filling," it may be said that there are only few other types known that have the same high AB coordination numbers, namely 12 and 6. These types are the AlB_2 and $ThSi_2$ types (Figs. 2 and 3),

* The $MgNi_2$-type structure is not considered here, as there are less than 25 two-component representatives known.

which have been discussed in detail elsewhere.[5] It may be added that no AB_2-type structures are known that have higher AB coordination numbers than 12 and 6, and it is very improbable that such structures could exist, for purely geometrical reasons (cf. at this point a paper by Dehlinger and Schulze[6]). Thus, considering the AB coordination numbers, there would be no difference among the types of MgX_2, AlB_2, and $ThSi_2$. However, a distinct difference exists if the coordination numbers of the homogeneous A connections and B connections are taken into account: In MgX_2 the average CN is $16/3$, against $14/3$ in AlB_2 and $ThSi_2$. In addition, MgX_2 has the very low Q value[5] of 0.95, against 1.01 and 1.04 for AlB_2 and $ThSi_2$. Moreover, the MgX_2 types are favored against the two others if one calculates the average CN considering both the heterogeneous *and* the homogeneous connections. Such a procedure was first suggested by Dehlinger and Schulze,[6–8] and corresponds to the Frank-Kasper[1,9] point of view of defining coordination numbers. The results are: MgX_2 has the average CN 13.3, whereas the corresponding values for AlB_2 and $ThSi_2$ are 12.7. Finally, the MgX_2 types have high symmetry and three-dimensional homogeneous connections, whereas the B connection of the AlB_2 type is only two-dimensional, and the $ThSi_2$ type has a lower symmetry (tetragonal) than the MgX_2 types.

Summing up: The MgX_2 types are geometrically the most distinguished as far as *space filling, symmetry*, and *connection properties* are concerned. Thus, it is less a question of why there are so many representatives known

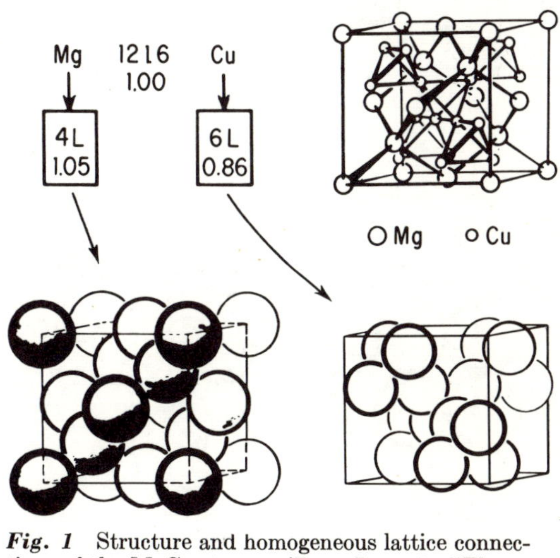

Fig. 1 Structure and homogeneous lattice connections of the $MgCu_2$ type. (*From Laves in "Theory of Alloy Phases,"* 1956; *courtesy of American Society for Metals.*)

than a question of why are more not known. For example: $InNi_2$ would have a perfect radius ratio to form an MgX_2-type compound, but it does not do so. It forms instead a filled-in NiAs-type structure.[3] To explain this one has to assume counteracting electrochemical factors.[3] In this respect, a discussion of *such* compounds which *should* crystallize in an MgX_2 type but do not do so may turn out to be especially fruitful.

To allow for good space filling the components of the MgX_2-type compounds should have a special radius ratio which is ideal when $R_A/R_B = \sqrt{3}/\sqrt{2} = 1.23$. Indeed, K and Na have the ratio 1.24 which approaches the ideal one very nearly, and this fact is considered as the main reason why the compound KNa_2 exists at all.[10]

On the other hand, there are many MgX_2-type compounds the components of which have R_A/R_B values that deviate rather widely from the ideal value 1.23. As such an example $LaNi_2$ may be mentioned. In the elementary state La and Ni have the radii 1.87 and 1.24, i.e., 1.87/1.24 = 1.50. In the $LaNi_2$ structure ($MgCu_2$ type) the La and Ni have the radii 1.57 and 1.28. To account for such differences it is plausible to assume that the components adjust their sizes to the space requirements, e.g., by

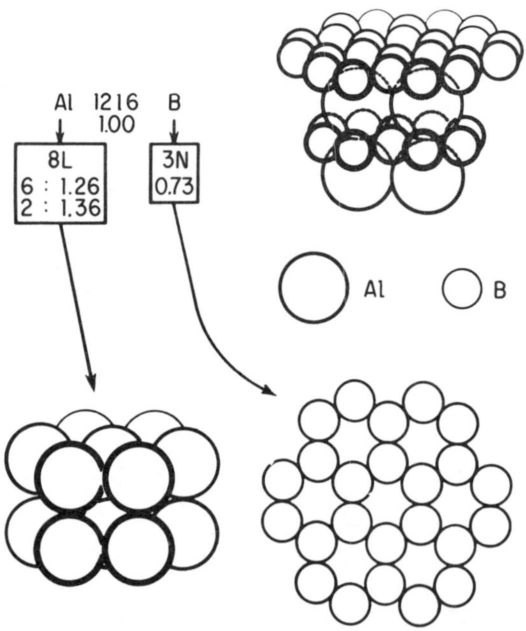

Fig. 2 Structure and homogeneous connections of the AlB_2 type. Note that B forms only a net connection. (*From Laves in "Advances in X-ray Analysis," vol.* 6; *courtesy of Plenum Press.*)

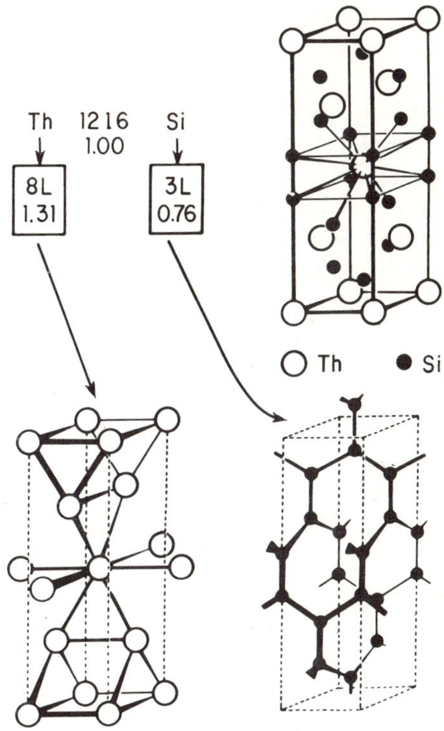

Fig. 3 Structure and homogeneous lattice connections of the $ThSi_2$ type. (*From Laves in "Theory of Alloy Phases," 1956; courtesy of American Society for Metals.*)

a suitable electron transfer. A more detailed discussion of such problems was given in 1956.[5]

ON THE FRANK-KASPER APPROACH TO THE PROBLEM OF SPACE FILLING AND COORDINATION NUMBER

Up to this point we have correlated good space filling with high CN, and we used the concept of CN in the classical sense; i.e., we took as the CN of an element the number of "nearest" neighbors (allowing for some small deviations from the *really* nearest distance). Analogously, we took as the CN of a compound A_mB_n the number of "nearest" neighbor B atoms an A atom has, and vice versa.

However, there is another point of view for defining coordination numbers which might be called the Frank-Kasper point of view; it was introduced by Kasper[11] and further developed by Frank and Kasper.[1,9]

As it is hard to visualize this point of view in three dimensions, let us discuss the principle in two dimensions. Figure 4 shows an AB arrangement. In the classical sense A has CN 4 and vice versa. In the Frank-

Kasper sense, A has CN 5 and B has CN 7 if chemical differences are disregarded; i.e., the average CN is 6.

If the Frank-Kasper approach is applied to the MgX_2-type structures, a CN $^{40}\!/_3 = 13.3$ results, whereas the classical approach would lead to an average CN of $(12 + 2 \times 6)/3 = 8$. In this connection it is interesting to note that the CsCl-type structure has CN 14 from the Frank-Kasper point of view and CN 8 from the classical point of view.

Both ways of looking at the problems have their advantages and demonstrate the high space filling realized in the MgX_2-type compounds. However, at this point it might be pointed out that though both ways of defining coordination numbers have their *advantages*, they may have also their *disadvantages*. Two examples may show the disadvantages:

1. Let us take the *classical point of view* in its rigid sense; i.e., we consider only those atoms as neighbors which have the very shortest distance between them. If we then look at the hexagonal close-packed structure,

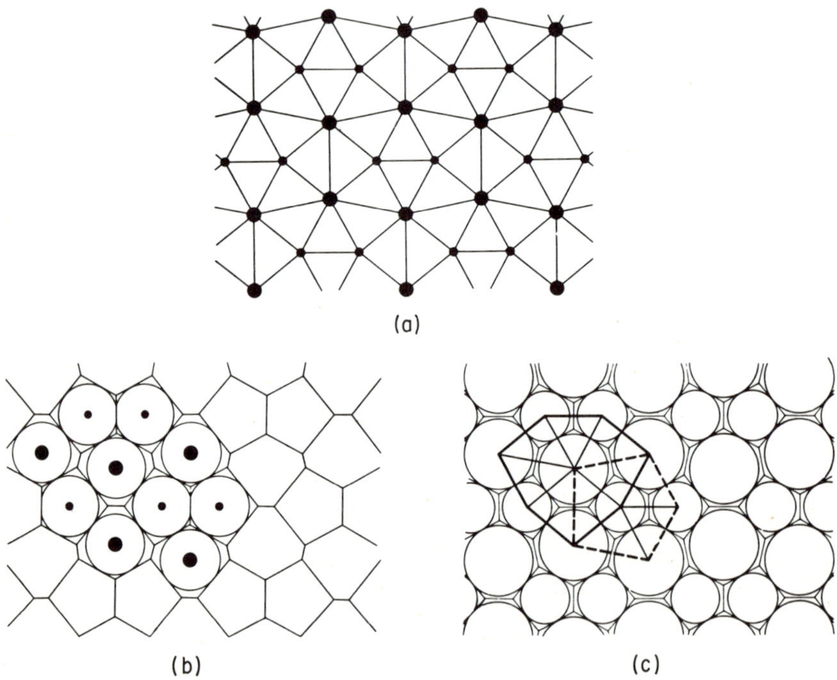

Fig. 4 Two-dimensional representation of the Frank-Kasper concept of coordination and coordination numbers: (*a*) a hypothetical AB structure, the small dots representing the A positions; (*b*) the "domains" as defined by Frank and Kasper; (*c*) a deformation of the "domains" in such a way that only A spheres have contact with B spheres and vice versa, corresponding to the classical point of view on coordination numbers. (*From Laves in "Beiträge zur Physik und Chemie des* 20 *Jahrhunderts," Friedr. Vieweg & Sohn, Brunswick, Germany,* 1959.)

most representatives would have CN = 6, since CN = 12 would only hold for the c/a ratio 1.63.

2. If we apply the *Frank-Kasper point of view* to the diamond type, CN = 16 would result, as the Dirichlet cells around the atoms are 16-faced polyhedra (see Fig. 5). Frank and Kasper call a Dirichlet cell a *domain* and define "the domain of an atom as the space in which all points are nearer to the center of that atom than any other."

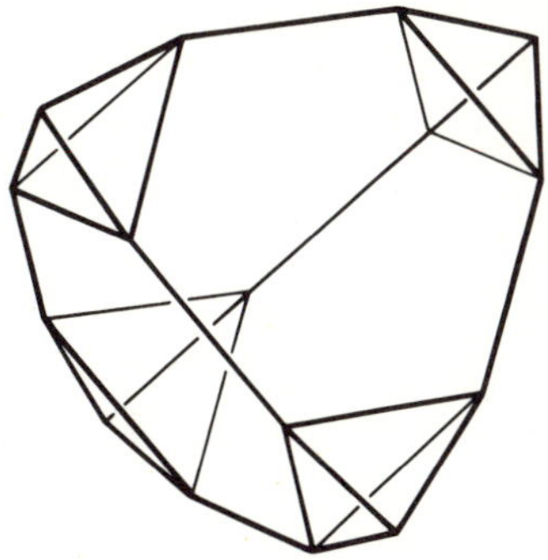

Fig. 5 Sixteen-faced polyhedron with which the space can be filled up completely. It corresponds to a "domain" (as defined by Frank and Kaspar[1]) of the diamond structure.

Thus the classical method *may* lead to senselessly small coordination numbers, and the Frank-Kasper method *may* lead to senselessly large ones. Therefore, which method is the most appropriate to use may frequently depend on the special problem to be discussed.

Surely, the Frank-Kasper approach is the best (if not the only one) when it comes to the problem of discussing intelligently the geometrical features of the so-called "electron compounds." As Kasper[11] found, there are four polyhedra that occur very frequently as coordination polyhedra in structurally complex electron compounds if the coordination number is defined in the Frank-Kasper sense. These polyhedra are shown in Fig. 6 and have the following characteristics:

1. They are convex.
2. They are bounded only by triangles.
3. They have corners in which at least five triangles meet. Table 6 lists their topological features.

As Frank and Kasper[1] could prove mathematically, the four polyhedra pictured in Fig. 6 and listed in Table 6 are the only possible ones that meet requirements 1 to 3 above.

In another paper[9] Frank and Kasper investigated the possibilities of in how many ways these polyhedra can be combined with one another and what the maximum average coordination number might be. The result appears to be 13.5, as realized, for example, in the σ-phase structure. Again, high space filling (high CN) appears to be a very important principle in this group of electron compounds.

It is interesting to note[9,11] that the MgX_2 types can be considered also as mixtures of two Kasper polyhedra (with the corner numbers 12 and 16) and with an average CN 13.3. Thus it is not surprising that occasionally electron concentrations appear to contribute to the stability of MgX_2-type

TABLE 6 ***Geometrical Features of the Kasper Polyhedra***

Coordination number, CN	Corners	Number of triangular faces	Edges
12	12 C_5	20	30
14	12 C_5, 2 C_6	24	36
15	12 C_5, 3 C_6	26	39
16	12 C_5, 4 C_6	28	42

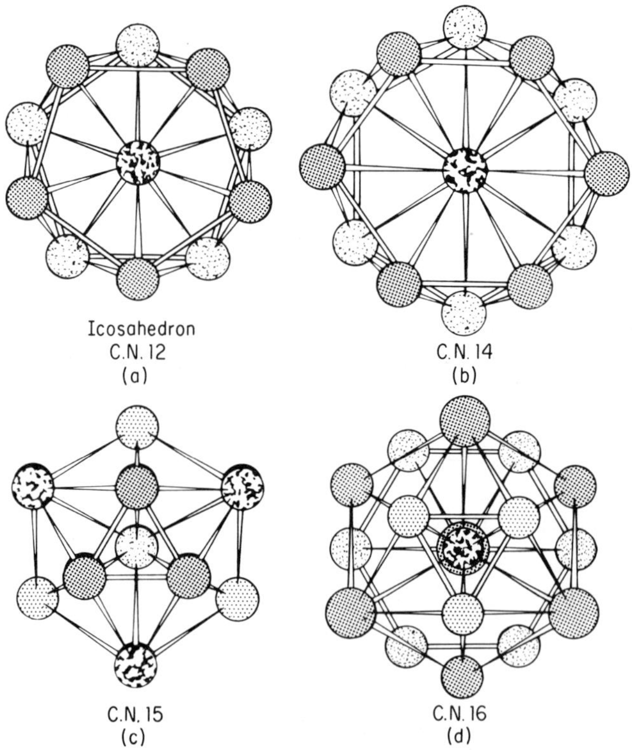

Fig. 6 The four "Kasper polyhedra," as first drawn by Kasper.[11] *(From Kasper in "Theory of Alloy Phases," 1956; courtesy of American Society for Metals.)*

compounds.[12] Furthermore, it is interesting to note that the MgX_2-type structures are very dense from *both* points of view, the classical one *and* the Frank-Kasper one.

SOME GENERAL CONSIDERATIONS

Having seen that good space filling can be correlated with high coordination numbers, the question arises which coordination numbers are the highest possible ones in A_mB_n compounds. If we knew the answer, the validity of the "space principle" could be stressed more rigorously.

In this respect it would help to know in how many ways space can be divided into equal polyhedra, i.e., polyhedra with the same number of faces, edges, and corners.

To illustrate the problem, let us discuss it for the two-dimensional case.

Given an infinite number of equivalent points in a plane, construct the Dirichlet cell for each point. In this way the plane will be divided into polygons of the same kind, i.e., having the same number of edges (and the same number of corners). It can be rigorously proved[13] that only 3-, 4-, 5-, and 6-sided convex polygons are possible. In addition, it can be rigorously proved that, from a topological point of view, 11 and only 11 kinds of plane partitions are possible using these four polygons. Having these 11 possibilities, we put the A atoms in the centers of the cells and examine how many ways B atoms can be put in the corners. Doing this,[13] we get all possible coordination numbers for all possible A_mB_n arrangements, and we can rigorously prove that there is a maximum of coordination numbers in the plane, e.g., 6 and 3 for a "compound" AB_2 with an average CN 4. Thus, the geometrical properties of the plane limit the number of possible coordination numbers.

In principle, the same must hold for the three-dimensional space, but we do not yet know whether the possible space partitions into polyhedra of the same kind proceed the same way as they did in the plane. In principle, it *could* be proved explicitly in a tedious time-consuming way. However, as far as the author is aware, nobody has bothered to do this, either in a systematic way or in an intelligent topological one. The author tried once to tackle the problem but became frustrated.

As an example of such space partition Fig. 5 may serve again, since space can be filled out completely with such 16-faced polyhedra. Notice that there are 4 + 12 corners. If we put Mg in the center of the cell and Cu on the 12 corners, we have the Cu coordination around Mg in $MgCu_2$. However, as mentioned, no systematic study has been carried out to find all possibilities of this sort. But, obviously, there must be a finite number of space partitions and, therefore, a maximum value of coordination numbers.

A first approach to solving this problem might be to find an answer to the following question: What is the average number of the faces, edges, and corners of convex polyhedra around "randomly" distributed points that fill out the space completely and fulfill the following conditions:

One face belongs to two polyhedra.
One edge belongs to three polyhedra.
One corner belongs to four polyhedra.

Among others, Meijering[14,15] and van der Waerden[16] worked on this problem. Unfortunately, their results came out differently: Meijering found as the average number of faces 15.6, whereas van der Waerden found 14. Who is right?

CONCLUDING REMARKS

As the purpose of this paper was to contribute to the question "which factors affect the stability of metallic phases," I tried to emphasize the influence of geometrical properties of our three-dimensional space. Exaggerating this aspect, I dare to make some rather trivial statements which, fortunately, cannot be checked by experiment; for example: (1) If space allowed us to have sphere packings with coordination numbers larger than 12, I would be sure that the Mg- and Cu-type structures would be realized only rarely; and (2) if space allowed the realization of AB_2 arrangements with coordination numbers larger than 12 and 6, I would be sure that MgX_2-type realizations would occur only rarely. My conclusion is: Space provides some structural-arrangement possibilities which are marked out as far as space, symmetry, and connection principles are concerned. However, it would be difficult to mold these qualitative points of view into a rigorous theory.

Finally, I apologize for talking so much about the geometrical features of the MgX_2-type structures. However, at present, they are the most convenient for demonstrating the influence of geometry on phase stabilization: The homogeneous connection allows for the development of covalent bonds; the heterogeneous connection allows for the development of polar bonds, and the dense packing allows for the development of metallic bonds, including the influence of electron concentration. Thus, for many metals such a structural arrangement may be a very convenient one for expressing their individuality, and analogous reasons will hold for other structure types realized by intermetallic compounds.

REFERENCES

1. Frank, F. C., and J. S. Kasper: *Acta Cryst.* **11**:184–190 (1958).
2. Laves, F.: *Z. Krist.* **73**:303 (1930).

3. Laves, F., and H. J. Wallbaum: *Z. Angew. Mineral.* **4**:17–46 (1941–1942).
4. Schubert, K.: "Kristallstrukturen zweikomponentiger Phasen," Springer-Verlag OHG, Berlin, 1964.
5. Laves, F.: In "Theory of Alloy Phases," American Society for Metals, Cleveland, Ohio, 1957.
6. Dehlinger, U., and G. E. R. Schulze: *Z. Metallk.* **33**:157–160 (1941).
7. Schulze, G. E. R.: *Z. Elektrochem.* **45**:849–865 (1939).
8. Dehlinger, U., and G. E. R. Schulze: *Z. Krist.* **102**:377–390 (1940).
9. Frank, F. C., and J. S. Kasper: *Acta Cryst.* **12**:483–490 (1959).
10. Laves, F., and H. J. Wallbaum: *Z. Anorg. Algem. Chem.* **250**:110–120 (1942).
11. Kasper, J. S.: In "Theory of Alloy Phases," *op. cit.*
12. Wallace, W. E., and R. S. Craig: This colloquium, p. 255 (look there for additional references).
13. Laves, F.: *Z. Krist.* **78**:208–241 (1931).
14. Meijering, J. L.: *Philips Res. Repts.*, 1953–1954.
15. Meijering, J. L.: *Acta Met.* **1**:607 (1953).
16. Meretz, W., and B. L. van der Waerden: *Naturwissenschaften* **53**:8–11 (1966).

Part Two

PURE METALS AND SUBSTITUTIONAL SOLUTIONS

PSEUDOPOTENTIALS APPLIED TO STABILITY OF STRUCTURES*

Volker Heine

Cavendish Laboratory, University of Cambridge, Cambridge, England

1 PSEUDISM

The following is a commentary on some recent theoretical work bearing on the structure of simple metals and alloys. We are interested in work which on the one hand is a direct development of the work of Jones on the role of reciprocal lattice vectors in relation to the Fermi sphere. At the same time we also select work with a fundamental approach, i.e., calculations of the electronic structure from a clear theoretical basis. Since we are interested in making certain points in the nature of commentary, no attempt is made at a complete survey.

At the center of recent understanding of simple metals lies the concept of the pseudopotential. By the term *simple* metals we exclude transition, rare-earth, and actinide elements, but include diamond and the group IV semiconductors as well as the group V semimetals. For some

* Supported in part by NASA.

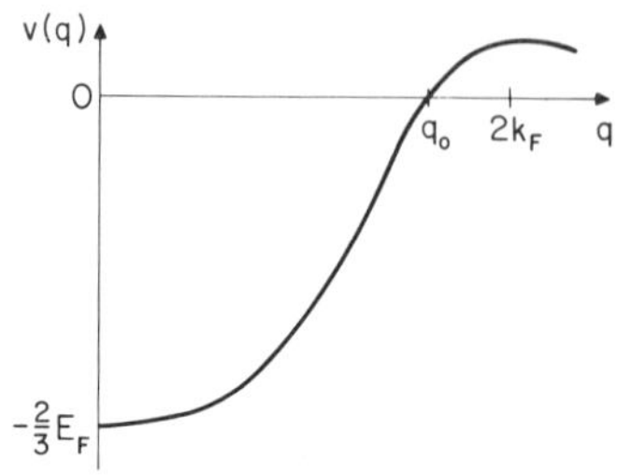

Fig. 1 Pseudopotential $v(q)$ for Al (schematic diagram). The (111) and (200) reciprocal lattice vectors fall between the zero q_0 and $2k_F$.

purposes the noble metals may be included, and we discuss them separately in Sec. 5. The general subject of pseudopotentials has been reviewed elsewhere.[2–5] Suffice it to note here that the band structures $E(\mathbf{k})$ of all simple metals conform to the nearly free electron (NFE) scheme. This is unexpected in view of the fact that the potentials of atoms are not weak perturbations but very strong. However, what is relevant is not really the potential but the effective scattering power of the atoms for electrons, as is clear if we view Brillouin zone boundaries in terms of Bragg reflections. This effective scattering power is found to be weak, and correspondingly we can represent the atom by a weak pseudopotential v. The form of the Fourier transform $v(q)$ is typically as shown in Fig. 1. At the same time we have pseudo wave functions ϕ, equal to the real wave functions ψ between the atomic cores, but having the rapid atomic-like oscillations of ψ inside the cores removed. Thus in lowest order of approximation the ϕ's become simple plane waves. All of this is not a qualitative swindle but an exact transformation of the Schrödinger equation.[7]

From the present point of view, the significance of the pseudopotential is threefold. Firstly, we can formulate all properties of simple metals by perturbation theory in terms of the nearly free electron gas: this includes the cohesive energy of metals and its change with structure. Secondly, since all properties are expressed in terms of $v(q)$, information about $v(q)$ obtained by fitting one set of experimental data may be applied to explain other properties. For example, the pseudopotentials of indium and antimony have been obtained[8] by fitting the band structure of the semiconductor InSb to its measured optical spectrum: the $v(q)$ for Sb is found to explain the band structure of Sb semimetal almost without change,[9] and that of In corresponds to what is needed to explain the tetragonal structure of this metal.[10] Thirdly, the formulation of terms of the free-electron gas allows us to calculate $v(q)$ from fundamentals,[11] including screening, self-consistency, and various correlation and exchange effects, for elements whose $v(q)$ has not yet been fitted to experiments. Such calculated $v(q)$ have, for instance, been used[10] in discussing the c/a ratios of the divalent hexagonal metals, and the variation of c/a in solid solution alloys of In.

2 BAND-STRUCTURE CALCULATIONS RELATED TO STRUCTURE

As is well known, the semimetals As, Sb, Bi have nearly a simple cubic structure. With 5 electrons per atom, they would be good metals if exactly simple-cubic. In accordance with the Jones argument,[1] they can lower their energy by structural changes which tend to make them insulators because this is the same as fitting electrons into low-energy states below a band gap. Cohen, Falicov, and Golin[14] have calculated the band structures, and shown that both the distortion in angle and the displacement of one atom relative to the other along the trigonal axis make the band structure less metallic. Moreover, both distortions together are needed to achieve a semimetal properly. Calculations have also been performed as a function of pressure, indicating a transition to a semiconductor as found experimentally.[14]

The band structures of Ca, Sr, Ba are being calculated as a function of pressure for the fcc and bcc structures.[15] Apparently[16] the results cannot be fitted in terms of a simple nearly free electron model with any constant pseudopotential, taking only lowest plane waves into account. The d states strongly perturb the bands, as is to be expected from the unusual shape of the calculated[12] $v(q)$ and the A_2 parameter in the model potential.[12] The effect increases with pressure because the region inside the model radius becomes a larger fraction of the atomic volume. This rules out any explanation of the interesting phase changes and electrical properties at high pressures, in terms of overlapping s and p bands only.[17]

Finally we note that diamond, Si, and Ge have small band gaps compared with the Fermi energy E_F, and thus come within the nearly free electron scheme. We take up in Sec. 4 the question of how far covalency and other chemical effects can be seen in band structure and cohesive energy calculations.

3 THE TOTAL ENERGY TO SECOND ORDER

We can write the total energy of the metal to sceond order in $V(q)$ for any structure, ordered or disordered.[18] Some contributions such as the Fermi energy and exchange and correlation energy of a free-electron gas are functions of volume only. The last is not well known, and we therefore restrict ourselves to structural changes at constant volume.[19] The energy for that consists of two terms. U_E is the electrostatic energy of point charges in a uniform negative charge. The band-structure contribution is approximately

$$U_b = \sum_g W(g)[v(g)]^2 \epsilon(g) f(g) \tag{1}$$

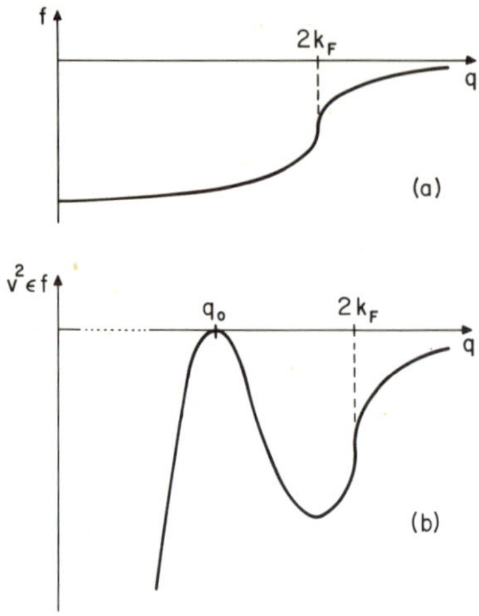

Fig. 2 The functions (schematic diagram only) (*a*) $f(q)$ and (*b*) $v^2\epsilon f$. Note that the scale in (*b*) between q_0 and $2k_F$ has been greatly expanded.

where ϵ is a screening factor which is near unity near $2k_F$ and may be ignored, and $f(g)$ is the negative function shown in Fig. 2*a*. The structural weight $W(g)$ is $|S|^2$ times the number of equivalent reciprocal lattice vectors **g**, where S is the structure factor. The combination $v^2\epsilon f$ is shown in Fig. 2*b*. The energy (1) includes both the sum of the occupied band-structure energies $E(\mathbf{k})$ and the electrostatic energy of the modulation of the electron density with wave vector **g**.

It is clear that there is little contribution to the binding energy from a reciprocal lattice plane if g falls near the zero q_0 of $v(q)$, i.e., if there is a small band gap. The energy is lowered by a structural change in two ways: (1) when a set of reciprocal lattice vectors moves further away from q_0 as a single set, and (2) when the set of g's splits, the energy lowering in this case being proportional to the second derivative of $v^2\epsilon f$ with respect to q. The c/a ratio of the divalent hexagonal metals; the distorted structures of Hg, Ga, and In; and the variation of the distortion in alloys of In may be understood qualitatively and in some cases semiquantitatively in these terms.[10]

The shape of $f(q)$ shows that other things being equal, i.e., $v(q)$ considered constant, a reciprocal lattice plane gives much more binding if g is just less than $2k_F$ rather than just greater. This is the well-known effect of Jones[1] as stated in Sec. 1. There are three comments to be made. Firstly, it is not correct just to add the one-electron energies $E(\mathbf{k})$, as assumed in some discussions, because this counts some of the electrostatic energy twice.[20] Secondly, if we fill the electrons beyond the zone plane, or move g well below $2k_F$, we do not lose this binding energy again but in fact keep gaining somewhat more (although the U_E also tends to increase then). Thirdly, it is not only the states at the zone boundary that contribute, but the second-order perturbation shift of all the occupied states. Thus there are contributions even in monovalent metals, as can be seen from the fact that $f(q)$ continues above $2k_F$ (Fig. 2*a*). They can be sizable, and can no longer be ignored as was done for example in the early work of Fuchs on the elastic constants.[21]

It is empirically well known that many structural properties of simple alloys correlate with the electron-per-atom ratio e/a. From the present discussion of Eq. (1), we see two factors as relevant: (1) the position of the g's relative to the zero q_0 of $v(q)$, and (2) the magnitude of the g's relative to $2k_F$. The latter certainly depends just on e/a. However, it is found that q_0 increases with valency z among the elements.[6,10] Thus the q_0 of the mean potential in alloys varies with concentration in much the same way as e/a does. Some of the apparent correlations of structure with e/a may therefore really be caused by the movement of q_0 rather than of $2k_F$. However, in some circumstances they have the opposite effect. For example in In alloys, the appropriate $g(200)$ falls near the zero q_0, and we have a tetragonal distortion splitting this set of vectors. The two factors, increase of q_0 and k_F on alloying with Pb say, work in opposite directions. In InPb and InSn the $2k_F$ effect wins: the distortion increases, but less so in InPb where the zero moves most. In InTl where there is only the change of q_0 (also increasing), the distortion decreases with concentration.[10]

The energy differences between structures are too small and the uncertainties in $v(q)$ and in the correlation and exchange effects too large at present for us to calculate energies blindly and expect to get the right result.[4]

There is considerable uncertainty about one highly technical point of great practical importance, namely the shape of $v(q)$ beyond the first zero and how quickly it cuts off at large q. Since there is a certain degree of arbitrariness[7] in the definition of the pseudopotential, there is in a sense no unique answer. However, what we want is that particular $v(q)$ which is in a certain sense a t matrix and cuts off most rapidly at large q so that the bands around E_F are decoupled to the maximum extent possible from the higher bands. Some of the calculated $v(q)$ do not cut off as rapidly at large q as one would like,[12] but there are several reasons for supposing that a suitable cutoff $v(q)$ exists. Of course, the decrease of $f(q)$ at large q helps in actual structural energy calculations (Fig. 2).

First we note that the higher q tend to describe the irregularities in the pseudopotential in and around the core of the atom. Such effects can be absorbed into an atomic t matrix, or from an alternative point of view cannot effect energy differences between two structures in neither of which there is significant overlapping of the cores. Second, it is found empirically in fitting band structures and optical spectra[8,9,22–24] that $v(q)$ can be taken as zero beyond the lowest few reciprocal lattice vectors. If the effects of higher $v(q)$ are zero in the one-electron energies $E(\mathbf{k})$, then they are not present in the total energy which is equal to the sum of occupied $E(\mathbf{k})$ apart from corrections such as due to counting some electrostatic energy twice. Third, it is always formally possible to decouple the bands and express the lower bands exactly in terms of a small finite secular equation.[25] The renormalized scattering matrix element for $\mathbf{k}$ to $\mathbf{k} + \mathbf{g}$ becomes to second order

$$v(\mathbf{g}) + \sum_{\mathbf{g}'} \frac{v(\mathbf{g}')v(\mathbf{g} - \mathbf{g}')}{E - E(\mathbf{k} + \mathbf{g}')} \tag{2}$$

The summation over $\mathbf{g}'$ excludes the $\mathbf{g}$ contained in the finite secular equation. The price that must be paid for having a finite secular equation is that the "renormalized" $v(\mathbf{g})$ in Eq. (2) are $\mathbf{k}$-dependent through the denominator in Eq. (2) (and through the numerator owing to the nonlocal nature of the pseudopotential). We therefore must make the finite equation large enough to reduce the $\mathbf{k}$ dependence to negligible proportions. The point is that once this has been done, we have probably also reduced the structure dependence of Eq. (2) to a negligible amount. For Eq. (2) is really a summation over all $\mathbf{k}'$ space, sampled at points $\mathbf{k} + \mathbf{g}'$. The sample depends on $\mathbf{k}$ and on structure through the $\mathbf{g}'$, the one variation not being radically different from the other. For large $\mathbf{g}'$ the different vectors $\mathbf{k} + \mathbf{g}'$ come close enough together in modulus, and the summation can be approximated to by a structure-independent continuous integral. As for the relatively low $\mathbf{g}'$, simple structures such as fcc, hcp, bcc are not really very different as regards shells of near neighbors in reciprocal space. Calculations[28] that have been made of the $\mathbf{k}$ dependence of Eq. (2) show that it is small, even when the bands of Pb are represented by a 4×4 secular equation. In any case we argue that if we can truncate $v(q)$ in fitting one band structure as found empirically, then this implies that we can use nearly the same $v(q)$ for all structures. This is indeed found for different semiconductors containing the same element,[8] and, as already mentioned, the In and Sb pseudopotentials from InSb[8] have been found to apply well to In metal[10] and Sb semimetal.[9] However, it is doubtful whether these can be used completely without change for calculations on highly localized defects such as vacancies, impurities, and surfaces.

4 COVALENCY

To what extent can we see some tendency toward covalency in the nearly free electron approach? Consider first a simple NFE band structure with $v(\mathbf{g})$ positive at the first band gap, as occurs for instance in Mg, Al, and several other metals. This means that the p-like state has the lower energy at the gap. It is a bonding state with $|\psi|^2$ a maximum halfway between the atoms, whereas the s-like state is antibonding with $\psi = 0$ there. In the first band as a whole there will be strong hybridization of s-like and p-like waves, all of bonding type tending to build up the charge density between the atoms. The effect is proportional to $v(\mathbf{g})$, and if the energy difference Δ_{sp} between the s and p levels in the free atom is small, this tends to make a positive hybridizing $v(\mathbf{g})$, as would be expected from a chemical point of view. All this appears to add up to a picture of some degree of covalency in the metal, and there is a sense in which this is so. However, it differs from molecular covalency in two important respects. First of all, the binding energy in our approximation is proportional to v^2, so that an antibonding gap [$v(\mathbf{g})$ negative, s below p state] of the same size gives just as much binding. Secondly, our theory is equivalent to calculating the charge density ρ to first order: i.e., the Fourier component $\rho(\mathbf{g})$ is proportional to $v(\mathbf{g})$. This means, because linearity implies superposition, that the total charge density is the sum of overlapping *spherical* pseudo atoms.[2] In other words we have no directed bonds. Moreover it cannot be argued that it is a serious fault to stop the perturbation expansion at this order, when the measure of higher-order corrections is the ratio $v(\mathbf{g})/E_F$, typically 0.1.

The situation in the group IV semiconductors (and group V semimetals) is different. Here q_0 has moved quite far to the right (in the sense of Fig. 1) because of the large valency.[10] The result is that v_{111} is not small, and in second order gives contributions to the band gap and to $\rho(\mathbf{g})$ comparable with the higher $v_{\mathbf{g}}$ in the first order.[29] We can illustrate the situation by calculating approximately the band gap (i.e., the "vertical" or "optical" gap in semiconductor terminology). We take plane waves $2\pi a^{-1}(1,1,0)$ and $2\pi a^{-1}(-1,-1,0)$ at the centers of opposite faces of the Jones zone.[30,31] The "renormalized" matrix element connecting them is to second order

$$\langle 110|V|\bar{1}\bar{1}0\rangle + \sum_{\mathbf{g}} \frac{\langle 110|V|\mathbf{k}+\mathbf{g}\rangle\langle \mathbf{k}+\mathbf{g}|V|\bar{1}\bar{1}0\rangle}{E_{110} - E_{\mathbf{k}+\mathbf{g}}} \tag{3}$$

In the summation we retain only terms giving v_{111}^2, i.e., intermediate states $(0,0,\pm 1)$. Putting in the structure factors, we obtain for the effective V_{220}

$$\begin{aligned} V_{220}(\text{effective}) &\equiv V(\text{direct}) + V(\text{2d order}) \\ &= v_{220} + v_{111}^2/(E_{111} - E_{001}) \end{aligned} \tag{4}$$

(We write V for the pseudopotential of the whole crystal, v for that of one atom.) Table 1 gives the two terms of Eq. (4) separately, the calculated band gap according to our simple approximation, i.e., twice Eq. (4), and the observed band gap at the point X in the Brillouin zone.[8,22] We see that the second term of Eq. (4) in fact dominates the first term. The sign of the interaction is such as to build up charge in the nearest neighbor bond directions (as is true of all states below the gap,[29] and also with the group V semimetals[14]), and the second term of Eq. (4) results in a large contribution which cannot be expressed as the sum of spherical pseudo

TABLE 1 ***Covalency in Semiconductors***

In rydbergs

Semiconductor	v_{220}	V, 2d order	Gap: Calculated	Gap: Observed	$(v_{111}/E_B)^2$
Si	0.04	0.12	0.32	0.30	0.26
Ge	0.01	0.15	0.32	0.28	0.37
Sn (gray)	0.00	0.1	0.22	0.23	0.26

NOTE: Values of v_{111}, v_{220} and the observed gap are from Cohen and Bergstresser.[8]

atoms. It has been calculated.[29,32] We might consider it the real degree of covalency, because it is a directed bond charge density. Its magnitude, expressed in dimensionless terms, is proportional to $(v_{111}/E_F)^2$. Unfortunately this quantity (Table 1) does not decrease monotonically from Si to Sn, as would be expected from the usual chemical picture. Actually the largeness of $(v_{111}/E_F)^2$ for Ge is to a considerable extent due to the decrease in E_F, a reflection of the fact that at lower mean charge densities it costs less kinetic energy to create a charge inhomogeneity.

The effects are clearly quite subtle, and we should undertake a proper analysis of the two-atom scattering terms in Bennemann's T-matrix formalism. The definition of degree of covalency is somewhat vague. Cohen[19] has shown how we can analyze the total energy in terms of two-atom, three-atom, etc., interactions. In an insulator we can, in principle, define the ionicity, number of electrons in a bond, etc., by the behavior of these functions at large separation.

Finally we inquire whether the distorted structures in Hg, Zn, and Ga can be understood in terms of a tendency to achieve certain bond distances or angles. The answer appears to be "no." We have already seen in Sec. 3 that these distortions are due to $v(q)$ being small,[10] nearly zero, at particular g: not to the potential being strong. Among the divalent metals the c/a ratio increases from Be to Cd with a completely distorted structure in Hg; i.e., distortion increases from Be to Hg, whereas the energy difference Δ_{sp} between the s and p levels in the free atom increases[33] (relative to E_F), and thus the tendency to sp hybridization decreases. Similarly among the trivalent metals, the distortion is at a maximum in the middle of the series Al, Ga, In, Tl (In-Tl alloys become fcc beyond 20 a/o Tl), whereas Δ_{sp}/E_F and other effects all change monotonically.[33] Also Hg, Ga, In are the lowest-melting and least-bound metals, hardly indicative of covalent binding. However, all these structural distortions have a natural explanation[10] in terms of the shape of $v^2\epsilon f$ near q_0, even down to such details as the switch with concentration in In-Pb and In-Sn alloys from $c/a > 1$ to $c/a < 1$. Because of symmetry, in lowest order the energy depends on $(c/a - 1)^2$ which does not distinguish between c/a greater or less than unity. It is the third-order term that distinguishes, and in a bond picture this would be given by the third derivative of the energy-separation curve. Now these curves always have the same Morse shape, steeper at small distances. The bond picture cannot therefore explain the switch.

5 THE NOBLE METALS

If we confine ourselves to considerations at and above the Fermi level, always retaining a full d shell, we are dealing with a nearly free electron band and can define a suitable pseudopotential.[34] However, it will not have quite the usual shape of Fig. 1. The $l = 0$ and 1 angular-momentum contributions to the pseudopotential will be "normal," but the proximity of the d band results in an exceptionally large $l = 2$ component. There are two terms.[26,35] The usual potential plus orthogonality term is net-repulsive because of the overcancelation from the spread-out d orbital.[6] In addition, there is the hybridizing term, also strong. All these terms can now be calculated for copper as a function of q from the work of Mueller.[35]

What can we say about the Jones explanation[36] of the Hume-Rothery rules, particularly for the α phase boundary? This theory, in terms of a free-electron sphere touching the zone faces at $e/a > 1.36$, has been under a cloud since the discovery that in pure Cu the Fermi surface is already so distorted as to touch the zone faces. The point is that this is irrelevant. We have in Eq. (1) the total energy to second order in $v(q)$. The factor f changes with e/a as k_F increases relative to g. Considering only the (111) reciprocal lattice vectors, U_b will appear as in Fig. 3, with the singularity where $2k_F$ *is equal to* g. Now in Cu at the zone faces the ratio $v(g)/E_F$ is approximately 0.3, large enough that third- and higher-order corrections will make some contribution but small enough that these do not fundamentally alter the shape of U_b. The effect will be to spread out the steep drop somewhat (Fig. 3). and add two small kinks from the critical points. However there should still be a region of strong upward curvature just beyond 1.36 (Fig. 3). In the alloy the kinks become smeared out because of the electron scattering. In any case the first kink at first touching the zone face does not concern us because we are already beyond

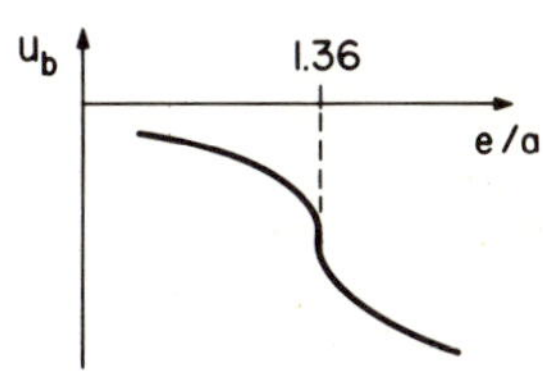

Fig. 3 Variation with electrons per atom ratio e/a of the band-structure energy U_b in Cu to second order (schematic diagram).

it in pure Cu, and the other is at too high an energy. Also the band gap decreases, making the simple second-order theory with its sharp curvature more nearly correct. The $l = 0, 1$ components of the pseudopotential will give a decreasing band gap for the reasons already argued elsewhere.[33] The two $l = 2$ terms, which are equally or more important, will also vary in a way to give a decreasing band gap: the potential plus orthogonality term because as the d shell of the solute is more tightly bound, there is less overcancelation; and the hybridizing term because E_F is rising and the mean E_d falling.

REFERENCES

1. Mott, N. F., and H. Jones: "Theory of the Properties of Metals and Alloys," chap. 5, Oxford University Press, London, 1936.
2. Ziman, J. M.: *Advan. Phys.* **13**:89 (1964).
3. Heine, V.: In J. G. Daunt et al., "Low Temperature Physics LT9," p. 698, Plenum Press, New York, 1965.
4. Harrison, W. A.: "Pseudopotentials in the Theory of Metals," Benjamin, New York, 1966.
5. See the introduction in Ref. 6.
6. Austin, B. J., and V. Heine: *J. Chem. Phys.* **45** (to be published).
7. See for example B. J. Austin, V. Heine, and L. J. Sham: *Phys. Rev.* **127**:276 (1962).
8. Cohen, M. L., and T. K. Bergstresser: *Phys. Rev.* **141**:789 (1966).
9. Falicov, L. M., and P. J. Lin: *Phys. Rev.*; **141**:562 (1966); P. J. Lin and J. C. Phillips: To be published.
10. Heine, V., and D. Weaire: To be published.
11. For some of the most recent calculated $v(q)$ see Ref. 12. For a comparison with experimentally fitted $v(q)$ see Ref. 13.
12. Animalu, A. O. E., and V. Heine: *Phil. Mag.*, **12**:1249 (1965).
13. Heine, V.: In F. Abelès et al., "Optical Properties and Electronic Structure of Metals and Alloys," p. 16. North Holland, Amsterdam, 1966.
14. Cohen, M. H., L. M. Falicov, and S. Golin: *IBM J. Res. Develop.* **8**:215 (1964). L. M. Falivoc: In C. T. Tomizuka et al., "Physics of Solids at High Pressures," p. 30, Academic Press, New York, 1965.
15. Animalu, A. O. E., and B. Vasvari: Private communication.
16. The author has so far only seen preliminary results for Ca.
17. Drickamer, H. G.: *Solid State Phys.* **17**:1 (1965).
18. Harrison, W. A.: *Phys. Rev.* **129**:2503 (1963); R. Pick and G. Sarma; *Phys. Rev.* **135**:A1363 (1964).
19. Cohen, M. H.: "Concentrated Solid Solutions," Proceedings of the special meeting of the ASM, Cleveland, 1963.
20. Reference 1, p. 46.
21. Reference 1, p. 149.
22. Brust, D.: *Phys. Rev.* **134**:A1337 (1964).
23. Ashcroft, N. W.: *Phil. Mag.* **8**:2055 (1963).
24. Anderson, J. R., and A. V. Gold; *Phys. Rev.* **139**:A1459 (1965).
25. See for example Ref. 26, also discussion in Refs. 23, 27.
26. Heine, V.: To be published.
27. Animalu, A. O. E.: *Phil. Mag.* **11**:379 (1965).

28. Animalu, A. O. E.: Unpublished calculation.
29. Kleinman, L., and J. C. Phillips: *Phys. Rev.* **125**:819 (1962).
30. Reference 1, p. 159.
31. The plane waves (1, −1, 0) and (−1, 1, 0) also belong to the same reduced wave vector (0, 0, 1). They have the same interaction (3). There is no interference between the two pairs of wave vectors, resulting in the double degeneracy of states at the band gap at X.
32. Bennemann, K. H.: *Phys. Rev.* **133**:A1045 (1964).
33. Cohen, M. H., and V. Heine: *Advan. Phys.* **7**:395 (1958).
34. See eq. (26) of Ref. 26.
35. Mueller, F. M.: *Phys. Rev.* (to be published).
36. Reference 1, p. 171.

DISCUSSION *on Paper Presented by V. Heine*

FRIEDEL: When Mott and Jones proposed their nearly free electron picture of diamond and Si, they had good experimental evidence from Skinner's X-ray emission data, which showed convincingly that the valence bandwidth of these elements was of the order of a rydberg.

KAUFMAN: Can you calculate the relative energies of the fcc, hcp, and bcc forms of Be, Al, Mg, Zn, and Cd?

HEINE: We do not yet know the magnitudes of the pseudopotentials and the correlation and exchange contributions to the total energy sufficiently well for such calculations to be meaningful at the present time. We are starting by understanding qualitative trends. All I can say is that fcc is obviously favored for group III metals and not favored for groups IA and 2B.

LAVES: What do you mean by "distortion" when you discuss the gallium structure?

HEINE: There is no unique way of viewing a three-dimensional structure. Mr. Weaire, who is working with me, has found that one can obtain the Ga structure from face-centered cubic by a *substantial* (15%) tetragonal distortion similar to that in In, followed by a *small* shift of all the atoms corresponding to a complex phonon of negative frequency squared. I was discussing only the large first effect.

KAUFMAN: High-pressure studies by J. Jamieson and H. Drickamer on the group IV elements Sn, Ge, Si and the III to V compounds have shown that these materials undergo a phase transformation from the semiconducting diamond structure to the metallic white Sn structure. The pressure at which these transformations occur at room temperature P multiplied by the volume change ΔV for these reactions is thus a first approximation of the energy difference between the diamond cubic and white Sn structures. Surprisingly, the product $P\,\Delta V$ has been found to approximate to one-half the semiconducting energy gap: $P\,\Delta V \approx E_G/2$. Would you care to comment or venture an explanation of this result?

HEINE: No, beyond the qualitative remark that the thermal band gap is of course some (complicated) measure of the strength of the pseudopotential. Actually, it did occur to me yesterday, as Professor Hume-Rothery was talking, that the zero of the pseudopotential in Sn does fall very near one of the reciprocal lattice vectors, which presumably has something to do with the transformation, analogous to the situation in Hg, Ga, and In.

LOMER: Is there any further evidence on the accuracy with which the pseudopotential for an atom can be transplanted from structure to structure?

HEINE: Yes, Al. Analysis of the Fermi surface gave $v(q)$ at $g(111)$ and $g(200)$. Theory was then used to put a continuous curve through the two points. This then gave good agreement with experiment for the resistance of liquid Al, the resistivity of solid Al due to phonons, and the phonon enhancement of the density of states at low temperatures. We also find in In and Ga that the $v(q)$ fitted from the semiconductors are closer to what is required for the metals than the $v(q)$ calculated purely theoretically.

BREWER: How is screening changed in going from InSb to Sb?

HEINE: It is assumed there is no change in the screening. This assumption is in practice not as bad as it appears because at q as large as the reciprocal lattice vectors, the screening factor is about 1.1 where the 0.1 comes from the screening electrons. A reasonable change in the latter, even 20% say, only gives a 2% change in $v(q)$. In the work on Sb, Lin and Falicov intended to use the $v(q)$ taken from InSb only as a first approximation. They found that they hit the right Fermi surface of Sb first shot, without any further adjustment.

THEORETICAL CONSIDERATIONS OF THE HUME-ROTHERY RULES

A. P. Blandin

Laboratoire de Physique des Solides associé au CNRS, Faculté des Sciences, Orsay, France

1 THE RIGID-BAND MODEL

In normal alloys (Cu-Zn for example), the Hume-Rothery rules[1] correlate the limit of stability of a phase to a critical value of the electron-per-atom ratio (1.4 for α phases of Cu alloys). Jones[2] explained this rule in the framework of the rigid-band model. Let $n_0(E)$ be the density of states per unit energy and per atom in the A metal; $n_0(E)$ presents kinks at G and H as shown in Fig. 1. (For comparison, the dashed curve represents the density of states for a free-electron gas.) These kinks occur when the surfaces of constant energy in **k** space are just touching or overlapping the Brillouin zone boundary.

In an alloy **A***B*, the first-order energy change ΔE for the one-electron state ψ_k is

$$\Delta E = \langle \psi_k | V_p | \psi_k \rangle$$

where V_p is the perturbing potential due to the B atoms. In the rigid-band model one takes ΔE independent of **k**. The density of states in the alloy is given by $n(E) = n_0(E - \Delta E)$; it keeps the same shape. If the B atoms have a greater valence than the A atoms, the point F representing the density of states for the Fermi energy is moving to the right.

In this model the total energy ϵ is calculated as the sum of one-electron energies:

$$\epsilon = \int^{E_F} E n_0(E)\, dE$$

A schematic curve for ϵ is given on Fig. 2 where for comparison the value of ϵ for a free-electron gas has been drawn. One sees on this figure that the energy of the phase is lowered compared to the energy of the free-electron gas. The decrease of the energy ϵ occurs mainly when the Fermi energy E_F is around E_D. Now, in a nearly free electron picture the point D represents very nearly the case for which the number Z_c of electrons per atom is such that the free Fermi sphere touches the Brillouin zone boundary. A given phase has a smaller energy than the free-electron gas for a number of electrons per atom Z near the critical value Z_c or greater than Z_c. This is the case when one vector τ of the reciprocal lattice becomes smaller than $2k_F$.

From this point, one can compare the stability of two different phases 1 and 2. The limits of stability of these phases will be controlled by the depletions of the total energy ϵ which occur around the critical values Z_{c1} and Z_{c2}. The limits of stability occur for values near the critical values Z_{c1} or Z_{c2}. If phase 1 is stable for the pure metal, it will thus be stable as long as the number of electrons per atom is smaller than Z_{c1}.

The argument remains valid if the true Fermi surface of the pure metal is in contact with the Brillouin zone boundary (F being to the right of G on Fig. 1). This is the case for Cu alloys.

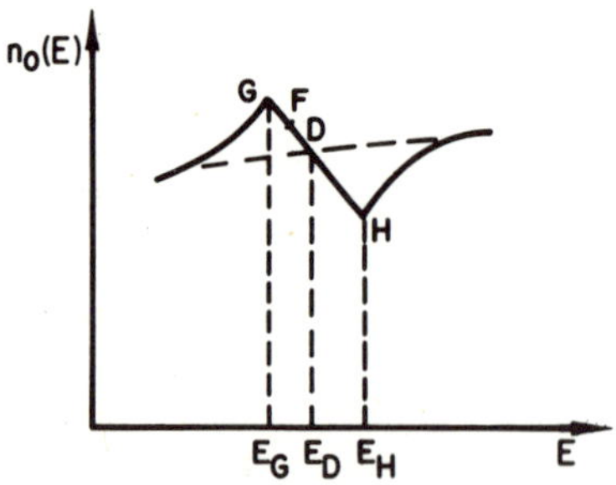

Fig. 1 Density of states. Solid line, in the metal A; dashed line, in a free-electron gas.

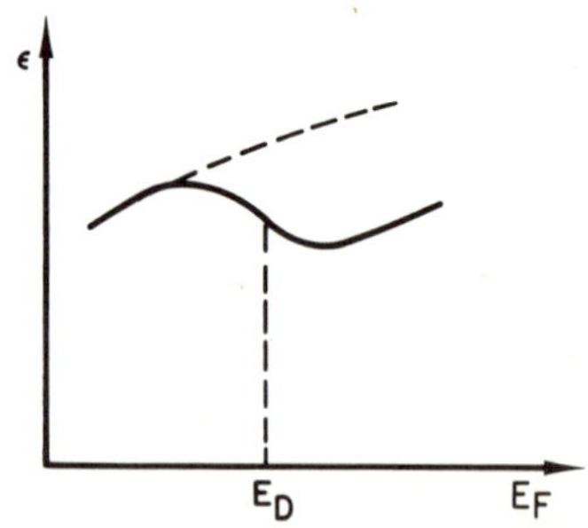

Fig. 2 Total energy of the alloy as a function of the Fermi energy. Solid line, corresponding to the density of states $n_0(E)$; dashed line, corresponding to the density of states of a free-electron gas.

But the rigid-band model can be criticized along the following lines:

1. ΔE may change with $\mathbf{k}$. Near the Brillouin zone boundary, states of definite and different symmetry have different ΔE values.

2. The calculation of the total energy ϵ takes care of only some second-order terms in the concentration of B atoms.

3. In the calculation of the total energy ϵ, the electron-electron interactions (which are implicitly contained in the unperturbed energies and in ΔE) are counted twice.

2 THE MODEL OF PAIR INTERACTIONS

Let us look now at a completely different way of approach, first in a simplified model of free electrons.

Pair Interactions in a Free-electron Gas

The Coulomb interaction between point charges Z_A and Z_B is

$$\frac{Z_A Z_B}{R} = \frac{Z_A Z_B}{2\pi^2} \int \frac{\exp i\mathbf{K}\cdot\mathbf{R}}{K^2}\, d\mathbf{K} \qquad \mathbf{R}_A - \mathbf{R}_B = \mathbf{R}$$

Owing to the screening by the electrons, the effective interaction of the charges in the free-electron gas is

$$W_{AB}(R) = \frac{Z_A Z_B}{2\pi^2} \int \frac{\exp i\mathbf{K}\cdot\mathbf{R}}{K^2 \varepsilon(\mathbf{K})}\, d\mathbf{K}$$

where $\varepsilon(\mathbf{K})$ is the dielectric constant for wave vector $\mathbf{K}$ and frequency 0. Owing to the screening, $W_{AB}(R)$ has no long-range Coulomb part—but it exhibits Friedel oscillations[4] for large R:

$$W_{AB}(R) \sim \frac{\cos 2k_F R}{(2k_F R)^3}$$

These oscillations in the real space are related to the logarithmic singularity of $\varepsilon(K)$ for K near to $2k_F$: $\varepsilon(K) \sim (K - 2k_F) \log |K - 2k_F|$ (Kohn anomaly[5]).

Pair Interaction in a Real Metal or Alloy

The method of orthogonalized plane waves,[6] which has been very useful in the understanding of Fermi surfaces of normal metals, can be used to make a self-consistent Hartree calculation of the energy of a metal[7,8] or alloy. One finds that the total energy in an alloy has a first term independent of the positions of the atoms, and a second term that is a sum of pair interactions: $W_{AA}(R)$, $W_{BB}(R)$, $W_{AB}(R)$ for AA, BB, and AB pairs, respectively. The effective interactions have also an oscillatory behavior as in the model of free electrons but with amplitudes α_{AA}, α_{AB}, and α_{BB} depending on the atoms A and B. The interactions $W(R)$ have been calculated for pure normal metals by Pick[8] (the method has not yet been applied to noble metals). In an alloy, no detailed calculations have been made so far, so that one does not know the amplitudes of the asymptotic form of $W(R)$. But one can say that they will vary smoothly with the concentration of B atoms; the main influence in alloying comes through the variation of the Fermi wave vector k_F.

The Hume-Rothery Rules

The method of pair interactions has been used with success to calculate the energy for different structures of the normal metals, and the phonon spectrum of these metals.[7,8] It is very helpful also for the understanding of the Hume-Rothery rules.

In a random alloy **A**B, the total energy of the alloy can be written

$$\epsilon = \epsilon_0 + \sum_{i<j} \bar{W}(R_{ij})$$

where ϵ_0 is a constant which varies smoothly with the concentration c of B atoms.

$\bar{W}(R)$ is an average interaction:

$$\bar{W}(R) = (1 - c)^2 W_{AA}(R) + 2c(1 - c)W_{AB}(R) + c^2 W_{BB}(R)$$

$\bar{W}(R)$ has an oscillatory behavior for large values of R:

$$\bar{W}(R) \simeq \alpha \frac{\cos 2k_F R}{(2k_F R)^3}$$

k_F is the Fermi wave vector of the alloy and thus varies with the concentration c of B atoms.

The second term in ϵ can be written

$$\epsilon_1 = \frac{N}{2}\left[\sum_{\tau} \Phi(\boldsymbol{\tau}) - \frac{\Omega}{(2\pi)^3}\int \Phi(\mathbf{K})\, d\mathbf{K}\right]$$

where N is the number of atoms, Ω the atomic volume, $\boldsymbol{\tau}$ the vectors of the reciprocal lattice, and $\Phi(\mathbf{K})$ the Fourier transform of $\bar{W}(R)$.

$W(R)$ exhibits Friedel oscillations, so that in $\Phi(\mathbf{K})$ there is a Kohn anomaly when $K \simeq 2k_F$: $\Phi(\mathbf{K}) \simeq \alpha|K - 2k_F| \log |K - 2k_F|$. This means that the derivative $d\Phi/dK$ of Φ has an infinite slope at $K = 2k_F$.

From this property, it can easily be seen that the total energy (in function of k_F) will have an infinite slope when one of the reciprocal lattice vectors τ is equal to $2k_F$. This is just the critical value of the rigid-band model that corresponds to the number of electrons per atom Z_c.

When α is positive, ϵ varies with k_F (or with the concentration c of B atoms) as given on Fig. 3. The energy is lowered when the Fermi wave vector is near the value $\tau/2$, or greater than $\tau/2$. This model gives results for ϵ that are very similar to those that were obtained in the rigid-band approach.

When α is negative, the energy ϵ is enhanced for $k_F > \tau/2$. This does not change the fact that the critical values for the limit of stability of phases will occur for the number of electrons per atom Z_c.

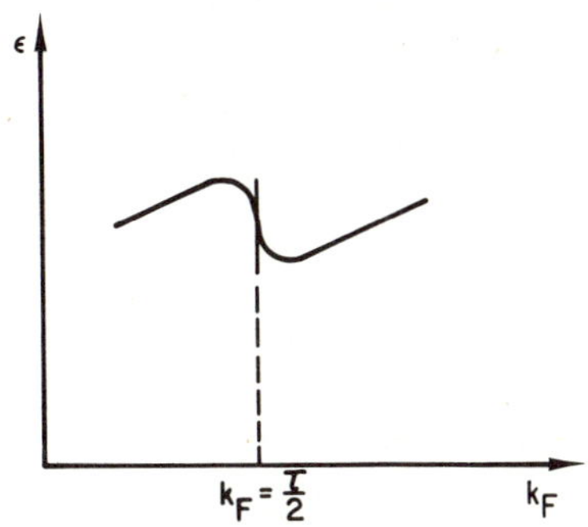

Fig. 3 Total energy of the alloy in the model of pair interactions.

These considerations are fairly qualitative, but one can conclude that the study of the effective interactions of atoms leads to similar conclusions to those for the rigid-band model (the only difference occurs when α is negative). In the model of the pair interactions, the conclusion follows directly from the existence of the Friedel oscillations (which are equivalent to the Kohn anomaly).

3 LONG-PERIOD SUPERLATTICES

In many ordered AB alloys, there exist, at temperatures below the critical temperature of order T_c, long-period superlattices.[10] Such structures occur in many instances, particularly in the Cu-Au system. A typical situation is the following: For planes of a given direction [1,1,1], for example, there is a modulation of the probability that a given plane is occupied by A (or B) atoms. This modulation has a period and a wave vector $\mathbf{K}_0$ which are not related to the ordinary lattice. The period and the wave vector $\mathbf{K}_0$ vary continuously with the concentrations of A or B and with temperature.

Two explanations have been given, and we shall see that they are complementary:

1. When a superlattice exists, new Brillouin zone boundaries exist. They are planes which are perpendicular to the wave vector $\mathbf{K}_0$ in the reciprocal space. With the help of the rigid-band model, one can say that a superlattice will be stable when one of the new Brillouin zone planes is in contact or will cut the free Fermi sphere. Sato[10] gets good agreement with experiments when the Fermi sphere is in contact with the new Brillouin zone. This explains well the variation of the period and of the vector $\mathbf{K}_0$ with the concentrations of A and B atoms.

2. The second approach is based on the model of pairs interactions. We shall give here the arguments for an alloy AB of equiatomic composition.[11] Taking care only of the long-range order, one can write the probability of having an atom A at $\mathbf{R}$: $P_A(\mathbf{R}) = [1 + \sigma(\mathbf{R})]/2$ and $P_B(\mathbf{R}) = [1 - \sigma(\mathbf{R})]/2 \quad -1 < \sigma(\mathbf{R}) < 1$.

The total energy of the alloy can be written in the following way:

$$\epsilon = \epsilon_0' + \sum_{\mathbf{R}<\mathbf{R}'} \sigma_{\mathbf{R}}\sigma_{\mathbf{R}'} V(\mathbf{R} - \mathbf{R}')$$

where $V(\mathbf{R}) = W_{AA}(\mathbf{R}) + W_{BB}(\mathbf{R}) - 2W_{AB}(\mathbf{R})$. The theory follows in the local Bragg-William approximation. For the equiatomic composition, the transition is a second-order one, and the critical temperature is given by

$$k_B T_c = v_0(\mathbf{K}_0)$$

$v_0(\mathbf{K}_0)$ is the minimum of the function $v_0(\mathbf{K})$:

$$v_0(\mathbf{K}) = \sum_{\mathbf{R}} \exp(i\mathbf{K}\cdot\mathbf{R}) V(\mathbf{R}) = \sum_{\boldsymbol{\tau}} v(\mathbf{K} + \boldsymbol{\tau})$$

where $v(\mathbf{K})$ is the Fourier transform of the function $V(\mathbf{R})$.

$V(\mathbf{R})$ is an oscillatory function of R so that $v_0(\mathbf{K})$ has a Kohn anomaly when $|\mathbf{K} + \boldsymbol{\tau}| = 2k_F$. The function $v_0(\mathbf{K})$ decreases near values of $\mathbf{K}$ such that $\mathbf{K} + \boldsymbol{\tau} = 2k_F$ and minima of $v_0(\mathbf{K})$ will occur near these values. If the absolute minimum of $v_0(\mathbf{K})$ is one of these minima, $\mathbf{K}_0 \neq 0$ or π, and one gets a superlattice which gives a new Brillouin zone boundary nearly in contact with the free Fermi sphere.

One sees that the two explanations that have been given are complementary. The occurrence of long-period superlattices is related to the presence of a new Brillouin zone boundary in the vicinity of the Fermi sphere. This is just a generalization of the Hume-Rothery rules.

4 STACKING-FAULT ENERGIES IN CLOSE-PACKED STRUCTURES[12]

Close-packed structures are characterized by the stacking of triangular planes of atoms in a given order: $ABABAB$ . . . for the hexagonal structure, $ABCABC$ for the face-centered cubic structure. A stacking fault is a fault in the stacking of the planes: $ABACAC$ . . . . In the model of pair interactions, one can calculate the energy of stacking faults for metals and alloys if one knows the effective interactions $W_{AA}(R)$, $W_{BB}(R)$, and $W_{AB}(R)$. For a random alloy it is sufficient to know the average interaction $\bar{W}(R)$. In the presence of stacking faults, the minimum distances of interest correspond to second neighbors, so that one can use the asymptotic form for $\bar{W}(R)$:

$$\bar{W}(R) \simeq \alpha \frac{\cos 2k_F R}{(2k_F R)^3}$$

In this case, the interaction is given when one knows α and the Fermi wave vector of the metal (or alloy) (or the number of electrons per atom).

Interactions between Planes

Using the pair interaction $\bar{W}(R)$, one can calculate first the change in the interaction energy of one atom with all the atoms j of a plane at a distance nd (d is the distance between planes) when a stacking fault is introduced. If $\mathbf{b}$ is the vector of the fault, this change of energy is

$$\Phi(nd) = \sum_j |\bar{W}(\mathbf{R}_i - \mathbf{R}_j + \mathbf{b}) - \bar{W}(\mathbf{R}_i - \mathbf{R}_j)|$$

Using the asymptotic value for $\bar{W}(R)$, one gets easily the following results:

Let λ be the first points of the reciprocal lattice of the triangular *plane*.

$$\lambda < 2k_F: \qquad \Phi(nd) \simeq -a\alpha \frac{\sin nk_\lambda d}{(2nk_F d)^2}$$

where a is a constant and $k_\lambda^2 = 4k_F^2 - \lambda^2$

$$\lambda > 2k_F: \qquad \Phi(nd) \simeq -a\alpha \frac{\exp(-k'_\lambda nd)}{(2nk_F d)^2}$$

where $k'^2_\lambda = \lambda^2 - 4k_F^2$.

The critical value $\lambda = 2k_F$ occurs when the average number of electrons per atom is $Z_1 = 1.14$. For $Z < 1.14$, $\Phi(nd)$ is exponentially decreasing. For $Z > 1.14$, $\Phi(nd)$ is oscillatory. One sees that the interaction function $\Phi(nd)$ depends essentially on the number of electrons Z.

Stacking-fault Energies

Knowing the energy of interaction between planes one can easily calculate the energies of stacking faults. The two parameters of importance are α, the amplitude of the oscillations, and Z, the number of electrons per atom.

The amplitude α has been calculated for some normal metals[8] and we shall use these values.

For monovalent metals, the hexagonal structure is stable, and the energies of the stacking faults 1Δ, 2Δ, and 3Δ are in the ratio 1:2:3 [this is due to the fact that $\Phi(nd)$ is nearly 0 for $n > 2$]. They are given in Table 1. The values are very small, especially for Na, but in this case α is very small and it is certainly necessary to use the exact function $\bar{W}(R)$.

TABLE 1 ***Energies of Stacking Faults in Alkali Metals***

In ergs per square centimeter

Element	fcc	hcp		
		$\Gamma_{1\Delta}$	$\Gamma_{2\Delta}$	$\Gamma_{3\Delta}$
Li	Negative values	2	4	6
Na	Unstable	(0.03)	(0.06)	(0.09)
K		0.4	0.8	1.2

These results are in agreement with experiment: one knows that at low temperature Li and Na have the hexagonal structure with many stacking faults.

For divalent metals, the hexagonal structure is stable. For trivalent metals, the face-centered cubic structure is stable. Numerical results are given in Table 2 for Mg and Al. Γ_m, Γ_i, and Γ_e are the energies of a twin, of an intrinsic, and of an extrinsic stacking fault in the face-centered cubic structure. The numerical values have the good order of magnitude, but there are no simple relations between them. This is due to the long-range nature of $\Phi(nd)$, the interaction between planes.

The Hume-Rothery Rules

This discussion can be generalized for alloys. We do not know α, but for normal metals (with the exception of the noble metals) all the calculated values are positive. (This is expected from the study of the screening in a free-electron gas.)

Knowing the behavior of $\Phi(nd)$ with n, one is able to calculate the energies of the faults. The sign of the energies will depend only on the values of the average number of electrons per atom—and the sign of α.

TABLE 2 ***Energies of Stacking Faults in* Mg *and* Al**

In ergs per square centimeter

Element	fcc			hcp		
	Γ_m	Γ_i	Γ_e	$\Gamma_{1\Delta}$	$\Gamma_{2\Delta}$	$\Gamma_{3\Delta}$
Mg	Negative values			60	120	195
Al	60	150	125	Negative values		

TABLE 3 ***Stability of Phases as Deduced for Stacking-fault Energies***

$\alpha > 0$

Z	1	1.14	1.27	1.36	1.66	2.10	2.29	3.53	4
hcp				Unstable			Unstable		
fcc		Unstable				Unstable			Unstable

One can thus get information on the stability of close-packed structures. The details of the calculations will not be given here: a detailed publication will be published soon.[12] Table 3 summarizes the results that are obtained when $\alpha > 0$; outside the classical limits given by the Hume-Rothery rules ($Z = 1.36$ for fcc structure, $Z = 1.66$ for hcp structures) there exist other limits of the phases ($Z = 1.27$ for example). We have thus a generalization of the Hume-Rothery rules.

Remark on Noble Metals

One should remark, in Table 3, that the noble metals Cu, Ag, and Au are anomalous. They have an fcc structure. Also, using the preceding formulas, one finds negative values for Γ_m, Γ_e, and Γ_i if they are taken for a positive value of α. These remarks lead to the conclusion that the model of pair interactions could be valid for the noble metals if α were negative. This may be due to the presence of the d band which has an energy not very small compared to the Fermi energy. This point must be clarified in order for one to understand the behavior of the noble metals and of their alloys.

CONCLUSION

This brief review shows that the Hume-Rothery rules have a very general meaning. The average electron-per-atom ratio is the critical parameter which controls many physical properties of alloys. The rigid-band model gives an interpretation of these rules. The more sophisticated method of pair interactions gives good informations on the origin of the rules. It can lead to quantitative results. The qualitative discussion of Secs. 2 and 3 of this paper has to be developed quantitatively. This can be done in the near future for simple cases. The calculations of Sec. 4 give some hope in regard to this development. The case of noble metals is more difficult from a theoretical point of view, but there is some hope that the situation could be also clarified.

REFERENCES

1. Hume-Rothery, W.: "The Metallic State," p. 328, Oxford University Press, New York, 1931.
2. Jones, H.: *Proc. Phys. Soc.* **49**:250 (1937).
3. Nozieres, P., and D. Pines: *Nuovo Cimento* **9**:470 (1958).
4. Friedel, J.: *Advan. Phys.* **3**:446 (1954).
5. Kohn, W.: *Phys. Rev. Letters* **2**:393 (1959).
6. Herring, C.: *Phys. Rev.* **57**:1169 (1940).
7. Harrison, W.: *Phys. Rev.* **129**:2503 (1963).
8. Pick, R.: Rapport CEA, thesis, Paris, 1965.
9. Blandin, A.: "Alloying Behaviour and Effects in Concentrated Solid Solutions," Proceedings of AIME, Cleveland, 1963, ed. by T. B. Massalski, Gordon and Breach, 1965.
10. Sato, H., and R. S. Toth: Proceedings of AIME, Cleveland, 1963, ed. by T. B. Massalski, Gordon and Breach, 1965.
11. Vallain, J.: *J. Phys. Radium* **24**:123 (1963).
12. Blandin, A., and G. Saada, *J. Phys.*, to be published.

THE LATTICE STABILITY OF THE TRANSITION METALS*

Larry Kaufman

ManLabs, Inc., Cambridge, Massachusetts

INTRODUCTION

Previous papers in this series have dealt with the relative stability of the body-centered cubic, hexagonal close-packed, and face-centered cubic configurations of metals. In particular, it was shown[1] that the difference in free energy between crystallographic forms of a metal made an important contribution in controlling alloying behavior, phase transformations, and stacking-fault energy of the metal over a wide range of temperature and pressure. Moreover, the difference in free energy between crystal forms, a quantity often much smaller than present errors in cohesive energy calculations, was shown to be relevant in interpreting specific kinds of activity data.[1] In view of present difficulties in making estimates of the free-energy difference and its temperature dependence from quantum-

* This research was sponsored by the Metallurgy Branch, Office of Naval Research, Washington, D.C., under Contract NONR 2600(00).

The author is indebted to Dr. Peter Rudman of Battelle Memorial Institute for several discussions of this subject and for furnishing references for a number of binary phase diagrams.

mechanical principles, a thermodynamic approach based on empirical data has been employed. In this way lattice-stability information for Ti,[1] Zr,[1] Zn,[2] Cu,[2] Ag,[2] and Fe[3] has been obtained. In addition, high-pressure studies[4,5] have been employed as a means for deriving lattice-stability information.

Recently, an analysis of the Fe-Ru system was performed using the regular-solution model and assuming that the lattice stability of hcp, fcc, and bcc Ru was similar to that of nonmagnetic Fe. The computed phase diagram for the Fe-Ru system, as well as the description of fcc $\rightleftharpoons$ hcp martensitic transformations at one atmosphere and high pressure resulting from these assumptions, was found to agree materially with observations.[6] In view of this finding, and the fact that transition elements of the second and third long periods exhibit a more regular pattern of crystal structure stability than their counterparts in the first period, it seems worthwhile to consider an analysis of the lattice stability of these elements. The absence of magnetic contributions to the second- and third-period counterparts of Cr, Mn, and Fe is undoubtedly the cause of the more regular progression from bcc to hcp to fcc stability across the transition series. In order to make a first approximation to this problem, it has been assumed that second- and third-row counterparts, i.e., Nb and Ta, Ru and Os, or Pd and Pt, can be characterized by identical lattice-stability parameters. Although this need not be true, it is introduced as a simplifying assumption.

On this basis, curves of the enthalpy and entropy differences can be estimated using the available results for hcp and bcc Zr[1]; hcp, bcc, and fcc Ru[6]; fcc and bcc Ag[2]; and hcp, bcc, and fcc Cd. The latter values were assumed similar to those for Zn.[2] With these values as fixed points, a series of curves can be drawn which are limited by observed stability relations: i.e., Re is hcp up to its melting point. In order to obtain better resolution, it became necessary to consider the binary phase diagrams between elements of the second- and third-row transition elements. These binary diagrams were chosen for those cases where the component elements have differing stable crystal structures: i.e., W-Ir or Mo-Re. In order to use these diagrams to advantage, the rather drastic assumption was made that the hcp, bcc, fcc, and liquid phases could be treated as ideal solutions. As a consequence, systems exhibiting miscibility gaps were eliminated from consideration. Surprisingly enough, as detailed below, a consistent series of lattice-stability parameters resulted. These parameters have been used to compute 30 binary phase diagrams which agree quite well with observations. In order to provide an additional evaluation, the calculations were extended to include the first-row transition elements V, Cr, Co, and Ni. Eighteen additional binary diagrams were computed assuming stability parameters for these metals identical to their second- and third-row counterparts. The only exception was hcp Co where a small change was made to reflect stability at low temperatures. As in the foregoing series, systems exhibiting miscibility gaps were not considered, and reasonable agreement was obtained. As a consequence, it is concluded that the aforementioned lattice-stability values constitute a good first approximation to defining the relative stability of these elements.

ENTHALPY AND ENTROPY DIFFERENCES AMONG THE HCP, BCC, AND FCC FORMS, AND THE COMPUTATIONS OF BINARY PHASE EQUILIBRIA

The lattice stability of the bcc (β), hcp (ϵ), and fcc (α) configurations of the second- and third-row transition elements is depicted in Figs. 1 and

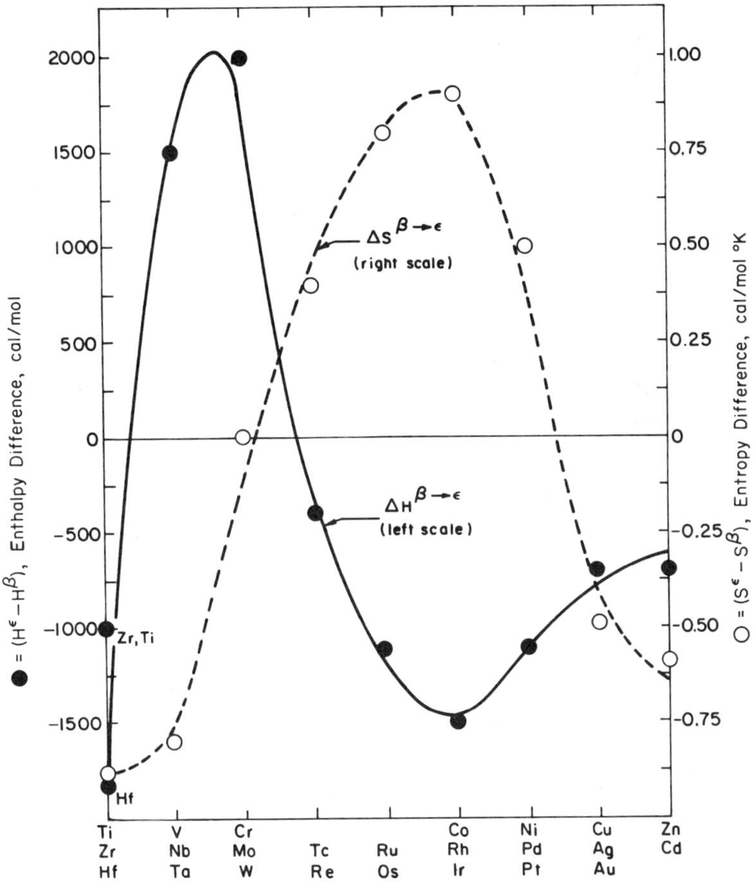

Fig. 1 Enthalpy and entropy differences between the hcp (ϵ) and bcc (β) forms of the transition metals.

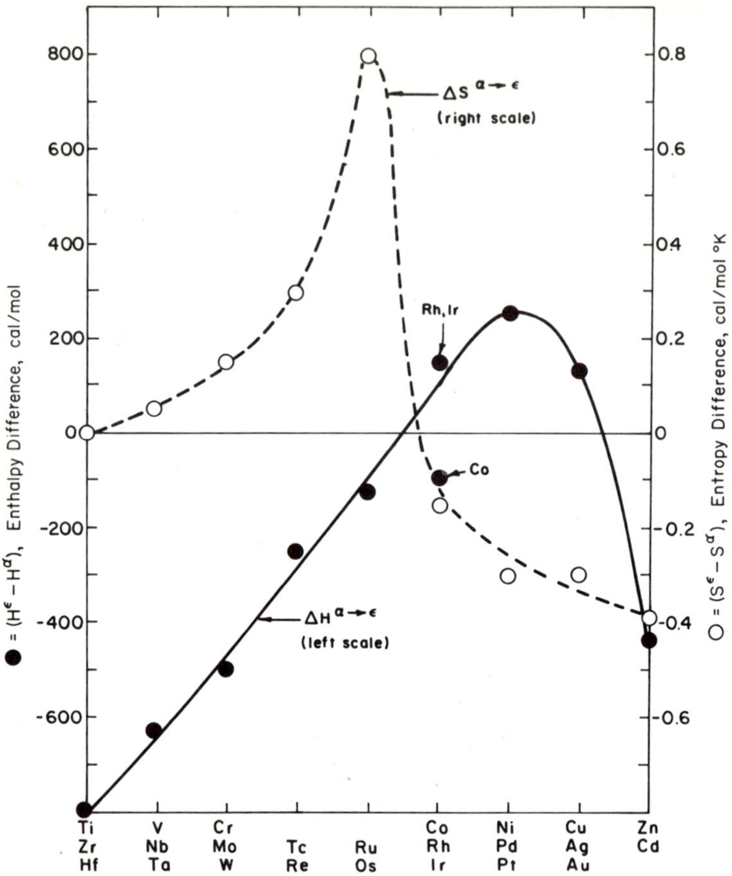

Fig. 2 Enthalpy and entropy differences between the hcp (ϵ) and fcc (α) forms of the transition metals.

2.* As indicated earlier, second- and third-row counterparts (i.e., Nb/Ta or Rh/Ir) are assumed to have identical lattice stabilities. This need not be true, as seen in the Zr/Hf case where the enthalpy difference $\Delta H^{\beta\to\epsilon}$ is more negative for Hf than for Zr. However, the assumption of equal parameters has been introduced for simplicity. It will be modified in later studies when the stability of these phases is considered in greater detail.

The equations for binary equilibria between two regular-solution phases ϵ and L at a fixed temperature, where the composition of element B at the $\epsilon/\epsilon + L$ boundary is x_ϵ and the composition at the $L + \epsilon/L$ boundary is

* These entropy and enthalpy differences refer to high temperatures far from the range influenced by third-law constraints.

x_L, can be expressed as[1]

$$\Delta F_A^{\epsilon \to L} + RT \ln \frac{1 - x_L}{1 - x_\epsilon} = x_\epsilon^2 E - x_L^2 L \qquad (1)$$

$$\Delta F_B^{\epsilon \to L} + RT \ln \frac{x_L}{x_\epsilon} = (1 - x_\epsilon)^2 E - (1 - x_L)^2 L \qquad (2)$$

In Eqs. (1) and (2) the free-energy differences $\Delta F_A^{\epsilon \to L}$ and $\Delta F_B^{\epsilon \to L}$ represent the differences between the hexagonal and liquid forms of elements A and B at the temperature in question. The regular-solution interaction parameters for the ϵ and L phases are E and L, respectively. If E and L are equal to zero, then the solutions are ideal, and the phase boundaries x_ϵ and x_L depend only on the free-energy terms for the pure metals. Analogues of Eqs. (1) and (2) can be written for β/L, α/L, ϵ/β, ϵ/α, and α/β equilibrium. These equations when coupled with the information contained in Figs. 1 and 2 permit direct computation of ideal-solution phase diagrams. The only additional information required is the melting point $\bar{T}$ and the entropy of fusion of the stable-form elements concerned. In the present case, the entropy of melting was taken to be 2.0 cal/(mole)(°K) in each of the following instances where experimental data are not available: Nb ($\bar{T}^\beta$ = 2740°K), Ta ($\bar{T}^\beta$ = 3270°K), Mo ($\bar{T}^\beta$ = 2900°K), W ($\bar{T}^\beta$ = 3650°K), Re ($\bar{T}^\epsilon$ = 3450°K), Ru ($\bar{T}^\epsilon$ = 2550°K), Os ($\bar{T}^\epsilon$ = 3300°K), Rh ($\bar{T}^\alpha$ = 2240°K), Ir ($\bar{T}^\alpha$ = 2750°K), Pt ($\bar{T}^\alpha$ = 2040°K), and Pd ($\bar{T}^\alpha$ = 1840°K). With the use of thermodynamic identities of the form

$$\Delta F_A^{\alpha \to L} = \Delta F_A^{\alpha \to \beta} + \Delta F_A^{\beta \to L} \qquad (3)$$

all the required free-energy differences can now be defined explicitly. As a consequence, the melting points of the unstable forms can be computed. These are shown for convenience in Table 1.

The computed phase diagrams are shown by the heavy lines in Figs. 3 to 17. These figures also show currently available experimental phase relations generated in a large measure as a result of recent research at Nuclear Metals Corporation of Concord, Massachusetts.[10,19] The calculations *do not* consider compound phases. These are shown as crosshatched regions on the phase diagram. As anticipated, the existence of compound phases will limit the range of stability of the β, ϵ, α, or L phases. In addition, where compound phases are dominant or numerous, no meaningful comparison of computed and observed behavior can be made. As a result, the binary-phase diagrams for either Zr or Hf with each of the elements Tc, Re, Ru, Os, Rh, Ir, Pd, and Pt are not presented. Similarly the binary diagrams for either Nb or Ta with each of the elements Rh, Ir, Pd, and Pt are not considered. Finally, in the binary systems between Zr and

TABLE 1 ***Summary of Computed Melting Points and Enthalpy and Entropy Differences***

Element	Melting points, °K		
	$\bar{T}^{\beta}$	$\bar{T}^{\epsilon}$	$\bar{T}^{\alpha}$
Nb	2740	1420	1170
Ta	3270	1800	1540
Mo	2900	1900	1530
W	3650	2650	2230
Re	2710	3450	2890
Ru	1420	2550	1780
Os	1960	3300	2310
Rh	930	2010	2240
Ir	1260	2490	2750
Pd	820	1470	1820
Pt	980	1670	2040

Element	Enthalpy and entropy differences*			
	cal/mole		cal/(mole)(°K)	
	$\Delta H^{\beta\to\epsilon}$	$\Delta H^{\alpha\to\epsilon}$	$\Delta S^{\beta\to\epsilon}$	$\Delta S^{\alpha\to\epsilon}$
Zr	−1030	−800	−0.90	0.00
Hf	−1830	−800	−0.90	0.00
Nb/Ta	+1500	−650	−0.80	+0.05
Mo/W	+2000	−500	0.00	+0.15
Tc/Re	−400	−250	+0.40	+0.30
Ru/Os	−1120	−120	+0.80	+0.80
Rh/Ir	−1500	+150	+0.90	−0.15
Pd/Pt	−1100	+250	+0.50	−0.30
Ag/Au	−700	+150	−0.50	−0.30
Zn/Cd	−700	−450	−0.60	−0.40

* See Figs. 1 and 2.

Hf with each of the elements Nb, Ta, Mo, and W, miscibility gaps dominate the binary systems. These cases will be treated at a later date when the ideal-solution approximation †(i.e., $A = B = E = L = 0$) is relaxed. Finally, the Tc and Re group is represented by Re because of the paucity

† The regular-solution parameters E and L are defined in Eqs. (1) and (2). The parameters A and B are corresponding terms for the α and β phases.

of information concerning Te. These diagrams can be easily computed for comparison with observed phase equilibria when the latter are available.

Figure 3 illustrates the intrusion of compound phases. However, the $\beta/\epsilon/L$ equilibria are in good accord with observation. This is equally true in Fig. 4 although the β/ϵ equilibrium in the Nb-Ru system is computed to be too near that of Nb. This is probably due to the fact that deviations in the β phase are more negative than in the ϵ phase (i.e., $B < E$). Figure 5 is similar to Figs. 3 and 4 showing the compound effect; nevertheless, the $\beta/L/\epsilon$ equilibria are in agreement with the computations. The W-Re and Mo-Re systems shown in Fig. 6 are similar to those cited above, as are the Mo-Ru and Mo-Os examples in Fig. 7. In the latter case, the β/ϵ equilibria are shifted too far toward Os (i.e., E is probably less than B). Good agreement is found in Fig. 8 for the W-Ru and W-Os examples. Figures 9 and 10 show the interesting situation in which the calculations actually predict the stability of the hcp ϵ phase. The larger range of stability of the ϵ phase observed is probably due to negative deviations (i.e., $E < 0$ and $E < A$, B, and L). In the Mo-Pt, Mo-Pd, W-Pt, and W-Pd examples, the hcp phase is almost, *but not quite*, stable according to the calculations. Interestingly ϵ appears in the Mo-Pt system (Fig. 11) and barely makes an appearance in the Mo-Pd system at high temperatures, but not in the W binary systems. The α/β equilibria in W-Pt are shifted farther toward the W-rich side because of negative deviations in the α phase ($A < 0$ and $A \leq B$, L). The latter condition would also account for the low computed peritectic. Similarly, the widths of the two-phase $\beta + L$ and $\alpha + \beta$ fields in W-Pd are probably due to deviations from ideality. Similar deviations are also seen in Fig. 13.

The Re-Pt example shown in Fig. 14 is in good agreement with observation, although in this case E is probably equal to A and both are less than L. A similar situation prevails in the Re-Pd system where positive values of the interaction terms are indicated by the width of the two-phase fields.

The computed Os-Ir phase diagram shown in Fig. 15 illustrates the poorest agreement between calculated and observed behavior. The difference could be due to the fact that A is less than E and L, or that the magnitude of the entropy difference between the α and ϵ phases previously computed for the Ru-Os case[6] is too large (see Fig. 2). The Ru-Pt, Ru-Pd, Os-Pt, and Os-Pd series shown in Figs. 16 and 17 indicate positive deviations in the Pd systems and only sketchy phase-diagram data in the Pt systems. However, the Os-Pt deviation is in the *opposite* direction from the Os-Ir deviation. Thus, if both systems were ideal, the Os-Pt deviation would demand larger values for $\Delta F_{\text{Os}}^{\epsilon\to\alpha}$ as opposed to the smaller values required by the Os-Ir deviation. As a result, the lattice-stability parameters for Ru/Os shown in Fig. 2 are probably correct. The discrepancies shown in the Os-Ir and Os-Pt examples can be rectified by having $A < E$ for Os-Ir and $E < A$ for Os-Pt.

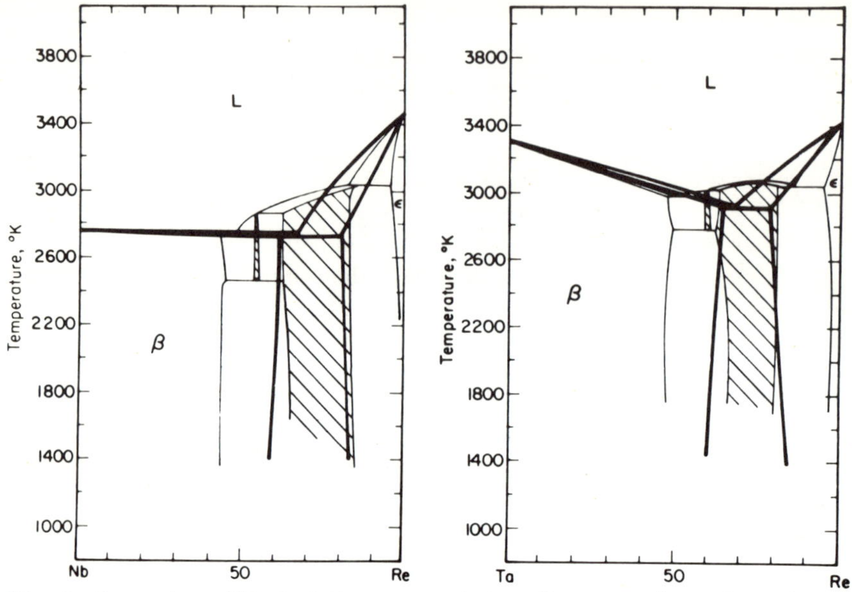

Fig. 3 Comparison of ideal-solution computed phase diagrams and experimental results.[7,8] (Volume differences: Nb-Re 20%; Ta-Re 21%.)

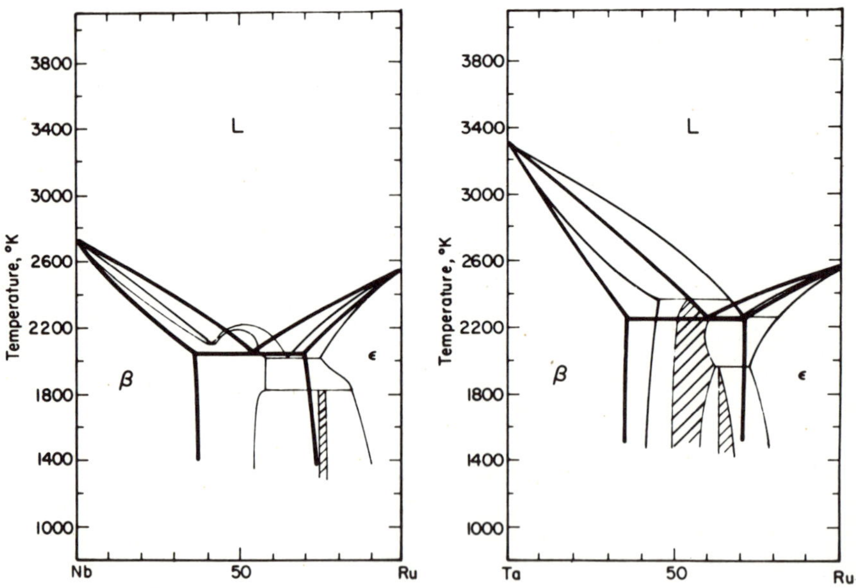

Fig. 4 Comparison of ideal-solution computed phase diagrams and experimental results.[9,10] (Volume differences: Nb-Ru 28%; Ta-Ru 28%.)

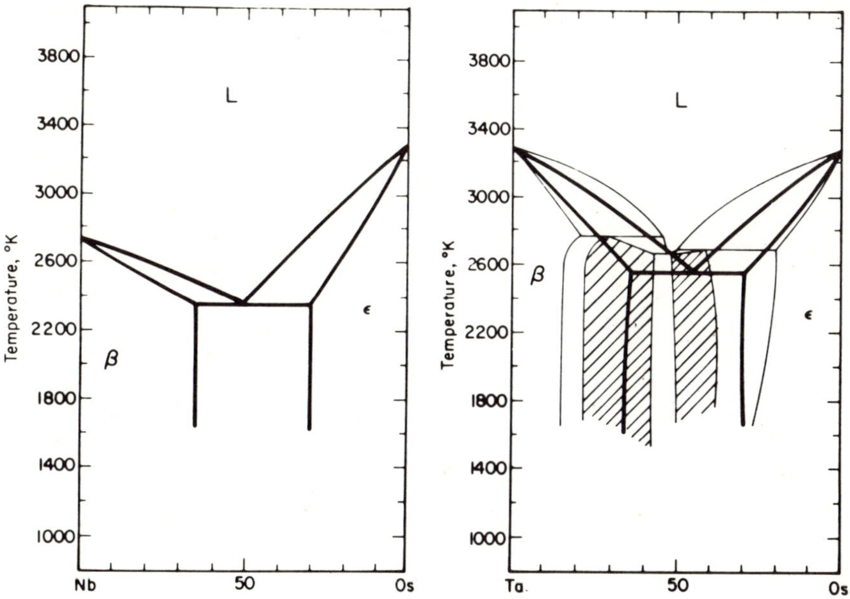

Fig. 5 Comparison of ideal-solution computed phase diagrams and experimental results.[10] (Volume differences: Nb-Os 25%; Ta-Os 25%.)

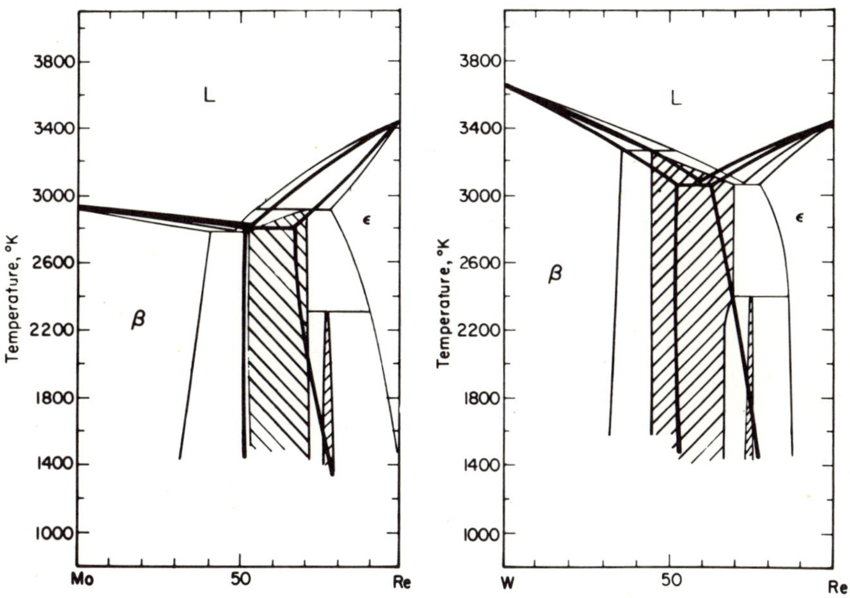

Fig. 6 Comparison of ideal-solution computed phase diagrams and experimental results.[11,12] (Volume differences: Mo-Re 6%; W-Re 8%.)

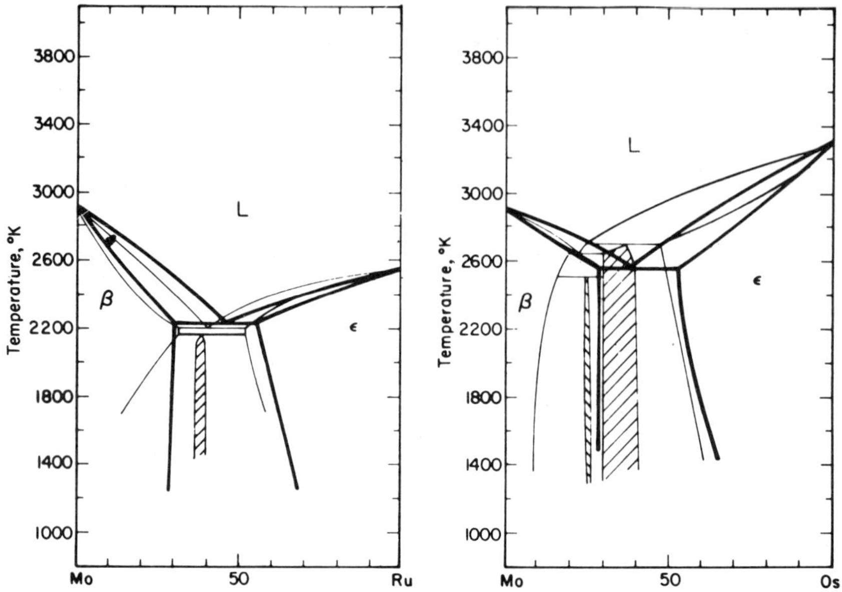

Fig. 7 Comparison of ideal-solution computed phase diagrams and experimental results.[13,14] (Volume differences: Mo-Ru 14%; Mo-Os 11%.)

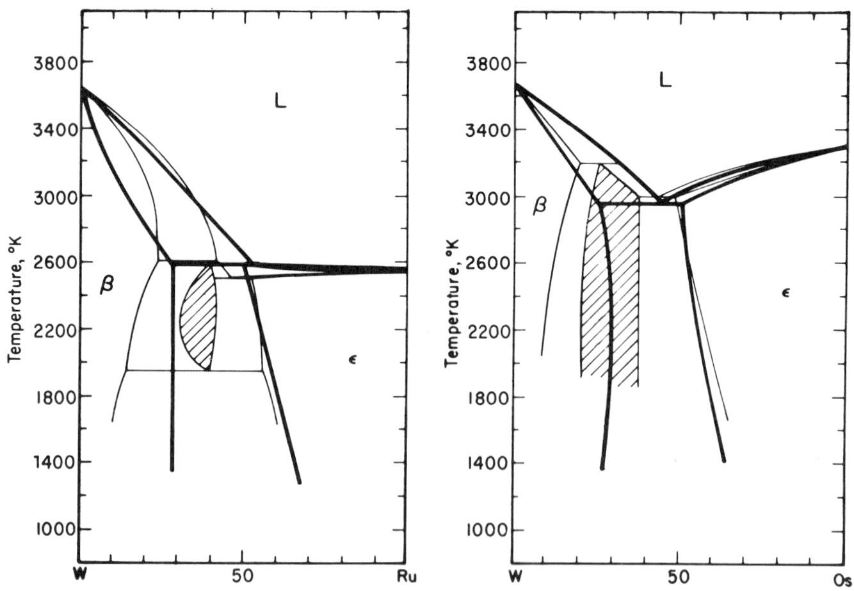

Fig. 8 Comparison of ideal-solution computed phase diagrams and experimental results.[15,16] (Volume differences: W-Ru 16%; W-Os 13%.)

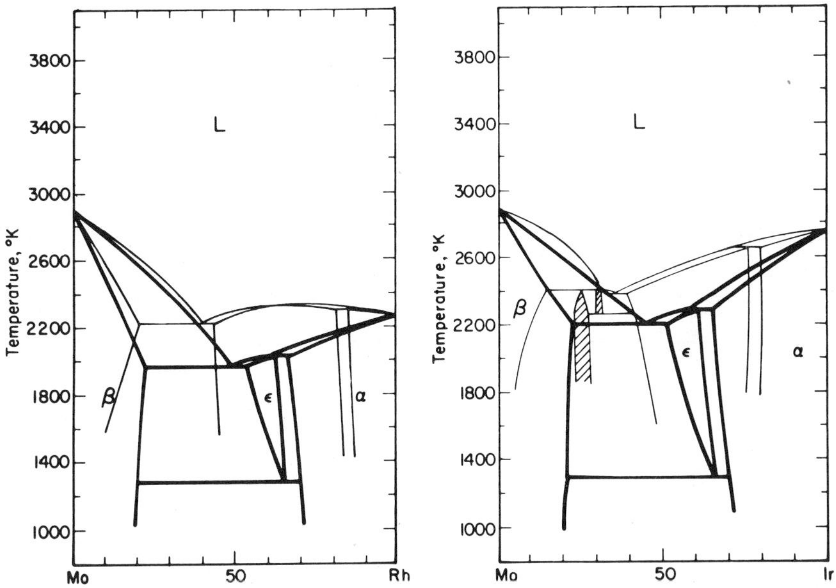

Fig. 9 Comparison of ideal-solution computed phase diagrams and experimental results.[17,18] (Volume differences: Mo-Rh 12%; Mo-Ir 9%.)

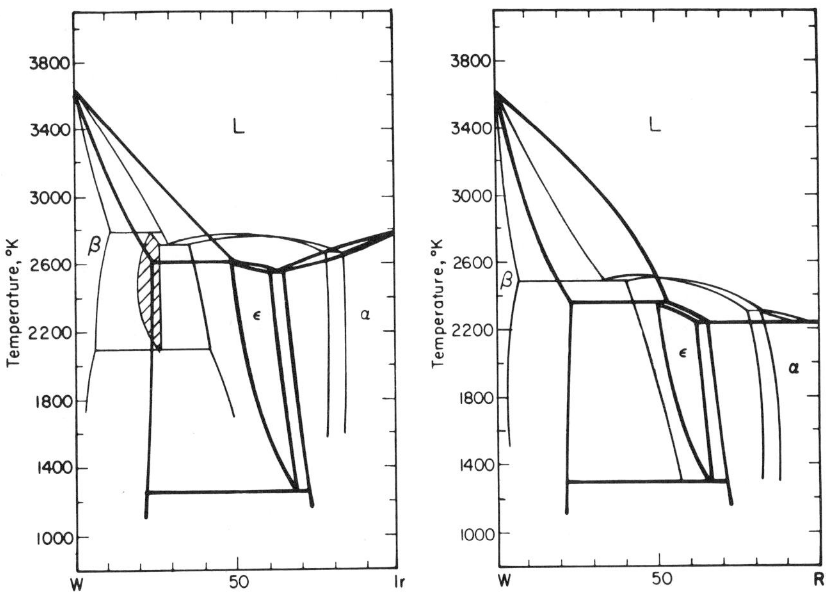

Fig. 10 Comparison of ideal-solution computed phase diagrams and experimental results.[19,19] (Volume differences: W-Ir 11%; W-Rh 14%.)

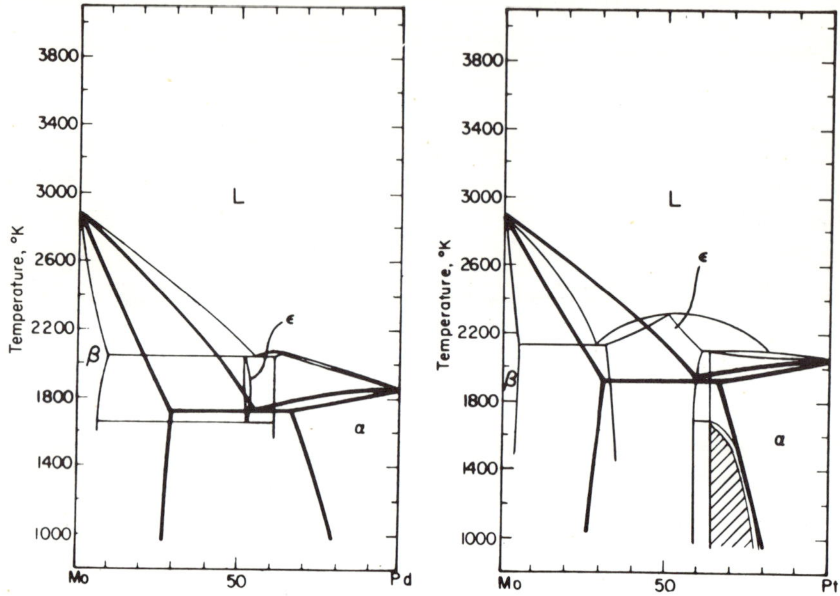

Fig. 11 Comparison of ideal-solution computed phase diagrams and experimental results.[20–22] (Volume differences: Mo-Pd 6%; Mo-Pt 3%.)

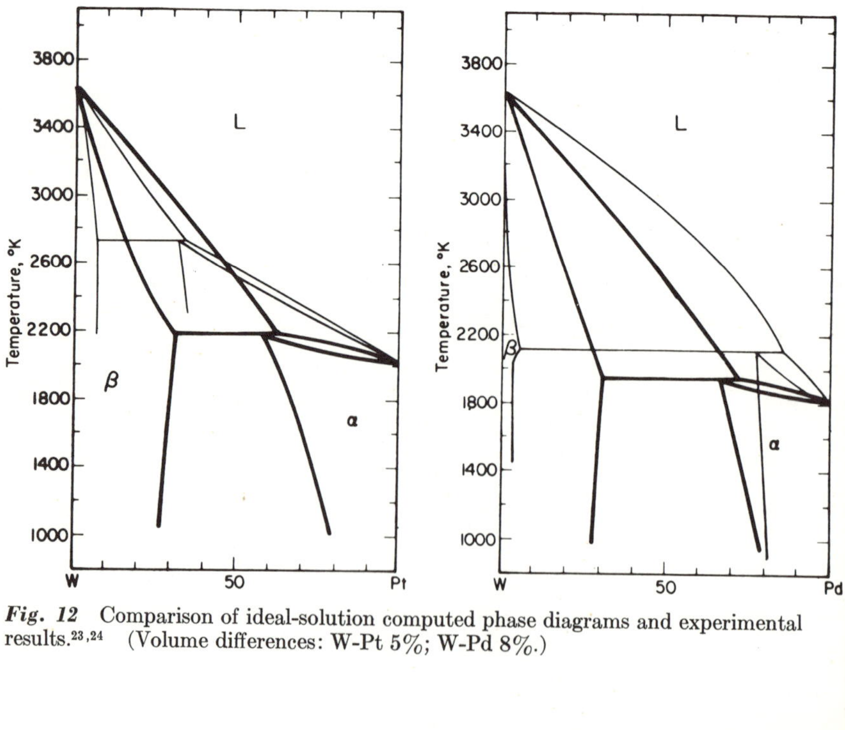

Fig. 12 Comparison of ideal-solution computed phase diagrams and experimental results.[23,24] (Volume differences: W-Pt 5%; W-Pd 8%.)

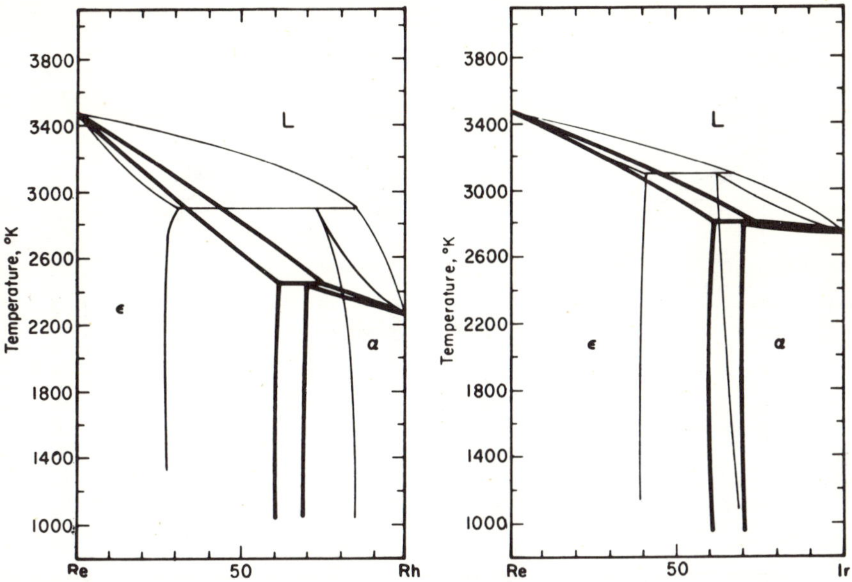

Fig. 13 Comparison of ideal-solution computed phase diagrams and experimental results.[25,25] (Volume differences: Re-Rh 6%; Re-Ir 3%.)

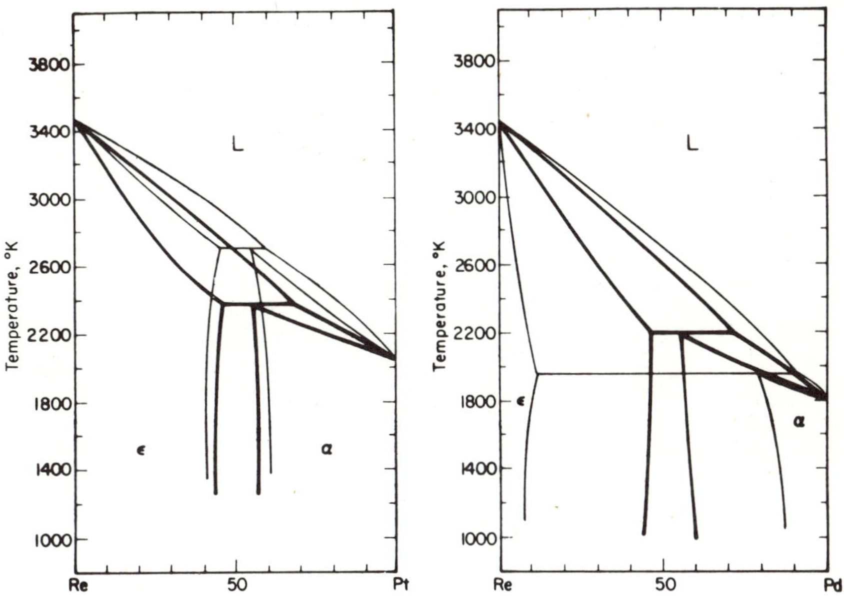

Fig. 14 Comparison of ideal-solution computed phase diagrams and experimental results.[26,25] (Volume differences: Re-Pt 3%; Re-Pd 0%.)

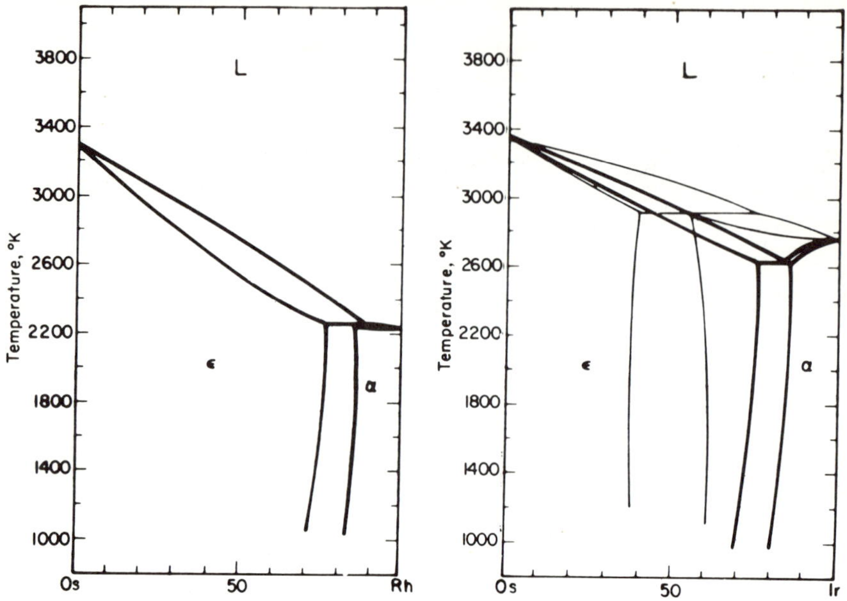

Fig. 15 Comparison of ideal-solution computed phase diagrams and experimental results.[27] (Volume differences: Os-Rh 1%; Os-Ir 1%.)

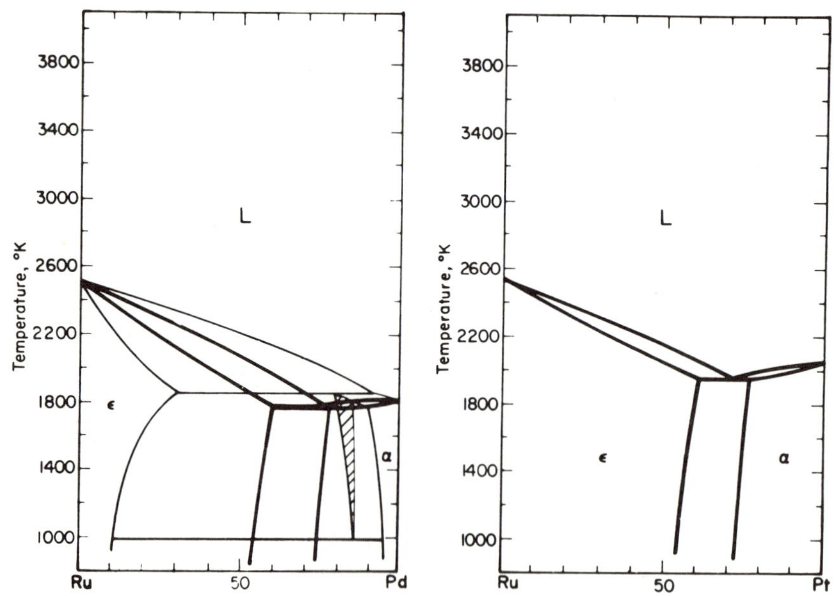

Fig. 16 Comparison of ideal-solution computed phase diagrams and experimental results.[28] (Volume differences: Ru-Pd 8%; Ru-Pt 10%.)

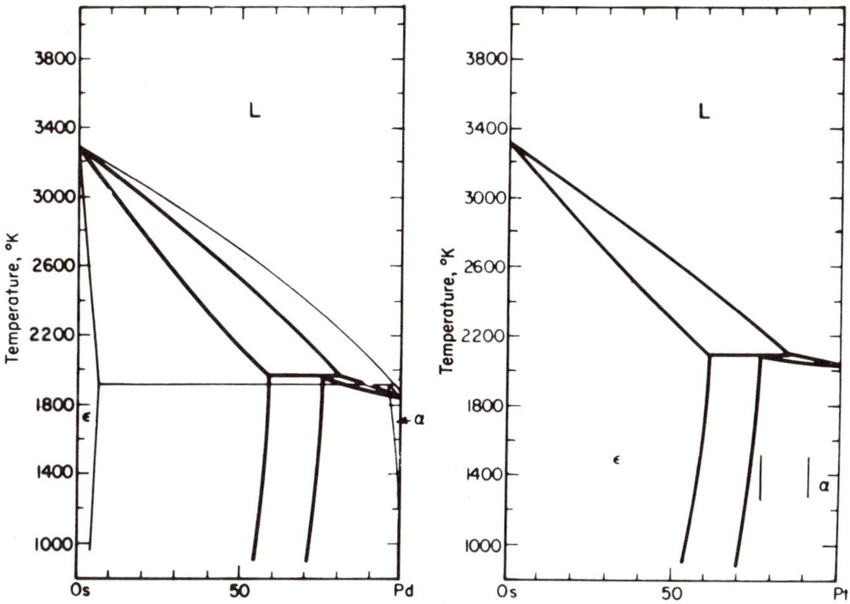

Fig. 17 Comparison of ideal-solution computed phase diagrams and experimental results.[29,26] (Volume differences: Os-Pd 5%; Os-Pt 8%.)

In order to extend the foregoing calculations to the first-row transition elements V, Cr, Ni, and Co, the free-energy differences among the β, ϵ, and α forms of these metals have been taken to be identical to their respective later-row counterparts in all cases save the α/ϵ case for Co. In the latter, only a small change from the α/ϵ Rh and Ir values was introduced to recognize the phase transformation exhibited by Co. In essence, these assumptions neglect the differences in magnetic free energy among the β, ϵ, and α forms of the foregoing first-row elements, which should be negligible for V, small for Cr, and significant for the β/α and β/ϵ stability in Co and Ni. Nevertheless, none of these situations are comparable to Fe[3,6] where the magnetic free-energy differences are dominant. The less said about Mn the better!

The lattice-stability parameters for the β, ϵ, and α configurations of V, Cr, Co, and Ni are given in Table 2, along with the computed melting points of the various crystal forms. The enthalpy and entropy differences are identical to those shown in Figs. 1 and 2, which were previously estimated for the second- and third-row counterparts, i.e., (ΔH and ΔS for V are the same as for Nb and Ta) except for α/ϵ Co. In the latter case, the Rh and Ir estimate of $\Delta H^{\alpha \to \epsilon}$ is $+150$ cal/mole while $\Delta S^{\alpha \to \epsilon}$ is approximately -0.15 cal/(mole)(°K). These parameters lead to an fcc structure stable at 0°K and becoming more stable with increasing temperature. In the case of Co, the fcc phase does become more stable as the temperature

TABLE 2 ***Enthalpy and Entropy Differences among the BCC (β), HCP (ϵ), and FCC (α) Forms of* V, Cr, Co, *and* Ni *and the Computed Melting Points***

Element	Enthalpy and entropy differences				Melting points, °K		
	cal/mole		cal/(mole)(°K)				
	$\Delta H^{\beta\to\epsilon}$	$\Delta H^{\alpha\to\epsilon}$	$\Delta S^{\beta\to\epsilon}$	$\Delta S^{\alpha\to\epsilon}$	$\bar{T}^{\beta}$	$\bar{T}^{\epsilon}$	$\bar{T}^{\alpha}$
V	+1500	−650	−0.80	+0.05	2180	1030	780
Cr	+2000	−500	0.00	+0.15	2175	1175	860
Co	−1500	−110	+0.90	−0.15	805	1705	1768
Ni	−1100	+250	+0.50	−0.30	885	1445	1725

increases, but the hcp phase is stable below 720°K with a $\Delta H_{\text{Co}}^{\alpha\to\epsilon} \approx -110$ cal/mole[30] or $\Delta S_{\text{Co}}^{\alpha\to\epsilon} = -0.15$ cal/(mole)(°K). This small difference can easily be explained in terms of small differences in magnetic free energy of α-Co and ϵ-Co.

The entropy of melting of fcc Co and Ni was taken as 2.32 and 2.44 cal/(mole)(°K), respectively,[30] while the entropy of melting of bcc V and Cr was estimated as 2.0 cal/(mole)(°K) as before. By employing the thermodynamic identities given by Eq. (3), all the required free-energy differences can now be defined explicitly. As a consequence, the melting points of the unstable forms can be computed. These are shown for convenience in Table 2. Utilization of the enthalpy and entropy differences given in Table 2 and Eqs. (1) and (2) permits computation of 18 additional phase diagrams. These are shown in Figs. 18 to 26 along with the observed phase diagrams.[26,31] As before, cases where miscibility gaps or diagrams completely dominated by compound phases have been eliminated from consideration.

Figure 18 shows the V-Re and V-Co systems. In the former, satisfactory agreement similar to that for Nb-Re and Ta-Re considered earlier is noted. In the latter, compounds dominate. However, the computed eutectic temperature is in good agreement with the published diagram. In the V-Ni and V-Pd examples shown in Fig. 19, compound formation is evident, indicating negative deviations from ideality in the α phase. Thus, the interaction parameter for the α phase, A, must be negative. If $A < L$, then the resulting L/α lines will be displaced to higher temperatures, and the β/α lines will be shifted toward the V side of the diagram. Restriction of α, β, and liquid phases by compound intrusion is as expected. Nevertheless, eutectic temperature agreement is satisfactory. The Cr-Re diagram shown in Fig. 20 is quite similar to the Mo-Re and W-Re diagrams in Fig. 6. In the present Cr-Re case, peritectic melting is correctly predicted at the proper level. Compound intrusion reduces solubility limits as in Fig. 6. In Cr-Ru (Fig. 20), the level of the eutectic is predicted properly. The unusual liquidus minimum near 10 a/o Ru shown in the observed diagram is worthy of reinvestigation.

The Cr-Pd and Cr-Pt diagrams shown in Fig. 21 are similar to the W-Pd and W-Pt cases discussed earlier. In the present case, the ordering tendencies in the α phase lead to stabilization of this phase at the expense of the β and liquid phases: i.e., $A < B$ and $A < L$. A similar effect is seen in the Cr-Ni example in Fig. 22 where ordering has been reported in the α phase. The computed Cr-Co diagram while suffering from compound intrusion and magnetic phenomena still shows the proper eutectic behavior.

Figure 23 shows the appropriate sequence from eutectic to peritectic behavior in the Mo-Ni and W-Ni examples. In addition, the invariant temperature levels are properly predicated. Peritectic behavior is correctly computed in the Ru-Ni and Re-Ni systems shown in Fig. 24. However, the widths of the two-phase fields are underestimated. The Co-Ru, Co-Os, Co.Re, and Co-Ni phase diagrams shown in Figs. 25 and 26 all exhibit the proper form, but the stability range of the α phase is underestimated in the first two and overestimated in the last two. These deviations can readily arise from the fact that the magnetic free-energy differences between the fcc and hcp forms of Ni and Co have been neglected.

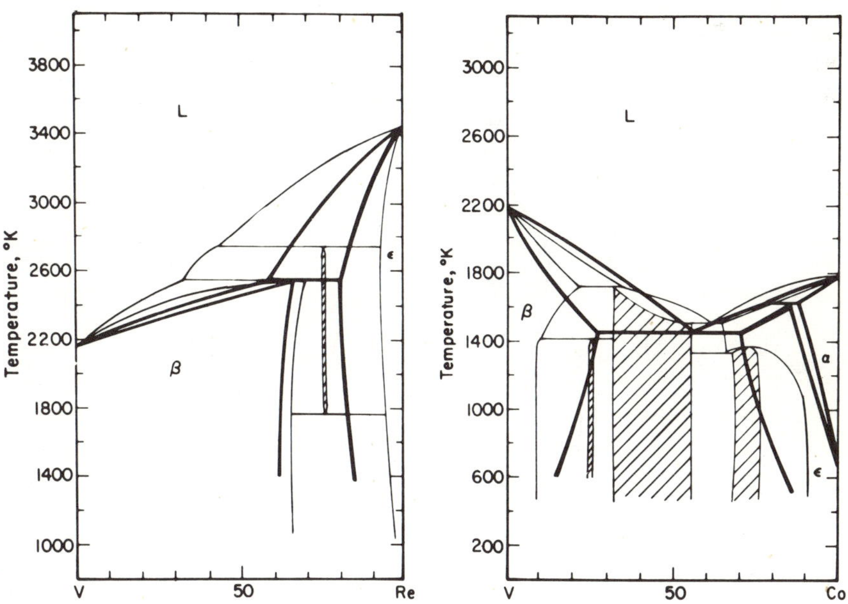

Fig. 18 Comparison of ideal-solution computed phase diagrams and experimental results.[31,26] (Volume differences: V-Re 6%; V-Co 22%.)

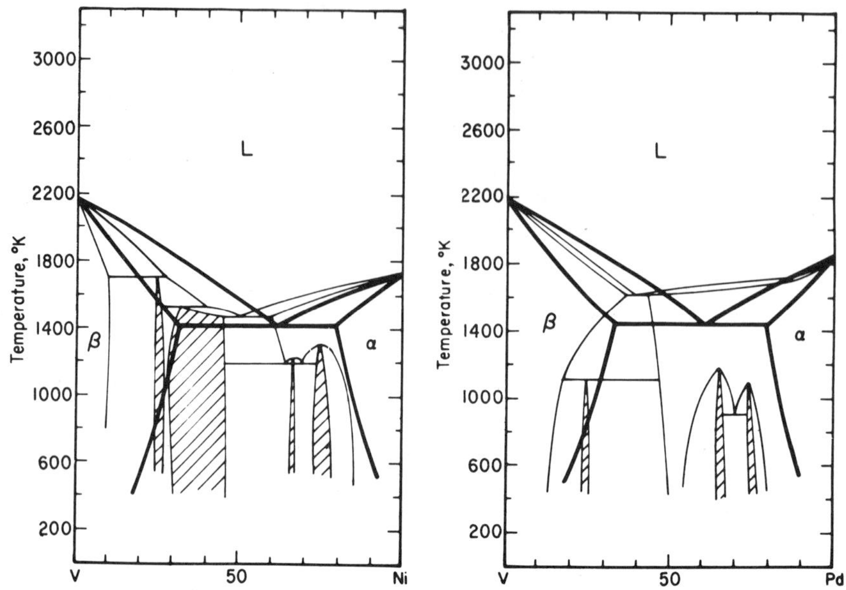

Fig. 19 Comparison of ideal-solution computed phase diagrams and experimental results.[26,31] (Volume differences: V-Ni 24%; V-Pd 6%.)

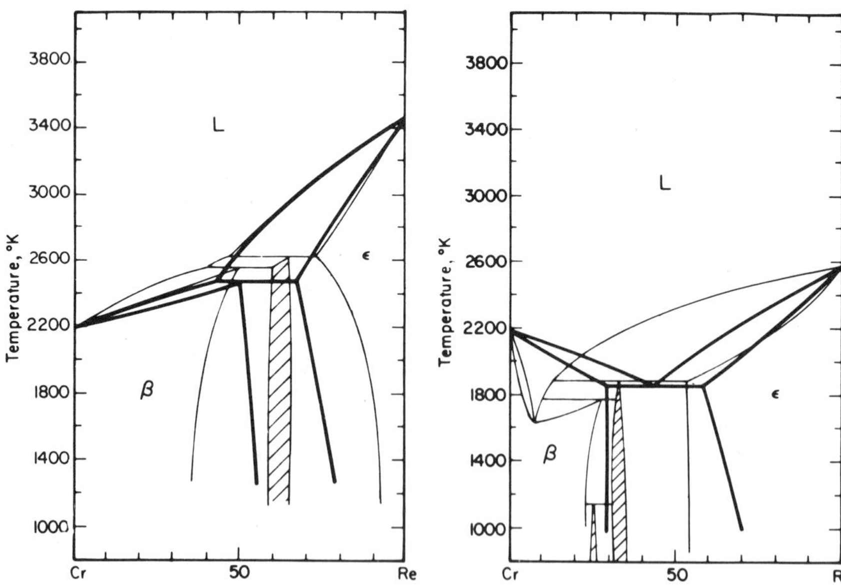

Fig. 20 Comparison of ideal-solution computed phase diagrams and experimental results.[31,31] (Volume differences: Cr-Re 20%; Cr-Ru 13%.)

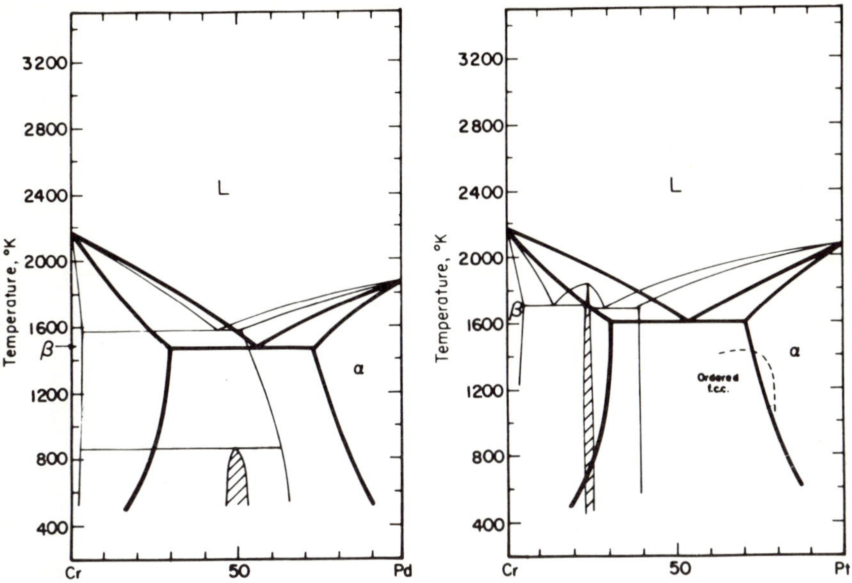

Fig. 21 Comparison of ideal-solution computed phase diagrams and experimental results.[26,26] (Volume differences: Cr-Pd 20%; Cr-Pt 23%.)

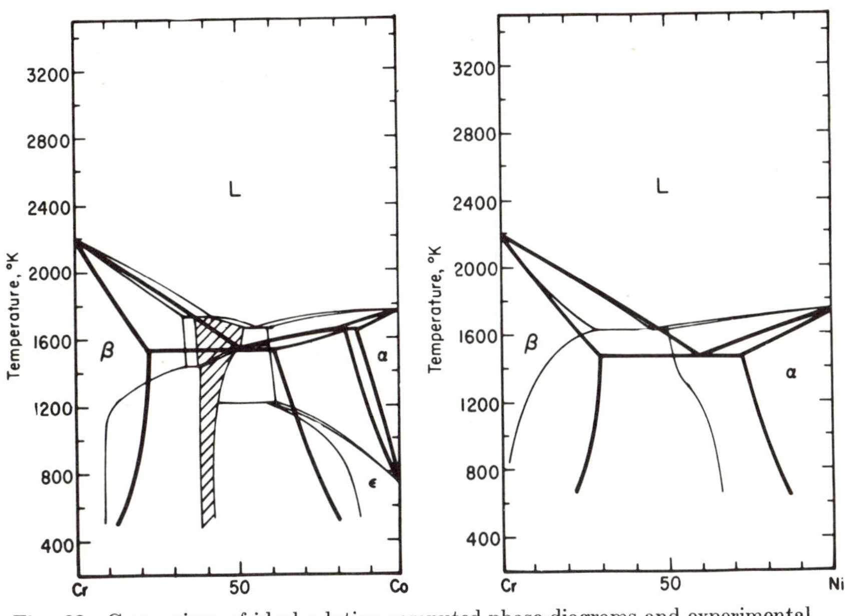

Fig. 22 Comparison of ideal-solution computed phase diagrams and experimental results.[26,26] (Volume differences: Cr-Co 8%; Cr-Ni 10%.)

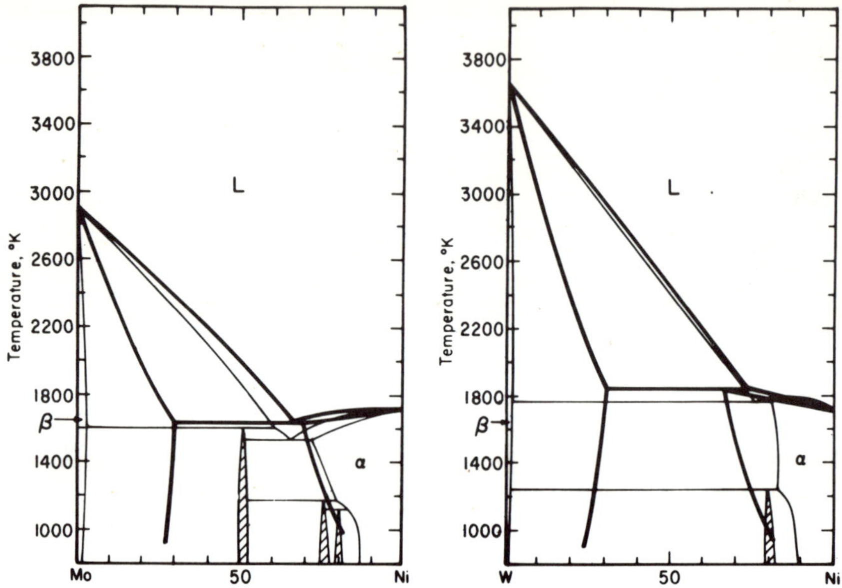

Fig. 23 Comparison of ideal-solution computed phase diagrams and experimental results.[26,26] (Volume differences: Mo-Ni 35%; W-Ni 37%.)

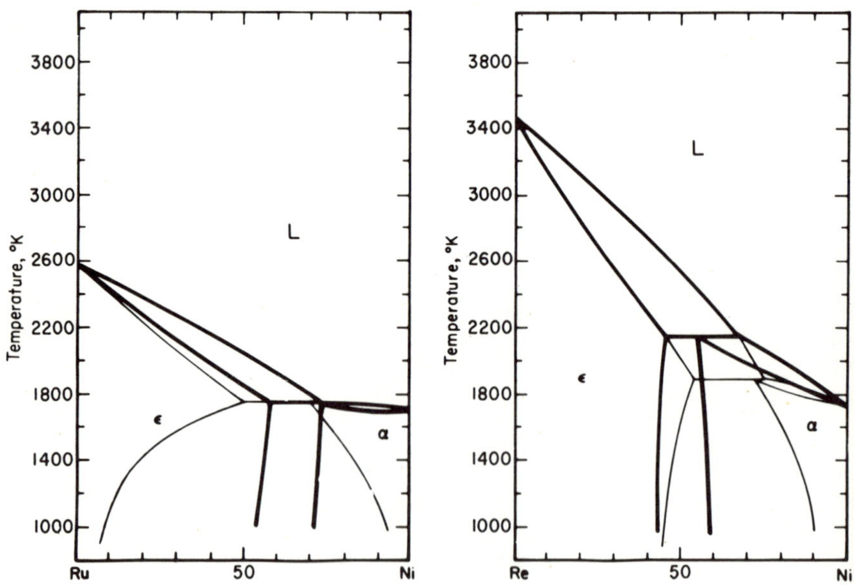

Fig. 24 Comparison of ideal-solution computed phase diagrams and experimental results.[31,31] (Volume differences: Ru-Ni 22%; Re-Ni 30%.)

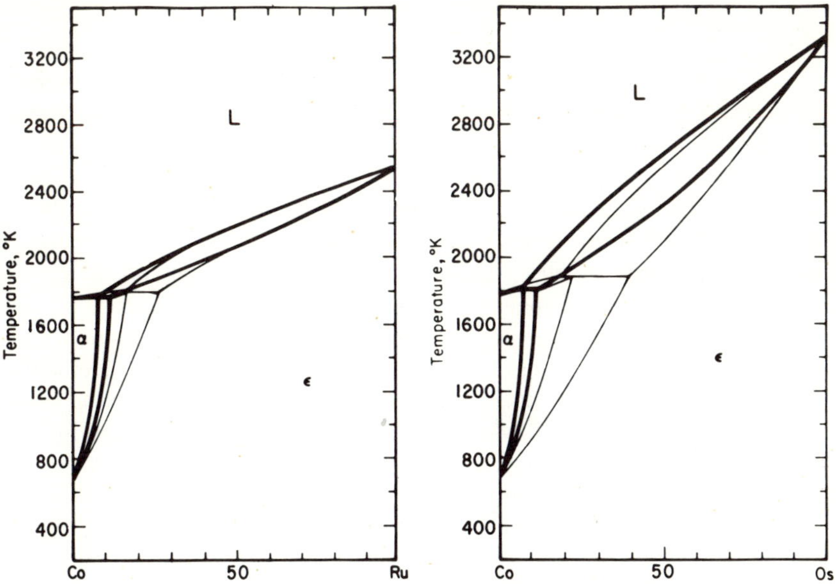

Fig. 25 Comparison of ideal-solution computed phase diagrams and experimental results.[31,26] (Volume differences: Co-Ru 20%; Co-Os 23%.)

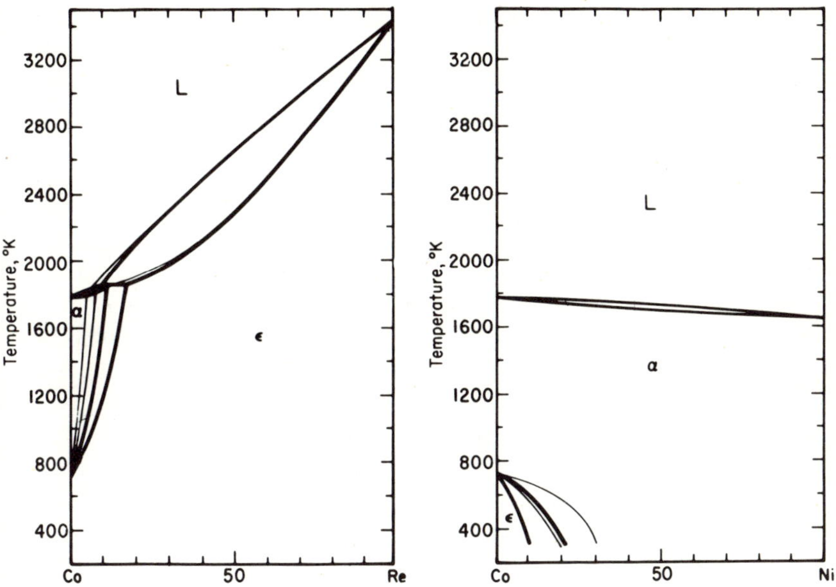

Fig. 26 Comparison of ideal-solution computed phase diagrams and experimental results.[26,26] (Volume differences: Co-Re 28%; Co-Ni 2%.)

SUMMARY

Sets of parameters representing the enthalpy and entropy differences between the bcc and hcp, and the fcc and hcp configurations of the transition metals have been estimated. These parameters, described as continuous functions of electron number, define the corresponding fcc/bcc differences. The free-energy differences have been used to compute a total of 48 ideal-solution binary phase diagrams. If each of the binary pairs is characterized by the third-row element, then the 96 end members consist of 10 Ta group elements (i.e., V, Nb, Ta), 22 W group elements (Cr, Mo, W), 12 cases where Re is considered, 18 cases where Os or Ru is involved, 14 Ir group elements (Co, Rh, Ir), and 20 Pt group elements (Ni, Pd, Pt). In each instance, *a single set of stability parameters for each group is used throughout.* The only exception is the small change due to Co α/ϵ discussed earlier. In spite of the oversimplified assumption of ideal solution behavior (which serves merely as a convenient baseline for all systems) and the fact that the computed phase diagrams are far from perfect, the overall comparison between predicted and observed behavior is quite satisfactory. Provisions are currently being made to reduce the phase-boundary calculations to computer format so that regular-solution phase diagrams can be readily generated from Eqs. (1) and (2). In order to reexamine the cases treated here in detail and to incorporate those systems dominated by miscibility gaps, such calculations would allow for size effects, solubility parameters, electronic interactions, etc., which contribute to the free energy of mixing, and control the interaction parameters A, B, E, and L. It should be pointed out that such factors must be *summed algebraically* in order to evaluate the sign and the magnitude of the regular-solution parameters.

It is a sad commentary, indeed, on the current status of solution theory, that the *sign* of the heat of mixing (or of the interaction parameters) for a solid or liquid phase cannot be explicitly predicted with confidence. Various factors including those listed above have been identified and discussed at length in the literature. However, the truth of the matter is that quantitative assessment of the component contributions *cannot* be performed with sufficient accuracy to permit reliable estimates of the algebraic sum. Moreover, individual factors such as the "size factor" cannot be used as a guide for predicting stability. In order to illustrate this point, the "percentage difference in volume" between the elemental partners in each of the binary systems considered herein has been computed, using the recent compilation of Rudman.[32] This difference (which is roughly three times the "size difference") is given in the legend of each binary system. Reference to these size differences shows little correlation with phase equilibria.

Emphasis on the various component contributions to the heat of mixing as the principal factors underlying phase stability suffers from the difficulty that relevant information is not commutative. Thus, such data on the W-Re system have little relevance to consideration of the W-Os or W-Ir systems. By contrast, however, information on the lattice stability of bcc, hcp, and fcc W is relevant to *all* alloy systems containing W.

REFERENCES

1. Kaufman, L.: *Acta Met.* **7**:575 (1959).
2. Kaufman, L.: *Bull. Am. Phys. Soc.* **4**:181 (1959); also Office of Technical Services Report PB 144220.
3. Kaufman, L., E. V. Clougherty, and R. J. Weiss: *Acta Met.* **11**:323 (1963).
4. Blackburn, L. D., L. Kaufman, and M. Cohen: *Acta Met.* **13**:533 (1965).
5. Kaufman, L.: "Condensed State Reactions at High Pressure," in W. M. Mueller (ed.), Energetics in Metallurgical Phenomena Series, Gordon and Breach, Science Publishers, Inc., New York, to be published.
6. Kaufman, L.: Proceedings of BISRA-ISI Conference on Physical Properties of Martensite and Bainite, Iron and Steel Institute Special Report 93, p. 48, London, 1965.
7. Giessen, B. C., R. Nordheim, and N. J. Grant: *Trans. AIME* **221**:1006 (1961).
8. Brophy, J. H., P. S. Schwartzkopf, and J. Wulff: *Trans. AIME* **218**:910 (1960).
9. Hurley, G. F., and J. H. Brophy: *J. Less-Common Metals* **7**:267 (1964).
10. Rapperport, E. J., and M. F. Smith: Refractory Metal Constitution Diagrams, Nuclear Metals Corp. WADD Technical Report 60–132, part 1, Concord, Mass., June, 1960.
11. Dickenson, J. M., and L. S. Richardson: *Trans. ASM* **51**:1055 (1959).
12. Dickenson, J. M., and L. S. Richardson, *Trans. ASM*: **51**:758 (1959).
13. Anderson, E. A., and W. Hume-Rothery: *J. Less-Common Metals* **2**:443 (1960).
14. Taylor, A., N. J. Doyle, and B. J. Kagle: *J. Less-Common Metals* **4**:436 (1962).
15. Rapperport, E. J., and M. F. Smith, *Trans. AIME* **230**:6 (1964).
16. Taylor, A., B. J. Kagle, and N. J. Doyle: *J. Less-Common Metals* **3**:330 (1961).
17. Anderson, E. A., and W. Hume-Rothery, *J. Less-Common Metals* **2**:19 (1960).
18. Michalik, S. J., and J. H. Brophy: *Trans. AIME* **227**:1047 (1963).
19. Rapperport, E. J., and M. F. Smith, Refractory Metal Constitution Diagrams, Nuclear Metals Corp. WADD Technical Report 60–132, part 2, Concord, Mass., March, 1962.
20. Anderson, E. A., *J. Less-Common Metals* **6**:81 (1964).
21. Raub, E.: *Z. Metallk.* **43**:23 (1954).
22. Knapton, A. G.: *Planseeber. Pulvermet.* **8**:2 (1959).
23. Jaffee, R. I., and H. P. Nielson: *Trans. AIME* **180**:603 (1949).
24. Goetz, W. K., and J. H. Brophy, *J. Less-Common Metals* **6**:345 (1964).
25. Tylkina, M. A., I. A. Tsyganova, and E. M. Savitskii: *Russ. J. Inorg. Chem.* **7**(8):990 (August, 1962).
26. Hansen, M., and K. Anderko: "Constitution of Binary Alloys," 2d ed., McGraw-Hill Book Company, New York, 1958.
27. Reiswig, R. D., and J. M. Dickenson: *Trans. AIME* **230**:469 (1964).
28. Rudnitsky, A. A., and R. S. Polyakova: *Russ. J. Inorg. Chem.* **4**(6): 634 (June, 1959).

29. Tylkina, M. A., V. P. Polykova, and O. Kh. Khamidov: *Russ. J. Inorg. Chem.* **8**(3):395 (March, 1963).
30. Hultgren, R., R. L. Orr, P. D. Anderson, and K. K. Kelley: "Thermodynamic Properties of Metals and Alloys," John Wiley & Sons, Inc., New York, 1963.
31. Eliot, R. P.: "Constitution of Binary Alloys," First Supplement, McGraw-Hill Book Company, New York, 1965.
32. Rudman, P. S.: *Trans. AIME* **233**:871 (1965).

DISCUSSION *on Paper by L. Kaufman*

HUME-ROTHERY: The calculations are of great interest, and come out surprisingly well in some cases.

In Fig. 23, the calculated solid solubility of Ni in Mo and W is very much greater than the relatively small values observed, although there are considerable solid solutions in Ni.

Do you not think this may be an electron orbital kind of restriction? The Mo and W structures require a very high proportion of *d* function. In Ni some of the orbitals have become atomic, and so the Ni atoms cannot give the high proportion of *d* function required by the W and Mo structures.

It is a general principle that in group VI to group VIIIC alloys, the solid solubility in the group VI metal is the smaller.

KAUFMAN: I cannot discuss the properties of the Mo-Ni and W-Ni systems in terms of electronic orbitals as suggested by Professor Hume-Rothery. As indicated in my paper, my computation is based on ideal solutions, and as such can be quite deficient for cases where substantial deviations from ideality exist. Within my framework, the differences between the computed and observed diagrams would be due to deviations in the β phase which are more positive than in the α phase. These deviations may be due to the electronic effects suggested by Professor Hume-Rothery or to other effects.

BREWER: Cannot your curves be equally represented by much larger energy differences (e.g., between bcc and hcp Mo), together with large negative deviations for intermediate compositions, when transition metals of the left side are mixed with metals of the right side?

KAUFMAN: The question raised by Professor Brewer is most interesting. At present, the values for the enthalpy difference between the bcc and hcp phases of the transition metals shown in Fig. 1 of my paper are approximately one-fifth of the values in Table 6 of his paper. It is noteworthy that although this difference in magnitude exists, the shape of Brewer's $\Delta H^{\beta\to\epsilon}$ vs. group number curve and Fig. 1 show very good agreement.

In response to this question however, it appears that very large difference between the enthalpies of the β and ϵ forms of the pure metals would lead to melting points for the unstable forms which would be near absolute zero (or at negative temperatures!). Thus, since the enthalpy of melting for Mo is about 5800 cal/mole, assigning an enthalpy difference of 6000 to 10,000 cal/mole for $\Delta H_{\mathrm{Mo}}^{\beta\to\epsilon}$ would lead to negative melting points. Apart from this difficulty, one could introduce departures from ideality (e.g., finite values of E, A, B, and L) to balance the large $\Delta H_{Mo}^{\beta \to \epsilon}$ term. However, under such circumstances, I believe that one would be very hard put to compute solidus-liquidus curves as shown in Figs. 3, 6, 8, and 12 of this paper and Fig. 9 of Agenda Discussion II which conform to observed behavior and also lead to a melting point for hcp Mo (hcp W or bcc Re) near 0°K.

FRIEDEL: I have no objection in principle to the results of your Fig. 1. However, the energy differences are small. You should therefore be careful in using them while neglecting effects such as the size factors which, as Professor Hume-Rothery pointed out yesterday, play a predominant role in the position of the liquidus and solidus in many binary diagrams.

KAUFMAN: The paper by R. A Buckley and W. Hume-Rothery (*J. Iron Steel Inst. London* **201**:227 (1963), which forms the basis of the conclusion that size effects play a *predominant* role in controlling the liquidus and solidus, completely ignores the lattice stability contribution! For purposes of evaluating the effects of size differences, Buckley and Hume-Rothery *assumed* that the melting points of the fcc, hcp, and bcc forms of the transition metals were equal (i.e., *the lattice stability differences were zero*). Under such conditions it is hardly surprising that they found size effects predominate.

In the opinion of the present author, the subject of size effects, etc., has had ample coverage in the literature. Indeed, the past 20 years have seen this subject discussed in almost agonizing detail! By contrast, the lattice-stability terms have been largely ignored or neglected. This situation surprisingly persists in spite of the fact that the lattice-stability term is a property of the metal and could be usefully employed for any binary or higher-order system in which the metal is alloyed, whereas the strain-energy terms would change from system AB to AC to AD, etc. Fortunately, however, the recent development of high-pressure technology has rescued the lattice-stability concept from oblivion. Thus, discussion of hcp Fe, fcc Tl, or bct Ge is no longer unusual. Indeed, the high-pressure studies have provided real information on the lattice stability of these structures. It is almost certain that continued research in the high-pressure and splat-cooling areas will provide even more quantitative data in the future.

As indicated in the present paper, the ideal-solution calculations have been presented with full realization that there are other effects such as strain-energy terms, vibrational terms, magnetic terms, electronic terms, etc., which contribute in the real case to the free energy of mixing. In the simplest case of a regular-solution description of these effects, their contribution might be included in the interaction parameters E, B, A, and L for the ϵ, β, α, and liquid phases. Nevertheless, the ideal-solution calculations have been presented to emphasize the lattice-stability contributions so that they might have a long overdue "day in court." Professor Friedel suggests that the lattice-stability differences are small. I agree, but hasten to point out that Professor Brewer thinks that they are large! The present state of affairs is that Figs. 1 and 2 present a set of enthalpy and entropy values which are in agreement with the known stability characteristics of 24 metals and which have been *explicitly* employed to compute 48 phase diagrams bearing a strong relation to reality. No effort has been made to exclude those diagrams that give poor results or to emphasize the case where good agreement is obtained. Although strain-energy terms and other factors could have been invoked for each specific system in order to *force-fit* the agreement, the author preferred to avoid such an exercise in order to evaluate the lattice-stability aspect. However, the present development does not preclude any future attempts to perform such an improvement. Indeed, future refinements are welcomed!

ELECTRONIC STRUCTURE OF ALLOYS OF 3d TRANSITION METALS WITH ALUMINUM*

T. M. Srinivasan, Helmut Claus, R. Viswanathan, and Paul A. Beck

Department of Metallurgy and Materials Research Laboratory, University of Illinois, Urbana, Illinois

D. I. Bardos

Metallurgy Division, Argonne National Laboratory, Argonne, Illinois

STRUCTURE OF THE 3d BAND OF BCC ALLOYS

Magnetic moment, as revealed in ferromagnets by saturation magnetization measurements and in antiferromagnets by neutron diffraction, is one of the most valuable sources of information on the structure of the 3*d* band. The well-known Slater-Pauling curve (Fig. 1) shows that for up to five electrons outside closed shells the population of the spin-up portions of the 3*d* band is equal to that of the spin-down portions. Spontaneous moment

* The work performed at the University of Illinois was supported by grants from the AEC, ARPA, and NSF. A portion of the work was performed at the Argonne National Laboratory under the auspices of the U.S. Atomic Energy Commission.

first appears in Cr with 6 outside electrons per atom, and it increases by approximately 1 μ_B for each additional outside electron up to about $e/a = 8.3$. Beyond that electron concentration it decreases at the same rate: 1 μ_B for each electron added. An interesting result indicated by the magnetic-moment data in Fig. 1 is the approximate validity of the rigid-band model over a rather wide range of alloys. For a given crystal structure, the magnetic moment is, roughly speaking, only a function of the electron concentration. This applies for the average moment of binary transition-metal alloys, excepting the fcc alloys of Co or Ni with transition metals having six or fewer electrons outside of closed shells.

Electronic specific-heat data are also very useful in studying the structure of the 3*d* band.[3] Recently it was recognized that the enhancement of the measured electronic specific heat due to the electron-phonon interaction may be quite appreciable.[4] In a few instances the enhancement factor has been estimated by using superconductivity data,[5] or phonon spectrum information obtained by inelastic neutron scattering.[6] In most cases, however, the enhancement factor is unknown, so that the experimentally determined electronic specific heat cannot be corrected. As a result, the band shape and the bandwidth derived from such experimental electronic specific-heat data are subject to some uncertainty, the extent of which is

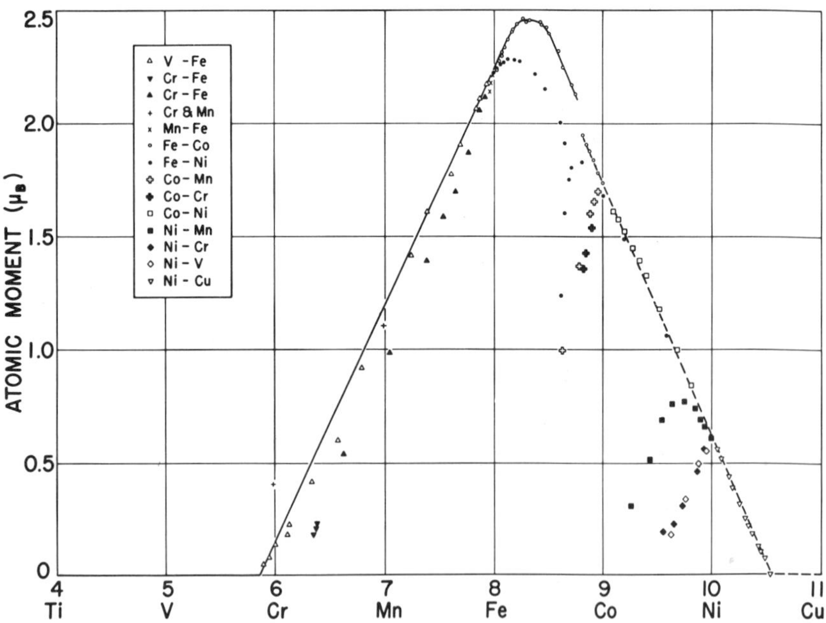

Fig. 1 Atomic moment in Bohr magnetons vs. number of outside electrons per atom for 3*d* metals and their bcc (continuous line) and fcc (dashed line) alloys—Slater-Pauling curve.

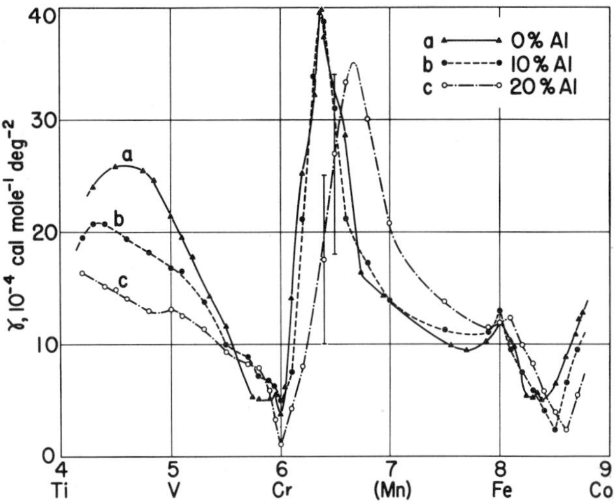

Fig. 2 Electronic specific-heat coefficients, in 10^{-4} cal/(mole) (deg^2), vs. electron concentration for (*a*) bcc alloys of 3*d* transition metals[7]; (*b*) similar ternary alloys with 10 a/o Al[11]; and (*c*) ternary alloys with 20 a/o Al.

difficult to evaluate. Nevertheless it is clear that the prominent features of the experimental electronic specific-heat vs. electron-concentration curves for bcc alloys of 3*d* transition elements (line *a*, Fig. 2) correlate well with those of the saturation magnetization (Fig. 1). Since the magnetic-moment values are presumably not subject to correction due to many-body effects, it appears that at least the location of the prominent features of the electronic specific-heat curves, for instance of the minima of line *a* in Fig. 2, is not altered by the enhancement due to electron-phonon interaction. Thus, the minimum of the electronic specific heat at $e/a = 6$ coincides with the first appearance of a spontaneous moment, and the second minimum at $e/a = 8.3$ coincides with the maximum value of the spontaneous moment. It appears, therefore, safe to conclude that the region of high electronic specific-heat values at $e/a < 6$ corresponds to a 3*d* sub-band with spin-up and spin-down electrons paired; the region of high electronic specific heat at e/a from 6 to 8.3 represents a second 3*d* sub-band with electrons in essentially one spin orientation; and the region of high electronic specific heat at $e/a > 8.3$ may be associated with a third 3*d* sub-band characterized by a spin orientation opposite to that in the second sub-band.[3] The electronic specific-heat data indicate, as do the magnetic-moment data, that the rigid-band model is, at least in a rough first approximation, rather widely applicable.[7] One may, therefore, conclude that, even without detailed knowledge of the enhancement factors arising from many-body effects, by combining the magnetic-moment data

(Fig. 1) with the electronic specific-heat results (line *a*, Fig. 2) one can form at least a semiquantitative picture of the structure of the 3*d* band, with a considerable degree of confidence.

EFFECT OF Al ON THE 3d BAND BCC ALLOYS

Figure 3 shows Arrott and Sato's magnetic-saturation data[8] for Fe-Al alloys. It may be noted that up to approximately 10 a/o Al the decrease of the saturation magnetization corresponds to the decreasing number of Fe atoms in the alloy owing to dilution with Al, so that the moment per Fe atom remains constant. The electronic specific heat of Fe-Al alloys, as shown in Fig. 4, is also very little affected by Al additions up to about 20 a/o Al.[9] The electronic specific-heat results for Fe-Si, Fe-Ge, and Fe-Sn alloys[10] are also similar. The simplest interpretation of the combined magnetic data and electronic specific-heat results is that, at least up to 10 a/o Al, the aluminum solute does not affect the 3*d* band of iron.[9,11,12] This conclusion is strongly supported by the electronic specific-heat data for ternary alloys of transition elements with 10 a/o Al (line *b*, Fig. 2). Comparison of lines *b* and *a* in Fig. 2 shows that at $e/a < 6$ the addition of 10 a/o Al decreases the electronic specific heat (an effect which is not well understood), but that at $e/a > 6$ the electronic specific heat remains essentially unchanged. It may be concluded that the absence of any appreciable effect of Al on the 3*d* band, as noted for Fe-Al alloys, in fact extends over both magnetic sub-bands of the bcc alloys.[11,12]

However, when the amount of the solute increases beyond 10 a/o Al, the magnetic moment is decreased below the level corresponding to simple dilution (Fig. 3). This decrease in the moment per Fe atom suggests the

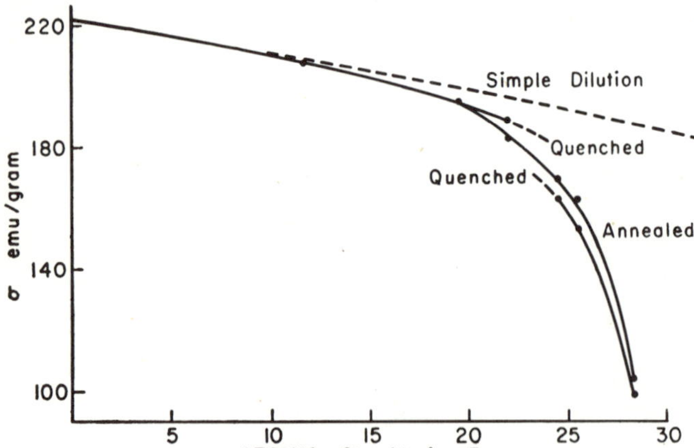

Fig. 3 Saturation magnetization for Fe-Al alloys, according to Arrott and Sato.[8]

beginning of a change in the 3*d* band of Fe. The results of recent electronic specific-heat measurements for a series of ternary alloys of transition elements with 20 a/o Al by Srinivasan and Claus[13] are shown in line *c* of Fig. 2. In this figure the electronic specific-heat data for the ternary alloys are plotted vs. electron-concentration values corresponding to the composition in transition elements only (i.e., in computing the "transition-element electron concentration" plotted in Fig. 2 the Al content is entirely disregarded). If the added Al in fact does not change the number of electrons in the *d* band, as in the case of the 10 a/o Al alloys with $e/a > 6$, then all the minima and maxima occur at the same transition-element electron concentration (lines *a* and *b*, Fig. 2). The fact that the prominent features of line *c*, representing the series of 20 a/o Al alloys, are displaced with respect to those for the alloys without Al (line *a*) indicates that the addition of 20 a/o Al introduces a change in the *d* band. Since the displacement took place toward higher electron concentrations, it is clear that the change corresponds to removal of electrons from the *d* band, so that, for instance, the filling of the third sub-band in the 20 a/o Al alloys begins at a higher transition-element electron concentration, i.e., at a higher Co/Fe ratio, than in the alloys without Al.

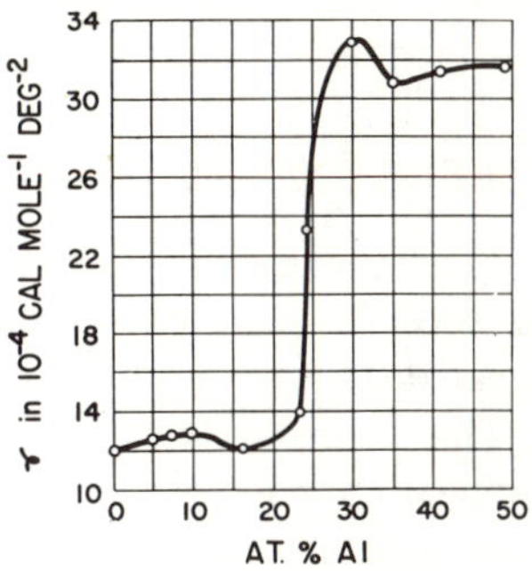

Fig. 4 Electronic specific-heat coefficients, in 10^{-4} cal/(mole) (deg^2), vs. electron concentration for bcc Fe-Al solid solutions.[9] The values given for more than 20 a/o Al comprise a large magnetic contribution.

The extent of the shift is an indication of the degree of depletion of the *d* band due to the addition of 20 a/o Al. For instance, the maximum of the electronic specific heat, which occurs at $e/a = 8$ (at Fe) in the alloys without Al addition, is displaced to approximately $e/a = 8.05$ to 8.1 in the alloys with 20 a/o Al. This would correspond to a loss of 0.05 to 0.1 electron from the *d* band. It is interesting to observe that the decrease in the moment per Fe atom resulting from the addition of 20 a/o Al is approximately 0.07 μ_B (Fig. 3). Thus, the decrease in the moment appears to correspond well with the decrease in the number of *d* electrons. These findings are consistent with the above interpretation of the Slater-Pauling curve (Fig. 1) in terms of the electronic specific-heat results (line *a*, Fig. 2). It may be noted in Fig. 2 that the minimum of γ at Cr (curve *a*) is not displaced in the 20 a/o Al alloy series (curve *c*). However, the lowering of this minimum (Fig. 2) may be a result of the shifting away from each other of the two sub-bands, which slightly overlap at Cr. That the two *d* sub-bands may be indeed essentially separated from each other is suggested by the lowering of γ in bcc Cr-Al alloys (at 20 to 25 a/o Al) to a value lower than that for Cu.[11] In such an alloy the electrons at the Fermi surface have presumably very little, if any, *d* character.

EFFECT OF Al ON THE 3d BAND OF FCC ALLOYS

The magnetic moments of fcc 3*d* metals and alloys are shown by the dashed descending line in Fig. 1. The corresponding electronic specific-heat data for the Fe-Ni alloys are shown by the solid thin line in Fig. 5.[14] (The dashed extension of this line to the left, toward higher Fe contents, appears to comprise a magnetic contribution, and it is, therefore, not to be regarded as representing electronic specific-heat data.)[14] The decrease in saturation magnetization on adding Cu to Ni (Fig. 1) and the corresponding decrease on adding Al (Fig. 6) have long been considered as a result of filling vacancies in the Ni 3*d* band. Recently, however, Beeby[15] proposed that the decrease in the moment in Ni-Al alloys may be due to a redistribution of the 3*d* vacancies between the spin-up and spin-down half-bands as a result of a decrease in exchange energy, rather than to filling of the 3*d* band. The thick solid line in Fig. 5 shows the electronic specific heat of ternary fcc Ni-Fe alloys with 10 a/o Al.[16] The direction of the displacement of this line with respect to that for the alloys without Al (thin solid line in Fig. 5) cannot be uniquely established in the absence of prominent features, such as the minima and maxima present in the corresponding curves for bcc alloys (Fig. 2). If one assumes a semirigid band model,* as suggested by the saturation magnetization data for the Al-free fcc alloys (Fig. 1), then the displacement of the two curves in Fig. 5 with respect to each other should be interpreted as a shift toward higher transition-element electron concentration upon the addition of Al. This would indicate not only that the *d*-band vacancies are not being filled by the valence electrons of the Al, but that, on the contrary, the effect is a removal of electrons from the *d* band, as in the case of the bcc alloys with 20 a/o Al.

The extent of the shift considered would be smallest near Ni, where it corresponds approximately to an electron loss of 0.3 from the *d* band as a result of adding 10 a/o Al. The corresponding increase in moment should be 0.3 μ_B, according to Fig. 1. Instead, a loss in moment of approximately 0.26 μ_B per Ni atom is observed (Fig. 6). At $Ni_{0.6}Fe_{0.4}$ the shift of the electronic specific-heat curve would suggest a loss of approximately 1.5 electrons upon adding 10 a/o Al; instead of the corresponding

* If the top of the filled half-band (spin up) in Ni is very near to the Fermi surface, as appears likely, the effect of decreasing the number of electrons (e.g., by alloying with Fe) in increasing the saturation magnetization (Fig. 1) can be interpreted as further depletion of the unfilled half-band (spin down), with the filled half-band shifting toward lower energies, so that it remains filled. This band model, in which each half-band may be assumed to be rigid, but the two are shifting with respect to each other, may be designated as a *semirigid* one.

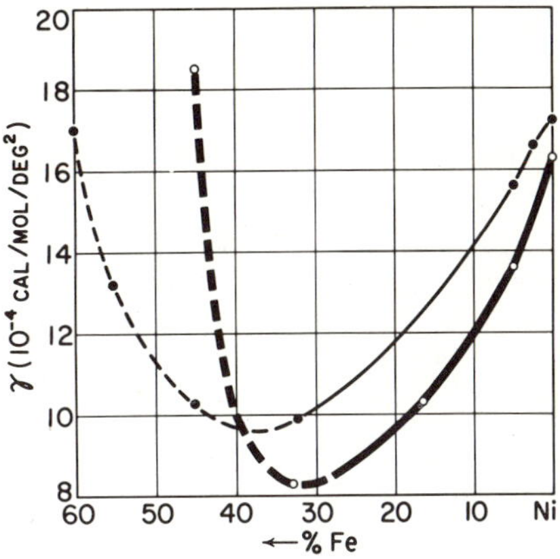

Fig. 5 Electronic specific-heat coefficients, in 10^{-4} cal/(mole)-(deg^2), for fcc Fe-Ni alloys[14] (thin line); and for similar ternary alloys with 10 a/o Al[16] (thick line); vs. transition-element electron concentration.

large increase, a decrease in moment of about 0.18 μ_B is observed (Fig. 6). The interpretation of these data is at present not clear. Further low-temperature specific-heat work is being done in an attempt to ascertain whether or not a semirigid-band model is appropriate for the Al-containing alloys.

MICROENVIRONMENT EFFECTS ON THE ATOMIC MOMENT

In addition to saturation magnetization and specific-heat data, which characterize average conditions in alloys and which one may attempt to interpret in terms of a band model, in recent years considerable information of various types accumulated relating to local conditions in atomic micro-environments in alloys. Nathans, Piggott, and Shull[17] showed by means of neutron diffraction (magnetic scattering) that in the ordered alloy Fe_3Al the Fe_I atoms, all eight nearest neighbors of which are Fe, have a spin moment of 2.18 μ_B or very nearly the same value as has bcc α-Fe, in the absence of Al. However, in the same ordered Fe_3Al structure the Fe_{II} atoms, having four Fe and four Al nearest neighbors, have a spin moment of only 1.5 μ_B. Thus, the spin moment of Fe is clearly shown to be sensitive to the nearest-neighbor microenvironment. On the other hand, the fact that Fe_I atoms in the Fe_3Al structure have nearly the same moment as have the atoms in pure α-Fe, suggests that the second-nearest Al neighbors have very little, if any, effect, since all six second-nearest neighbors of Fe_I in ordered Fe_3Al are Al atoms.

It is clear that the effect of nearest neighbor Al atoms in decreasing the spin moment of an Fe atom does not vary linearly with their number. It is possible to estimate the variation of this effect with the number of nearest-neighbor Al atoms on the basis of the known change of the saturation moment with Al content[8] (Fig. 3) by assuming a random distribution of the atoms in Fe-Al solid solutions up to 15 a/o Al. In this model the fraction of Fe atoms with four Al nearest neighbors is 0.0185. Each of these Fe atoms may be assumed to have 32% lower spin moment than the 2.2-μ_B value for Fe atoms in the absence of Al, as found for Fe_{II} in Fe_3Al. This portion of the loss of average moment per Fe atom in the 15 a/o Al alloy, due to those Fe atoms with four Al nearest neighbors, amounts to 0.59% or almost half the measured loss of average moment of 1.25% per Fe atom. Assignment of the remaining 0.66% loss must be made so that the condition of no loss in the average moment per Fe atom for alloys with 10 a/o, or less, Al will be satisfied as nearly as possible. This can be done by maximizing the contribution to the remaining loss in the 15 a/o Al alloy by Fe atoms with five or more Al nearest neighbors, since the fraction of such atoms decreases very rapidly at lower Al contents. However, even if a 100% loss of moment is assumed for each Fe atom with five or more Al nearest neighbors, a loss of 0.38% in the bulk moment of the 15 a/o Al alloy remains unaccounted for, and this loss must be attributed to Fe atoms with three Al nearest neighbors. Since the fraction of such Fe atoms in the random model of the 15 a/o Al alloy is 0.084, the loss of moment for each of them must be 4.5%. This assignment of the loss of moment to the Fe atoms with different numbers of Al nearest neighbors leads to an average loss of moment per Fe atom for the 10 a/o Al alloy of 0.34%, a value somewhat larger than allowed by the probable error in the saturation magnetization measurements. The discrepancy can be corrected by modifying the model for the 15 a/o Al alloy to include a certain degree of short-range order, instead of ideal randomness. Short-range order will increase the average number of Al nearest neighbors around Fe atoms. Because of the increase in the fraction of Fe atoms with four and more Al nearest neighbors, the contribution of these Fe atoms to the overall loss of moment in the 15 a/o Al alloy becomes larger, and the remaining loss, which must be assigned to Fe atoms with three Al nearest neighbors, becomes smaller. As a result, the loss of average moment calculated for alloys with 10 a/o and less Al decreases. In any case, it is clear that the constancy of the average moment per Fe atom in alloys up to about 10 a/o requires an assignment of very little, if any, loss of moment to Fe atoms with one or two Al nearest neighbors. By far the largest part of the decrease in spin moment with an increasingly Al-rich nearest-neighbor shell evidently takes place quite abruptly in the range between three and five Al nearest neighbors. (An even more abrupt decline in the spin moment of Fe with changing composition of the nearest-neighbor shell of atoms was indicated for Nb-Mo solid solutions of varying composition, containing 1 a/o Fe.[27])

Interestingly, practically identical conclusions were reached previously by Friedman and Nicholson[22] on the basis of their data on Mössbauer hyperfine splittings in ordered Fe-Al alloys, even though they made use of

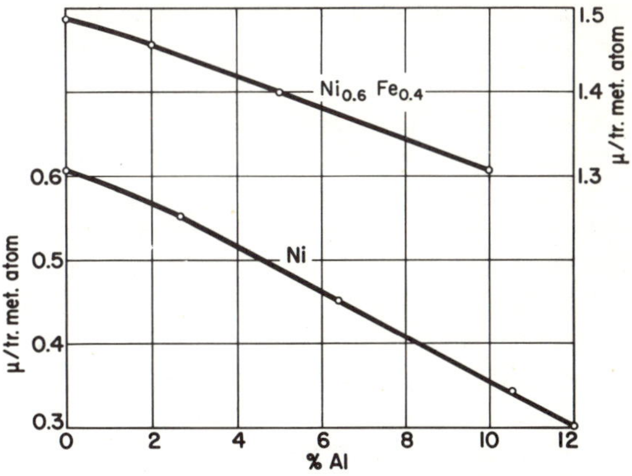

Fig. 6 Magnetic moment in Bohr magnetons per transition-metal atom vs. Al content, for fcc Ni-Al alloys[20] and for fcc $Ni_{0.6}Fe_{0.4}$-Al alloys.[21]

an assumption now known to be incorrect, namely that the hyperfine field at the nucleus of an Fe atom is proportional to the spin moment on that atom. The fact that this assumption is incorrect is clearly shown, for instance, by the linear decrease in the average hyperfine field at Fe^{57} nuclei with increasing Al content[23,24] in the alloy composition range where the spin moment per Fe atom is independent of Al content.[8]*

In contrast to the Fe-Al alloys, in Ni-Al alloys the moment per Ni atom decreases with increasing Al-content even at low Al concentrations[20] (Fig. 6). Accordingly, in the Ni-Al solid solutions one must assume a considerable decrease in spin moment even on those Ni atoms with only one Al nearest neighbor.

The spin moments of both Fe and Ni in fcc Fe-Ni solid solutions were studied by Shull and Wilkinson[18] by means of neutron magnetic-disorder scattering measurements, using also the known bulk moment of these alloys. Their data indicate an increase in the spin moment of iron with increasing Ni content. In a later similar investigation by Collins, Jones,

* In view of this linearity, at low Al contents the rate of decrease of the average hyperfine field with increasing Al content should be correlated with the increase in the number of Fe atoms, which have one of their nearest Fe neighbors replaced by Al. From this correlation the magnitude of the loss of hyperfine field at Fe nuclei with only seven Fe nearest neighbors may be calculated to 6.1% (using the data from Johnson et al.[23]) or to 7.6% (using the data from Stearns[24]). The average of these two values is in excellent agreement with the 7.2% loss derived from the ferromagnetic NMR satellite line corresponding to Fe nuclei with an Al nearest neighbor in dilute Fe-Al solid solutions.[25] That the effect of a nearest-neighbor Al atom here consists mainly in the elimination of the spin moment of the nearest-neighbor Fe atom that it replaces is suggested by observations[26,25] giving very similar losses in the hyperfine field of Fe by a large variety of nonmagnetic nearest-neighbor solute atoms.

and Lowde,[19] which took into account short-range order effects, this increase was not so clearly recognizable. However, both investigations were in quite good quantitative agreement as to the value near 3 μ_B for the spin moment of Fe in ordered $FeNi_3$. Also, the bulk moment of this alloy increases by 5% on ordering.[28] Since in ordered $FeNi_3$ all Fe atoms have only Ni nearest neighbors, it appears reasonable to assume that in the highly Ni-rich solid solutions, in which the Fe atoms also have practically only Ni nearest neighbors, the spin moment of the Fe atoms is similarly high. In Fig. 7, which shows the data points for fcc Fe-Ni solid solutions from both neutron diffraction studies, straight line I, representing the approximate variation of the Fe moment with the alloy composition, was drawn accordingly. If it is assumed that the $FeNi_3$ specimens used in the two investigations were not completely ordered (certainly true for the one used by Collins et al.,[19] since it contained 30% Fe), it is possible that the spin moment of Fe with *only* Ni nearest neighbors, as in very dilute Ni-base alloys, is even higher than 3 μ_B. Accordingly, an alternative approximate variation of the spin moment of Fe with composition is shown in Fig. 7 by straight line II.

The results of recent saturation magnetization measurements with ternary fcc solid solutions of compositions of the type $Ni_{0.6}Fe_{0.4}$-Al are given in Fig. 6. The decrease of the average moment per transition-metal atom with increasing Al content in these alloys is less pronounced than the corresponding decrease for the binary Ni-Al alloys (Fig. 6). The average slope of the moment per transition-metal atom vs. Al content curve for the Ni-Al binary alloys $dp/dc = -2.52$, while the corresponding average slope for the ternary alloy series $dp/dc = -1.83$. The saturation-moment data for the ternary alloys can be satisfactorily accounted for on the basis of the following assumptions: The decrease in the average Ni spin moment in each ternary alloy is the same as that for the binary Ni-Al alloy of the same Al content (same average number of Al nearest neighbors), and the decrease in the spin moment of the Fe atoms in each ternary alloy corresponds to the decrease in the number of Ni nearest neighbors due to their partial replacement by Al atoms. If the calculation is based on the increase of spin moment of Fe by Ni shown by line I in Fig. 7, the calculated rate of decrease of the spin moment per transition-metal atom with increasing Al content $dp/dc = -1.70$, while the relationship expressed by line II in

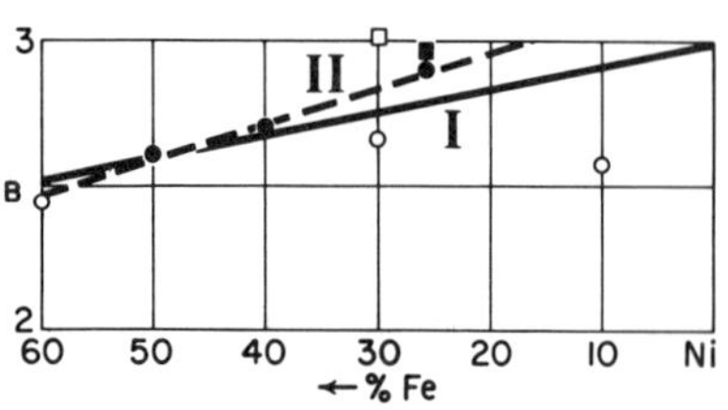

Fig. 7 Atomic moment associated with Fe atoms μFe in Bohr magnetons in fcc Fe-Ni alloys vs. composition. Circular symbols refer to disordered alloys; square symbols to ordered ones. Data from Collins et al.,[19] empty symbols; from Shull and Wilkinson,[18] filled symbols.

Fig. 7 leads to $dp/dc = -1.80$ for the $Ni_{0.6}Fe_{0.4}$ alloys, in good agreement with experiment. These considerations are suggestive of the importance of the microenvironment in influencing the magnitude of the magnetic moments in alloys of 3*d* metals.

REFERENCES

1. Slater, J. C.: *J. Appl. Phys.* **8**:385 (1937).
2. Pauling, L.: *Phys. Rev.* **54**:899 (1938).
3. Gupta, K. P., C. H. Cheng, and P. A. Beck: *J. Phys. Radium* **23**:721–726 (1962).
4. Clogston, A. M.: *Phys. Rev.* **136A**:8 (1964).
5. Clogston, A. M.: Personal communication.
6. Krebs, K.: *Phys. Letters* **6**:31–33 (1963).
7. Cheng, C. H., C. T. Wei, and P. A. Beck: *Phys. Rev.* **120**:426–436 (1960).
8. Arrott, A., and H. Sato: *Phys. Rev.* **114**:1420–1426 (1959). Similar data also in D. Parsons, W. Sucksmith, and J. E. Thompson; *Phil. Mag.* **3**:1174 (1958).
9. Cheng, C. H., K. P. Gupta, C. T. Wei, and P. A. Beck: *J. Phys. Chem. Solids* **25**: 759–765 (1964).
10. Gupta, K. P., C. H. Cheng, and P. A. Beck: *J. Phys. Chem. Solids* **25**:1147–1149 (1964).
11. Pessall, N., K. P. Gupta, C. H. Cheng, and P. A. Beck: *J. Phys. Chem. Solids* **25**:993–1003 (1964).
12. Beck, P. A.: Proc. International Conference on Magnetism, Nottingham, pp. 178–181, Institute of Physics and Physical Society, London, 1964.
13. Srinivansan, T. M., and H. Claus: To be published.
14. Gupta, K. P., C. H. Cheng, and P. A. Beck: *J. Phys. Chem. Solids* **25**:73–83 (1964).
15. Beeby, J.: *Phys. Rev.* **141**:781–788 (1966).
16. Viswanathan, R.: To be published.
17. Nathans, R., M. T. Pigott, and C. G. Shull: *J. Phys. Chem. Solids* **6**:38 (1958).
18. Shull, C. G., and M. K. Wilkinson: *Phys. Rev.* **97**:304–310 (1955).
19. Collins, M. F., R. V. Jones, and R. D. Lowde: *J. Phys. Soc. Japan* **17** (Suppl. B-III): 19–25 (1962).
20. Crangle, J., and M. J. C. Martin: *Phil. Mag.* **4**:1006 (1959).
21. Bardos, D. I.: To be published.
22. Friedman, E. A., and W. J. Nicholson: *J. Appl. Phys.* **34**:1048 (1963).
23. Johnson, C. E., M. S. Ridout, and T. E. Cranshaw: *Proc. Phys. Soc.* **81**:1079 (1963).
24. Stearns, M. B.: To be published in *Phys. Rev.*
25. Rubinstein, M., G. H. Strauss, and M. B. Stearns: *J. Appl. Phys.*, **37**:1334–1335, 1966.
26. Wertheim, G. K., V. Jaccarino, J. H. Wernick, and D. N. E. Buchanan: *Phys. Rev. Letters* **12**:24 (1964).
27. Jaccarino, V., and L. R. Walker: *Phys. Rev. Letters* **15**:258 (1965).
28. McKeehan, L. W., and E. M. Grabbe: *Phys. Rev.* **55**:505 (1939).

DISCUSSION *on Paper Presented by P. A. Beck*

BREWER: When you say 20 a/o Al removes *d* electrons from the *d* band of the Fe, would this also be consistent with the statement that unpaired *d* electrons are removed by adding *d* electrons to produce internally paired *d* electrons?

BECK: Yes, if one drops the rigid-band model, the decrease in the number of electrons in the "magnetic" portion of the d band (at $e/a \geq 6$) may be due to their transfer into the "nonmagnetic sub-band" (at $e/a > 6$) upon the addition of 20 a/o Al. In order that the nonmagnetic sub-band may accept additional electrons in spite of the decrease in the density of states from a to c in Fig. 2, this sub-band must become wider.

FRIEDEL: One can easily show, using a tight binding approximation, that the rigid-band model is *not* valid for analysis of the γ values in Ni-Co and Ni-Fe alloys. The essential reason is that the d band is nearly full in Ni, and the number of positive holes per atom (about 0.5) is too small compared with the minimum excess nuclear charge that they have to screen.* Indeed, using a reasonable form for the density of states in the d band of Ni, one can explain why γ decreases both when Cu and Co (or Fe) are added. In these "strong perturbation" computations, the slope of the γ versus electron-per-atom ratio plot has *no* direct relation with the slope of $n(E)$ near the Fermi level of Ni.

These deviations from the rigid-band model are of course expected to be smaller, the nearer the matrix is to the middle of the transitional series. There the rigid-band model is expected to work very well, as all your work shows.

Turning to Ni-Fe-Al, I would think that the effects of Fe and Al on γ are essentially *additive* in the limit of small Fe and Al concentrations. It is therefore not surprising that Al reduces γ in Ni-Fe as it does in pure Ni. Both effects are consistent with the idea that Al fills the d band of Ni, but that this effect is *local*.†

BECK: Professor Friedel's results in regard to the deviation from rigid-band behavior in Ni-Fe-Al alloys are most interesting. Since the measured γ (e/a) curves shown in Fig. 5 of the paper do not have maxima and minima, the displacement of which one could observe (as in the case of the bcc alloys, Fig. 2), we wondered whether the effect of alloying with Al should be interpreted as a displacement of the curve toward higher transition-element electron concentrations (corresponding to loss of electrons from a "semirigid" d band) or toward lower density of states (corresponding to Professor Friedel's model). In order to answer this question experimentally, Dr. Viswanathan, who obtained the data for the Ni-Fe-Al alloys, is at present extending his investigation to Ni-Cu-Al alloys as well. On the semirigid-band model one may expect the maximum of γ, which in the Al-free alloys is at Ni, to be displaced toward some Ni-Cu composition. Thus, on adding Cu at a constant Al content, the value of γ should initially *increase*. On the other hand, on Professor Friedel's model the γ maximum may be expected to stay at Ni and, thus, on adding Cu γ should *decrease*. The experimental results will be available shortly.

HEINE: I would like to comment on Beck's results. First let me refer to some recent work by Geballe and collaborators on niobium nitride. They find that the Fermi level E_F appears to lie in a pure sp band, the d band being pushed above E_F. Anderson has suggested an explanation along the following lines. The $2p$ states of the N atoms, which lie well below the Nb bands, can overlap to some extent

* Gomes, *J. Phys. Chem. Solids*, 1966; Friedel, Gautier, and Lenghart, *Slater Festschr.*, 1966.

† Friedel, *ibid.*

onto Nb atoms. The overlap is then of bonding type. Similarly the *d* states of Nb overlap to some extent with the N, and have to have a corresponding antibonding character. This shifts the *d* band upward sufficiently to empty it completely.

We can apply somewhat similar arguments to the Fe-Al and other alloys. Let us think of the Al atom as one cell of Al metal: its occupied states are spread over a wide range of 1 rydberg, more than covering the Fe *d* band. The lower bonding states of the *d* band will interact in a bonding sense with the Al and are lowered in energy. The upper *d* states have an antibonding interaction and are moved upward in energy. Thus, at electron concentrations less than Cr, there results some filling of the *d* band due to the effect of the Al, and beyond Cr some emptying, exactly as on Beck's slide.

A REVIEW OF THE PHYSICAL PROPERTIES OF SOME PRIMARY SOLID SOLUTIONS, AND THEIR INTERPRETATION

John Stringer

Metal Science Group, Battelle Memorial Institute, Columbus Laboratories, Columbus, Ohio

INTRODUCTION

It follows from the early results of Hume-Rothery, and the later observations of Raynor, Massalski, and Beck among others, that the electronic contribution to the free energy of a metal is frequently dominant in determining the stability of phases. Theoretical description of the electrons in disordered metallic phases has, however, proved extremely difficult; although the early model of Jones, designed to rationalize the experimental results of Hume-Rothery, was based on a very simple picture of a metal and is clearly incorrect in detail, most subsequent descriptions have failed to produce significant improvements.[1]

Pippard's[2] determination of the Fermi surface in Cu using the anomalous skin effect first showed that the Jones model was invalid, although it had been suggested earlier on indirect grounds (by Olson and Rodriguez[3] among others) that the Fermi surface in Cu already touched the {111} faces of the Brillouin zone. Heine and Cohen[4] thereupon suggested, instead of the "rigid band" assumed by Jones, that the band structure itself was a function of alloying. On the basis of this model, they made some specific predictions of the shape of the Fermi surfaces in pure Au and Ag, which were subsequently proved to be incorrect by the de Haas–van Alphen studies of Shoenberg.[5]

Friedel[6] has developed an idea originally due to Mott,[7] who pointed out that the excess charge on a solute atom of differing valency would be effectively shielded by a local distortion of the conduction band, so that in effect the additional electrons (or holes) introduced by the solute are localized. This is different from the idea of a bound state, which may also exist around a solute atom; in the free-electron theory one may regard the completely ionized solute atom as introducing a local "well" in the otherwise uniform lattice potential; this well may contain one or more bound states, but of course these states lie in an otherwise forbidden energy band, normally below the conduction band.

At large alloy concentrations, Friedel has demonstrated using a Thomas-Fermi method that the effect of the alteration in lattice potential due to the solute atoms may be approximately regarded as a displacement of the whole band without change of shape; this is not valid at the bottom of the band, but is probably valid at the Fermi level. The effect of the displacement is to *reduce* the change in Fermi level (relative to vacuum) from the value to be expected if the position of the band remained unchanged. Further calculation showed that at small concentrations the density of states in the alloy is always smaller than that expected by rigid-band theory.

Raimes[8] has made some calculations for a simple cubic lattice of touching square wells treating a tight-binding approximation with first-order perturbation theory. The wells were made deep enough to contain two bound levels, giving rise to two bands. Even for the "pure metal"—all the wells the same depth—the energy surfaces produced by this model were considerably distorted from the spherical. Where the impurity wells were *shallower* than the matrix, the energy surfaces became more spherical; but where the impurity wells were deeper, the distortion of the energy surfaces increased. Surprisingly, the smaller the energy gap between the two bands, the less spherical were the energy surfaces, contrary to the predictions of Heine and Cohen. Raimes notes that this seems to be a result of using touching square wells.

It is of interest to consider whether in fact the concepts of a Fermi surface, a Brillouin zone, and so forth have any meaning for a concentrated disordered alloy. This question was posed by Heine[9] recently, who concluded that the Fermi surface was probably still sharp to within approximately

0.1 ev or so. A number of calculations have been made of the energy spectra in disordered lattices, but these calculations are exact enough to be of immediate interest only for the one-dimensional case. In this context it is extremely dangerous to extrapolate results obtained for one-dimensional models to a three-dimensional real situation. Makinson and Roberts[10] determined the spectrum for a linear chain of equal delta functions whose spacing was varied according to a variety of statistical functions, and found that the band structure was sensitive to short-range order rather than to long-range order. More recently, Agacy and Borland[11] have studied the one-dimensional problem for a diatomic linear lattice, and find that prominent peaks in their computed density-of-state curves could be associated with atom clusters. These authors note that any energy band forbidden for a perfect lattice of either type of atom is also forbidden in the disordered case, a conclusion first stated by Luttinger.[12] They also stress that the fact that the (site occupation) disorder in a solid lattice is discrete, as opposed to the continuous (spacing) disorder possible in liquid metals, is important in producing peaks in the spectrum, and suggest that this part of the conclusion is valid for the three-dimensional case. Physically, one may interpret the peaks in the density-of-state curves as electrons in states that are effectively localized in regions of the lattice whose local structure permits such a state, but cannot propagate into neighboring parts of the lattice.

Studies of three-dimensional disordered lattices, because of the difficulty of the problem, have so far not produced results of such obvious physical relevance. Edwards[13] and co-workers, and also Jones,[14] have developed a technique based on an investigation of the properties of Green's functions of electrons in disordered systems.

Jones[15] comments that the energy band is broadened in a dilute disordered alloy, with the maximum effect on the line width being at the center of the band, although the displacement at this point is a minimum. If the Fermi level is near the center of a band, line broadening will decrease the density of states at the Fermi level, whereas if the Fermi level is near the band edge, the density of states will increase. Short-range or long-range order may produce an effect on the density of states which may not be strictly parallel with the residual resistivity. Interestingly, Jones's calculations showed no singularities either within or below the pure metal band; but of course this may well be a consequence of the method of calculation.

Beeby[16] has extended the work of Edwards, and has applied the results to a description of alloys; however, his discussion is mainly confined to the properties of the *d* bands in transition metals.

The singularities in the one-dimensional models which one may associate with local fluctuations in composition have in effect been predicted earlier.

Flinn[17] discussed the correlation between short-range order and the total electron energy, and by implication the correlation suggests a variation of density of states with local order from point to point in the alloy. Smoluchowski[18] also suggested that anomalies in the electronic specific heat might be removed by summing a suitably weighted average of the density of states over the whole solid.

In summary, it appears that in alloys one may distinguish a number of effects:

1. Local perturbations of the conduction band due to screening of the additional charge of isolated solute atoms.
2. Virtual bound states arising from resonance between bound states about an impurity atom and the band; this effect will be marked for d bands, but may well be present also for s and p bands.
3. Local perturbations of the band and the density of states due to local variations in composition.

Because of the complexity of the effects in alloys, and the absence of any very convincing theory for even relatively simple cases, it is important to measure experimentally the relevant properties of the electron distribution. Unfortunately this is far from simple. The experiments that give clear unambiguous information, such as de Haas–van Alphen measurements, are not applicable to alloys. One is therefore restricted to indirect techniques, and the theory relating these effects to the properties of the electrons is again not well understood. In the next section, some of the experimental results for the primary solid solutions of Cu-Zn and Cu-Ge have been collected, and a discussion of the state of the theory will be made with reference to these data.

EXPERIMENTAL INVESTIGATIONS OF Cu–Zn AND Cu–Ge ALLOYS

The various data are shown in Fig. 1. In all diagrams, the circles refer to Cu-Zn; the squares to Cu-Ge. The sources of the data are identified in the legend to the figure.

Electronic Specific Heat

The low-temperature specific heat of a metal is usually of the form

$$C = \gamma T + A \left(\frac{T}{\Theta}\right)^3 \tag{1}$$

where γ is the electronic contribution to the specific heat, and Θ is the Debye temperature.

According to the free-electron theory, the electronic specific heat is related to $g(E_F)$, the density of states at the Fermi level, by the expression

$$\gamma = \tfrac{1}{3}\pi^2 k_B{}^2 g(E_F) \tag{2}$$

where k_B = Boltzmann's constant. However, as Jones[31] pointed out, this must be modified because of exchange interactions, and because of electron-phonon interactions. One form of electron-phonon interaction was discussed by Buckingham and Schafroth[32], and a second, appreciable only when the electron energy departs significantly from the free-electron value, by Jones. Jones predicted a very rapid rise in the apparent value of γ immediately before the Fermi surface touched a Brillouin zone face, followed by an abrupt drop back to the free-electron value. In a later paper, Jones[33] recalculated to take account of the fact that the Fermi surface in Cu already touches the Brillouin zone walls, and to allow for the randomness of the solid solution. In this case,

$$\gamma = \gamma_0(1 + bc)$$

where γ is the electronic specific heat of the alloy, γ_0 is that of the pure metal, c is the fractional concentration, and b is a constant which depends on the scattering power of the solute atom, and is of order unity.

Boardman and March[34] have also developed a theory of the specific heats of dilute alloys, but as yet a comparison between their results and experiments is not possible.

The upper curve shown in Fig. 1a is due to Rayne.[19,20] Although there is some scatter, the Cu-Zn and Cu-Ge results superimpose well, and show a marked rise in dilute solutions. Since it is known that the density of states should fall in dilute Cu alloys as the electron concentration increases, Heine[35] remarked that this effect might be due to ferromagnetic impurities in the alloys, and indeed small amounts of impurities can give a large contribution to the low-temperature specific heat which may be linear with temperature. Rayne[36] analyzed the Cu-Ge alloys, and showed that the Fe content was in fact very low, of the order of 1 or 2 ppm (parts per million by weight). Veal and Rayne[21] repeated the experiments on Cu-Zn using very high purity starting materials, and were careful to avoid any chance of picking up ferromagnetic impurities. Again, the curve initially rose, although the values were much smaller than in the earlier paper, suggesting that in fact some ferromagnetic impurity had been present. This did not, however, explain the anomalous rise. Veal and Rayne, following Hume-Rothery and Roaf,[37] suggested that the density-of-state curve will have two peaks, one corresponding to overlap of the {111} faces, the second to overlap of the {200} faces, of the Brillouin zone. If the energy gap across the {200} faces is appropriately chosen, the density of

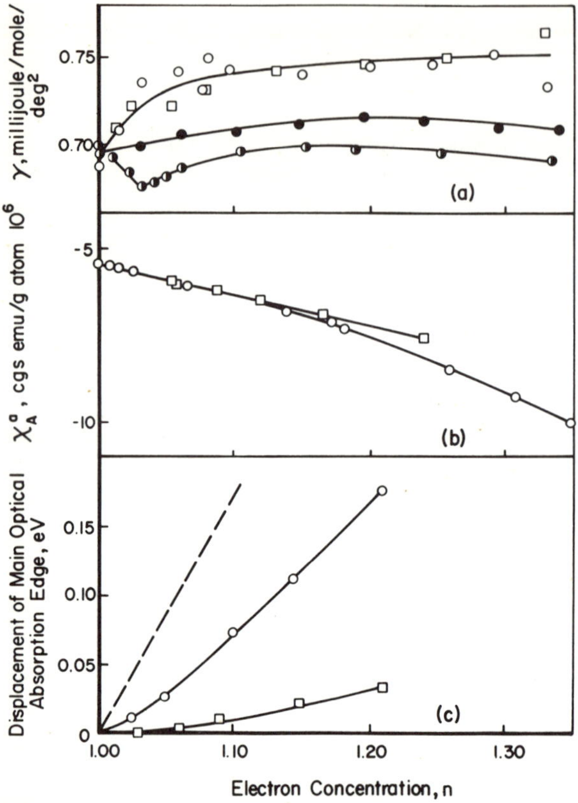

Fig. 1 The variation of some physical properties with electron concentration for Cu-Zn and Cu-Ge alloys. In all cases the circles refer to Cu-Zn alloys; the squares to Cu-Ge alloys. (*a*) Empty circles, from Rayne[19]; empty squares, from Rayne[20]; filled circles, from Veal and Rayne[21]; half-filled circles, from Isaacs and Massalski.[22] (*b*) Circles and squares, from Henry and Rogers.[23] (*c*) Circles, from Biondi and Rayne[24]; squares, from Rayne.[25]

states does in fact rise between the two peaks; but the energy gap has to be about 12 ev: a more likely value is 5.7 ev.[38]

Recently, Massalski and Isaacs[22] repeated the measurement of the electronic specific heat of Cu-Zn alloys and obtained the significantly different values shown in the lower curve. The authors are unable to suggest a reason for the initial decrease, or for the difference between their results and those of Veal and Rayne, but remark that in both sets of results the variation of the specific heat is very small, only a little over 3% in each case.

It is important to note that there is no rapid rise in the specific heat before the phase boundary, as would be expected from any of the theories in which the phase boundary is assumed to occur immediately after a Brillouin zone face is contacted. Further, any modification of the Heine-Cohen model in which alloying produces "spheroidization" of the Fermi surface and consequent loss of contact with the {111} faces of the Brillouin zone should also produce (accepting Jones) a large positive anomaly. Further, the "Jones anomaly" due to scattering of electrons by the impurities giving effective contributions from electrons near the peak in density of states below the Fermi surface cannot (as suggested by Massalski and Isaacs) explain the rapid rise in the Cu-Ge results at low concentrations, since it should be followed by a rapid fall as the peak becomes submerged. Finally, the theories that suggest that there is some shielding of solute atoms would imply little change in very dilute alloys, with a gradually increasing effect as the shielding breaks down at higher concentrations. In fact, the change appears to be most rapid for very dilute alloys, although of course this is where the two sets of data are most in conflict.

It will appear later that the scattering produced by impurities in Cu is anisotropic. Deaton and Gavenda,[39] using an ultrasonic attenuation method, determined relaxation times for different parts of the Fermi surface at 4.2°K. At this temperature, the relaxation they were observing was *probably* impurity relaxation, and the relaxation time at the {111} necks was some eight times as great as that at the belly of the surface. This variation of the relaxation produces (following Jones) greater contributions from "submerged" electrons at the belly than at the necks. Kip[40] et al. have determined effective masses at various points on the surface, using cyclotron resonance techniques, and find that

$$\frac{m_b}{m_0} \simeq 1.39$$

$$\frac{m_n}{m_0} \simeq 0.6$$

It follows therefore that the density of states at the belly is greater than that at the necks, and hence this represents a larger contribution to the specific heat. The anisotropic impurity relaxation increases the effective contribution of the belly, and therefore increases γ. Plainly, the effect will only be important for the first 1 or 2%, and will not be very large, since anyway the contribution of the necks to the specific heat will not be great.

Magnetic Susceptibility

Another property that is generally regarded as closely connected with the density of states at the Fermi level is the magnetic susceptibility. The problem is complicated by the large number of contributions to the susceptibility. The ion cores give a diamagnetic contribution, which in the case of Cu alloys is large because of the full d shell. The electrons in the unfilled shells have a susceptibility composed of a Pauli paramagnetism and a Landau diamagnetism; in the simple case the Landau term is one-third of the Pauli term, but this depends on the effective mass of the electrons; in fact

$$\chi_E = \mu_B^2 \left[1 - \frac{1}{3} \left(\frac{m}{m^*} \right)^2 \right] g(E_F) \tag{3}$$

The result is that, even neglecting effects due to the lattice disorder, determination of the density of states may be difficult.

The data shown in Fig. 1b are from Henry and Rogers.[23] The authors consider that the susceptibility of the alloy $\chi_A{}^a$ may be divided into the susceptibility of the ions χ_{Cu^+} and χ_s, the net paramagnetic contribution of the band $\chi_E(x)$, and the susceptibility χ_e of the additional electrons contributed by the solute elements, which they regard as being localized around the impurity atoms, but not in bound states. The experimental curve is approximated by a straight line; then

$$\frac{\chi_A{}^a - \chi_{\mathrm{Cu}}}{x} = (\chi_s + \chi_e) - \chi_{\mathrm{Cu}^+} + \frac{d\chi_E(x)}{dx} \tag{4}$$

where χ_{Cu} is the susceptibility of pure Cu.

The first two terms on the right-hand side are calculated from self-consistent fields available in the literature; the third is calculated by assuming that the density of states decreases linearly with x. Allowing for the volume change on alloying,

$$\frac{d\chi_E(x)}{dx} \simeq 2\zeta\chi_E(0) - (5.2 \times 10^{-6})Z \tag{5}$$

where ζ is the proportional change in lattice parameter for unit concentration, and $\chi_E(0)$ is the observed susceptibility of Cu minus χ_{Cu^+}. This expression gives quite good agreement with experiment for a number of Cu alloys, including the two shown here.

The authors critically examine the not dissimilar theory of Friedel,[6] and conclude that it does not give satisfactory agreement with experiment. Although the criterion suggested by Mott[41] and Friedel for the existence of a bound state is satisfied,

$$E_2{}^s < E_1 + H_1 + \tfrac{3}{2}E_F \tag{6}$$

(where $E_2{}^s$ is the second ionization potential of the solute, E_1 is the first ionization potential of the solvent, H_1 is the heat of sublimation of the solvent, and E_F is the Fermi energy), there is no evidence on the basis of the susceptibility results of any bound states. Again, there are no anomalies that would be expected as a result of {200} contact, or spheroidization.

Optical Absorption

The optical absorptivity for Cu-Zn and Cu-Ge alloys was measured by Biondi and Rayne[24] and by Rayne.[25] The displacement of the main optical absorption edge on alloying is shown in Fig. 1*c*. This edge is associated with the transfer of electrons from the top of the *d* band to the Fermi level, and the dashed line is the expected change in terms of rigid *d* and conduction bands. Very recently these data have been reanalyzed by Lettington,[42] who, using the band scheme of Segall,[38] calculated the change in Fermi level relative to the bottom of the conduction band from the optical data. The results agree quite well with what one would expect for a rigid parabolic band with unit effective mass. It appears therefore that in agreement with Friedel, the conduction band moves down in energy in such a way as to keep E_F roughly constant, the *d* band remaining stationary. However, Lettington modified the rigid-band model slightly, by allowing the band gap to vary as a result of interaction between the *d* band and the conduction band, thus moving some way toward the model of Heine and Cohen. Interestingly, the calculations suggest that the limit of α-phase solubility coincides with contact of the Fermi surface with the {200} Brillouin zone faces. The model does not suggest any initial spheroidizing of the surface.

Of the available experiments, the optical absorption data seem to be easiest to interpret in terms of the existing models, and to give the least ambiguous answers. However, the experiment is very difficult, and without implying any defects in the technique of Biondi and Rayne, it would seem highly desirable that the measurements should be repeated, and extended to higher alloy contents.

Soft X-ray Spectra

Clift, Curry, and Thompson[43] have determined the soft X-ray emission M spectrum of a Cu–30 a/o Zn alloy. The Zn emission spectrum was exactly the same in the alloy as in pure Zn. The authors conclude from this that there is a different conduction band associated with each atomic species, and that this is virtually unaffected by alloying. Beeby in effect reached a similar conclusion on theoretical grounds, but for *d* bands only. This is much easier to understand, because of the more localized nature of the *d* states; it is difficult to understand for the extended states in *s* and *p* bands. In contrast, Bearden and Friedman[44] determined the K emission spectra for the whole range of Cu-Zn alloys, and found progressive changes in both the Cu and Zn K bands on alloying. Although it is difficult, as

Clift et al. point out, to compare K and M spectra, it is equally difficult to reconcile these two results.

The change in the width of the K bands is less than that required by the free-electron theory, and the shift on the high-frequency side of the $CuK\beta_2$ is only of the order of the experimental error, suggesting that relative to the K level the Fermi energy changes much less than a rigid-band model would require in agreement with the Friedel model. An extremely interesting point is that the emission spectra are apparently almost insensitive to crystal structure. A comparison of the $CuK\beta_{2,5}$ and $ZnK\beta_{2,5}$ shapes indicates that there are an increased number of high-energy electrons around Cu, and a corresponding decrease around Zn, but not nearly as large as a complete sharing would predict. Again, this is in line with the picture of localization of the additional electrons around the more positively charged solute ions, but does demonstrate also that the shielding is incomplete.

Lattice Parameter

Figure 1*d* shows the variation of lattice parameter with electron concentration for both alloy systems. The data, from a number of investigators, are taken from Pearson;[26] the various results agree very closely. The relationship between lattice parameter and electronic effects has been discussed by Massalski and King.[1] Although a number of attempts have been made to correlate deviations from Vegard's law with valence effects, these have not been satisfactory. Further, there is some doubt as to the exact value of the atomic radius of the solute atom for a comparison with Vegard's law. Pauling calculates the radius for a coordination number of 12; and his radii are[45]

$$Zn_p{}^{12} = 1.339$$

$$Ge_p{}^{12} = 1.444$$

For comparison, the closest distance of approach of Zn atoms in pure Zn corresponds to a radius

$$Zn_a = 1.3322$$

A number of authors have suggested using atomic volume as a parameter, and Rudman[46] has recently tabulated atomic volumes for a large number of metal atoms; from this, an equivalent radius may be calculated:

$$Zn_R = 1.3902$$

$$Ge_R = 1.4708$$

Yoshioka and Beck[47] have recently discussed the apparent volume of Ge in a number of structures, and note that it is much smaller in combination with transition elements. Its apparent value in an fcc Co solution is 12.8 A^3, whereas in an fcc Al-based solid solution its value is 18.7 A^3.

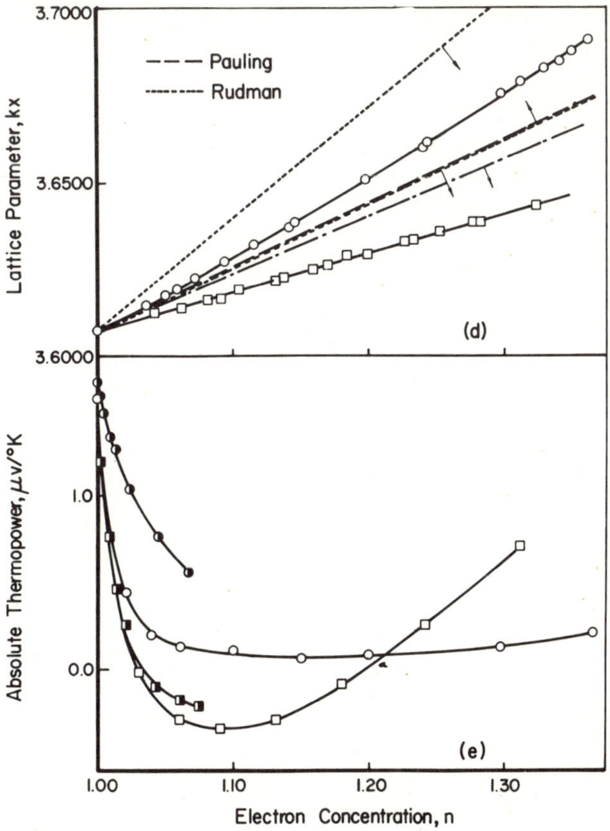

Fig. 1 *(continued)* Further properties of Cu-based alloys. (*d*) Various sources, tabulated by Pearson.[26] (*e*) Empty circles and squares, from Köster and Rave[28]; half-filled circles and squares, from Crisp and Henry.[27]

Finally, Rudman[48] has examined the deviations to be expected from Vegard's law on the basis of elastic constraints; although the elasticity of metallic Ge is not known, this criterion predicts a negative deviation from Vegard's law for both solutes. This is in agreement with experiment, provided one uses the Rudman radius (dotted line) and not the Pauling radius (dashed line).

Absolute Thermopower

The absolute thermopower is given by

$$S = \frac{\pi^2 k_B{}^2 T}{3e} \left\{ \frac{\partial \ln \sigma(E)}{\partial E} \right\}_{E=E_F} \tag{7}$$

However, this expression takes no account of phonon drag, and, as Ziman[49,50] points out, this may give a contribution 10 to 100 times the purely electronic value.

The experimental values shown in Fig. 1*e* are from Crisp and Henry,[27] who determined the values at 290°K, and Köster and Rave,[28] who determined them at 296°K. It can be seen that both curves have the same shape, but the absolute values are rather different, particularly for the Cu-Zn alloys. Crisp, Henry, and Schroeder[51] had earlier determined the thermoelectric powers of a number of alloys over a wide range of temperature. At low temperatures—around 60°K—there is a very marked phonon-drag peak, but at room temperature Crisp and Henry conclude that the phonon-drag contribution is of the order of 0.1 μv/°K in pure Cu and negligibly small in the alloys. Earlier data (of Blatt and Kropschot[52] and of Norbury[53]) agree approximately with the results of Crisp, but lie rather higher, an effect which Crisp identifies with ferromagnetic impurities. The materials used in this investigation were carefully prepared from very high purity materials, and probably contained less than 1 ppm Fe. One cannot therefore attribute the difference between the results shown in Fig. 1*e* to impurities.

Henry and Schroeder[54] found that at high Zn contents the thermopower was very sensitive to thermal history: quenched alloys had smaller thermopowers than slow-cooled ones. This effect was attributed to short-range ordering, following Clareborough[55] et al. Köster and Rave found that specimens cold-worked 90% gave *higher* thermopowers than recrystallized specimens. For comparison, according to Henry and Schroeder, Cu–10.22 a/o Zn had a thermopower of 0.32 μv/°K when slow-cooled, and 0.24 μv/°K when quenched, from 575°C. According to Köster and Rave,[28] an alloy of this composition had a thermopower of 0.36 μv/°K cold-worked 90%, and 0.08 μv/°K recrystallized. Since cold working should also destroy short-range order, it is clear that this explanation of the results must be treated with some suspicion.

Both these effects were small in dilute alloys, increasing until the Zn content reached about 10 a/o, and then remaining sensibly constant. By contrast, the alloys containing Ge were much less sensitive to heat treatment and to cold working.

According to the simple theory, the thermopower should vary linearly with composition. Crisp's curves are only slightly different from straight lines for dilute solutions, and since the corresponding curves for Ag-based alloys are rectilinear, he attributes the curvature to the small phonon-drag term. However, this plainly cannot explain the results of Köster and Rave. These points will be discussed in more detail in the section on Hall effect.

Stacking-fault Energy

From the point of view of a discussion of phase stability, the stacking-fault energy is an extremely interesting parameter. Since a stacking fault in a face-centered cubic structure may be regarded as a thin layer of close-packed hexagonal structure, the stacking-fault energy can be regarded (with some reservation) as a measure of the energy difference between the fcc and cph structures. Unfortunately, it is not always easy to measure the stacking-fault energy. The most direct method is by measuring the geometry of extended nodes in the electron microscope, after Whelan, and Howie and Swann.[56] This technique has been refined by Thornton[29] et al. and more recently by Lothe[57] et al. In addition, Seeger[58] has estimated the stacking-fault energy from the temperature and strain-rate dependence of the stress at the onset of stage III deformation. Smallman[59] et al. have estimated it from rolling textures, and a number of authors[60] have determined it (or more properly the stacking-fault probability) from X-ray line-shift measurements. Unfortunately, there is a wide discrepancy among the absolute values obtained, although the general variation of the energy with composition is generally the same. The data points in Fig. 1*f* are taken from Thornton et al.; Massalski's X-ray data[60] are in good agreement with these values. The value for pure Cu is from Dillamore,[59]

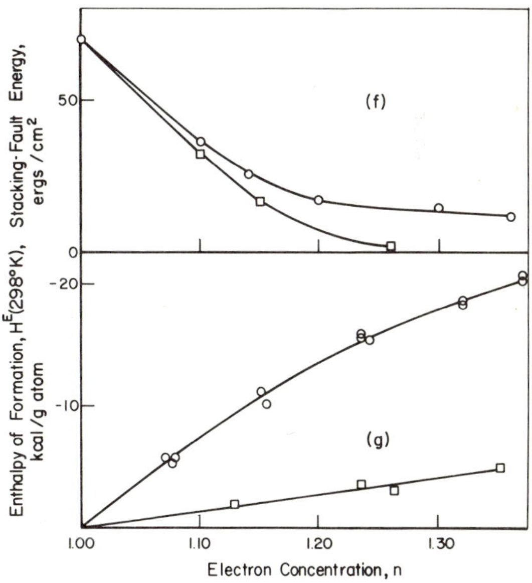

Fig. 1 (*continued*) Further properties of Cu-based alloys. (*f*) Circles and squares, from Thornton et al.[29] (*g*) Circles and squares, from Kleppa and King.[30]

although this is probably too high. It is interesting to note that the energy approaches zero at the phase boundary for Cu-Ge, but remains positive for Cu-Zn. This may be correlated with the fact that the neighboring phase in the Cu-Ge system is the cph ζ phase, whereas in the Cu-Zn system it is the bcc β phase; however, the difference in energy between this and a hexagonal phase is slight, since β-brass can be transformed martensitically to a hexagonal phase at low temperatures.

Accepting that this quantity is in fact a direct measure of the stability of the fcc phase relative to an hcp phase of the same composition,[61] it is interesting to note that, contrary to the predictions of the Jones–Hume-Rothery theory, there is no rapid change in relative stability in the vicinity of the phase boundary. In particular, it would be anticipated that the energy-versus-composition curve should be convex with respect to the composition axis; but this is not so. In this context it is interesting to note that Vassamillet and Massalski[62] have studied stacking-fault energy in isoelectronic noble-metal alloys. The energy falls up to about 1 a/o of solute, but thereafter remains substantially constant; it appears therefore that size and disorder factors are relatively unimportant for more concentrated alloys.

Enthalpy of Formation

Figure 1g shows thermodynamic data from Kleppa and King.[30] Orr and Argent[63] have recently also determined the heat of formation, and have reviewed the data from the literature. In the main, their results agree with those shown in the figure, but there is some evidence of an abrupt drop in H^E (increase in the sense of 1g) in the range 31 to 36 a/o Zn. However, the authors doubt that this is real. The free-energy data of Argent and Wakeman[64] show no such change, but vary smoothly up to the phase boundary.

Short-range Order

Because of the closeness of the elements in the periodic table, it is difficult to measure short-range order by conventional X-ray techniques. However, there is strong evidence of short-range order in α-brasses containing more than about 8 a/o Zn. Clareborough, Hargreaves, and Loretto[55] observed specific-heat anomalies and associated anomalies in electrical resistivity, and Childs and Le Claire[65] observed an internal friction peak at 573 to 623°K for alloys containing more than 8 a/o Zn, which they attributed to short-range order. Mention has already been made of the sensitivity of the absolute thermopower to heat treatment and to cold working, which the respective authors also attribute to short-range order. Keating[66] attempted to study order in α-brass using neutron diffraction, but was unable to detect any significant amount.

For Cu-Ge the situation is less clear. The thermopower (and also the electrical resistivity) is far less sensitive to heat treatment and to cold work than for Cu-Zn, but this may well be a consequence of the lower absolute concentrations involved; one would anticipate if anything a stronger tendency to ordering in Cu-Ge than in Cu-Zn. Nakajima[67] et al. observed that the stacking-fault energy of Cu-Ge alloys increased on low-temperature annealing, which they attributed to the formation of domains of short-range order on the dislocations; the effect was not due to Suzuki locking.

Elasticity

Few data are available on the elasticity of Cu alloys, and apparently no single-crystal data. Figure 2 shows results for a number of Cu alloys from Neighbours and Smith,[68] expressed in terms of a parameter Z_E. This parameter is calculated by assuming that the change in the elastic constants on alloying can be attributed to ionic and electrostatic factors. Writing $C = C_{44}$ and $C' = (C_{11} - C_{12})/2$, the changes on alloying may then be written

$$\Delta C = C_E(Z_E^2 - 1) + C_i \alpha x$$

$$\Delta C' = C'_E(Z_E^2 - 1) + C'_i \alpha x \tag{8}$$

where the subscripts E and i refer to the electrostatic and ionic contributions, respectively; α is a constant, and x is the atomic concentration of

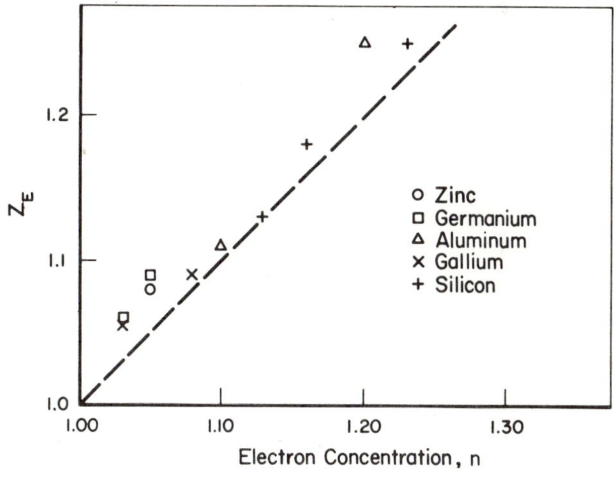

Fig. 2 Neighbours-Smith parameter Z_E versus electron concentration n. Z_E is calculated from elastic data, and in a simple model $Z_E = n$ (shown as a dashed line).

solute. A simple free-electron theory would result in

$$Z_E = n$$

shown as the dashed line in Fig. 2. A metal in which the excess charge is shielded should give negative deviations from this line; however, as can be seen, all the points lie on the positive side. In view of the weight of evidence in support of a "shielded solute" model, it must be concluded that the theory is oversimplified. Plainly it would be valuable to reexamine the elasticity problem, both experimentally and theoretically.

HALL EFFECT AND CONDUCTIVITY

These classical transport properties have been left until the end, since they lead to the second part of the paper. Figure 3*a* shows the results of a number of workers. The agreement is very good, and Köster[28] has furthermore shown that the Hall coefficient in the Cu-Zn system is not very sensitive to cold work. In the Cu-Ge system the Hall coefficient of the two most concentrated alloys is increased by 90% cold work; for example, an alloy with 10.3 a/o Ge ($n = 1.31$) has a Hall constant of -4.07×10^{-5} cm^3/A-sec in a recrystallized condition, and -3.72×10^{-5} in the cold-worked condition. In more dilute alloys, however, the effect was negligible.

Figure 3*b* shows the reciprocal Hall coefficient as the "effective number of electrons per atom" n^*:

$$n^* = \frac{1}{NecR_H}$$

where N is the number of atoms per unit volume, and c is the velocity of light.

Figure 3*c* shows the conductivity at room temperature as a function of n in the two alloy systems. Again, there is fairly good agreement among the various investigators; as might be expected, Ge is a more effective scatterer than Zn.

Coles[71] first suggested that the Hall effect might provide useful information on the electronic structure of alloys, and replotted the early data of Dorfman and Zhukova[72] for Ag alloys as n^* versus n. A similar plot with results for Cu-, Ag-, and Au-based alloys, taken from Ziman,[49] is shown in Fig. 4. The results show that instead of lying along the line $n = n^*$ as the free-electron theory would predict, or on a single smooth curve as a rigid-band model would suggest, the curves are all very different. There is a marked cusp at the pure metal composition, and (at least on the $n > 1$ side) a marked minimum in most of the curves. Clearly, this is very difficult to understand in terms of a simple band theory. Cooper and Raimes[73] introduced the idea that it was necessary to consider a relaxation process composed of a phonon relaxation and an impurity relaxation, both of which might be anisotropic with different anisotropies. A calculation based on this concept, together with a picture of the Fermi surface as a distorted sphere, produced curves of the right shape, but quantitatively the agreement was not good. Ziman,[50] recognizing that the Fermi surface in the noble metals already touched the zone walls, reexamined the problem.

First, the Fermi surface is approximated by eight cones, and a parameter u is defined which is related to the magnitude of the energy gap across the $\{111\}$ Brillouin zone faces. For a spherical Fermi surface, therefore, $u = 0$; for Cu, $u = 0.20$ if the band gap is 3.5 ev. Now, if the relaxation time is regarded as isotropic, a family of curves of n^* versus n can be calculated for different values of u. These do not, however, exhibit a cusp at $n = 1$, nor do they exhibit a minimum. However, they do show that the

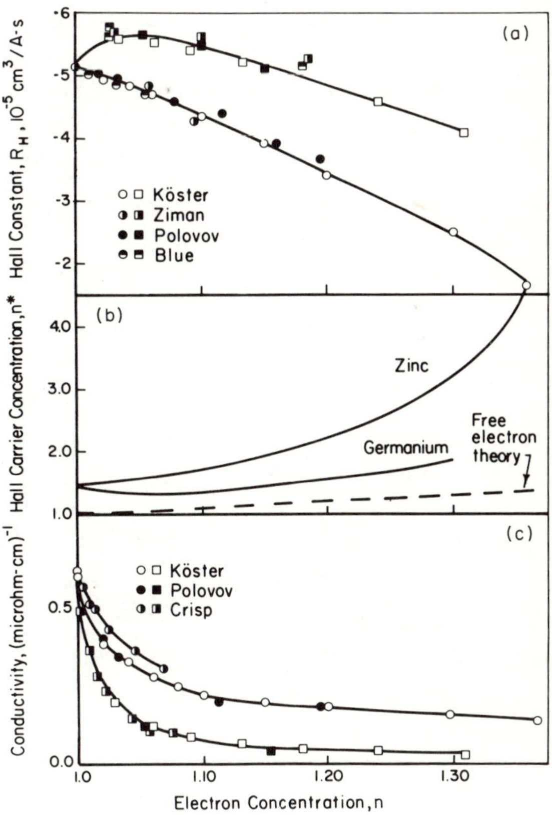

Fig. 3 Hall effect and resistivity data for Cu-Zn and Cu-Ge solid solutions.

Hall coefficient will be rather sensitive to the detailed shape of the Fermi surface, and in the case of Cu alloys is more sensitive to the necks than to the belly of the surface. The calculated value of n^* for the pure metal is 2.05, which compares rather poorly with the experimental value of $\sim$1.45. These disagreements demonstrate that one cannot in fact regard the phonon relaxation as being isotropic. Ziman suggested that the relaxation time on the necks is significantly different from that on the belly, and that, in fact, if τ_n is greater than τ_b, n^* will be increased. If the functional form of the impurity relaxation differs from that due to the phonons, there will be a rapid change in n^* as the alloy content increases and the residual resistance due to the impurity becomes dominant over the phonon resistance. Thereafter, n^* varies along a line characteristic of the new (impurity) relaxation as the Fermi surface swells or contracts while electrons are added to or removed from the alloy. Ziman comments that the minima in the curves shown in Fig. 4 come at about the point where the impurity resistivity equals the phonon term.

This theory has been widely quoted in the last few years in discussions of Hall-effect results, but there has been no quantitative development; the quantitative theory of Cooper and Raimes cannot be modified to be compatible with the real Fermi surface. Hagmann and Ricker[74] further developed the Ziman eight-cone model, assuming (as Ziman suggested) an

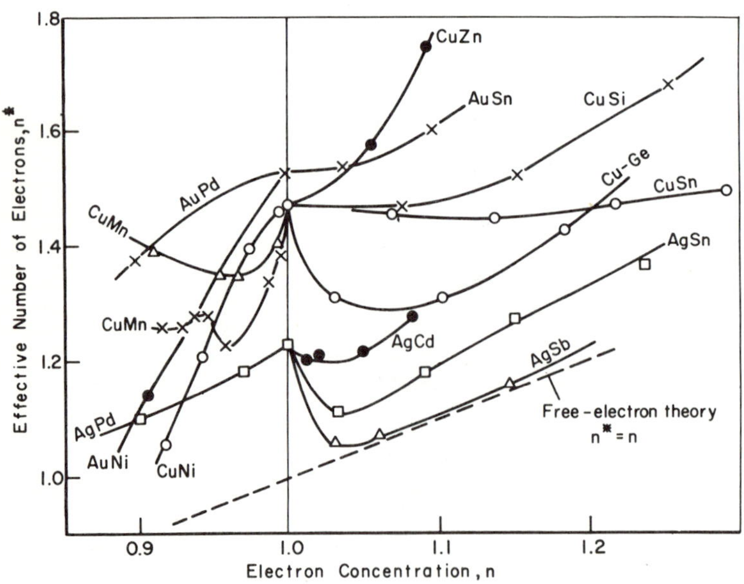

Fig. 4 The effective number of free electrons in the noble-metal alloys. (*After Ziman*[49].)

anisotropic impurity relaxation time, and produced a "two-band" type result:

$$R_H = -\frac{1}{eN}\left(\frac{m_a}{m_1}\frac{n_1}{n}\frac{\langle\tau_1^2\rangle}{\langle\tau\rangle^2} - \frac{m_a}{m_2}\frac{n_2}{n}\frac{\langle\tau_2^2\rangle}{\langle\tau\rangle^2}\right) \tag{9}$$

Here, the angular brackets indicate averages over the Fermi surface, and the subscripts 1 and 2 refer to the two bands, so that $\langle\tau_1^2\rangle/\langle\tau\rangle^2$ is a measure of the anisotropy of the relaxation in the first band. (It should be recalled that this differs from the two-band model in detail, since the two relaxation times are associated with impurity and phonon relaxation.) This may be rewritten as

$$R_H = -R_1t_1 + R_2t_2 \tag{10}$$

where the terms t_1 and t_2 may be regarded as "anisotropy factors." Above the Debye temperature, thermal scattering is approximately isotropic.

Again, this model suggests a close connection between the Hall coefficient and resistivity, a point that had originally been made by Coles. In fact, Hagmann and Ricker show that there is a linear relationship between the change in n^* and the change in the resistivity ρ for an electron concentration $n = 1.05$ in Ag alloys. This lends weight to the hypothesis that the initial fall in the Hall coefficient is due to impurity scattering.

Hurd[75] has analyzed the Hall-effect data for the Au-Ag system obtained by Köster[76] et al. The Hall effect is lower for Ag than for Au, which is in agreement with the observation that the ratio of neck-to-belly radius is 1.9% for Ag and 3.1% for Au. Hurd defines a contribution to n^* due to impurity scattering as

$$n_i = n^* - n_0 = n^*\frac{R_0 - R}{R_0} \tag{11}$$

where R_0 is the Hall coefficient for the pure solvent. Apparently, n_i correlates well with the impurity scattering as measured by the electrical resistivity; but it was not possible to allow for lattice-parameter changes, since the data were not available.

That this result is not due to an anomalous variation in the Fermi surface in Ag-Au alloys can be demonstrated by reference to the work of McAlister[77] et al. who measured Faraday effect in Ag-Au alloys and concluded that the Fermi surface varied smoothly from one end of the binary system to the other.

Heine[78] has examined Hurd's empirical theory, and notes that impurity scattering should increase with concentration, but retains its own anisotropy; and since it is the anisotropy which controls the value of the Hall coefficient, once the impurity scattering dominates the phonon scattering, the Hall effect should show no variation with composition. However, since $\tau_{n,i}/\tau_{b,i} \cong 8$, where the suffix i refers to impurity scattering, it is possible that impurity scattering may dominate on the belly, but phonon scattering may still dominate at the necks, so that in the alloy

$$\tau(\text{belly}) \cong \tau_{b,i}\alpha[C(1 - C)]^{-1}$$

$$\tau(\text{neck}) \cong \tau_{n,p}\alpha \frac{1}{T}$$

(The suffix p refers to phonon scattering.) Hence,

$$\left(\frac{\tau_n}{\tau_b}\right)_{\text{alloy}} \approx \alpha C \frac{1 - C}{T} \tag{12}$$

which is roughly in agreement with the experimental variation of R.

However R actually depends on τ^2. Heine rationalizes this in terms of the alteration in curvature of the necks as alloying increases, and a semi-quantitative theory on this basis gives approximate agreement with Hurd's curves.

Taylor[79] has calculated the relaxation time in the noble metals, and has plotted the results for various directions in the Brillouin zone. At low temperatures ($T/\Theta \sim 0.1$) there is considerable anisotropy, with the relaxation time along [100] being between two and three times that along [111]; however, as $T \rightarrow \Theta$, the anisotropy becomes less and is negligible at $T = 0.7\Theta$. This is qualitatively in agreement with the predictions of Ziman. It is interesting to note (comparing this result to the experimental result of Deaton and Gavenda[39]) that phonon and impurity relaxations have opposite anisotropies. Polovov and Sumin[69] also discussed their experiments in terms of the Ziman theory. They comment that the anisotropy of impurity scattering in Cu-Zn alloys is considerably less than that of phonon scattering, and instance Frank's[80] results on the temperature dependence of the Hall coefficient in α-CuZn in support. In addition, they suggest that the anisotropy of impurity scattering increaset as the valence difference between the solute and solvent increases.

Experimentally therefore both Ziman and Polovov must invoke a significant anisotropy in phonon scattering to account for the room-temperature results, and even to explain Frank's results at 600°C. However, qualitative reasoning would suggest, and Taylor's calculations confirm, that any anisotropy in phonon-electron scattering vanishes as $T \rightarrow \Theta$. For Cu, $\Theta \sim 340$°K, so that room temperature is 0.88Θ, and the phonon scattering should be entirely isotropic.

The Hall-effect data for the Cu-Zn and Cu-Ge systems do not suggest any discontinuous change in the properties of the Fermi surface. For example, there is good reason to believe that the relaxation time is significantly different in the neck region of the Fermi surface compared with that in the belly. It would appear therefore that either the elimination of the necks as suggested by Heine and Cohen or the appearance of new necks on the {100} faces as suggested by Hume-Rothery and Roaf would result in an important change in the anisotropy of the relaxation time averaged over the Fermi surface, with an attendant alteration in the Hall coefficient. Further, either of these events would result in quite sharp changes in the average *curvature* of the Fermi surface, and this also should affect the Hall coefficient. The data would seem to indicate (at least for Cu-Zn alloys) that there is no *sharp* change in the overall properties of the Fermi surface.

The importance of short-range order has been mentioned a number of times already. In an early study, Flanagan and Averbach[81] suggested that the deviations from the predictions of the free-electron theory in their studies of the Hall effect in the alloy systems Cu-Ni, Au-Ag, and Cu-Au could be interpreted in terms of short-range order. This, of course, was prior to the studies that have resulted in the Ziman theory. The relative insensitivity of the Hall effect to cold work revealed by Köster's study suggests that in fact this is not so; there is reasonably good evidence that there is short-range ordering in Cu-Zn alloys, and that this can be partially destroyed by cold work; however, the Hall effect is almost completely unchanged.

Hall-effect Studies in Polyvalent Alloys

Although the idea of using the reciprocal Hall constant as an effective electron concentration was reasonable when it was applied to alloy systems in which it was thought that the Fermi surface was a distorted sphere lying in a single zone, it becomes less obviously suitable for an alloy in which the Fermi surface does make contact with the zone walls. For example, for a spherical or distorted spherical Fermi surface where the curvature is everywhere positive, the quasi-particle wave packets are composed of electron-like wave functions; and to talk of an "effective number of electrons" is not unreasonable. Similarly, for a nearly full zone, with negative curvature everywhere, one may speak of an effective number of holes. As the band fills, the situation becomes less clear; whereas quasi-particles on the belly in Cu, for example, are plainly electron-like, those at the necks are neither hole nor electron-like, being composed of wave functions whose curvature differs in sign. Plainly, as the zone fills, one will pass through a point at which the Hall coefficient vanishes, since a nearly empty zone gives a negative Hall coefficient and a nearly full one a positive Hall coefficient. This would give a value of $n^* = \infty$. This trend can clearly be seen in Fig. 3*b* for Cu-Zn alloys, where n^* approaches 4.5 at the zone boundary; and it becomes even more noticeable in polyvalent metals, where the Fermi surface may lie in a number of zones. Plainly, the Hall coefficient itself is more accurately related to the real physical situation than the "effective number of carriers" n^*. Relatively few data are available for the Hall coefficient in polyvalent metal alloys. Most work has been done for Mg based alloys, but these are hexagonal close-packed, so that there are *two* components of the resistivity and the Hall effect. Determination of these two components requires single-crystal specimens, although some idea of their relative magnitude can be obtained from textured polycrystals (Lane, Huglin, and Stringer[82]). In fact, all measurements have been made on polycrystals for which no texture determinations were made. The datum point is thus some unknown average of the two components. However, it is probable that the effect of alloying on the texture developed by a given fabrication procedure is not great, and that the observed variation in properties is due to the effect of the solute on the Fermi surface and the relaxation processes, and not to texture effects.

Mg-based Alloys. Raynor[83] made accurate lattice-parameter measurements of a number of Mg primary solid solutions, and observed what he regarded as a discontinuous change in $d(c/a)/dn$ at $n \cong 2.005$ where c/a is the axial ratio of the hexagonal lattice. Jones[84] had suggested that a discontinuity of this type might be expected if the Fermi surface overlapped the (0002) faces of the Brillouin zone (more properly, the Jones zone). However, subsequent study has cast some doubt on the reality of this discontinuity. Figure 5 shows c/a versus n for a number of alloys. The curves are drawn through experimental points from a number of sources, tabulated by Busk.[85] If (following Massalski and King) one regards the axial ratio as a measure of the electron concentration, it would appear that some, at least, of the solutes are incompletely ionized or have some electrons in bound states. The difference, however, is not great.

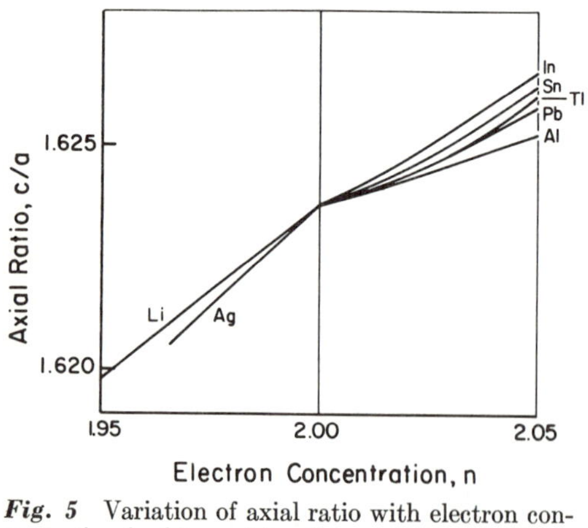

Fig. 5 Variation of axial ratio with electron concentration for binary Mg alloys. (*After Busk*[85].)

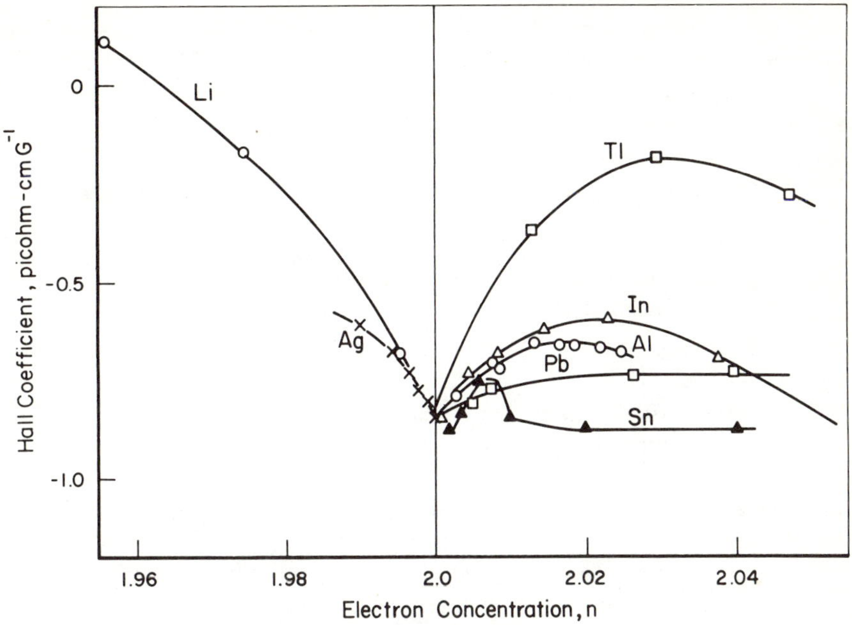

Fig. 6 Hall effect in binary Mg alloys. (*From Salkovitz*[86].)

The Hall effect was measured by Salkovitz, Schindler, and Kammer.[86] They interpreted the data in terms of Raynor's theory, but as Fig. 6 shows, they are more easily interpreted in terms of a set of curves similar to those obtained for group I metals. These authors also determined the resistivity and thermopower of the alloys. All the elements increase the resistivity linearly with atomic percent of solute; Cd, Li, and Ag have approximately the same effect on the resistivity; Al and In rather more; Tl more; and Sn and Pb the most. The effect on the thermoelectric power is in the reverse order.

It is interesting to note that this order bears little relation to the Hall effect; notably Sn and Pb produce very similar effects on the resistivity and the thermopower, but quite different changes in the Hall effect.

Hedgcock and Muir,[87] using some of the same alloys as Salkovitz et al., studied the temperature dependence of the resistivity, and concluded that an exact agreement with Matthiessen's law could be obtained provided an appropriate value of the Gruneisen constant θ was chosen for each alloy. For example, θ for pure Mg is 340°K; for 0.29 a/o Al, θ = 328°K; for 0.80 a/o Al, θ = 314°K; and for 2.41 a/o Al, θ = 300°K. In general, the Gruneisen θ should be equal to the Debye temperature Θ, but experimentally it is often a little different. Rayne et al. determined from specific-heat data that Θ for Cu-Zn and Cu-Ge fell on alloying; this would appear to imply that the most important characteristic effect of the alloy elements is their effect on the electron-phonon relaxation. This is slightly different from Ziman's suggestion, that the electron-phonon relaxation process remains substantially unchanged as the alloy content increases. However, it is also clear that the resistivity data cannot be used to explain the Hall-effect results.

Elasticity data for Mg-alloy single crystals have been obtained by Long and Smith,[88] using an acoustic technique at room temperature. Changes are evident only in C_{44}, which decreases slowly as n increases. Theoretical reasoning suggests that $(C_{11} - C_{12})/2$ should show an abrupt change of slope versus n when overlap takes place (Leigh[89]), but there is no such cusp. This suggests that Raynor's suggested overlap at $n \simeq 2.005$ does not in fact take place.

Zn Alloys. Stringer and Lane[90] have measured Hall effect in three Zn-based alloys; the results are shown in Fig. 7. The metallurgical condition of the alloys was carefully controlled, and pole figure determinations were made on the alloys; it is certain in this case that none of the effects observed can be attributed to orientation differences. Although the experiments did not specifically determine resistivity, some approximate resistivity data were obtained, and are shown in Fig. 8. Lattice-parameter measurements on the alloys agreed well with the data in the literature,[91,92] shown in Fig. 9.

It is clear that the electronic contributions of Ag and Cu, as revealed by the axial ratio data, are very similar. However, the Hall-effect results are very different, with the Cu alloys showing a much more rapid initial change,

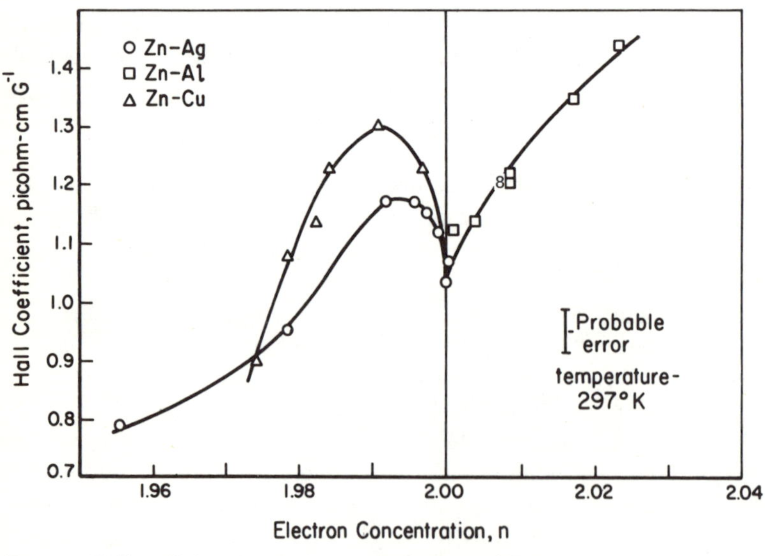

Fig. 7 Hall coefficient in solution-treated alloys of Zn[90].

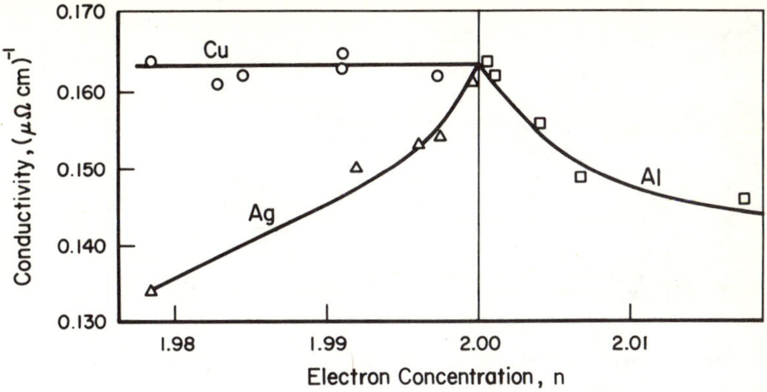

Fig. 8 Conductivity of solution-treated Zn-based alloys at room temperature[90].

a much deeper maximum, and a sharper fall after the minimum. On the basis of Ziman's theory, this would imply a greater scattering by the Cu atoms, and a greater anisotropy in the solute scattering. However, the resistivity data suggest the reverse; apparently Ag is a much more effective scatterer in Zn (as one might expect) than Cu is.

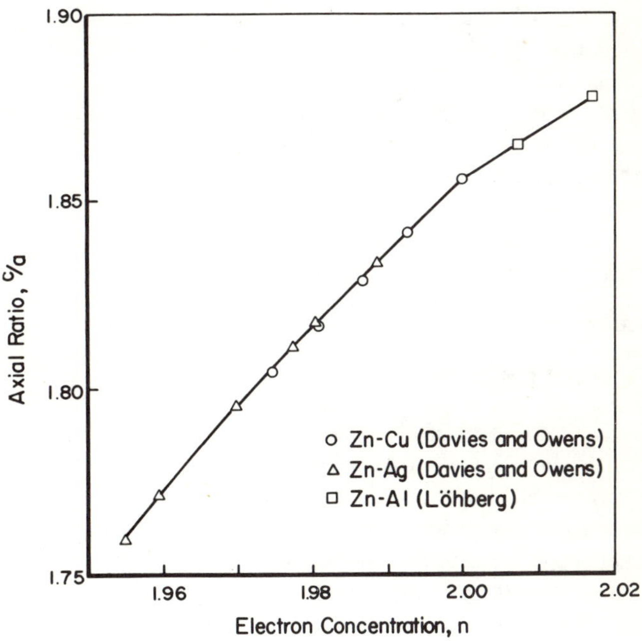

Fig. 9 Axial ratios of primary solid solutions of Zn-based alloys.

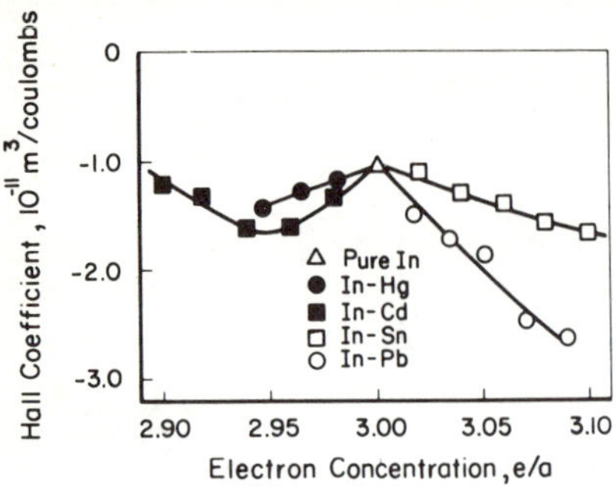

Fig. 10 Hall coefficient in In-based alloys. (*After Yonemitsu et al.*[93])

In Alloys. Recently, Yonemitsu et al.[93] have studied transport properties in primary In solid solutions. Their Hall-effect results are shown in Fig. 10. The Hall effect in these alloys is very small, and the changes produced by alloying are not great. The curves, however, have the same general appearance as the earlier data, although only the In-Cd curve has a minimum. All the solutes produce very little effect on the resistivity, and once again there is no evident correlation between the effect on the resistivity and the change in the Hall effect. Thermopower measurements were also made, with results fairly similar to the Hall data. Lead produced the largest change, and the In-Cd curve has a minimum.

Axial ratio data are also available for these alloy systems; Pb produces a very small increase in axial ratio, Sn a significantly larger increase. Both Hg and Cd reduce the axial ratio, with the effect of Cd being significantly greater.

Yonemitsu et al. discuss their results in terms of a kind of Ziman theory, in which one divides the Hall effect into an "electronic" effect and a "scattering" effect. The electronic effect decreases with increasing n, in a manner similar to Ziman's curves of n^* versus n for an intermediate value of u. The scattering effect decreases linearly with $|n - 3|$. Summing these two effects gives the sort of curves one observes: a cusp at the pure metal composition, and a minimum on one side.

CONCLUSIONS

The stability of electron compounds, which correlates in so many cases with the electron-to-atom ratio, has been explained in terms of a Fermi surface–Brillouin zone interaction which produces a rapid increase in the electron energy immediately before the phase boundary. Data have been collected for a large number of property variations with electron-to-atom ratio for the primary α-solid solution of the binary systems Cu-Zn and Cu-Ge, which are regarded as representative examples of this type of phase.

The results very strongly suggest that there is *no* discontinuous change in the shape or properties of the Fermi surface anywhere in the α-phase field. Specifically, it seems extremely unlikely that there is any "spheroidizing" of the surface sufficient to cause contact between it and the {111} faces of the Brillouin zone to be lost.

None of the models for metal alloys seems entirely compatible with the data, but some modification of the Mott-Friedel picture seems best; the solute atoms are partially screened by localized unbound electrons, and the bottom of the band is displaced in such a way that the Fermi energy increases considerably more slowly with n than the rigid-band model would suggest. It is not possible to estimate whether there are any other changes in the band structure, but it seems likely that the alteration in density of states at the Fermi level is not large.

The difficulty of performing and interpreting many of the experiments is obvious. Of those available, one of the more reproducible, and less sensitive to metallurgical and other variables, is the Hall effect. The current view of the Hall effect in noble-metal alloys is given, and some defects in the model are pointed out. Finally, data are presented for Hall-effect measurements in polyvalent metals, and the results are compared with a qualitative extension of the existing theory. This strongly suggests that in fact some important modifications may have to be made.

REFERENCES

1. The early work has been reviewed in some detail. See for example T. B. Massalski and H. W. King: *Progr. Mater. Sci.* **10**(1):1 (1961), which also describes Massalski's approach. A representative recent paper of Beck's is D. I. Bardos, R. K. Malik, F. X. Spiegel, and P. A. Beck: *Trans. AIME* **236**:40 (1966).
2. Pippard, A. B.: *Phil. Trans. Roy. Soc.* **A250**:323 (1957).
3. Olson, R., and S. Rodriguez: *Phys. Rev.* **108**:1212 (1957).
4. Cohen, M. H., and V. Heine: *Advan. Phys.* **7**:395 (1958).
5. See for example D. Shoenberg: In W. A. Harrison and M. B. Webb (eds.), "The Fermi Surface," p. 74, John Wiley & Sons, Inc., New York, 1960.
6. Friedel, J.: *Advan. Phys.* **3**:446 (1954).
7. Mott, N. F.: *Proc. Cambridge Phil. Soc.* **32**:281 (1936).
8. Raimes, S.: *J. Phys. Radium* **23**:639 (1962).
9. Heine, V.: In "The Fermi Surface," *op. cit.*, p. 279.
10. Makinson, R. E., and A. P. Roberts: *Australian J. Phys.* **13**:437 (1960).
11. Agacy, R. L., and R. E. Borland: *Proc. Phys. Soc.* **84**:1017 (1964).
12. Luttinger, J. M.: *Philips Res. Rept.* **6**:303 (1951).
13. See for example S. F. Edwards: *Proc. Roy. Soc.* **A267**:139 (1962).
14. See for example H. Jones: *Proc. Roy. Soc.* **A285**:461 (1964).
15. Jones, H.: In "Quantum Theory of Atoms, Molecules, and the Solid State," to be published by Academic Press Inc., N. Y.
16. Beeby, J. L.: *Phys. Rev.* **135**:A130 (1964).
17. Flinn, P. A.: *Phys. Rev.* **104**:350 (1956).
18. Smoluchowski, S.: *Phys. Rev.* **84**:511 (1951).
19. Rayne, J. A.: *Phys. Rev.* **108**:22 (1957).
20. Rayne, J. A.: *Phys. Rev.* **110**:606 (1958).

21. Veal, B. W., and J. A. Rayne: *Phys. Rev.* **130**:2156 (1963).
22. Isaacs, L. L., and T. B. Massalski: *Phys. Rev.* **138**:A134 (1965).
23. Henry, W. G., and J. L. Rogers: *Phil. Mag.* **1**:237 (1956).
24. Biondi, M. A., and J. A. Rayne: *Phys. Rev.* **115**:1522 (1959).
25. Rayne, J. A.: *Phys. Rev.* **121**:456 (1961).
26. Pearson, W. B.: "A Handbook of Lattice Spacings of Metals and Alloys," pp. 579, 619, Pergamon Press, London, 1958.
27. Crisp, R. S., and W. G. Henry: *Phil. Mag.* **10, 11**:841 (1965).
28. Koster, W., and H. P. Rave: *Z. Metallk.* **55**:750 (1964).
29. Thornton, P. R., T. E. Mitchell, and P. B. Hirsch: *Phil. Mag.* **7**:1349 (1962).
30. Kleppa, O. J., and R. C. King: *Acta Met.* **10**:1163 (1962).
31. Jones, H.: *Proc. Roy. Soc.* **A240**:321 (1957).
32. Buckingham, M. J., and M. R. Schafroth: *Proc. Phys. Soc.* **67**:828 (1954).
33. Jones, H.: *Phys. Rev.* **134**:A958 (1964).
34. Boardman, A. D., and N. H. March: *J. Phys. Chem. Solids* **25**:1435 (1964).
35. Heine, V.: In "The Fermi Surface," *op. cit.*, p. 287.
36. Rayne, J. A.: In *ibid.*, p. 316.
37. Hume-Rothery, W., and D. Roaf; *Phil. Mag.* **6**:55 (1961).
38. Segall, B.: *Phys. Rev.* **125**:109 (1962).
39. Deaton, B. C., and J. D. Gavenda: *Phys. Rev.* **129**:1990 (1963).
40. Kip, A. F., D. N. Langenberg, and T. W. Moore: *Phys. Rev.* **124**:359 (1961).
41. Mott, N. F.: *Progr. Metal Phys.* **3**:102 (1952).
42. Lettington, A. H.: *Phil. Mag.* **11**:863 (1965).
43. Clift, J., C. Curry, and B. J. Thompson: *Phil. Mag.* **8**:639 (1963).
44. Bearden, J. A., and H. Friedman: *Phys. Rev.* **58**:387 (1940).
45. Pauling, L. "The Nature of the Chemical Bond," p. 403, Cornell University Press, Ithaca, N. Y., 1960.
46. Rudman, P. S.: *Trans. AIME* **233**:864 (1965).
47. Yoshioka, T., and P. A. Beck: *Trans. AIME* **233**:1788 (1965).
48. Rudman, P. S.: *Trans. AIME* **233**:872 (1965).
49. Ziman, J. M.: In "The Fermi Surface," *op. cit.*, p. 296.
50. Ziman, J. M.: *Advan. Phys.* **10**:1 (1961).
51. Crisp, R. S., W. G. Henry, and P. A. Schroeder: *Phil. Mag.* **10**:553 (1964).
52. Blatt, F. J., and R. H. Kropschot: *Phys. Rev.* **118**:480 (1960).
53. Norbury, A. L.: *Phil. Mag.* **2**:1188 (1926).
54. Henry, W. G., and P. A. Schroeder: *Can. J. Phys.* **41**:1076 (1963).
55. Clareborough, L. M., M. E. Hargreaves, and M. H. Loretto: *Proc. Roy. Soc.* **A261** 500 (1961).
56. Howie, A., and P. R. Swann: *Phil. Mag.* **6**:1215 (1961).
57. Jossang, T., M. J. Stowell, J. P. Hirth, and J. Lothe: *Acta Met.* **13**:279 (1965).
58. Seeger, A., R. Berner, and H. Wolf: *Z. Phys.* **155**:247 (1959).
59. Dillamore, I. L., R. E. Smallman, and W. T. Roberts: *Phil. Mag.* **9**:517 (1964).
60. See for example L. F. Vassamillet and T. B. Massalski: *J. Appl. Phys.* **34**:3398 (1963).
61. de Wit, R., and R. E. Howard: *Acta Met.* **13**:655 (1965).
62. Vassamillet, L. F., and T. B. Massalski: *J. Appl. Phys.* **34**:3402 (1963).
63. Orr, R. L., and B. B. Argent: *Trans. Faraday Soc.* **61**:2126 (1965).
64. Argent, B. B., and D. W. Wakeman: *Trans Faraday Soc.* **54**:799 (1958).
65. Childs, B. G., and A. D. Le Claire: *Acta Met.* **2**:718 (1954).
66. Keating, D. T.: *Acta Met.* **2**:885 (1954).
67. Nakajima, K., J. J. Slade, and S. Weissmann: *Trans. ASM* **58**:14 (1965).
68. Neighbours, J. R., and C. S. Smith: *Acta Met.* **2**:591 (1954).
69. Polovov, V. M., and V. I. Sumin: *Fiz. Metal. i Metalloved.* **17**:184 (1964).

70. Blue, M. D.: *J. Phys. Chem. Solids* **11**:31 (1959).
71. Coles, B. R.: *Phys. Rev.* **101**:1254 (1956).
72. Dorfman, Y. G., and P. N. Zhukova: *Zh. Eksperim. i Teor. Fiz.* **9**:51 (1939).
73. Cooper, J. R. A., and S. Raimes: *Phil. Mag.* **4**:145, 1149 (1959).
74. Hagmann, D., and T. Ricker: *Z. Metallk.* **55**:789 (1964).
75. Hurd, C. M.: *Phil. Mag.* **12**:47 (1965).
76. Köster, W., D. Hagmann, and K. E. Saeger: *Z. Metallk.* **54**:19 (1963).
77. McAlister, A. J., E. A. Stern, and J. C. McGroddy: *Phys. Rev.* **140**:A2105 (1965).
78. Heine, V.: *Phil. Mag.* **12**:53 (1965).
79. Taylor, P. L.: *Proc. Roy. Soc.* **A275**:209 (1963).
80. Frank, V.: *Dan. Mat. Fys. Medd.*, vol. 30, no. 4, 1955.
81. Flanagan, W. F., and B. L. Averbach: *Phys. Rev.* **101**:1441 (1956).
82. Lane, G. S., A. S. Huglin, and J. Stringer: *Phys. Rev.* **135**:A1060 (1964).
83. See for example W. Hume-Rothery and G. V. Raynor: *Proc. Roy. Soc.* **A177**:27 (1940); G. V. Raynor, *Proc. Roy. Soc.* **A180**:107 (1942).
84. Jones, H.: *Proc. Roy. Soc.* **A147**:400 (1934); *Phil. Mag.* **41**:663 (1950).
85. Busk, R. S.: *Trans. AIME* **188**:1460 (1950).
86. Salkovitz, E. I., A. I. Schindler, and E. W. Kammer: *Phys. Rev.* **105**:887 (1957).
87. Hedgcock, F. T., and W. B. Muir: *Phys. Rev.* **136**:A561 (1964).
88. Long, T. R., and C. S. Smith: *Acta Met.* **5**:200 (1957).
89. Leigh, R. S.: *Phil. Mag.* **42**:139 (1951).
90. Stringer, J., and G. S. Lane: To be published.
91. Davies, D. A., and E. A. Owen: *Brit. J. Appl. Phys.* **15**:1309 (1964).
92. Löhberg, K.: *Z. Metallk.* **32**:86 (1940).
93. Yonemitsu, K., K. Takano, Y. Hishiyama, T. Matsuda, and I. Shiozaki: *J. Phys. Soc. Japan*, **20**:1797 (1965).

ORDER-DISORDER PHENOMENA AND SOLID SOLUTIONS

Gérard Fournet

Laboratoire Central des Industries Electriques, Fontenay-aux-Roses, France

INTRODUCTION

We shall show how the use of Yvon's theory[1] provides a better understanding of order-disorder phenomena, short-range order in solid solutions, magnetic effects, and so on.

Basis of Yvon's Theory

As usual in the relevant theories of binary alloys, Yvon's work is based on the following assumptions:

1. The electronic structure of the material is not known explicitly.
2. The lattice is considered perfect, it does not depend on the temperature, and thermal vibrations are neglected.
3. Only thermal excitation is considered, namely, the possibility for atoms to exchange their position.
4. The lattice obeys classical thermodynamic statistics.

5. The total cohesion energy of the system is the sum of the interaction energies of ions taken two by two. If we denote by A and B the two types of atoms, the mere knowledge of the function

$$W_p = W(r_p) = W_{AA}(r_p) + W_{BB}(r_p) - 2W_{AB}(r_p) \tag{1}$$

is sufficient; here r_p is the distance between two atoms in the position of pth neighbors (we are considering not only the first neighbors).

6. Each atom, besides the interaction energies, can be subjected to an exterior field of forces, and we consider

$$U(\mathbf{x}_i) = U_A(\mathbf{x}_i) - U_B(\mathbf{x}_i) \tag{2}$$

where $\mathbf{x}_i$ gives the position of node i.

SUMMARY OF YVON'S THEORY

A New Formulation

With the preceding assumptions it is sufficient to determine the position of atoms of type A to obtain at the same time the position of B atoms. The statistical problem that is to be solved now is analogous to the problem offered by the kinetic theory of fluids. Thanks to some results of the latter theory obtained from a method[1] devised by Ursell, Yvon[2] can immediately evaluate the probability $n_i = n_{1,A}(\mathbf{x}_1)$ of finding an atom of type A at node i, the probability $n_{ij} = n_{2,AA}(\mathbf{x}_i, \mathbf{x}_j)$ of finding an A atom at node i and another A atom at node j at the same time, the probability $n_{3,AAA}(\ .\ .\ .\ )$, and so on. Yvon gets the logarithm of n_i in the form of a series,

$$\log n_i - \sum_j g_{ij}n_j - \frac{1}{2}\sum_j\sum_k g_{ij}g_{jk}g_{ki}n_jn_k + \ .\ .\ .\ = K - \frac{U_i}{kT} \tag{3}$$

after noting that

$$g_{ij} = \exp\left(-\frac{W_{ij}}{kT}\right) - 1 \tag{4}$$

where W_{ij} is $W(|\mathbf{x}_i - \mathbf{x}_j|)$. K is an integration constant and depends on the conditions of the problem. In the second term of the series in Eq. (3) appears a sum of groups of two nodes (one being node i and the other any node). In the third term appears a sum of groups of three nodes among which stands necessarily the node i. In the same way Yvon has obtained

$$n_{ij} = n_in_j \exp\left(-\frac{W_{ij}}{kT}\right)\left(1 + \sum_k g_{ik}g_{jk}n_k + \cdots\right) \tag{5}$$

In Eqs. (3) and (5) the nodes in a group may be distinct or not, and this structure lets us avoid tedious numerical calculations. The best proof of this is that information only on the node i which is considered appears in each term of Eq. (3): in the second, by the group ii; in the third, by the group iii, etc.; all this information gives

$$\log n_i + n_i + \tfrac{1}{2}n_i^2 + \cdots \tag{6}$$

since g_{ii} is equal to -1, W_{ii} being considered as infinite.

The basic idea of Yvon's method[2] is to transform Eqs. (3) and (5) into other expansions where the terms are ordered according to the number of distinct nodes they contain. The equations will turn into

$$f_i = \log A_i + \sum_{j\neq i}{}' \log A_{ij} + \sum_{j\neq i}{}' \sum_{k\neq i}{}' \log A_{ijk} + \cdots = K - \frac{U_i}{kT} \tag{7}$$

$$\log n_{ij} = \log B_{ij} + \sum_{k\neq j}{}' \log B_{ijk} + \cdots \tag{8}$$

where the $A_i\ldots$ and $B_{ij}\ldots$ must be calculated.

Calculation of the $A_i\ldots$

We can get the expression of $\log A_i$ by two different approaches.

The first method consists of adding all the information concerning node i and coming from the different terms of Eq. (3). As, by definition, there is no term A_{ii}, A_{iii}, etc.,

$$\log A_i = \log n_i + n_i + \tfrac{1}{2}n_i^2 + \cdots \tag{9}$$

The second method, suggested by Yvon, originates from the following situation. The expression of an $A_i\ldots$ is independent of the lattice type; it is a function of the n_i and of the g_{ij} which does not change whether the lattice is fcc or bcc, or even if the lattice is reduced to a few nodes. To calculate A_i, for instance, let us consider a lattice reduced to a single node. We can then write Table 1.

For this reduced lattice, the probability of finding an A atom in i is $n_i = r/(1 + r)$, whereas Yvon's general expression gives $A_i = r \exp K$,

TABLE 1

Occupancy	Energy	Relative probability
1 A atom in i	U_i	$r = \exp(-U_i/kT)$
0 A atom in i	0	1

TABLE 2

Occupancy	Energy	Relative probability
0 atom	0	1
1 A atom in i	U_i	r
1 A atom in j	U_j	$s = \exp(-U_j/kT)$
1 A atom in i and 1 A atom in j	$U_i + U_j + W_{ij}$	rsw_{ij}

which can be still written as

$$A_i = \frac{n_i}{1 - n_i} \exp K \tag{10}$$

This solution agrees with the expansion of Eq. (9) provided that we have made the choice $K = 0$, as can be seen by calculating the first terms of the expansion of Eq. (10) in powers of n_i. We now see the great superiority of a calculation that considers a single-node lattice over the general calculation that only gives the expansion of Eq. (9).

To obtain A_{ij} we must consider a two-node lattice defined by Table 2, where w_{ij} (hereafter written w for simplicity) $= \exp(-W_{ij}/kT)$. Direct analysis of this lattice gives

$$n_i = r \frac{1 + sw}{1 + r + s + rsw} \tag{11}$$

$$n_j = s \frac{1 + r}{1 + r + s + rsw} \tag{12}$$

$$s = n_j \frac{1 + r}{(1 + rw)(1 - n_j)} \tag{13}$$

$$(1 - n_i)wr^2 + [1 - n_i - n_j + w(n_j - n_i)]r - n_i = 0 \tag{14}$$

The only physical solution of Eq. (14) is positive. On the other hand, in the two-node lattice, the general expression [Eq. (7)] gives $A_iA_{ij} = r \exp K$, so that we find

$$A_{ij} = r \frac{1 - n_i}{n_i} \times \text{constant} \tag{15}$$

and from Eq. (3),

$$A_{ij} = \frac{n_i + n_j - 1 + w(n_i - n_j) + \sqrt{[(n_i + n_j - 1) + w(n_i - n_j)]^2 + 4n_i(1 - n_i)w}}{2wn_i} \tag{16}$$

If two types of sites exist (i.e., α sites and β sites for an alloy undergoing the order-disorder transformation below its critical temperature), we shall write $A_{\alpha\alpha}(w)$, $A_{\alpha\beta}(w)$, and so on.

Calculation of the B_{ij}...

The two-note lattice can be used to calculate B_{ij}. In this case we have in fact

$$B_{ij} = n_{ij} = \frac{rsw}{1 + r + s + rsw} \tag{17}$$

and eliminating first s, then r, we get

$$B_{ij} = \frac{1+(w-1)(n_i+n_j) - \sqrt{[(n_i+n_j-1)+w(n_i-n_j)]^2+4n_i(1-n_i)w}}{2(w-1)} \tag{18}$$

We note that $B_{\alpha\alpha}(w)$, $B_{\alpha\beta}(w)$, . . . , B_{ijk} (and A_{ijk}) could be calculated from a lattice reduced to three nodes.

Case with No Long-range Order

In some problems, all sites are identical (order-disorder alloys above the critical temperature) and n_i can take only one value n; if only this type of problem is to be solved, the two-site lattice can be settled with $U_i = U$ as shown in Table 3. This leads to expressions of A_{ij} and B_{ij} which we denote respectively as $A_2(w)$ and $B_2(w)$; for example [cf. Eq. (18)],

$$B_2(w) = \frac{1 + (w - 1)2n - \sqrt{(1 - 2n)^2 + 4n(1 - n)w}}{2(w - 1)} \tag{19}$$

and $B_2(1) = n^2$. Calculations get even simpler where, the alloy composition being AB, the probability $n = 0.5$; in this case

$$B_2(w) = \frac{w - \sqrt{w}}{2(w - 1)} = \frac{\sqrt{w}}{2(1 + \sqrt{w})} \tag{20}$$

TABLE 3

Occupany	Energy	Relative probability
0 atom	0	1
1 atom in i	U	r
1 atom in j	U	r
1 atom in i and 1 in j	$2U + W_{ij}$	r^2w_{ij}

When the value of n is known, B_{ijk} depends only on w_{ij}, w_{jk}, and w_{ki}. The series of Eq. (8) shows us that w_{ij} plays a different part, and we thus note $B_3(w_{ij},w_{jk},w_{ki})$; our results are given in Refs. 3 and 4.

ORDER-DISORDER PHENOMENA

First, we shall study long-range order and, second, short-range order.

Long-range Order

At a temperature lower than the critical temperature, there exists long-range order. The parameter s is used to define probabilities n_α and n_β of finding A atoms on the two types of sites α and β. For an alloy of equiatomic composition and in which both types of sites are equally present (e.g., Cu-Zn, Cu-Pt), we know that

$$n_\alpha(s) = n_{1,A}(\alpha) = \frac{1+s}{2}$$

$$n_\beta(s) = n_{1,A}(\beta) = \frac{1-s}{2} \tag{21}$$

while in the case of Au-Cu_3 alloys we have (A atoms being Au atoms)

$$n_\alpha(s) = \frac{1+3s}{4} \qquad n_\beta(s) = \frac{1-s}{4} \tag{22}$$

Since the exterior field of forces does not interfere in the actual lattice, the parameter s is obtained by writing that the summation f_i [cf. Eq. (7)] is constant whether node i is an α or a β node. In f_i the terms are grouped according to a single criterion: the number of nodes which appear in each term. We shall now consider a second criterion: the various energies $W(r_1)$, $W(r_2)$, and so on, the knowledge of which is needed to calculate each term. Then we can display the various terms of f_i in the following way [cf. Eq. (23)]: In the first column we put the terms in which a single node occurs; in the second column terms with two nodes; and so on. In the first line we put the terms in which only W_1 occurs; in the second line terms with only W_1 and W_2 except those of the first line; and so on. We thus obtain

$$\begin{aligned} f_i = \log\frac{n_i}{1-n_i} &+ \sum_{z1} \log A_{ij}(w_1) + \sum \log A_{ijk}(w_1,w_1,w_1) + \cdots \\ &+ \sum_{z2} \log A_{ij}(w_2) + \cdots \\ &+ \sum_{z3} \log A_{ij}(w_3) + \cdots \end{aligned} \tag{23}$$

The general term is denoted by $f_{pq,i}$ [$w_p = \exp(-W_p/kT)$, q nodes].

This arrangement enables us to study the history of the various theories of long-range order. Bethe[5] only considered the first two terms of the first line f_{11} and f_{12}, and the corresponding equation for long-range order is

$$f_{11,\alpha} + f_{12,\alpha} = f_{11,\beta} + f_{12,\beta} \tag{24}$$

We know from Fowler and Guggenheim[6] that the classical theory of Bragg and Williams[7] assumes that in Eq. (24) the number of nodes z_1 located at a distance r_1 from the node they consider is infinite. In his numerical calculation, Yvon,[2] neglecting the effect of $W_2, W_3, \ldots$, considers the three- and four-node terms of the first line, and his equation for long-range order is

$$f_{11,\alpha} + f_{12,\alpha} + f_{13,\alpha} + f_{14,\alpha} = f_{11,\beta} + f_{12,\beta} + f_{13,\beta} + f_{14,\beta} \tag{25}$$

Yvon's and Bethe's results display little difference; the agreement of either theory with experiment is not always excellent.

Nearly all numerical developments, except Chang[8] and Pines,[9] of the long-range order equation and the various theories (except Yvon's) assume that W_2 and the other W_p of higher order vanish. Our aim was to avoid this assumption. By comparing Yvon's and Bethe's results, we have been able to neglect the terms with three or four nodes. We shall only examine the role of W_2 so that the long-range order equation becomes

$$f_{11,\alpha} + f_{12,\alpha} + f_{22,\alpha} = f_{11,\beta} + f_{12,\beta} + f_{22,\beta} \tag{26}$$

Cu-Zn Alloys.[10] In the Cu-Zn bcc lattice, each node (α or β) possesses eight first neighbors located at a distance $r_1 = a\sqrt{3}/2$ (these eight first neighbors have a common index β or α, different from the one of the central atom), and six second neighbors at a distance $r_2 = a$ (these six second neighbors have a common index identical to the one of the central atom that we are considering. We thus write the long-range order equation in the form

$$\log \frac{1+s}{1-s} + 8 \log A_{\alpha\beta}(w_1) + 6 \log A_{\alpha\alpha}(w_2) = \log \frac{1-s}{1+s} + 8 \log A_{\beta\alpha}(w_1) + 6 \log A_{\beta\beta}(w_2) \tag{27}$$

which, after all intermediate calculations, yields

$$\varphi(s,T) \equiv \log \frac{\sqrt{(1-s^2)/w_1 + s^2} + s}{\sqrt{(1-s^2)/w_1 + s^2} - s} - \frac{3}{4} \log \frac{1+s}{1-s} + \frac{3}{4} \log \left[\frac{\sqrt{w_2(1-s^2) + s^2} + s}{\sqrt{w_2(1-s^2) + s^2} - s} \frac{1-s}{1+s} \right] = 0 \tag{28}$$

If we suppose that W_2 vanishes, then w_2 becomes equal to unity and the last term in Eq. (28) vanishes. In this manner we find again the equation for long-range order of Bethe's theory or of Fowler and Guggenheim's quasi-chemical method (cf. eq. 1317 of Fowler and Guggenheim[6]).

If we want to study the effect of introducing W_2 in the equation for long-range order, we must consider the function $F(s,T)$, i.e., the configurational free energy. At a given temperature T_0, matter tends to minimize $F(s,T_0)$, and, therefore, the equation giving the *possible* values of s (at T_0) is

$$\frac{\partial F}{\partial s}(s,T_0) = 0 \tag{29}$$

We said possible values, for among the solutions of Eq. (29) we shall have to choose the one that gives the smallest minimum of $F(s,T_0)$. It is obvious that we should be able to identify any equation for long-range order with Eq. (29), except that a multiplicative factor and an additive term (which is independent of s but may depend on T) can be associated with $F(s,T_0)$. The equation for long-range order which we have established [Eq. (28)] can then be written in the form

$$\frac{\partial H}{\partial s}(s,T) = 0 \tag{30}$$

where

$$H(s,T) = f_1(T)F(s,T) + f_2(T) \tag{31}$$

and we can thus determine

$$h(s,T) = H(s,T) - H(0,T) = \int_0^s \varphi(s,T)\, ds \tag{32}$$

By setting $w_2 = 1$ we have been able to return to the function $F(s,T)$ of the quasi-chemical method, writing $f_1(T) = 2NkT$. This confirms our

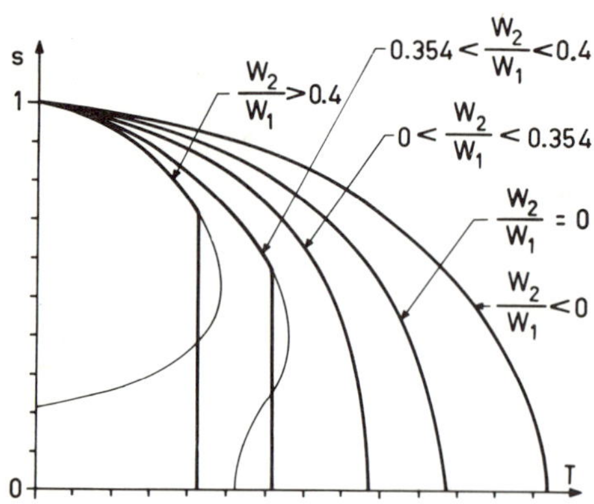

Fig. 1 Long-range order parameter s in cubic-centered lattice versus T for constant W_1 and various values of ratio W_2/W_1.

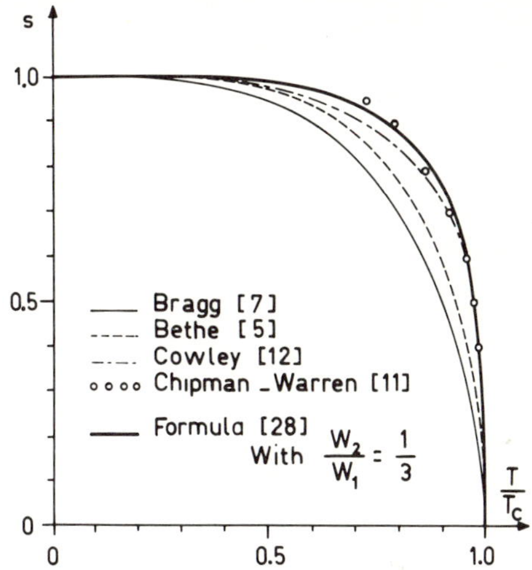

Fig. 2 Long-range order parameter s in β-brass versus T/T_c according to various theories and experiment.[11]

previous considerations [we also could determine $F(s,T)$ by a method similar to that of Fowler and Guggenheim by first deriving[10] Eq. (32) for the average configurational energy]. Consideration of the function $h(s,T_0)$ can be sufficient to solve entirely the problem of determining the long-range order at temperature T_0. In Fig. 1 we give $s(T)$, W_1 being constant and W_2/W_1 variable. We have plotted experimental[11] and theoretical[5,7,10,12] results in Fig. 2.

Cu-Pt Alloys. We have undertaken the theoretical study[13] of Cu-Pt because the order-disorder phenomenon of this alloy is very peculiar. The identity of the first neighbors is the same in the perfectly ordered state and in the perfectly disordered state. Thus there is no noticeable change in the number of first neighbors, and the classical theories can give no explanation of the phenomenon. The experimental studies were made by Johansson and Linde,[14] Linde,[15] and Walker.[16] Walker has shown that:

1. The variation of the long-range order parameter s as a function of temperature can be schematically represented as in Fig. 3, the critical point being defined by $s_0 = 0.77$ and $T_c = 810°C$.
2. Above the critical temperature, there exists a short-range order.

In the disordered state the lattice is fcc. From the disordered state we can get to the ordered state by putting Cu and Pt atoms alternately on (111) planes: therefore the cubic symmetry cannot remain, and the ordered alloy possesses the rhombohedral symmetry. It must be noticed, however, that the positions of the nodes of the lattice change very little in going from disorder to order: the angle θ (cf. Table 4) is very close to 90°. We can describe precisely the neighborhood of each node in the lattice by Table 4.

TABLE 4

Neighborhood	Disordered state	Ordered state
First neighbors.........	12 atoms at $r_1 = r_2/\sqrt{2}$	6 similar atoms at $l_1' = r_2 \sin(\theta/2)$ 6 different atoms at $l_1'' = r_2 \cos(\theta/2)$
Second neighbors.......	6 atoms at r_2	6 different atoms at r_2

Then, with Eqs. (21) and (26), the long-range order equation is

$$\log \frac{n_\alpha}{1 - n_\alpha} + 6 \log A_{\alpha\alpha}(w_1) + 6 \log A_{\alpha\beta}(w_1) + 6 \log A_{\alpha\beta}(w_2)$$

$$= \log \frac{n_\beta}{1 - n_\beta} + 6 \log A_{\beta\beta}(w_1) + 6 \log A_{\beta\alpha}(w_1) + 6 \log A_{\beta\alpha}(w_2) \qquad (33)$$

where we have assumed that $W(l_1') = W(l_1'') = W_1$. When all calculations have been performed, we get

$$\varphi(s, T) \equiv \log \frac{s + \sqrt{s^2 + w_1(1 - s^2)}}{-s + \sqrt{s^2 + w_1(1 - s^2)}}$$

$$+ \log \frac{s + \sqrt{s^2 + (1 - s^2)/w_1}}{-s + \sqrt{s^2 + (1 - s^2)/w_1}} + \log \frac{s + \sqrt{s^2 + (1 - s^2)/w_2}}{-s + \sqrt{s^2 + (1 - s^2)/w_2}}$$

$$- \frac{8}{3} \log \frac{1 + s}{1 - s} = 0 \qquad (34)$$

A detailed study of $h(s, T)$ [cf. Eq. (32)] shows that in our model the 1.3 value of the ratio W_2/W_1 separates the two possible cases:

1. For $W_2/W_1 < 1.3$ the disordered state is stable; the alloy is always a solid solution.
2. For $W_2/W_1 > 1.3$, order-disorder transformations exist; the $s = f(T)$ curves always show a discontinuity at the critical temperature. The experimental observations relative to the critical point can be explained by choosing

$$W_1 = 763k \qquad \text{and} \qquad W_2 = 1244k \qquad (35)$$

With these values, we have established the curve for $s = f(T)$ in Fig. 3. The experimental points of Walker (personal communication) are plotted on the same figure. The agreement is very good.

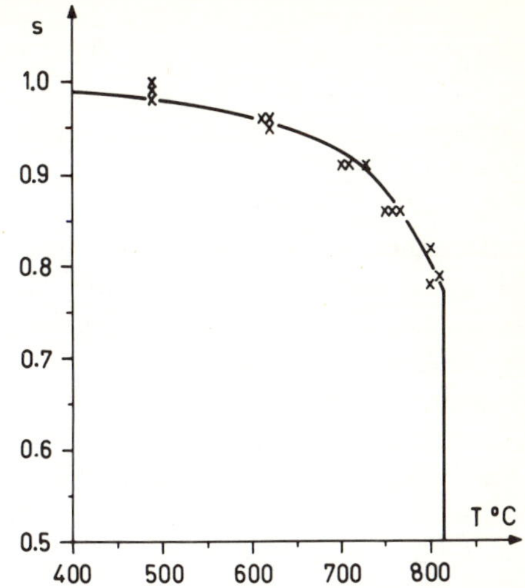

Fig. 3 Long-range order parameter s in Cu-Pt alloy versus T; theoretical curve in full line and experimental data.[16]

Au-Cu$_3$ Alloys. The lattice of Au-Cu$_3$ is an fcc lattice in the ordered state; the α nodes, located at the corners of the elementary cube, are occupied by Au atoms (or A atoms) and the β nodes, at the centers of the faces, by Cu atoms (or B atoms). A detailed analysis of the lattice gives

$$\log \frac{n_\alpha}{1 - n_\alpha} + 12 \log A_{\alpha\beta}(w_1) + 6 \log A_{\alpha\alpha}(w_2) = \log \frac{n_\beta}{1 - n_\beta}$$
$$+ 4 \log A_{\beta\alpha}(w_1) + 8 \log A_{\beta\beta}(w_1) + 6 \log A_{\beta\beta}(w_2) \qquad (36)$$

The calculation of $h(s,T)$ is not straightforward, and we can refer to Ref. 17. The complete discussion gives the theoretical curves of Fig. 4. Our

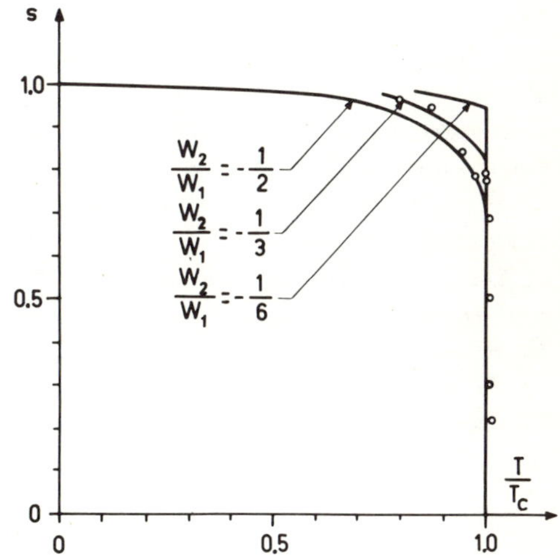

Fig. 4 Long-range order parameter s in Cu-Au$_3$ alloy versus T; theoretical curves in full line and experimental data.[18]

curve corresponding to $W_2/W_1 = -0.5$ predicts the correct value for s at the critical temperature, but there are divergences of shape between this curve and the experimental results.[18] We have needed two arbitrary constants in our theory: on the one hand, we choose the value $-\frac{1}{2}$ for the ratio W_2/W_1; on the other hand, the comparison of the theoretical value of $w_{1,c}(0.40)$ and of the measured critical temperature (388°C) determines $W_1 = 608k$. The general agreement is not as good as we obtained under the same conditions for the order-disorder phenomenon in α-brass. We think that this difference can be explained at least partly by the following fact: in bcc CuZn the terms A_{ijk} in which the three nodes appear, all being first neighbors of one another (i.e., bound by W_1 energies), do not exist; it is not the same for the fcc $AuCu_3$. The same approximation (A_1 and A_2 terms, W_1 and W_2 energies) can thus yield a different value according to the cases considered, CuZn and $AuCu_3$, owing to the nature of the first neglected terms.

Short-range Order

The existence of this order appears through several phenomena: specific-heat anomaly, X-ray scattering, and others.

Specific Heat. In Yvon's[2] approach where one considers only the $N_A A$ atoms to which N positions are offered, the energy is defined by

$$E = \tfrac{1}{2} \sum_i \sum_j{}' W(|\mathbf{x}_i - \mathbf{x}_j|) \tag{37}$$

where the sum is obviously extended only to the nodes occupied by A atoms and where W is defined by Eq. (1). To calculate the configurational specific heat we just differentiate E with respect to T; the general calculation is painful with this method, and we shall limit ourselves to calculating the specific heat for temperatures $T_c - 0$ and $T_c + 0$ immediately lower and higher than the critical temperature. For Cu-Zn alloys, with $W_2/W_1 = \frac{1}{3}$, our results[10] are, on the one hand, $C(T_c - 0) = 5.8k$, which is in good agreement with the experimental value of $5.1k$, whereas Bethe's theory gave $1.8k$, and Cowley's infinity. On the other hand, we obtain $C(T_c + 0) = 0.23k$, in fair but better agreement with experiment ($0.8k$) than Bethe's results.

X-ray Scattering. X-ray scattering appears to be an important, and the only direct, means of observation of short-range order. When there is no long-range order, i.e., $n_{1,A}(\mathbf{x}_i) = n$, the expression for that part of the intensity which varies slowly with the angle (in comparison to the "peaks") has already been given.[19–21] In studying the effect produced by a single crystal we shall use the expression

$$I(h) = I_e(h)Nn(1 - n)[F_A(h) - F_B(h)]^2[1 + \sum_{r \neq 0} \alpha_p \cos(\mathbf{h}\cdot\mathbf{r}_p)] \tag{38}$$

where $\mathbf{h}$ is the vector $2\pi(\mathbf{s} - \mathbf{s}_0)/\lambda$, $\mathbf{s}$ and $\mathbf{s}_0$ being the unit vectors directed along the scattered and incident rays [the absolute value h of $\mathbf{h}$ is $(4\pi \sin \theta)/\lambda$, 2θ being the scattering angle]; $I_e(h)$ the intensity diffused by an electron in the conditions we consider; $F_A(h)$ the structure factor of the atom of type A; and N the *total* number of atoms. The expression of α_p is

$$\alpha_p = 1 - \frac{n_{2,AB}(\mathbf{r}_p)}{n_{1,A}n_{1,B}} = \frac{n}{1-n}\left[\frac{n_2(\mathbf{r}_p)}{n^2} - 1\right] \tag{39}$$

In Eqs. (38) and (39), the $\mathbf{r}_p$ are the vectors joining the centers of the atoms to each other; the p index gives the rank of the neighbors. In the fcc system for instance, there are 12 first neighbors of any atom and therefore 12 factors of α_1 in Eq. (38), each of these being of the form $\cos(hr_1 \cos \Phi)$ where Φ is the angle between $\mathbf{h}$ and one of the $\mathbf{r}_1$ vectors. For a polycrystalline material Eq. (38) becomes

$$I(h) = I_e(h)Nn(1-n)[F_A(h) - F_B(h)]^2\left(1 + \sum_p z_p\alpha_p \frac{\sin hr_p}{hr_p}\right) \tag{40}$$

where z_p is the number of pth neighbors of an atom. In all cases the calculation of the n_2 by means of the $B_{ij\cdots}$ [cf. Eq. (41)] allows us to obtain the scattered intensity theoretically.

We have used the expression

$$\log n_{ij} = \log B_{ij} + \sum_k{}' \log B_{ijk} \tag{41}$$

reduced to the two- and three-node terms, under the following conditions:

1. The long-range order is supposed to have completely vanished at $T > T_c$.
2. We take into account only the energies W_1 and W_2 relative to the first and second neighbors. We notice that if we take into account terms of higher order than B_2, *the field of existence of short-range order is broader than the field of existence of the potential energies between the atoms.* If, supposing that only W_1 does not vanish, we should take into account an infinite number of terms: B_2, B_3, B_4, . . . , the result would then be the existence of "short-range order" between two nodes i and j infinitely distant from each other, this "order" being infinitely weak but different from zero. We have developed[3,4] the calculations of all classes of the $B_3(w_{ij}, w_{jk}; w_{ki})$ coefficients. We now know the B_3 coefficients that may occur—that is, those that are different from unity—and we ought, for each value of the distance r_p, to write Eq. (41) in more detail by indicating the classes of B_3 that we need.

Cu-Pt Alloys.[13] Let us carry out this work for first neighbors in disordered Cu-Pt alloys, i.e., apply Eq. (41) to the probability $n_2(r_1)$.

The B_2 term to be considered is the one corresponding to the same distance r_1 as is $B_2(w_1)$. To find the B_3 that occur in $n_2(r_1)$ let us consider Fig. 5 where we have represented in perspective a part of the fcc lattice; Cx, Cy, Cz are three coordinate axes, rectangular and parallel to the axes of the cubes. The nodes C and D are in first-neighbor position, and we look for the B_3 coefficient that binds these two nodes. Node E of coordinates $a/2$, 0, and $a/2$, first neighbor of C and D, introduces coefficient $B_3(w_1;w_1,w_1)$; three similar terms occur from points $(a/2,0,-a/2)$, $(0,a/2,a/2)$, and $(0,a/2,-a/2)$ which bring to 4 the coefficient of the term $\log B_3(w_1;w_1,w_1)$. Let us now consider point F which is the second neighbor of point C and the first neighbor of point D. The existence of the triangle CDF involves a coefficient $B_3(w_1;w_1,w_2)$. Three other such terms exist from points $(a,0,0)$, $(a/2,-a/2,0)$, and $(-a/2,a/2,0)$. The $B_3(w_1;w_2,w_2)$ does not exist in the lattice we are considering; all the B_3 coefficients of the kind $B_3(w_i;1,w_j)$ are equal to unity, so that in the approximation chosen we get

$$\log n_2(r_1) = \log B_2(w_1) + 4 \log B_2(w_1;w_1,w_1) + 4 \log B_3(w_1;w_1,w_2) \tag{42}$$

We shall give no details of how the various probabilities $n_2(r_p)$ are established. It should be noticed, however, that, on the one hand, coefficients B_3 are independent of the lattice in which they can occur, whereas, on the other hand, the numbers affecting these B_3 coefficients in the expression giving the probabilities n_2 depend on the lattice we are considering. For instance, in the expression relative to $n_2(r_1)$ these numbers equal 4 for $B_3(w_1;w_1,w_1)$ and 4 for $B_3(w_1;w_1,w_2)$; the number corresponding to $B_3(w_1;w_1,w_1)$ is 0 for the simple cubic lattice or square plane lattice. For the fcc lattice the results are:

$$\begin{aligned}
\log n_2(r_1) &= \log B_2(w_1) + 4 \log B_3(w_1;w_1,w_1) + 4 \log B_3(w_1;w_1,w_2) \\
\log n_2(r_2) &= \log B_2(w_2) + 4 \log B_3(w_2;w_1,w_1) \\
\log n_2(r_3) &= \log B_2(1) + 2 \log B_3(1;w_1,w_1) + 2 \log B_3(1;w_1,w_2) \\
\log n_2(r_4) &= \log B_2(1) + \log B_3(1;w_1,w_1) + 2 \log B_3(1;w_2,w_2) \\
\log n_2(r_5) &= \log B_2(1) + 2 \log B_3(1;w_1,w_2) \\
\log n_2(r_6) &= \log B_2(1) \\
\log n_2(r_7) &= \log B_2(1) \\
\log n_2(r_8) &= \log B_2(1) + \log B_3(1;w_2,w_2)
\end{aligned} \tag{43}$$

All the probabilities $n_2(r_p)$ $(p > 8)$ are equal to $B_2(1) = n_1^2 = 0.25$, which shows that in our approximation the short-range order extends to the eighth neighbors. We have summarized our results by drawing, for

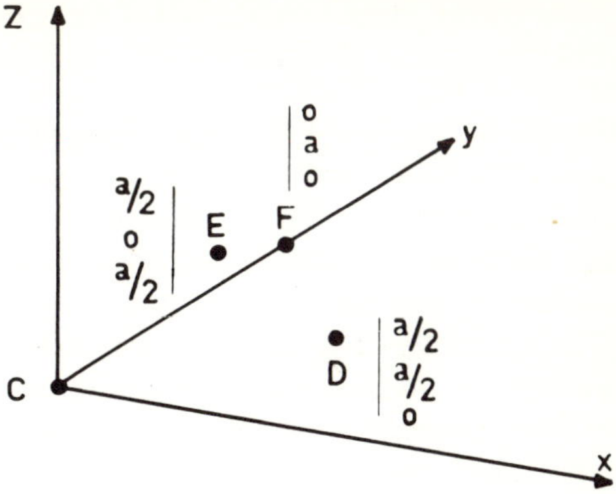

Fig. 5 Perspective view of a part of the fcc lattice.

$p = 1, 2, 3, 4, 5,$ and 8, the isoprobability $4n_2(r_p)$ curves in the plane with coordinates w_1 and W_2/W_1. These curves (not reproduced) allowed us[13] to determine for the alloy defined by conditions (35) the following values relative to a temperature of 890°C ($w_1 = 0.52$):

$$4n_2(r_1) = 1.148 \qquad 4n_2(r_5) = 1.088$$
$$4n_2(r_2) = 0.844 \qquad 4n_2(r_6) = 1.000$$
$$4n_2(r_3) = 1.150 \qquad 4n_2(r_7) = 1.000$$
$$4n_2(r_4) = 1.178 \qquad 4n_2(r_8) = 1.065$$

It is now possible to obtain the absolute value, in electron units, of the intensity of the scattered rays per atom for an alloy kept at 890°C. In Fig. 6 we give the experimental points of Walker (private communication), the experimental curve as drawn by Walker[16] (dashed line), and our theoretical curve (full line). We also give the Laue intensity curve: $I_e(h)Nn(1 - n)[F_A(h) - F_B(h)]^2$. The general agreement between experiment and our prediction is good; in particular the position and the value of the two maxima are correctly predicted.

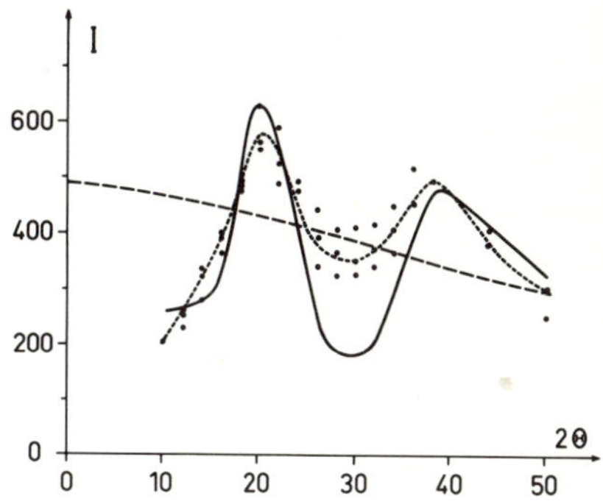

Fig. 6 Intensity of X-ray scattering in electron units versus angle, theoretical curve, and experimental data.[18]

Au-Cu Alloys.[3] Our equations (43), obtained for an fcc lattice, are still valid. The proportion $n = \frac{1}{4}$ of A atoms (Au) implies that $B_2(1) = \frac{1}{16}$. We have then plotted the isoprobability $16n_2(r_p)$ curves in the plane $(w_1, W_2/W_1)$. The experimental value[22] for 405°C of α_p, with Eq. (39), thus implies the relation $w_1 = f_p(W_2/W_1)$.

If our theory correctly predicted the experimental facts, all the corresponding curves should have a common point defined by $w_{1,0}$ and $(W_2/W_1)_0$. Before concluding, let us notice that Cowley[22] claims an absolute uncertainty of 0.010 for his α_p which yields an uncertainty of 0.030 for the parameter $16n_2(r_p)$. In such conditions, in the plane $(w_1, W_2/W_1)$ we must wonder whether some bands centered on the previously drawn curves would show common points. When we study the figure (not reproduced here) we see that there exists a point common to bands $p = 1, 3, 4$, and 5, but that the band for $p = 2$ stands at a small distance from that point. The interesting fact is that the common point corresponds to $W_2/W_1 = -\frac{1}{2}$, a value that gave satisfactory results in the study of long-range order. The agreement for the value of w_1 at the common point is not so good. We find $w_{1,0} = 0.295$ when in fact $w_{1,c} = 0.40$ for 388°C would impose the value $w_1 = 0.41$ for 405°C. Cowley also studied the Au-Cu_3 alloys at temperatures of 460 and 550°C. We have repeated the same work for these temperatures. There is a common area for bands with $p = 1, 3, 4$, a point $W_2/W_1 = -0.50$; the corresponding $w_{1,0}$ is 0.335 for 460°C and 0.380 for 550°C. If we choose $W_1 = 807k$, the absolute temperatures corresponding to the $w_{1,0}$ are 673, 740, and 835°K instead of 678, 733 and 823°K which Cowley had observed. Table 5 shows that there is a very good agreement except for the results for $n_2(r_2)$ between Cowley's observation and our theoretical predictions for short-range order.

TABLE 5

p	405°C		460°C		550°C	
	Experimental	Theoretical	Experimental	Theoretical	Experimental	Theoretical
1	0.544	0.520	0.556	0.559	0.607	0.607
2	1.588	1.880	1.516	1.760	1.315	1.635
3	1.027	1.059	1.057	1.052	1.078	1.046
4	1.285	1.248	1.204	1.206	1.135	1.166
5	0.841	0.855	0.853	0.880	0.904	0.903
6	1.075	1.000	1.021	1.000	0.973	1.000
7	0.952	1.000	0.976	1.000	0.991	1.000
8	1.144	1.045	1.126	1.035	1.057	1.025

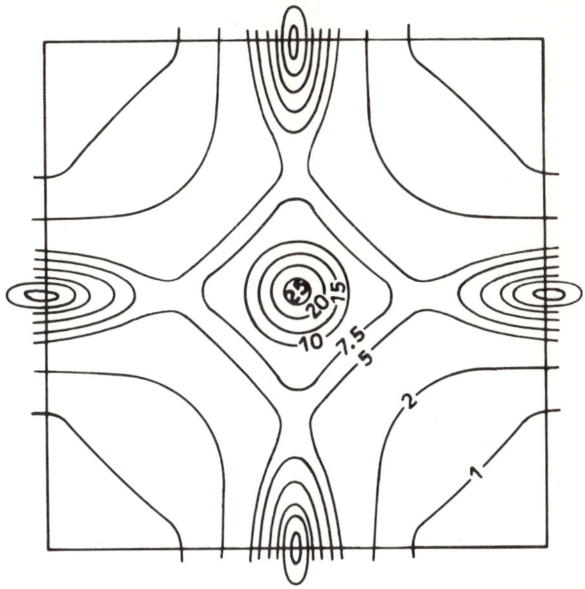

Fig. 7 Experimental equal diffusing-power curves coming from the short-range order in the plane (100) of the reciprocal lattice at 405°C.[22]

Fig. 8 Theoretical equal diffusing-power curves coming from the short-range order in the plane (100) of the reciprocal lattice at 405°C.

In the first column we have indicated the rank p of the neighbors considered, and in the next columns we wrote the values of $16n_2(r_p)$ (the experimental values of Cowley's and the theoretical ones). We recall that the accuracy of Cowley's measurements as estimated by himself corresponds to an absolute error of 0.030 on the $16n_2(r_p)$. In agreement with Cowley's results we found that α_3 was positive. This fact ought to be stressed as α_3 is negative in the ordered structure. In Figs. 7 and 8, we have represented the equal-diffusing power curves—experimental and theoretical—coming from the short-range order, in the plane [1.0.0.] of the reciprocal lattice for 405°C. Figure 7, from Cowley's paper, is drawn in arbitrary units: the maximum scattering power corresponds to 27 units. In Fig. 8, derived from our theoretical results, we have multiplied our bulk results so that the maximum intensity also gives 27 units. The general agreement of Figs. 7 and 8 is very satisfying.

SOLID SOLUTIONS

Alloys that undergo order-disorder transformations show short-range order above the critical temperature. The calculations on this subject are general, and, therefore, we have mentioned[4] that *the short-range order phenomenon is always present in metallic solid solutions*, thus extending the ideas of Hume-Rothery[23] and Barrett.[24]

The calculation of n_2 by means of B_2 and B_3 always enables us to estimate the intensity of the scattered X rays. As an example we considered Al-Ag alloys in which the atomic concentration of Ag is 6 a/o. The curves (Fig. 9) represent our predictions[25] for 515°C (full line) and 610°C (dashed line) when choosing $W(r_1) = -550k$ and $W(r_2) = 1{,}100k$, $W(r_3) = W(r_4) = \cdots = 0$. Accounting for the experimental dispersion (repre-

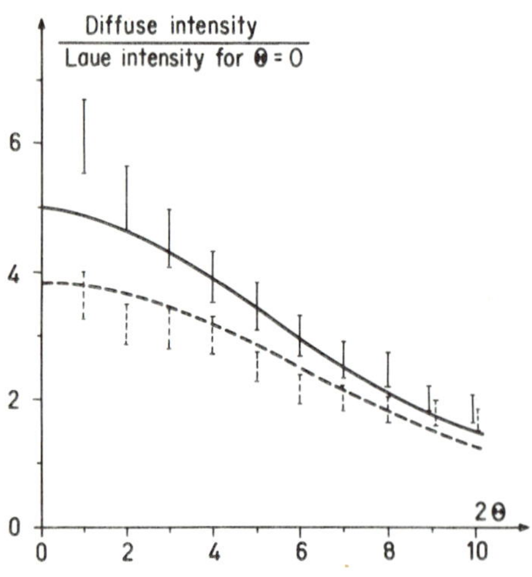

Fig. 9 Intensity of X-ray scattering versus angle for Al-Ag alloy (6 a/o in Ag); theoretical and experimental data for 515°C (full line) and 610°C (dashed line).[26]

sented by the vertical segments), our theory is in good agreement with the experimental results.[26] There is also good agreement for the absolute measures if we multiply the experimental absolute values, determined with a 30% accuracy, by a factor 0.83.

For alloys that do not undergo the order-disorder modifications, A and B atoms have no tendency to form pairs, and our configurational energy E should be negative, although some W_p might be positive. The expansion of Eq. (38) or (40) in terms of n, when the concentration of A atoms tends to zero, shows[4] that the value of the intensity is larger than Laue scattering for alloys that do not undergo the order-disorder transformation (W_p negative as average) and smaller than this scattering in the opposite case (W_p negative). Experiments for $AuCu_3$ and AlAg confirm this fact.

MAGNETIC PHENOMENA

We have also used Yvon's formalism[27] to study the Curie temperature for different kinds of lattices. We shall call A the ions that are magnetized parallel to the magnetic field and B those that are antiparallel to it. To be able to compare our determinations with those that have already been obtained, we shall assume that only the energy W_1 for first neighbors does not vanish. This energy W_1 is negative, and the parameter

$$x = \exp \frac{W_1}{2kT} \tag{44}$$

is introduced. Yvon gets the equation for the spontaneous magnetization by noticing that, without any exterior field, the probability of observing a magnetization in one direction (state I) is the same as the probability of observing it in the opposite direction (state II). Then values such as $n_{\mathrm{I}} + n_{\mathrm{II}} = 1$ must give [cf. Eq. (7)] $f_{\mathrm{I}} = f_{\mathrm{II}}$; by noting that $n_{\mathrm{I}} = (1 + s)/2$, we obtain $f_{\mathrm{I}}(s) = f_{\mathrm{I}}(-s)$, and the Curie temperature which corresponds to $s = 0$ is defined by

$$\left[\frac{\partial f_{\mathrm{I}}(s)}{\partial s}\right]_{s=0} = 0 \tag{45}$$

By introducing terms of the type

$$a_{ij} = \frac{1}{A_{ij}} \frac{\partial A_{ij}(s)}{\partial s} \tag{46}$$

Eq. (45) becomes[27]

$$a_i + \sum_j{}' a_{ij} + \sum_j{}' \sum_k{}' a_{ijk} + \cdots = 0 \tag{47}$$

The study of each type of lattice permits us to obtain the expression

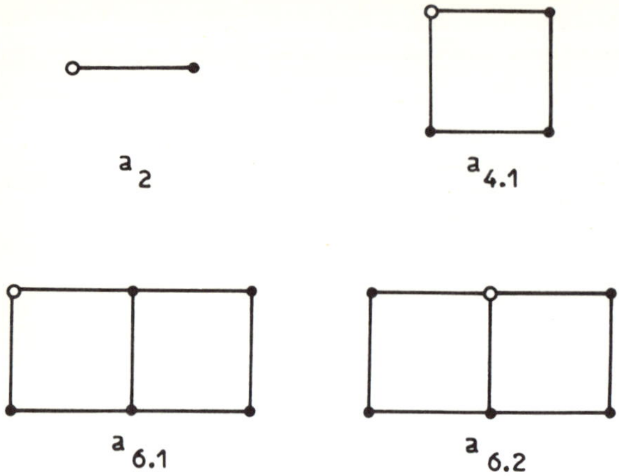

Fig. 10 Graphs defining the various $a_{q,n}$.

corresponding to Eq. (47). For example, the square plane lattice gives

$$a_1 + 4a_2 + 4a_{4,1} + 8a_{6,1} + 4a_{6,2} + \cdots = 0 \qquad (48)$$

The first index q of $a_{q,n}$ gives the number of nodes in a graph in which each pair of first-neighbor nodes is joined by a line; the node around which we carry the statistics is noted by a circle. In Fig. 10, we give graphs for the square plane lattice. Studying such a lattice (Fig. 11), we see that the 0 and A nodes introduce an a_2 term; the square $OGHA$ introduces an $a_{4,1}$ term; $OGVWHA$ or $OGHYIA$, an $a_{6,1}$ term; $COGHAB$, an $a_{6,2}$ term.

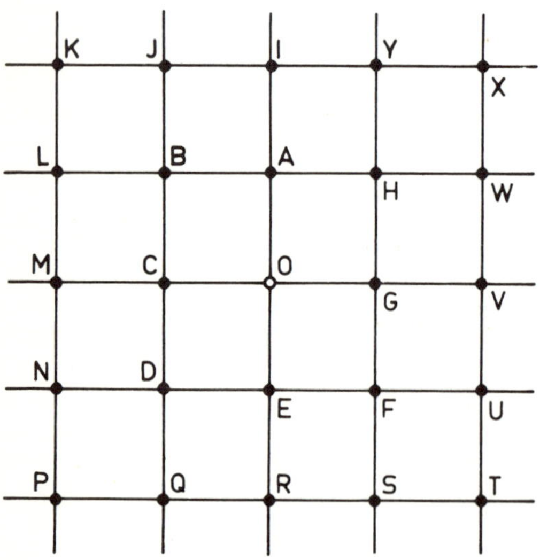

Fig. 11 Square-plane lattice.

The calculation of a_1 can be carried out, either from the expression of the A_1 or by a direct calculation using the analysis of the lattice reduced to one node. We find [cf. Eq. (10)]

$$a_1 = \left[\frac{1}{r}\frac{\partial r}{\partial s}\right]_{s=0} = 2 \tag{49}$$

by considering $n_I = (1 + s)/2 = r/(1 + r)$. This direct method allows us to calculate from place to place all the a_i... and in particular

$$\begin{aligned} a_1 &= 2 \\ a_2 &= x - 1 \\ a_{4,1} &= \frac{(x-1)^4}{2(x^2+1)} \\ a_{6,1} &= \frac{(x-1)^6(x+1)^2}{2(x^2+1)(3x^2+1)(x^2+3)} \\ a_{6,2} &= \frac{(x-1)^6(x^5+x^4+6x^3+x-1)}{(3x^2+1)^2(x^2+3)(x^2+1)} \end{aligned} \tag{50}$$

We can now calculate the Curie point relative to the square plane lattice. The first approximation corresponds to $a_1 + 4a_2 = 0$; then $x_{c,2} = 0.500$. The second approximation $a_1 + 4a_2 + 4a_{4,1} = 0$ gives $x_{c,4} = 0.467$; the third, implying terms in $a_{6,1}$ and $a_{6,2}$, gives $x_{c,6} = 0.453$. We have found it interesting to consider the curve $x_{c,q} = f(1/q)$, where q is the number of nodes contained in the last term taken into account to calculate $x_{c,q}$. Then the exact value x_c corresponds to $f(0)$. The three known points of the curve [cf. Eq. (27)] can, by means of a parabolic extrapolation, give the approximate value $x_c = f(0) = 0.421$, which is a very good result, as can be seen from the following table:

Bragg-Williams[7]	$x_c = 0.657$
Bethe[5]	$x_c = 0.500$
Yang[28]	$x_c = 0.486$
Kikuchi[29]	$x_c = 0.439$
Our value $f(0)$	$x_c = 0.421$
Exact value[30,31]	$x_c = 0.414$

We have given similar data for the plane triangular lattice, hexagonal plane lattice, simple cubic lattice, bcc, and fcc in another paper.[27]

CONCLUSIONS

We hope to have shown the power of Yvon's method. We believe that the advantages of the method are:

1. The parameters A, B, a can be calculated without any particular lattice being specified.
2. The evaluation can be as accurate as necessary, although the length of the calculations increases. The number q of nodes contained in the last term which is accounted for gives the possibility of classifying the various approximate values.
3. The method could be used to account for the interactions between second, third, etc., neighbors.
4. The length of the necessary calculations is much less than in the other methods.

However, it is likely that the potential energies are not as simple as the summation ΣW_{ij} which has been assumed, and improvements should be considered along this line. Nevertheless our results satisfactorily exhibit all the proper consequences of the assumptions. Moreover, we have indicated[32] how Yvon's method can be extended to the case where more than two types of atoms are present.

REFERENCES

1. Yvon, J.: "Fluctuations en densité," Herman & Cie., Paris, 1937.
2. Yvon, J.: *Cahier Phys.* **28**:1 (1945); **31**:12 (1948).
3. Fournet, G.: *J. Phys.* **14**:226 (1953).
4. Fournet, G.: *Acta Met.* **1**:383 (1953).
5. Bethe, H. A.: *Proc. Roy. Soc.* **150A**:552 (1935).
6. Fowler, R. H., and M. A. Guggenheim: "Statistical Thermodynamics," Cambridge University Press, London, 1939.
7. Bragg, W. L., and E. J. Williams: *Proc. Roy. Soc.* **145A**:699 (1934); **151A**:540 (1935); **152**:231 (1935).
8. Chang, T. S.: *J. Chem. Phys.* **9**:174 (1941).
9. Pines, B. Ya.: *Zh. Eksperim. i. Teor. Fiz.* **11**:725 (1941).
10. Fournet, G.: *J. Phys.* **13**:14A (1952).
11. Chipman, D., and B. E. Warren: *J. Appl. Phys.* **21**:696 (1950).
12. Cowley, J. M.: *Phys. Rev.* **77**:669 (1950).
13. Fournet, G.: *Bull. Soc. Franç. Min. Crist.* **77**:711 (1954).
14. Johansson, C. H., and J. O. Linde: *Ann. Phys.* **82**:449 (1927).
15. Linde, J. O.: *Ann. Phys.* **30**:1951 (1937).
16. Walker, C. B.: *J. Appl. Phys.* **23**:118 (1952).
17. Fournet, G. Colloque international sur les changements de phase, p. 199, Paris, 1952.
18. Keating, D. T., and B. E. Warren: *J. Appl. Phys.* **22**:286 (1951).
19. Wilchinsky, Z.: *Phys. Rev.* **63**:223 (1943); *J. Appl. Phys.* **15**:806 (1944).
20. Smoluchowski, R.: *Phys. Rev.* **64**:257 (1943).

21. Fournet, G.: *Bull. Soc. Franç. Min. Crist.* **74**:37 (1951).
22. Cowley, J. W.: *J. Appl. Phys.* **21**:24 (1950).
23. Hume-Rothery, W.: "Atomic Theory for Students of Metallurgy," p. 110, Institute of Metals, London, 1946.
24. Barrett, C. S., and T. B. Massalski: "Structure of Metals," 3d ed., McGraw-Hill Book Company, New York, 1966.
25. Fournet, G.: *Compt. Rend.* **237**:75 (1953).
26. Walker, C. B., J. Blin, and A. Guinier: *Compt. Rend.* **235**:254 (1952).
27. Fournet, G.: *J. Phys.* **18**:663 (1957).
28. Yang, C. N.: *J. Chem. Phys.* **13**:66 (1945).
29. Kikuchi, R.: *Phys. Rev.* **81**:988 (1951).
30. Kaufmann, B.: *Phys. Rev.* **60**:532 (1942).
31. Onsager, L.: *Phys. Rev.* **65**:117 (1944).
32. Fournet, G.: Third International Congress of Crystallography, Paris, 1954.

A SELF-CONSISTENT O.P.W. CALCULATION OF PAIR-INTERACTION FUNCTIONS IN SOME NORMAL METALS

R. M. Pick

Service de Physique du Solide
Centre d'Etudes Nucléaires de Saclay

INTRODUCTION

We shall summarize a method, originally due to W. Harrison,[1] which enables one to obtain a pair-interaction function for normal metals. This method is based on a self-consistent perturbation calculation of the energy of a metal in which the perturbative potential is an operator which automatically orthogonalizes the wave functions of the conduction electrons to those of the inner-core electrons.

In order not to enter into the mathematical details, we shall first present a simple calculation which shows how the classical calculation of the total energy of a metal by a perturbation expansion leads to a result which can be reexpressed as a sum of pair interactions between the ions. We shall show what are the principal features of this pair-interaction force and what kinds of approximations are involved in such a procedure. After having briefly mentioned the main differences between this simple model and the actual one, we shall give some results concerning this pair-interaction function as well as the density of the conduction electrons, and the relative stability of different crystallographic phases for some normal metals.

1 THE HARTREE MODEL

This is the simplest model with which the total energy of a metal can be computed self-consistently up to second order in perturbation. In this model, the ions are schematized as points with charge Z. The bare potential, $V(\mathbf{r})$, is thus simply that originated by those point charges plus a compensating distribution of free electrons. Its Fourier transform is related to the Fourier transform of the total potential, $W(\mathbf{q})$ by $W(\mathbf{q}) = V(\mathbf{q})/\epsilon(\mathbf{q})$ where $\epsilon(\mathbf{q})$ is the dielectric constant of the system, and consequently

$$V_s(\mathbf{q}) = V(\mathbf{q})\left(\frac{1}{\epsilon(\mathbf{q})} - 1\right)$$

is the potential due to the rearrangement of the conduction electrons under the influence of $V(\mathbf{q})$.

One can easily show that the total energy of the system can be written as

$$E_{\text{tot}} = E_w + E_p^\circ \tag{1}$$

where E_w is the self-energy of the system: ion plus free electrons. E_w can clearly be written as

$$E_w = Z^2 e_w$$

where e_w depends only on the crystalline structure of the metal and its density. E_p° is the sum of the energy of the actual electrons in the metal (measured from the mean value of W) minus the self-energy of the charge associated with V_s:

$$E_p^\circ = \sum_{k<k_F} [E_{\mathbf{k}} - W(0)] - \frac{1}{2}\int \rho_s(\mathbf{r}) V_s(\mathbf{r})\, d^3\mathbf{r}$$

Furthermore when $E_{\mathbf{k}}$ is expressed as a function of W, in a perturbation expansion, one gets

$$E_p^\circ = \sum_{k<k_F} \left[E_{\mathbf{k}} + \sum_q{}' \frac{W(\mathbf{q})V(\mathbf{q})}{\epsilon_{\mathbf{k}} - \epsilon_{\mathbf{k}+\mathbf{q}}} \right] \tag{2}$$

where $\epsilon_{\mathbf{k}}$ is the corresponding free-electron energy. The summation on k

is easily performed and leads to

$$E_p^\circ = \frac{L^3}{\pi^2}\frac{k_F{}^5}{10} - \frac{L^3}{4\pi}\frac{k_F}{\pi}\sum_q{}' W(\mathbf{q})V(\mathbf{q})g\left(\frac{q}{2k_F}\right) \tag{3}$$

with

$$g(x) = 1 + \frac{1-x^2}{2x}\log\left|\frac{1+x}{1-x}\right| \tag{4}$$

These formulas are independent of $\epsilon(\mathbf{q})$. In order to obtain its value, one proceeds also to the perturbation expansion of the wave function $\Psi_\mathbf{k}$ of the conduction electrons. An expansion to first order only is sufficient to obtain the density $\rho_{s(q)}$ of the electrons displaced by the potential $V_{(\mathbf{q})}$. Using the Poisson formula, one then gets

$$\epsilon(\mathbf{q}) = 1 + \frac{2k_F}{\pi q^2}\, g\left(\frac{q}{2k_F}\right) \tag{5}$$

It is clear that (1) and (2) lead to a formulation of the total energy as a sum of a volume-dependent constant plus a sum of pair interactions. Indeed, we have

$$V(q) = \frac{1}{L^3}\int\sum_R \frac{Z}{|\mathbf{R}-\mathbf{r}|}\exp i\mathbf{q}\cdot\mathbf{r}\, d^3\mathbf{r} = \frac{4\pi}{L^3}\frac{S(\mathbf{q})Z}{q^2} \tag{6}$$

where $S(\mathbf{q})$ is the structure factor $\sum_R \exp i\mathbf{q}\cdot\mathbf{R}$. Thus $W(\mathbf{q})$ is also the product of $S(\mathbf{q})$ by a function of q. The same is true for E_w, so that one finally gets

$$E_{\text{tot}} = \text{const.} + \frac{1}{2}\frac{4\pi}{L^3}\sum_q \frac{|S(\mathbf{q})|^2 Z^2}{q^2 + (2k_F/\pi)g(q/2k_F)}$$

which can be interpreted as a sum of pair interactions, the interaction function being

$$f(\mathbf{r}) = \frac{1}{2\pi^2}\int \frac{Z^2 \exp i\mathbf{q}\cdot\mathbf{R}}{q^2 + (2k_F/\pi)g(q/2k_F)}\, d^3\mathbf{r} \tag{7}$$

This pair-interaction function depends then only on the charge of the ion, and on the value of k_F, i.e., of the specific volume.

The two asymptotic forms of $f(\mathbf{r})$ for small and large r can be predicted at once.

1. For $r \to 0$, $f(\mathbf{r}) \simeq Z^2/r$, as $f(\mathbf{q}) \simeq Z^2/q^2$ when $q \to \infty$.

2. For $r \to \infty$, $f(\mathbf{r})$ should tend exponentially to zero, because $f(\mathbf{q})_{q=0}$ is finite, if $g(q/2k_F)$ had not a logarithmic singularity for $q = 2k_F$.

This singularity leads in fact to

$$f(\mathbf{r}) \rightarrow \frac{\pi Z^2}{k_F(1 + 2\pi k_F)^2} \cos \frac{2k_F r}{r^3} \qquad r \rightarrow \infty \tag{8}$$

In most practical cases $f(\mathbf{r})$ reaches already its asymptotic form for the third neighbor, as shown in Fig. 1 relative to Li.

We have thus seen how the total energy of a metal can be expressed as a sum of pair interactions between the ions, the conduction electrons having formally disappeared in this treatment, and what are the general features of the pair-interaction function. Before explaining how the same kind of calculation can be undertaken with a more realistic potential, some remarks on the approximations involved are in order.

1. It is not self-evident that a perturbation expansion of the energy E_k leads to a correct value of E_{tot} as the individual E_k diverges in the vicinity

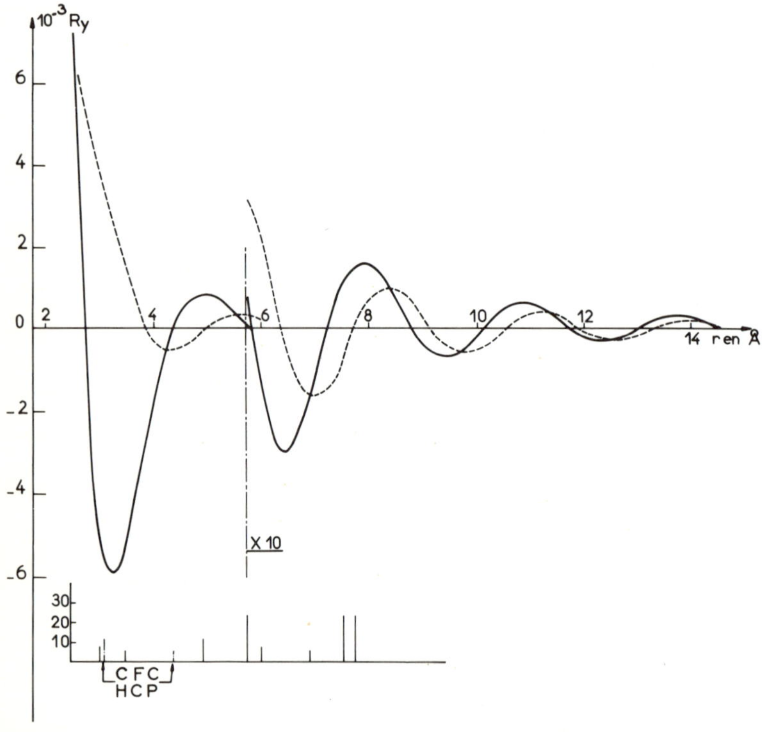

Fig. 1 Pair interactions in Li. Results with the pseudopotential shown as a full line; results in the Hartree model shown as a dashed line. The number and position of neighbors are also shown in the lower part of the diagram: the full lines refer to a bcc lattice, and the dashed lines to fcc and cph lattices.

of Brillouin zones. This can nevertheless be proved to be rigorous for a second-order perturbation expansion.[2]

2. The occurrence of a pair-interaction function is the consequence of the second-order perturbation expansion—a third-order one would have also given three body forces, etc.

3. The occurrence of the logarithmic singularity is the consequence of the discontinuity in the electron occupation function at 0°K. At finite temperature, this singularity then disappears. But, as the usual temperatures are orders of magnitude smaller than the Fermi one, the oscillations of the pair-interaction function will only be damped out at extremely large distances.

4. This pair-interaction function obtained with the Hartree model does not lead to correct physical results.* This comes from the fact that we have taken a too simple bare potential. Should we have chosen a more sophisticated real potential (i.e., a local pseudopotential), all formulas up to (5) would still be valid—the principal modification in (7) would be replacement of Z by $q^2V(q)L^3/4\pi$. Thus any local pseudopotential will lead also to asymptotic oscillatory forces.

2 THE OPERATIONAL PSEUDOPOTENTIAL

One of the main troubles with the Hartree model is that the wave functions of the conduction electrons are not correctly described. In the Hartree approximation, the wave functions of all the electrons, those belonging to the ions as well as the conduction ones, are the eigenfunctions of the same hamiltonian. They are then orthogonal to each other. This leads to a complicated form of the wave functions of the conduction electrons inside and in the vicinity of the ions, thus to a high kinetic energy, which must be compensated by an equivalent potential energy if these electrons have a quasi-free-electron-like behavior (see the paper by V. Heine[3]). We shall first set up, from here up to Eq. (10), a pseudopotential which automatically leads to a correct wave function, and will then discuss the main differences between this treatment and the former one.

Let $|\, c\,\rangle$ represent a core wave function, which is an eigenvalue of

$$H^\circ = -\frac{\nabla^2}{2} + W_0(\mathbf{r})$$

where $W_0(\mathbf{r})$ is the self-consistent potential, and let $\Psi_{\mathbf{k}}$ be the wave function of a conduction electron. Because of the orthogonality relationship be-

* For example, the most stable phase found for K would be hexagonal close-packed with c/a approximately 2.85.

tween $|c\rangle$ and Ψ_k, whatever will be the set of function $\lambda_c(\mathbf{r})$, Ψ_k will also be an eigenfunction (with the same eigenvalue E_k) of

$$H_1{}^t = -\frac{\nabla^2}{2} + W_0(\mathbf{r}) + \sum_c |\lambda_c(\mathbf{r})\rangle\langle c|$$

As $H_1{}^t$ is not a hermitian operator, the adjoint operator

$$H_1 = -\frac{\nabla^2}{2} + W_0(\mathbf{r}) + \sum_c |c\rangle\langle\lambda_c(\mathbf{r})|$$

will have the same eigenvalues but its eigenfunctions φ_k will be different, and will depend on the functions λ_c. It is then possible to choose these λ_c in such a way that φ_k might be more easily computed; in our case, we shall require that they might be obtained by a perturbation expansion as in Sec. 1.

It can be proved that:

1. Such a requirement is approximately satisfied[4] if λ_c has the form $-W|c\rangle$.

2. Whatever is λ_c, the relationship between $|\varphi_k\rangle$ and the corresponding electronic wave function $|\Psi_k\rangle$ is[5]

$$|\varphi_k\rangle = (1 - P)|\Psi_k\rangle \tag{9}$$

where $P = \sum_c |c\rangle\langle c|$.

One is then led to look for the eigenvalues and eigenfunctions of

$$H = -\frac{\nabla^2}{2} + (1 - P)W_0(\mathbf{r}) \tag{10}$$

where the bare potential V_0, from which W_0 has to be deduced, explicitly takes into account the actual charge distribution of the ions. In (10) the real potential $W_0(\mathbf{r})$ has now been replaced by the operational pseudopotential

$$W' = (1 - P)W_0(\mathbf{r}) \tag{11}$$

As W' is an operator, the simple procedure, briefly described in Sec. 1, which leads from the bare potential to the total energy of the metal and the pair interaction cannot be used as such. The mathematical machinery is a little more complex, but the general scheme remains unchanged, and so do the principal results. We shall briefly sum up the main modifications.

1. The total energy may still be written as a sum of an electrostatic term and a perturbation one, but:

(*a*) In E_w, the electrostatic term, the real charge Z is replaced by Z^*, an effective charge always bigger than Z. (The physical reason for this change will be given in Sec. 3.)

(b) As W' is no longer a real potential, the second term in formula (2) must be replaced by

$$\frac{1}{2}\sum_{k<k_F}\left(\frac{\sum_q{}'\langle\mathbf{k}\mid W'\mid\mathbf{k}+\mathbf{q}\rangle\langle\mathbf{k}+\mathbf{q}\mid W'-V_s\mid\mathbf{k}\rangle}{\epsilon_{\mathbf{k}}-\epsilon_{\mathbf{k}+\mathbf{q}}}+\text{c.c.}\right)$$

(c) V_s is still that potential which is created by the conduction-electron charge distribution (minus the constant represented by the equivalent free-electron charge distribution), but the relationship between the total potential matrix element $\langle\mathbf{k}\mid W\mid\mathbf{k}+\mathbf{q}\rangle$ and the corresponding effective bare potential $\langle\mathbf{k}\mid W'-V_s\mid\mathbf{k}+\mathbf{q}\rangle$ can no longer be expressed easily through a dielectric constant formalism. Nevertheless, as in Sec. 1, $V_s(\mathbf{q})$ is readily obtained, from V_0 and P, through the perturbation expansion of the electronic charge distribution.

2. The pair-interaction formulation of the total energy is still valid. Its Fourier transform has now an awkward form, but its principal characteristics are unchanged:

(a) It is proportional to q^{-2} for large q.

(b) It is finite for $q \to 0$ but has a logarithmic singularity for $q = 2k_F$. Its asymptotic form for $r \to \infty$ can then be written

$$f_p(\mathbf{r}) = \frac{\pi}{k_F(1+2\pi k_F)^2}\,Y^2\,\frac{\cos 2k_F r}{r^3}$$

(c) The relationship between $|Y|$ and $\langle-\mathbf{k}_F\mid W\mid\mathbf{k}_F\rangle$ is nearly the same as that existing between Z, the ionic charge, and $W(2k_F)$.

The operational pseudopotential theory thus parallels exactly the local potential one. To test its validity, we shall take a very simple form for the local bare potential, namely, the simple ionic charge distribution plus a Slater exchange potential between the core and the conduction electrons. This leads to two restrictions for its application:

1. We need the core wave functions $|c\rangle$. We choose them to be the same as in the free ion. We can then only study those metals in which both are nearly the same, i.e., the normal metals. Transition and noble metals must be excluded.

2. The cohesive energy of a metal contains many-body effects which have not been taken explicitly into account in E_{tot}. Though it is approximately possible to account for them by choosing special forms of the pseudopotential (see L. J. Sham[6] for instance), we have not used this procedure. Nevertheless, many-body effects are essentially functions of the electronic density, i.e., k_F and Z. k_F will have then to be considered

as a fixed parameter,* and this theory will not be able to give the atomic volume of the metal, or to explain experiments involving a bulk or local change of electronic density.

3 RESULTS

We shall present now the results of some of our calculations on three alkaline metals, Li, Na, and K, as well as two other metals of the second row, Mg and Al. We give first the form of the interaction function in these metals, then discuss the distribution of the conduction electrons, then proceed to the study of the relative stability of different phases at 0°K, and end by a quick study of the internal stability of the body-centered phase of the alkaline metals and of Al.

Form of the Pair-interaction Function

As already stated, the asymptotic form of the pair-interaction function is given by

$$f_p(\mathbf{r}) = \frac{\pi}{k_F(1 + 2\pi k_F)^2} Y^2 \frac{\cos 2k_F r}{r^3}$$

which is the formula obtained in the Hartree model except for the replacement of the ionic charge Z by Y. The ratio Y/Z is given in Table 1.

TABLE 1. ***The Ratio Y/Z***

Li 1.02		
Na 0.161	Mg 0.535	Al 0.633
K 0.745	Ca 0.520	

We see that, except for Na, the intensity of the asymptotic form (which is already valid for the third or fourth neighbor of an ion) is not much smaller than that obtained with the Hartree model. On the contrary, for the first or second neighbor the results are quite different, as shown on Figs. 1 and 2 relative to Li and Al. They exhibit a huge minimum, always in the immediate vicinity of the first neighbor, whose intensity might be an order of magnitude bigger than for the second one, in contradistinction with the Hartree model (dashed lines on the same figures).

* No other parameters enter this theory as the wave functions $|c\rangle$ are obtained through a first-principles calculation involving only Z.

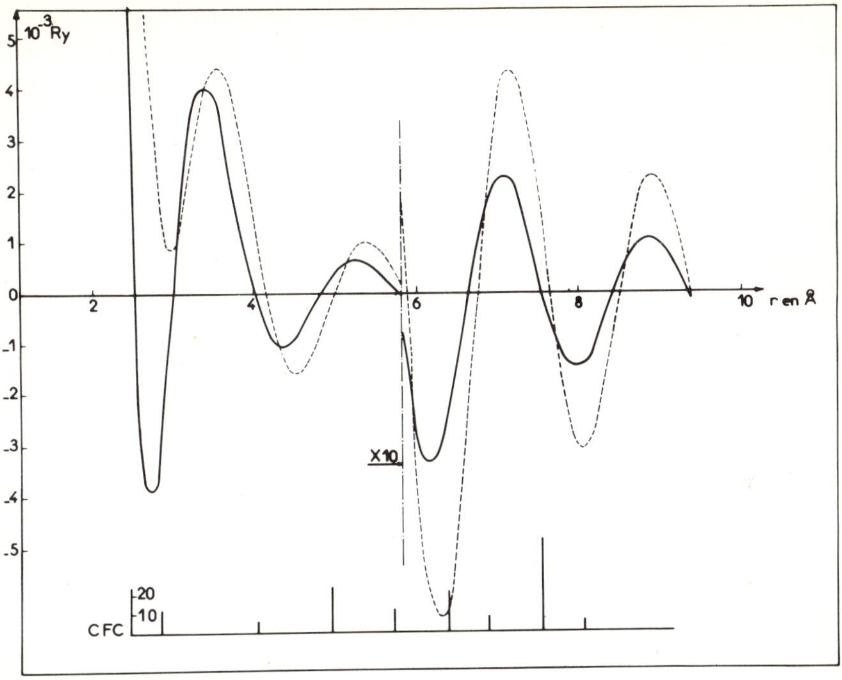

Fig. 2 Pair interactions in Al. Results with the pseudopotential shown as a full line; results in the Hartree model shown as a dashed line. In the lower part of the figure are shown the position and the number of neighbors in the fcc lattice.

The fact that the ratio between first and more distant neighbor interaction is much bigger with alkaline metals than with Al is substantiated by phonon spectra analysis[7,8] which shows that one needs a large number of pair interactions to explain the Al spectrum, though this is not the case for K[9] or Na.

But direct comparisons between these pair-interaction functions and those deduced by Johnson and others[10] from X-ray and neutron diffraction data on liquid metals do not show a good agreement, except for the position of the first minimum as shown in Table 2.

TABLE 2. ***Position of the First Minimum of the Pair Interaction***

	Li	Na	K	Al
This calculation	3.2	3.4	4.6	2.7
Johnson, et al.[10]	3.3	3.7	4.8	3.1

The main disagreement is the size of the first minimum in the alkaline metals, which appear to be three or four times bigger than those of Johnson and others.[10] This might come from the fact that our minima are so sharp that at the actual temperature of measurement, they are partly cut off by thermal movements. In Al, on the other hand, where our computed minimum is not relatively as important as in the alkaline metals, the Johnson result and ours agree to within 20%.

Distribution of the Conduction Electrons

To see to what extent these metals have a free-electron-like behavior, we looked at the distribution of the conduction electrons. This distribution is more or less uniform except in the vicinity of the ions. This last effect comes from the orthogonalization of the conduction-electron wave functions to those of the core electrons. It appears as a kind of repulsion between the two types of electrons, and thus increases the density of the conduction

TABLE 3. ***Values of*** α

Li 0.080		
Na 0.078	Mg 0.084	Al 0.079
K 0.138	Ca 0.171	

electrons in the rest of the crystal. This effect is the origin of the replacement of Z by Z^* in the Ewald energy of Sec. 2. The value of α, where

$$Z^* = Z(1 + \alpha)$$

is given in Table 3, which shows that this effect is not negligible.

Except for this effect, the distribution of conduction electrons is mostly uniform as has been verified by Harrison[11] in Al. This is illustrated in Fig. 3 which gives this distribution for the $\{110\}$ plane for this metal.

Relative Stability of the Alkaline Metals at 0°K

As is well known, some alkaline metals undergo a transformation from a cubic body-centered phase at normal temperature to an ideal hexagonal close-packed one at very low temperature. Moreover, when hexagonal close-packed Li is cold-worked, it does not transform into body-centered but into face-centered cubic lattice. Then in Li and Na which are the two metals in which this transformation happens,[12] at 0°K, the sequence of

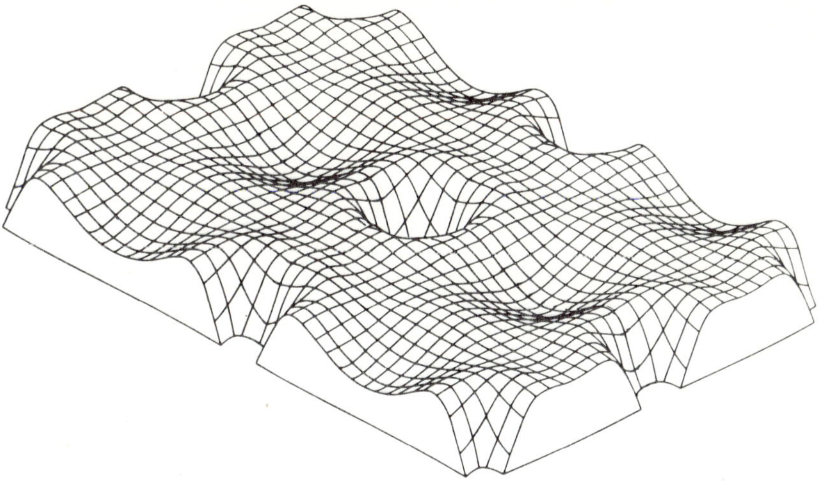

Fig. 3 Conduction-electron density on a {110} plane in Al. This is a projection of a three-dimensional plot; if viewed from above, the lines would form a square grid.

phases is: hexagonal close-packed more stable than cubic face-centered, which in turn is more stable than the cubic body-centered phase.

We checked our pair potential against those experimental facts. The results are given in Table 4, where the difference in energy between the different phases is measured in degrees Kelvin, taking our zero of energy as the calculated most stable phase. The measured temperature of transformation is also given. One sees that

1. The hexagonal close-packed structure is the most stable in our calculation, with a c/a ratio near ideal, as is the case for Li and Na.
2. The order of magnitude of the difference in energy between the phases is correctly given as well as their sequence.
3. The results for K are qualitatively equivalent to those for Li and Na, contrary to experiment. Nevertheless the difference in energy among the three phases is smaller than in the two other cases.

TABLE 4. ***Relative Phase Stabilities in the Alkali Metals***

	Li	Na	K
Tc_{exp}	77°	40°	
bcc	39°	63°	25°
fcc	22°	22°	8°
hcp $c/a=\sqrt{8/3}$	6°	4°	1.6°
c/a_{min}	1.597	1.580	1.613

Internal Stability of Phases

We have looked at the phonon spectrum of the cubic body-centered alkaline metals, and of the cubic face-centered Al. Though there is never an excellent detailed agreement between the computed spectra and the measured ones, the overall shape is always correct, and the disagreement is never more than 20% of the measured frequency.*

As an example, the computed elastic constants of Li and K are compared with their measured values in Table 5. The agreement is very good for Li but, again, poorer for K.

TABLE 5. ***Elastic Constants in 10^{10} dyne/cm^2***

	$\frac{1}{2}(c_{11}-c_{12})$	c_{44}
Li_{exp}	1.21*	11.5*
Li_{cal}	1.24	14.4
K_{exp}	0.377	2.86
K_{cal}	0.395	4.35

* Extrapolated to 0°K.

In conclusion, these pair-interaction functions seem to give rather good agreement with experiment, whenever comparisons are possible. The agreement is in fact worse, the higher the number of the row in which is the metal. A similar conclusion has been obtained by V. Heine[13] with a model potential which does not give as good results for K and Ca as for lighter metals.

REFERENCES

1. Harrison, W.: *Phys. Rev.* **129**:2503, 2592 (1963).
2. Pick, R., and A. Blandin: *Phys. Kondens. Mater.* **3**:1 (1964).
3. Heine, V.: This volume, p. 103.
4. Pick, R., Thesis: Rapport CEA, R. 2820.
5. Austin, B. J., V. Heine, and L. J. Sham: *Phys. Rev.* **127**:276 (1962).
6. Sham, L. J.: *Proc. Roy. Soc.* (*London*) **A283**:33 (1965).
7. Woods, A. D. B., B. N. Brockhouse, et al.: *Phys. Rev.* **128**:1112 (1962).

* This is only true for transverse waves, as longitudinal phonon frequencies cannot be predicted accurately, because of the variation of cohesive energy associated with such phonons.

8. Yarnell, J. L., J. L. Warren, and S. M. Koenig: *J. Phys. Chem. Soc. Suppl. I*:57 (1965).
9. Woods, A. D. B.: Private communication.
10. Johnson, M. D., P. Hutchinson, and M. N. March: *Proc. Roy. Soc.* (*London*) **A282:**283 (1964).
11. Harrison, W.: *Phys. Rev.* **136A:**1107 (1964).
12. Barrett, C. S.: *Acta Cryst.* **9:** 671 (1956).
13. Heine, V.: Private communication.

Agenda Discussion:
PURE METALS AND SOLID SOLUTIONS

chairman **Jacques Friedel**

Service de Physiques des Solides, University of Paris, Orsay, France

secretary **John Stringer**

Metal Science Group, Battelle Memorial Institute, Columbus, Ohio

This discussion ranged on the stability of pure metals and solid solutions. The papers discussed were in principle those of the first two days, but the discussion centered actually on two main topics:

1. The validity and use of effective potentials in describing "normal" metals and alloys—i.e., those classified as group I by Hume-Rothery and where the valence electrons behave in many ways as nearly free
2. The Brewer-Engel theory, especially as applied to transitional metals and alloys

For the first kind of topic, the discussion was animated, but led on the whole to a general agreement so that only the main protagonists and conclusions will be reported. It has been thought more advisable, on the contrary, to report more nearly verbatim the discussion on the Brewer theory, to which a written report by Lomer and some comments by the chairman have been added.

1. NORMAL METALS AND ALLOYS

Long-range Oscillations in the Effective Potential

There was a general agreement that Heine's treatment fails to include the long-range oscillations necessarily present in the self-consistent effective potential $v(r)$, owing to its "clothing" by the conduction electrons; these oscillations are considered in Pick's treatment. It was agreed that the difference between Heine's and Pick's treatments did not much affect the pair-interaction potential $F(r)$ at distances of the order of the nearest neighbors, where it had usually a strong minimum, so that the general order of magnitude of the cohesive energy was not affected. It was also agreed that the long-range oscillations of $v(r)$ present in Blandin's and Pick's treatments were essential when comparing two states which only involved a change in the arrangement of fairly distant neighbors: e.g., the stacking-fault problem or the relative stability of the fcc and hcp phases. It was also stated by Pick that, for more than second neighbors, the use of the asymptotic form was excellent.

On the relative stability of phases involving a change in nearest neighbors, the meeting was more divided. Heine pointed out that ultimately it is only $w(q)$ at the reciprocal lattice vectors that matters, and these do not usually lie at $2k_F$. He wondered therefore whether all this emphasis on the long-range oscillations in real space, related to the precise form of the singularity just at $2k_F$ in reciprocal space, was not somewhat misplaced. Blandin remarked that Heine himself had invoked in his discussion the rapid variation of $w^2(q)\epsilon(q)$ in the neighborhood of $2k_F$, owing to the logarithmic anomaly of the dielectric constant $\epsilon(q)$; a more careful study of this variation was therefore in order. Friedel remarked that the oscillatory behavior of $w(r)$ at fairly close range (a few interatomic distances) was actually dominated by the part in the logarithmic anomaly which was not so close to $2k_F$, and where changes in $w(q)$ due to the anomaly were not so small. One should therefore think twice before dropping the long-range oscillations altogether. Lomer was skeptical on that score.

Validity of Perturbation Methods. Phase Shift in the Effective Potential

On a question by Lomer, Pick pointed out that in the perturbation treatmeat used by everybody, the *wave functions* were valid so far to an order of perturbation lower than the energy. This did not satisfy Heine, who thought that, knowing the pseudopotentials, one could solve the wave equation by machine computations, thus nearly exactly.

Lomer inquired whether the choice of trial function was important. But Heine said that one uses the pseudopotentials and plane waves, which give one the pseudowave functions. One then turns them into real-wave functions by putting back the "wiggles" in the core which one had previously cut out. This can be done, because the pseudowave function and real-wave function will be identical outside the core. The orthogonalizing procedure is not useful in nuclear calculations, but, fortunately, it works quite well for atoms. It is possible that some properties—the Knight shift for instance—may depend very strongly on the form of the wave function in the core; one may then have to calculate a Wigner-Seitz wave function, and use it to calibrate the wave function. This is, however, perfectly feasible.

It seems to the chairman that here Heine misses the point of self-consistency; so far, the best that has been done to "clothe" the effective potential is to use an essentially free-electron dielectric constant. Independently of the objections that can be raised about such a procedure in the atomic cores, it is clear that to go higher in the order of approximation would require a very complex correction of the dielectric constant.

In a postmeeting written comment, Heine points out that there are already in the literature numerical calculations on diamond-type semiconductors, where these effects are very important, which have included (1) k-dependent correlation and exchange based on the extension of Hubbard's theory, (2) the actual band structure, not free-electron theory, and (3) higher orders of perturbation by calculating directly a self-consistent t matrix.

Also, as pointed out by Seeger, one could expect a *phase shift* to appear in the asymptotic oscillations of the effective potential. Friedel explained that the whole concept of pair interactions or second-order perturbations as developed so far in this field was only valid if the phase shift was small. Furthermore, to include it in the interaction was to take into account third-order terms, of the same strength therefore as three-body interactions. Seeger was not quite convinced.

The Noble Metals

The phase-shift question then centered on noble metals. Seeger pointed out that to reverse the sign of α in the asymptotic interaction $\alpha r^{-3} \cos 2k_F r$ for noble metals, as suggested by Blandin, was equivalent to introducing a phase shift of π. His feeling was therefore that the whole concept broke down for these metals, perhaps because of the large size of their inner (d) shells. Heine answered that the presence of a d band not much below the Fermi level produced indeed a significant contribution to the pseudopotentials for the Fermi electrons, which his group was in the process of computing. This was not finished, and he could therefore give no firm answer yet, but he was confident that the pseudopotential concept would still hold and give results in agreement with experiment, especially about such things as the anisotropy of relaxation time. Commenting further on Stringer's paper on the Hall effect in Ag-Au alloys, he remarked that the difference in relaxation time on the neck and belly of the Fermi surface of copper may well be due to the effect of the *d* band on the pseudopotential. The states on the neck have pure *p*-like symmetry and so do not see at all the *d* contribution to the pseudopotential and the scattering. Data were now being produced for the Fermi surface of alloys, some on Cu-Zn being presented in Pearson's paper. The parameters calculated for the copper atoms from these data should be another test of the theory.

Later on in the discussion, Brewer and Hume-Rothery commented on the special stability of the noble metals compared with other "normal" metals, which make them look much as if they belonged to the end of the transitional series. Hume-Rothery thought this might be due to a correlation effect owing to the polarization of the not very stable *d* shells (van der Waals interaction), while Brewer liked to think of a *dsp* hybridization which could in first approximation be described by an atomic configuration with a partly filled *d* shell.

These points gave rise to heated discussions, and the chairman would like to make at this stage the following considered comment:

1. The concept of a van der Waals interaction between *d* shells, for which he was initially responsible, does not seem to him to hold much promise: it must be very sharply reduced by screening through the conduction electrons. There is, however, a possible increase in *s-d* correlation compared with free atoms, because the *d* band of metallic copper is more easily excited, thus more polarizable, than the *d* shell of a free atom. This interatomic effect has been partly taken into account in the band computation of Brooks' pupils. It seems to lead to an appreciable but insufficient increase in stability. More work in this field could use our present knowledge of plasmon excitations and might be of some interest.

2. From the band point of view, another possible factor is *s-d* mixing. It seems the most direct equivalent, in this language, to Brewer's point of view. It seems also that it can lead to a long-range pair interaction. It will therefore be sketched here on a simplified model, where one neglects the intrinsic width of the *d* band. There is no doubt that similar but more complex conclusions could be drawn for a *d* band of finite width.

Let us start therefore with a metal like Zn where the *d* band is definitely below the conduction band (Fig. 1). To go now to pure copper, we have to apply a repulsive potential V which shifts the *d* band into the energy range of the conduction band (Fig. 2). One knows also that the two bands will mix so as to give rise to bands which avoid one another. This is the *s-d* mixing which gives some *s* character to the full band and some *d* character to the conduction band.*

* Figure 2 has been drawn neglecting the orbital degeneracy of the *d* band. This would only introduce complexities in the neighborhood of energy E_d; the main conclusions would, however, still hold.

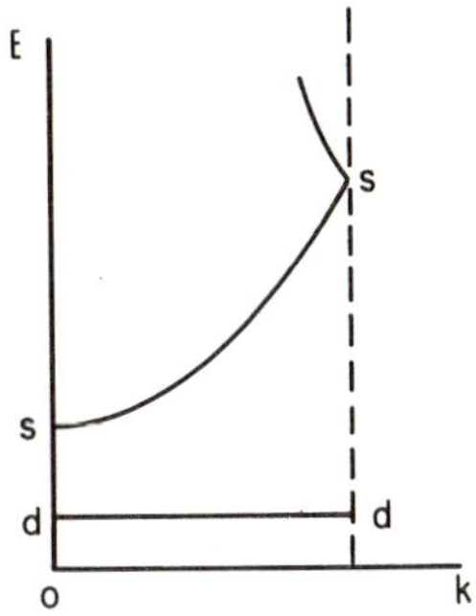

Figure 1

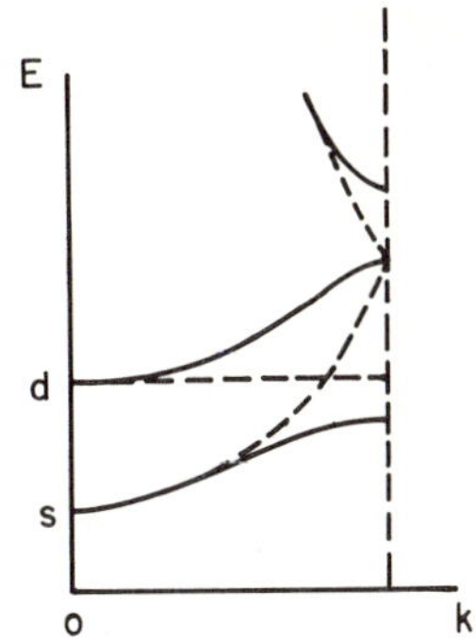

Figure 2

The mixing will be large over a region of the order of the matrix element V_{sd} around the crossing point C.

As always in such resonance effects, it will split the band structure without changing much its average energy. If therefore, as in copper, the Fermi level falls above the E_d but within a range of the order of V_{sd}, the *s-d* mixing will definitely stabilize the system. Values of V_{sd} of the order of an electron volt are likely, leading to appreciable stabilization. If now a Brillouin-zone limit $k/2$ is near to the crossing point C, the *s-d* mixing contributes to the band gap. If for instance, as pictured in Fig. 1, the *d* band is initially flat and the conduction band free-electron-like, the *s-d* mixing introduces by itself a band gap of the order of $V_{sd}^2/(E_s - E_d)$. This is pictured in Fig. 2 for the simple case where no other matrix element of V contributes to the band gap. The fact that the conduction band joins on the free-electron curve at the lower edge of the gap indicates that the presence of a Brillouin-zone limit below the Fermi level stabilizes the structure further.

Finally a *d* band with no width can be considered as a collection of nonoverlapping *d* shells. We could then consider the effect of *s-d* mixing on each atomic shell before treating their interaction. The *s-d* mixing transforms each *d* level into a "virtual bound level" of width V_{sd} in energy. As shown by Caroli in another context (Thèse 1966, Orsay), these virtual bound levels have a pair interaction with an asymptotic form such as const$\cdot\sin^2\delta\cos(2k_Fr + 2\delta)/r^3$. Here δ is a phase shift related to the position of the Fermi level to E_d. When, as in copper, the Fermi level is at a distance from the *d* band comparable to or larger than $|V_{sd}|$, the phase shift is known to be near to $\pi(\delta/\pi$ gives essentially the relative part of *d* character above the Fermi level).

The long-range interatomic coupling should then have the same type of spatial variation as for the more normal effective potentials discussed before. For metals such as copper ($E_m \gg E_d$), its strength should be proportional to the square of the part of *d* character which is not occupied, owing to *s-d* mixing. Because the phase shift δ is near to π, the scattering due to *s-d* mixing is small, and this validates again the concept of pair interaction. But contrary to Blandin's prediction, the amplitude of the long-range oscillations does not seem to change sign when δ changes from small values as assumed in the perturbation scheme to values near to π as in this model.

Electronic Structure of Calcium and Lead

Questions were put to Heine about the deviations from free-electron behavior in these two metals. Brewer was glad to hear that preliminary calculations of the band structure as a function of volume, carried out in Cambridge, suggested that the usual methods to compute pseudopotentials were not good enough, and that probably the *d* band should be brought in.

Concerning lead, Heine thought that Hume-Rothery's viewpoint that it was divalent was in a way consistent with the fact that its pseudopotential was attractive, contrary to most normal metals. As a result, one has *s*-like states below the gap, *p*-like ones above. The first band has therefore a nearly pure *s* character, while the states at the Fermi surface are mostly of *p* character. It seems to the chairman, however, that the nearly spherical nature of the Fermi surface can only be fitted with a total volume of 4 valence electrons per atom. Everybody seemed also to agree that as far as the spatial distribution of the electronic density was concerned, the question was probably meaningless.

Dynamic Effects

Choquard remarked that all the calculations are performed for a static lattice. Can one now take the lattice dynamics into account, by for example integrating the phonon spectrum with effective force constants in metals and nonmetals, including the structure factor, and calculating the effect of the atomic motion on the effective pseudopotential? Blandin suggested that one could take account of it by altering the dielectric constant, but that for low-frequency phonons this change would be small. Choquard agreed, but commented that he was thinking more of a Debye-Waller type effect, and that while the effect on the tail of the effective potential would be small, the nearest-neighbor interactions would clearly be modified, the effect being to smear out the effective potential. Pick commented that if one looks at the logarithmic derivative of the ratio of the elastic constants with the lattice parameters as a function of either temperature or pressure, for aluminum the results are more or less the same, suggesting that the pair-interaction functions will be sufficient, since one knows that the phonon spectrum that one gets from such a model is more or less correct. However, for a low-melting-point metal such as lithium or sodium, there is a clear difference between these two quantities. This implies that dynamical effects are important only very near the melting point. If one needs to calculate dilatation, the sample pair interactions are insufficient, and one needs to consider third- and fourth-order terms.

α-Phase Boundaries in the Alloys of Noble Metals

Except for an intervention of Massalski on Brewer's theory for β-brasses, which will be reported below, the discussion on alloys of normal metals concentrated mostly on this topic. For a constant band gap and using

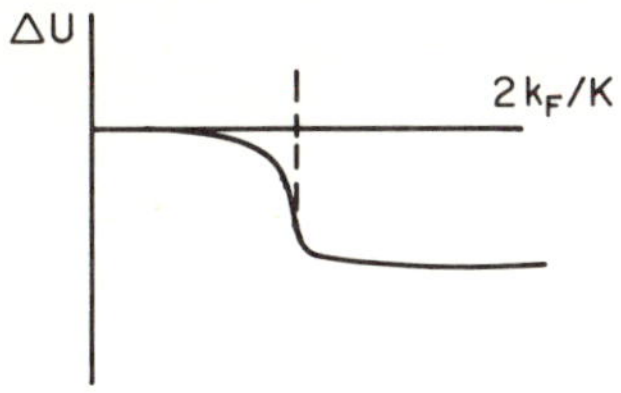

Figure 3

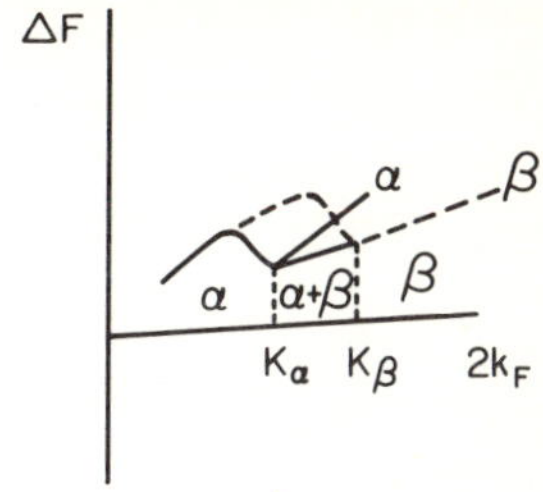

Figure 4

second-order theory, one finds that the lowering of energy ΔU due to a Brillouin-zone limit drops quickly at $k = 2k_F$, and has no real minimum: the further in the Fermi sphere the Brillouin-zone limit is, the stronger the stabilization (Fig. 3).

If one uses better than second-order techniques, one introduces small kinks in $\Delta U(k)$ but without changing the general form of $\Delta U(k_F)$. If one now compares the stability of various phases α, β . . . , one expects the boundary of the α phase to be on the right of $k_F = k_\alpha/2$ (Fig. 4). Thus the fact that the Fermi surface touches the Brillouin-zone limit in the pure noble metals does not invalidate Jones's explanation of the Hume-Rothery rules.

Seeger remarked that the phase boundary in Au-based alloys always comes before that of Ag- and Cu-based alloys; this seemed difficult to explain on the basis of the models proposed.

On a question by Pearson, Heine guessed that the Fermi surface did not touch the (200) faces in the α-phase field.

Finally, Perry cast some doubts on the vacancy effect mentioned by Heine in Sn-In alloys. This work was originally due to Raynor,[1] but recent work by Ridley[2] has failed to confirm it.

2 TRANSITIONAL METALS AND ALLOYS—BREWER'S THEORY

This section was meant to compare the relative merits of the traditional *d*-band— or more complex *dsp*-band—theories with Brewer's Heitler-London mode of approach.

As not much was said of the band point of view, it is perhaps worth recalling its main points in the simplest form. A contribution by Lomer then analyzes Brewer's point of view critically. A series of interventions by various people including Brewer himself are then recorded.

The Stability of Transitional Metals from the Band Point of View (Friedel)

From the band point of view, what seems to characterize these metals is the existence of a band of essentially *d* character of measurable width

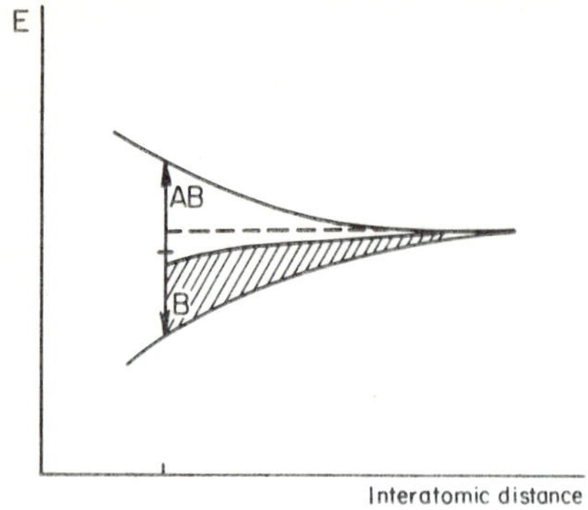

Figure 5

(about 5 ev from X-ray emission spectra) but narrow enough for a tight binding computation to be valid. There would also be a broader conduction band, mostly of s and p character, which provides most of the conductibility but with few electrons (perhaps 0.5 to 2 per atom) and thus not able per se to give more stability than the low-melting mono- or divalent normal metals. Of course, this is a simplification. There is mixing between s, p, and d bands. As discussed earlier for the noble metals, this can add somewhat to the stability.

The essential point about a tight binding approximation is that the broadening of the band when the atoms come closer is nearly symmetrical and larger than the general energy shift (Fig. 5).

In this "molecular orbital" approach, the stability of an incompletely filled d band then comes from the fact that the average energy of the occupied part necessarily is smaller than the energy for isolated atoms.

Physically, this lowering in energy comes from the fact that only the Bloch functions B with "bonding" character will be occupied, i.e., those where the electronic density between neighboring atoms is larger than the mere addition of atomic densities—the unstable antibonding states AB are correlatively empty.

It is then easy to show that when the d band is progressively filled, as occurs for a transitional series, the cohesive energy should vary nearly parabolically with the atomic number. Its maximum should be for a nearly half-filled d band, thus near to Cr, Mo, and W. This maximum value depends on the exact form of the density of states $n(E)$ for the d band; but, for a d band, it should be of the order of its width, thus about 5 ev/atom.

Along this line of approach, the relative stability of various phases for a given transitional element could be computed if one knew well enough the form $n(E)$ of the d band. This is unluckily not yet the case, except for the bcc structure. This is known, experimentally and theoretically, to be fairly sharply divided into bonding and antibonding halves of roughly equal weight, by a well-marked minimum in the density of states. The bcc structure is thus expected to be especially stable in the middle of the series, as is observed experimentally. The origin of this division of the band is to be found in the fact that in the bcc structure all the d functions can be chosen with their lobes pointing toward first or second nearest neighbors.

One can also make the following remarks:

For a given element, whatever the (fairly close-packed) structure, the *d* band will have a similar width, and the cohesive energy will thus be of the same order of magnitude. This even applies to transitional liquids, where a tight binding approach is possible (Part IV) or to diatomic molecules, where the molecular-orbital treatment would give rise to bonding and antibonding states separated by energies given by the same overlap integrals.

It can be shown that, for a crystal, the pure *d* bonding described here is necessarily metallic, if the band approximation is valid. This is because the bonding and antibonding parts of the *d* band cannot be separated by a band gap. It is tied up with the fact that each *d* orbital is used for more than one interatomic bond.

Of course, exchange and correlation corrections should be made to this very crude picture, especially to take into account the magnetic moment of the free transitional atom (and eventually magnetism in the metallic phase). However, these corrections are known to be definitely smaller than the cohesive energy. It might therefore seem better to do them *after* the band computation. Also, in the simple-minded Hartree scheme just outlined, one counts twice the electrostatic interaction between *d* electrons. In the tight binding approximation, the correction one should introduce is, however, practically independent of the lattice structure.

Critical Analysis of Brewer's Theory (Lomer)

On reflection of a few hours, and after conversation with Brewer, it seemed to Lomer that Brewer's idea is essentially this: "The binding energy for each structure of the solid is a smooth function of the number of binding electrons." The terms in this sentence are to be understood as follows:

1. The binding energy is to be measured from the lowest atomic-energy level of a suitable configuration. The erratic variation of this level removes much of the erratic variation of solid stability. (This might be altered if one used a "mean of multiplet," but data are difficult for this.)

A configuration is "suitable" if it has one *s* electron, or two *s,p* electrons, or three sp^2, and as many unpaired *d* electrons as possible.

2. The number of *s,p* outer electrons correlates with the structure

1	——	bcc
2	——	hcp
3	——	fcc

3. The cohesive energy depends smoothly on the number of *s,p* electrons and the number of *d* electrons.

The correlation proposed is well defined, and either works well or not.

The claim for high success must stand or fall on its own merits, not on the idea that one prefers to start the estimation of binding energy on another basis. Because of the large number of data incorporated empirically, a very high success rate must be demanded.

The normal-band theorist's approach is rather different. Starting with a bare ion lattice, one adds band electrons. The atom is built up *without net spin* so that its energy in some way corresponds to the energy of the center of gravity of a spectroscopic multiplet rather than to the *lowest* component of a multiplet. The structure of the bands depends on the assumed crystal potential, itself determined by the assumed division between d and s,p bands (see Mattheiss, MIT Quarterly Report 48, 1962, for work on V d^3s^2 and d^4s); self-consistency can in principle be applied (Waber at Los Alamos is in the midst of a good program of this work). In practice, one finds that the configuration for bcc transition metals is $d^{n-1}s$, and recent work by Loucks at Iowa on Y and Ho, and Altmann at Oxford on Sc and Y, suggests that a completely filled pair of bands may be assigned largely to $s\,p$, giving roughly $d^{n-2}\,s\,p$ for the hexagonal structure. Work by Waber on Cu-Ni bands confirms all previous views that any reasonable potential for fcc gives a configuration between $d^{n-0.5}s^{0.5}$ and s^1 for fcc structures.

The problem of rationalizing the Engels-Brewer correlation then appears to reduce to the following two questions:

1. Why should one be interested in the high-multiplicity atomic states, where exchange energy is very favorable, if one intends immediately to use this prepared state to make bonds in a large resonating system, where the bonding resonance breaks down the spin correlations, and screening alters the outer-electron states beyond recognition?
2. How can one believe any correlation based on three outer electrons fcc when it is so far from the final description we need for noble metals?

Brewer, as an empiricist, is entitled to ignore these questions. Band theorists who accept that the statistics of "success" presented by Brewer exceed those for other methods of correlation must, Lomer thinks, wonder whether his use of the lowest multiplet is significant, and whether the coincidence of band-theory estimates of the mean configuration of bcc and hcp transition metals and Brewer's criteria is entirely accidental. The impossibility of calculating the resonance energy of the degenerate bond system leads Lomer to suppose that Brewer's correlation can never be the basis of a quantitative theory: that in itself does not make it a useless predictor, but means it cannot lead to new understanding of stability parameters. There would seem little evidence for the correlation of three outer electrons ↔ fcc, and it may not be so easy in practice to separate the structures as is suggested.

Discussion

Hume-Rothery said that it would probably be of great interest if the present meeting would discuss the Engel-Brewer correlations, and say whether they are sound or not. He is naturally very sympathetic to any attempt which will forecast alloy structures, and that the theory has had some success one cannot dispute. In commenting on it therefore it would be better to concentrate on the fundamental assumptions on the one hand and some very serious objections on the other hand, and not to haggle over minor points. The fundamental assumption is that one can associate numbers of electrons with crystal structures: 1 e/a with bcc, 2 e/a with hcp, and 3 e/a with fcc. He didn't see how one could reconcile this last with copper, silver, and gold, since one must surely accept the Fermi-surface results which produce a surface containing just one electron per atom. He can, however, reconcile this with the model which attributes the higher bonding with van der Waals forces arising from the polarization of the *d* electrons. One must, of course, accept that there is additional bonding in these metals than can be provided by the single electron. The justifications for this assumption appear to be that in the second row of the periodic table one has Na (bcc) with 1 e/a; Mg (hcp) with 2 e/a; and Al (fcc) with 3 e/a. However, in fact at low temperatures Na is not bcc but close packed. This close-packed hexagonal phase is a genuine stable structure, according to Barrett. One cannot envisage finding another two electrons in sodium. The later elements of group II are fcc, not hcp; so that the basic assumption seems invalid.

Massalski referred to Zener's curve of the stability of the β-brasses, and commented that one must always take into account hexagonal phases in these systems. In all the systems formed by a noble metal with zinc or cadmium, with the exception of Ag-Zn, a martensitic transformation is observed (Fig. 6), in most cases the martensite being hcp or fcc with many faults. In fact, the equilibrium hexagonal phases appear to be stable in many of these systems, and they occur in the whole range of electron concentration from 1.2 to 1.8. He suggested that this was incompatible with the Engel-Brewer theory. If one assumes (with Brewer) that copper is 3-valent, then some valence transition would have to occur between the pure metal and the solid solutions, and then also one should get opposite effects on alloying, for example, tin (3-valent) with silver, and cadmium (2-valent) with silver, and this should show, for example, in the electronic specific heat; but it does not.

Brewer remarked that in pure copper two electrons were promoted from the *d* band. As one adds zinc, this promotion becomes increasingly difficult, with the effect that the apparent valency of copper falls.

Massalski does not entirely accept this, because according to Engel the hcp structure is only stable for electron-to-atom ratios of slightly less

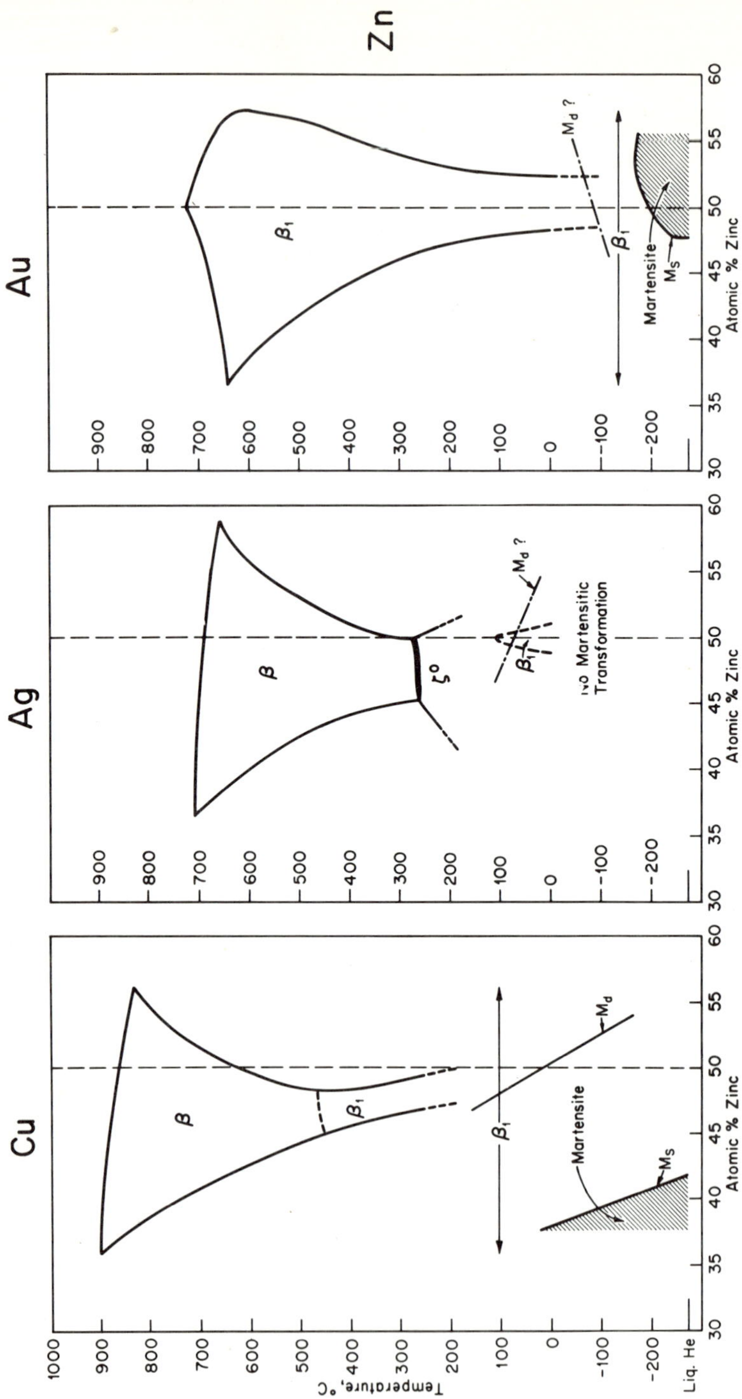

Cu
Ag
Au
Zn
Temperature, °C
1000
900
800
700
600
500
400
300
200
100
0
-100
-200
Liq. He
30
35
40
45
50
55
60
Atomic % Zinc
β
β_1
β_1
Martensite
M_s
M_d
ζ^o
M_d ?
No Martensitic Transformation

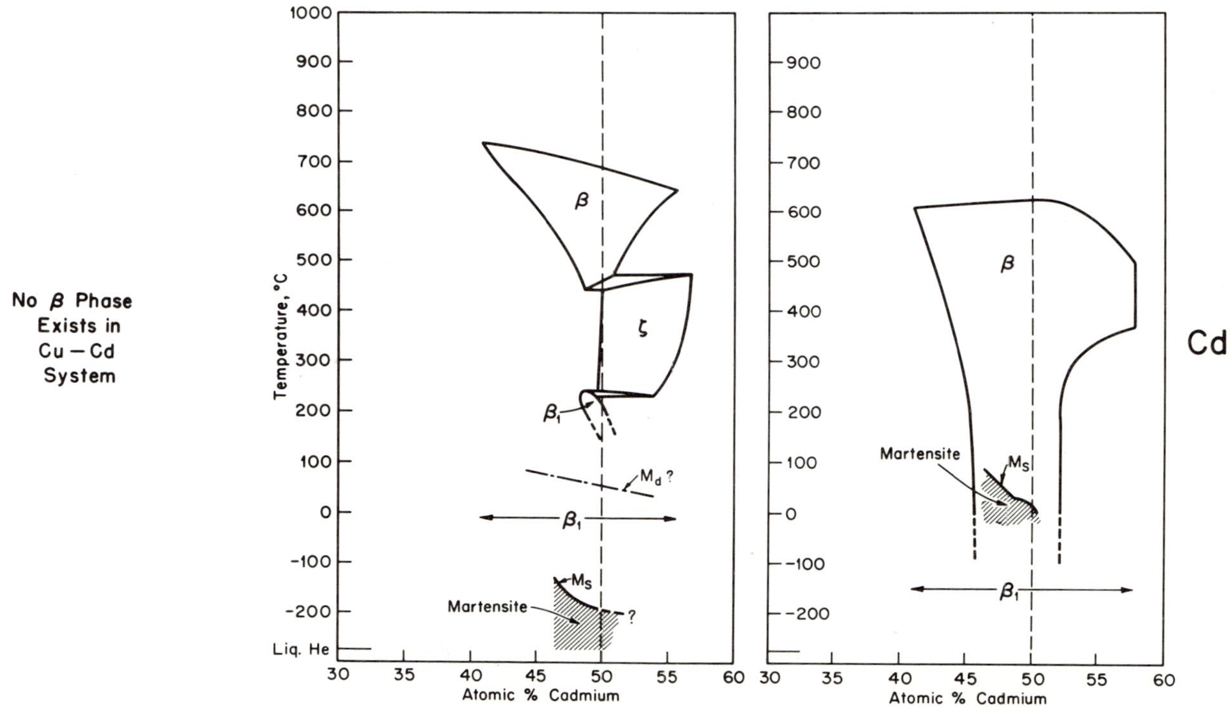

Fig. 6 Martensite transformations in the β phase of alloys of the noble metals with zinc and cadmium.

than 2. Further, there are examples in alloys where all three structures are stable at the same composition—for example in 20.2 a/o Cu-Ga, which is bcc at high temperatures, fcc at intermediate temperatures, and hcp at the lowest temperatures.

Pearson said that the main problem of matter in the solid state is that one does not know how the electron levels of the atoms (known for the free atoms from spectroscopic data) interact as the atoms condense in the solid state. Crystal structure gives the resultant of these complex interactions, and it is therefore the first and most important experiment of the solid state. If one can understand crystal structure, and distortions of crystal structures, then one has a potential source of direct information on electron interactions. Brewer's development of Engel's observations, to the extent that he says that the electronic basis for the occurrence of at least the three fundamental metallic structures is understood, therefore calls for careful consideration and scrutiny.

Brewer asked first what is a reasonable range of values for energies of different structures. Many people have implied that they must all be close together, apparently based on the fact that all the listed heats of transformation are very small. However, since the free energy must be the same at the transformation temperature and pressure, and the entropies cannot differ very much, it follows that the heats of transformation also cannot differ very much. That does not imply that the heats of the structures that are not found must be small also. If one thinks about strongly bonded materials, like for example the transition metals with bonding energies of 70 to 200 kcal, the majority of them have only one structure, even though they have been studied over wide ranges of temperature and pressure. There is no evidence from their alloying behavior that any other structure will be stable. There are interactions in these strongly bonded metals very much stronger than the quantities that have been discussed so far. For example, the atomization energies are up to 200 kcal, but the heat of fusion is only about 5 to 10 kcal; so the major bonding is also present in the liquid, and is thus not calculable from theories which depend on a consideration of the crystal state. On the basis of these and other considerations, he feels that there is no justification for the assumption that all structures will have closely similar energies.

Pearson made the point that the experimental Fermi surface is a one-electron surface. A second point concerned the *d*-electron configuration in the second and third periods. One knows from the chemistry of sulfides and selenides of these elements that in an octahedral ligand field there is a very large splitting, Δ, of the energies of the two sets of *d* electrons. This Δ is so large for the second and third periods that one never does get the high spin state. One is also led to believe that in the octahedral field of the metal this Δ is going to be of the same order of magnitude, and so one does not expect in fact to have a high spin state in the metal. If one did, one would have ferromagnetism as in the first long period, where the splitting is very much smaller. Consider for example iridium, which is fcc. Brewer would regard this as d^6sp^2. The cohesion is then discussed in terms of the four unpaired d electrons and the s,p electrons. In view of the absence of the high spin state, surely one must regard the electrons as paired. Further there is some evidence that one can't always mix one's d electrons: for example, consider the PtMn phase which has the Co-Au-I superstructure. In this, one has layers of magnetic Mn atoms alternating with layers of Pt atoms. This phase has a range of stability from about 32 to 66% Pt, and studies show (lattice parameter, magnetic susceptibility, and neutron diffraction), that the magnetic manganese atom has four unpaired spins. The Mn-Mn distance in the manganese layers is 2.8 A, and in analogous nickel compound (NiMn) only 2.6 A, giving very strong coupling involving one of the five manganese d electrons, which leads to antiferromagnetism. Two of the manganese d electrons point toward the adjacent platinum layer. If there was strong interaction between the d states in the platinum and the manganese (which are about 2.8 A apart), one would not of course have one's four spin moments. There may be two explanations of why there is no interaction: There may be a filled Pt shell, or antibonding states at the top of the band; but there is not, in general, mixing, as one would perhaps envisage from normal alloy chemistry. Also in compounds where one has a shell model where the transition metals don't interact, again one can, as expected, give examples where there has been no d mixing. For example $(Fe_{1/2}Ni_{1/2})As_2$ must come out as an *average* d^5 configuration. However, all the true d^5 compounds have the monoclinically distorted $CoSb_2$-type structure, but $(Fe_{1/2}Ni_{1/2})As_2$ does not: it has the undistorted marcasite structure. Hence one must have a mixture of d^4 iron and d^6 nickel.

Brewer replied to Hume-Rothery. In fact, the basis or justification for the theory is much wider than Hume-Rothery stated. One example he quoted is this: All 18 metals of the first six groups have the body-centered cubic structure. This comes directly from the Engel theory, but Brewer challenged anyone else who could produce a theory which gave the same result. Friedel asked if this involved classifying Na as a bcc metal. Brewer said yes, and Friedel remarked that it certainly was not at 0°K. Brewer remarked that in some other cases—zirconium for example—a cph phase supersedes the bcc at low temperatures. From his point of view, this was not important, since (as appeared before) this means that the structures must have very nearly the same stability: it is enough that the elements exhibit the structure at all. Further, his later studies have been confined to strongly bonded metals, since it is obvious for weakly bonded metals that the energies of the various structures are going to be much the same. With regard to copper, there seems to have been a concensus of opinion that, as long as Brewer did not attribute a magnetic moment to his d holes, the question was, in band terms, how much admixture of d character in the unoccupied part of the conduction band an s-d mixing or a polarization of the d shells actually meant. There was a general agreement of ignorance. The band protagonists, however, stuck to the fact that there should be a full (and nearly all d) band and a Fermi surface enclosing one nearly free electron per atom. This point has been already discussed (Sec. 1) above.

Kaufman disagreed with Brewer's earlier point about the relative stability of phases, illustrating this by projection of phase boundaries (Fig. 7). Brewer said that his previous discussion had shown that he did not really feel that because some energy differences are small (which everyone agrees with), it follows that all are. The fact that the solidus and liquidus run together on the phase diagram merely makes the point that the interactions in the liquid are as strong as those in the solid, but it tells one nothing about relative stability. Kaufman said an enthalpy difference of atomization which is larger than the enthalpy of fusion of the pure metal implies a melting point of the unstable phase which is below 0°K.

Drowart remarked that the comparison of the heats of sublimation of the transition and rare-earth metals with the first excited levels of the atoms gives correlations very similar to those of Brewer. The same correlations

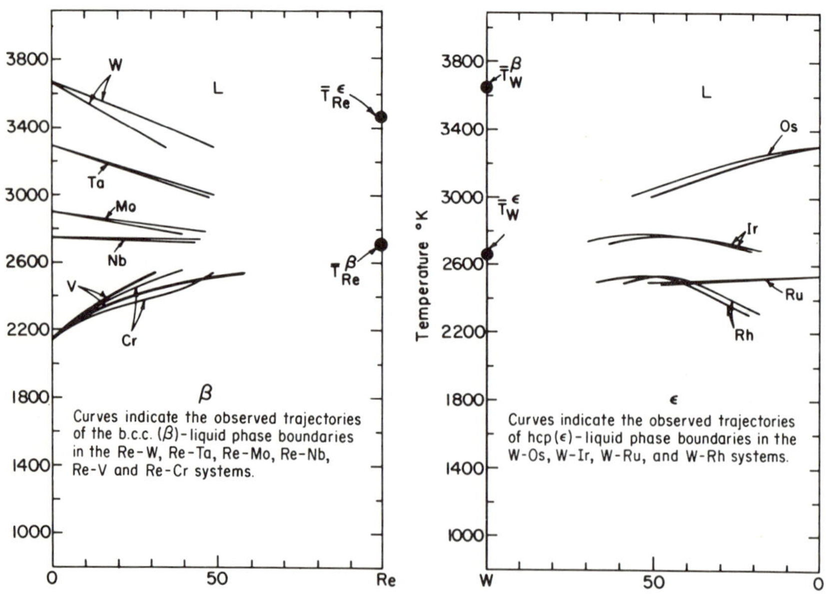

Fig. 7 Comparison of computed melting temperatures of bcc Re and hcp W with β/L and ϵ/L equilibria.

hold in the gaseous oxides of the transition and rare-earth metals. With reference to previous comments about ligand field effects, a recent paper by Jørgensen shows that for the gaseous oxides and halides of the transition metals the ligand field and molecular orbital treatment taking spin-orbit coupling into account are related and lead to analogous conclusions. In the comparison of the dissociation energy of the diatomic molecules of the metals with their heat of sublimation a great regularity appears which extends to all usual metals including the transition elements. Wherever the heat of sublimation is high, the dissociation energy is large. The dissociation energy of the Cu_2, Ag_2, and Au_2 molecules being three to five times larger than those of the alkali metals, it seems likely that the former are not single-electron metals.

POSTMEETING WRITTEN DISCUSSION: SEEGER

I should like to state my viewpoint regarding the validity of the pseudopotential and pseudo-atom approach more clearly than was possible during the oral discussion.

I have no objections to the idea of the pseudopotential as a physical concept; with the present art of computing being what it is, the use of the pseudopotential concept in order to take into account the effect of inner-core electrons is probably a necessity in any quantitative treatment of metals of high atomic number Z. My misgivings concern in particular the way in which the application of this concept is tied together with the use of second-order perturbation theory in most treatments, e.g., those reported in Harrison's recent book.[3]

The two basic ideas of these treatments are the following: (1) The pseudo-atom idea: In a metal each atom carries with it, irrespective of its position relative to its neighbors, a screened potential with long-range oscillations, the so-called pseudopotential. (2) The pseudopotential is assumed to be so weak that it may be treated by second-order perturbation theory.

It appears to me that these two assumptions are, as a rule, incompatible with each other. To see this, consider a monovalent metal. In order to permit the linear superposition of the pseudopotentials of neighboring atoms, the main part of the pseudopotential must be well concentrated within the atomic cell. This is the assumption usually made in the free-electron calculation for vacancies (or interstitial atoms) in monovalent metals (see, e.g., A. Seeger[4]). From these treatments, based on a variety of potentials, it is well known that second-order perturbation theory does not suffice for reliable calculations. This is simply a consequence of the Friedel neutrality condition, which says that the potential must be strong enough to repel (in the case of the vacancy) or attract (in the case of a pseudopotential or an interstitial) one electron per atom. In the example (vacancy in Cu) treated by Stehle (see Ref. 4), the constants in the asymptotic electron density

$$\rho_{el} = -\frac{\alpha_F}{2\pi^2 r^3} \cos (2k_F r + \varphi_F) \tag{1}$$

come out to be $\alpha_F = -0.478$, $\varphi_F = -\pi/2 + 0.198 = -1.313$. Mann[5] has generalized Eq. (1) to anisotropic energy surfaces and has treated explicitly the following case which corresponds particularly well to the concept of the "pseudo atom": The crystal has a conduction band with one electron per atom. The conduction electrons are characterized by their Wannier function $a(r)$. A perturbative potential $V(r)$ is assumed to be so localized that the matrix elements

$$V_{ij} = \int a(r - p_i) V(r) a(r - R_j) d\tau \tag{2}$$

are different from zero only for $i = j = 0$. In this case an equation similar to Eq. (1) holds, with α_F replaced by $-\sin \pi\alpha(E_F)$ and φ_F replaced by $-\pi\alpha(E_F)$. Here $\alpha(E_F)$ is a generalized phase shift at the Fermi energy E_F. In a monovalent metal the pseudopotential must be such as to attract a screening cloud containing just one electron. Mann[5] shows that this is equivalent to $\alpha(E_F) = -\frac{1}{2}$. This means that in Eq. (1) we have $\phi_F = 1$ and $\varphi_F = \pi/2$. The usual second-order perturbation-theory treatment gives, however, $\varphi_F = 0$ and is applicable only if $\alpha_F \ll 1$.

We see that in these examples the conditions for the applicability of the perturbation theory are grossly violated. The situation is even worse for multivalent metals, where the pseudopotential must be even stronger in order to attract the correct screening charge. The only way to make φ_F small (apart from the accidental cancellation of terms in the expression for φ_F, which, however, does not ensure the validity of the perturbation theory) is to make the potential so extended that all partial wave phase shifts at the Fermi surface become small. Such a potential, however, would have to be so extended that it would have a substantial magnitude outside the atomic cell and would therefore be unsuited as a pseudopotential. The limiting case which might still be considered as a possibility (although a completely unrealistic one) for a pseudopotential is an attractive square-well potential with a radius equal to the Wigner-Seitz radius. For such a potential the phase shifts at the Fermi surface were given by Seeger and Mann[6] for a monovalent fcc free-electron metal. Computing α_F and φ_F from these values gives us $\alpha_F = 0.159$ and $\varphi_F = -0.733 \approx -\pi/4$. We see that even in this extreme case φ_F is not negligible, although α_F happens to be small (on account of accidental cancellations). For realistic pseudopotentials the situation is aggravated by the change of sign of the potential within the atomic cell.

The result of the preceding discussion is that the two basic assumptions formulated above cannot be fulfilled simultaneously. This means that we have no right to expect the second-order perturbation treatment of pseudopotentials to give results which agree with experiment. This is particularly true if the results depend on the phase φ_F of the long-range oscillations, which is, in general, given quite wrongly by the perturbation treatment. Unfortunately this applies to almost all the defect calculations, e.g., those on the interaction between solute atoms and vacancies and the calculation of stacking-fault energies. I am inclined to consider in this field the agreement with experiment, which has occasionally been reported, as fortuitous.

Furthermore, our result means that the usual justification for using pair interactions between atoms in metals is invalid (compare the contribution by Pick). It remains to be seen whether a valid justification for the use of pair interactions can be found.

REFERENCES

1. Lee, J. A., and G. V. Raynor: *Proc. Phys. Soc.* **B67**:737 (1954).
2. Ridley, N.: *J. Inst. Metals* **92**:123 (1963–1964).
3. Harrison, W. A.: "Pseudo-potentials in the Theory of Metals," W. A. Benjamin, New York, Amsterdam, 1966.
4. Seeger, A.: *J. Phys. Radium* **23**:616 (1962).
5. Mann, E.: *Phys. Stat. Sol.* **13**:293 (1966).
6. Seeger, A., and E. Mann: *J. Phys. Chem. Solids* **12**:326 (1960).

Part Three

INTERMETALLIC PHASES

ELECTRON CONCENTRATION AND PHASE STABILITY IN LAVES PHASES AND Ni-Al ALLOYS*

W. E. Wallace and R. S. Craig

Department of Chemistry, University of Pittsburgh, Pittsburgh, Pennsylvania

INTRODUCTION

The studies described in the present paper fall into three categories:

1. Electronic specific-heat studies of four pseudobinary Laves-phase systems
2. Electronic specific-heat studies of β-phase Ni-Al alloys
3. Phase relationships in the pseudobinary Laves-phase systems $LnCo_2$-$LnAl_2$, in which Ln represents Gd or Er

Although the three groups of studies are independent of one another, they share a common feature. In each case there are one or more structural or constitutional features that relate or can be related to the electron concentration.

* This work was assisted by a contract with the U.S. Atomic Energy Commission.

The factors responsible for the stability of alloy systems have only been adumbrated in the several decades since the advent of quantum theory. Of the several important factors involved, the electron concentration is the one whose role has perhaps been elucidated most fully. However, even in this instance the issue is far from clear. This is especially true for systems containing transition metals, for reasons that are amplified below.

The present work has involved transition-metal systems in the main. It is the purpose of this paper in part to show that by combining magnetic, calorimetric, and structural data in the fashion described below some of the approaches and procedures applied earlier to non-transition-metal systems can be applied almost equally well to systems having one or more transition-metal components. It now appears that, given sufficient information, the role of the electron concentration in transition-metal systems is only slightly more obscure than in systems composed entirely of non-transition metals.

The work reported in this paper was, in the largest sense, planned so as to test at a few points the rather shaky and very incomplete theoretical framework currently used in discussing stability of alloys. To anticipate that which follows it has been found that a rather wide variety of phenomena (involving mainly transition-metal systems) are consistent with the general concepts of alloy stability developed by Mott, Jones, Hume-Rothery, Laves, Witte, and others several decades ago.

ELECTRONIC SPECIFIC-HEAT STUDIES OF PSEUDOBINARY LAVES-PHASE SYSTEMS

General Remarks

A number of systems of this nature have received attention as regards their electronic specific heats in our laboratory to date. Of these, four have been studied to the extent that results for them are essentially complete. These are the following: $MgCu_{2-x}Al_x$, $MgCu_{2-x}Si_x$, $Zr_{1-x}Ti_xFe_2$, and $ZrFe_{2-x}Co_x$. In the first two of these the electron concentration (e.c.) as customarily defined (valence electrons per atom) is clearly varying. It is instructive, as will be discussed further below, to study the variation of the electronic specific coefficient γ with e.c. The variation in the e.c. in them is more difficult to determine—a complication which they share with other systems containing $3d$ transition metals. In fact a rigorous, unambiguous determination of their e.c. is not possible at present, although a suggestion, which appears reasonable, for determining it is presented below. These complications diminish the probability of an unambiguous interpretation of results for intermetallics containing transition metals. However the challenge heightens interest and studies of them concurrent with non-transition-metal systems improves chances of meaningful interpretation of the results obtained.

The $MgCu_2$-base Systems

Before discussing the results on the two $MgCu_2$-base systems it is appropriate to review briefly the significance of e.c. in Laves-phase systems. The stability of these phases is usually ascribed to geometrical factors.[1–3] The relative sizes of the two components is such that packing is particularly efficient.[4] In the three prototype Laves phases—the cubic $MgCu_2$ (C_{15}), the hexagonal $MgZn_2$ (C_{14}), and the hexagonal $MgNi_2$ (C_{36}) structures—the average coordination number is $13\frac{1}{3}$. If contacts are regarded as bonds, then the Laves phases are "more bonded" than their components, which are in 8- or 12-fold coordination. Disregarding varying strengths of these "bonds," we expect the compounds to have lower energy than, and hence be stable with respect to, the equivalent mixture. However, these considerations hold for each of the three prototype structures. On this basis each would be equally stable, whereas it is well known that a given substance usually adopts a particular structure. Clearly at least one other secondary factor is involved. Laves and Witte[1] in their classic work in the '30's clearly identified this, at least for non-transition-metal systems, as the e.c. The e.c. makes its influence felt through the interaction of the Fermi surface with the Brillouin zone boundaries. The shape and size (in terms of electron capacity) of a particular zone are of course dependent on the structure. The size of the Fermi volume is determined by the e.c. It follows therefore that at a given e.c. the Fermi surface–zone boundary (FSZB) interaction is structure-dependent, and hence one of the Laves-phase structures may be energetically favored over the others—or so the argument has run.

To make the argument of the preceding paragraph more explicit we review the work done on $MgCu_{2-x}M_x$ systems, where M = Zn, Al, Si, and other polyvalent metals. Laves and Witte showed[1–4] that replacement of monovalent Cu by the polyvalent metals could be accomplished at first without a structure change, but with increasing x the cubic C_{15} structure gave way to either the C_{14} or C_{36} hexagonal structures. They found the C_{15} structure to be unstable above an e.c. of about 1.7 and interpreted this in terms of Brillouin zone filling—a type of FSBZ interaction. As filling of the zone approaches completion, the density of states $N(E)$ drops so that a further increase in e.c. is energetically unfavorable and an alternate structure is adopted in order that electrons can be accommodated at a lower energy. This postulate of zone filling was supported by several lines of evidence: magnetic susceptibility,[5] hydrogen solubility,[6,7] and semi-empirical calculations treating the valence electrons as nearly free.[8,9]

Several systems were studied by Klee and Witte[5] and were observed to be

Pauli paramagnets. The susceptibility of the electron gas, which provides a rough measure of $N(E)$, can be computed from the measured susceptibility. However, in this case where the zone is nearly full, the orbital contribution to the susceptibility is anomalous, and the estimate of the trend of density of states from susceptibility work is particularly difficult. Nevertheless Klee and Witte showed that the susceptibility behavior generally supported the concept that zone filling was responsible for the destabilization of the $MgCu_2$ structure.

The same conclusion was reached on the basis of H solubility measurements.[6,7] Implicit in the argument is the concept that H ionizes as it dissolves and raises the e.c. Hence as the zone is filled, the H solubility decreases. This has been confirmed by Lieser and Witte for several appropriate systems.

In 1957 Bode[8] on the basis of quantum-mechanical considerations established a trend for the density of states for e.c. between 1.48 and 1.68. These were extended by Schnabl[9] in 1964 to lower e.c. These calculations not only showed the expected decline in density of states near the phase boundary but also revealed structure in the density-of-states–e.c. curve at lower e.c. which appeared in the experimental results as well.

Thus magnetic susceptibilities, H solubilities, and theoretical work were all consistent with the viewpoint that the limit of the cubic phase is brought on by zone filling. However, each of these pieces of evidence is developed either using assumptions of questionable validity or employing corrections that cannot be rigorously and reliably made. Direct confirmation of the expected decline in $N(E)$ near the phase boundaries has not been forthcoming until recently. One of the authors (R.S.C.) and his associates have recently presented determinations of $N(E)$ for the $MgCu_{2-x}Al_x$[10] and $MgCu_{2-x}Si_x$ systems.[11] These have been established from the electronic specific-heat coefficient γ, obtained by specific-heat determinations in the liquid He range.

Results for the $MgCu_{2-x}Al_x$ system are summarized in Fig. 1. In the upper portion of the diagram are shown the $N(E)$ values obtained in the course of the calorimetric work. The sharp fall in $N(E)$ expected near the $MgCu_2$ phase boundary if the zone is nearly filled is clearly apparent. The susceptibility (χ) values obtained by Klee and Witte are shown in the lower part of Fig. 1, together with the values computed from the relationship

$$\chi = \tfrac{2}{3}\mu^2 N(E) \tag{1}$$

using $N(E)$ values computed from the measured γ's. Equation (1) gives the "free electron" susceptibility of the system (ignoring of course the diamagnetism of the ion cores) except that the measured density of states is used instead of the value for a free-electron gas. Differences between the measured χ values and those computed for Eq. (1) are evident.

The $N(E)$ values provide no indication of the fall in χ shown in Fig. 1 at low Al concentrations. The reality of this fall was, however, validated by measurements in this laboratory, which duplicated the results of Klee and Witte.

Results for the $MgCu_{2-x}Si_x$ system are shown in Fig. 2. The measurements are again consistent with the concept of a nearly filled zone at the phase boundary.

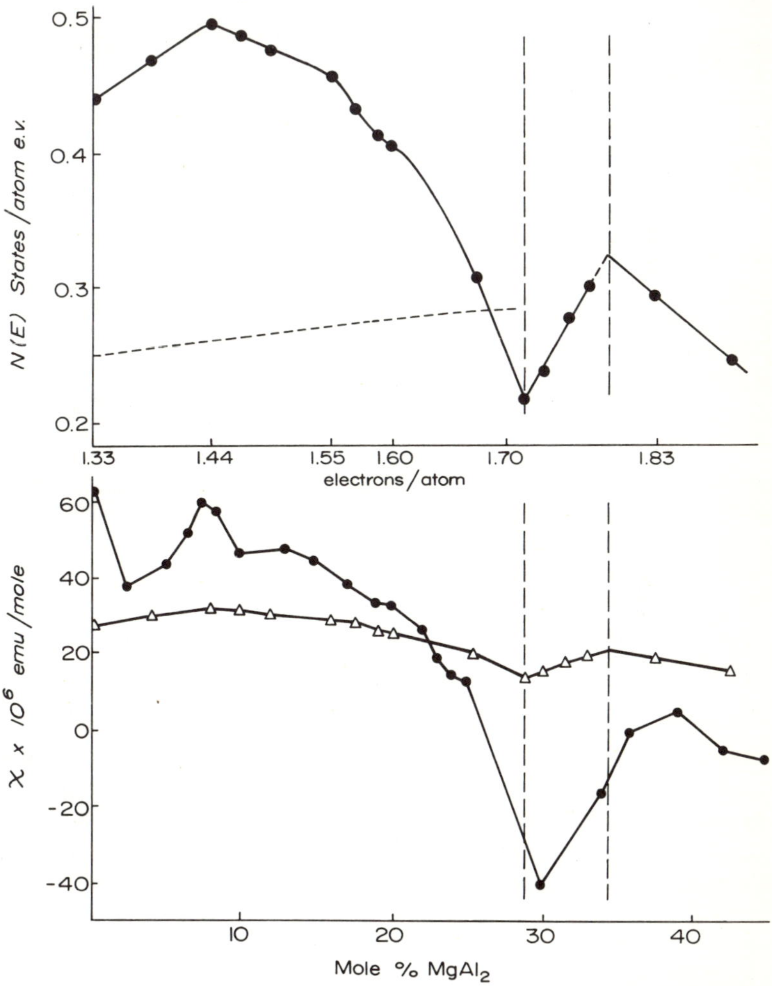

Fig. 1 Density of states $N(E)$ and susceptibility χ versus electron concentration and composition for the $MgCu_{2-x}Al_x$ system. The curve through the triangles is computed from Eq. (1) and represents the "free-electron" χ except that the measured $N(E)$ is employed instead of the free-electron $N(E)$. The dashed curve is the free-electron $N(E)$. [*From P. I. Slick, C. W. Massena, and R. S. Craig, J. Chem. Phys.*, 43: 2792 (1965).]

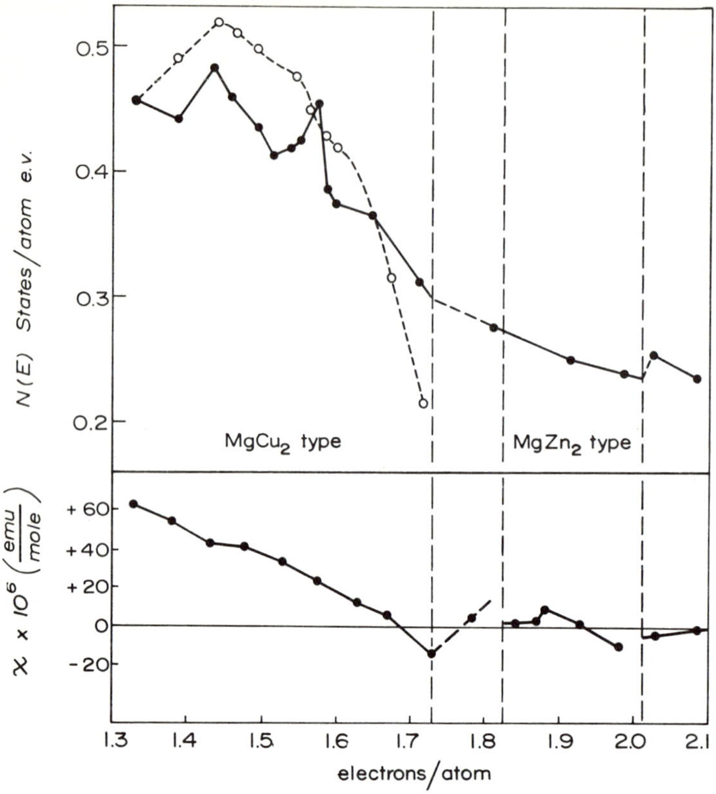

Fig. 2 $N(E)$ and χ versus electron concentration for the $MgCu_{2-x}Si_x$ system (solid curve). $N(E)$ versus electron concentration for the $MgCu_{2-x}Al_x$ system is also included (dashed curve). [*From P. I. Slick, C. W. Massena, and R. S. Craig, J. Chem. Phys.*, 43: 2792 (1965).]

Comparison of the measured density-of-states curves with those calculated are shown in Fig. 3. Agreement between computed and observed results at the higher electron concentrations is striking. It is clear that the calorimetric measurements fully support the concept of a nearly filled zone at the phase boundary in these two $MgCu_2$-base alloy systems, and the notion that FSZB interactions constitute the primary destabilizing factor for the $MgCu_2$ phase.

The $Zr_{1-x}Ti_xFe_2$ System

Difficulties in establishing the e.c. of transition-metal systems have been alluded to earlier. As a working hypothesis, we employ the following practice: (1) If the substance exhibits a local moment, we exclude the electrons involved in establishing that moment from the e.c. (2) If the

TABLE 1 ***Contributions to the Electron Concentration***

Intermetallics	3*d* shell, electrons per atom			Contribution to the sea of delocalized electrons, per atom
	+ spin	− spin	Total	
Zr				4
Ti				4
Fe	4.6	2.4	7.0	1
Co	5.0	3.2	8.2	0.8
Ni	5.0	4.4	9.4	0.6
Fe in $TiFe_2$	5.0	4.6	9.6	−1.6
Fe in $ZrFe_2$	5.0	3.4	8.4	−0.4
Co in $ZrCo_2$	5.0	5.0	10.0	−1
Co in $GdCo_2$	5.0	4.0	9.0	0
Co in GdCoAl	5.0	5.0	10.0	−1

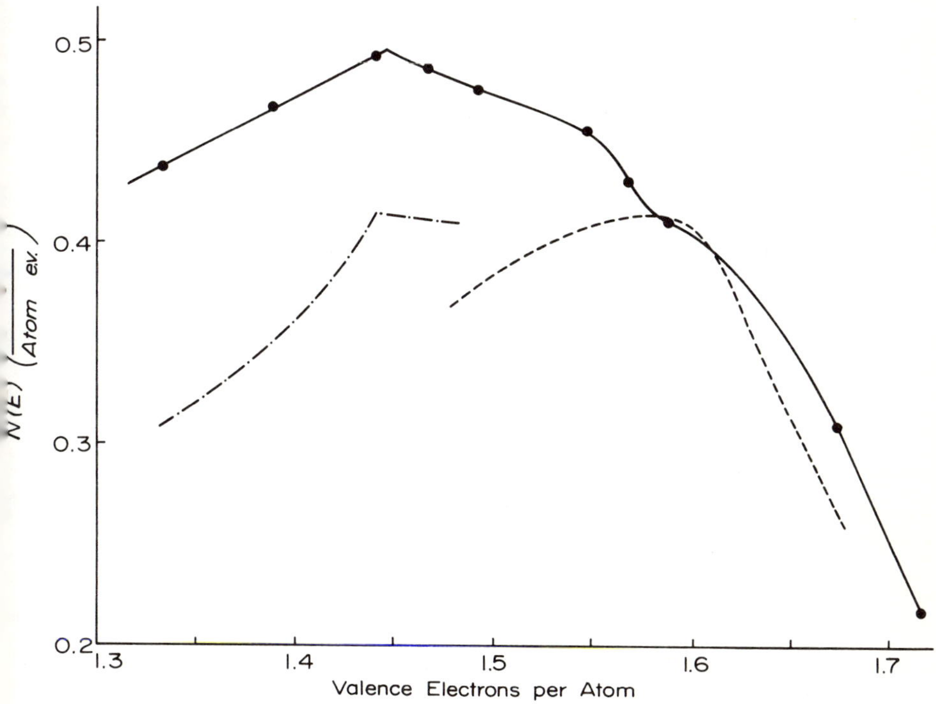

Fig. 3 $N(E)$ versus electron concentration. The solid curve gives the experimental values for the $MgCu_{2-x}Al_x$ system.[10] Results obtained theoretically by Bode[8] and by Schnabl[9] are also shown.

atom in the element possesses a local moment, but because of electron transfer this moment has been reduced to zero in the intermediate phase, we assume the *d* electrons to have remained localized and exclude them from consideration. (3) If the material is a Pauli paramagnet, we include in establishing the e.c. all electrons beyond those in the appropriate inert-gas atom—Ar for the 3*d* metals. On this basis the situation in Zr, Ti, Fe, Co, Ni and for Fe and Co in $ZrFe_2$, $TiFe_2$, $ZrCo_2$, $GdCo_2$, and GdCoAl is as set forth in Table 1. The difference in the + and − spin populations is established from the observation saturation moment in each case. The moments of the elements are of course well known. The moment in $ZrFe_2$ and $ZrCo_2$ was taken from the work of Piegger and Craig.[12] That for $TiFe_2$ was estimated by Wallace[13] on the basis of Mössbauer measurements. The moments of Co in GdCoAl and in $GdCo_2$ are from work of Oesterreicher and Wallace[14] and Farrell and Wallace,[15] respectively.

TABLE 2 ***Electron Concentration in Several Intermetallics***

In electrons per atom

$TiFe_2$	0.27
$ZrFe_2$	1.07
$ZrCo_2$	.67
$GdCo_2$	1.00
GdCoAl	1.67

The number of 3*d* electrons in Ni is 9.4, as indicated by magnetic measurements. Theoretical work[16] and evidence based on Mössbauer measurements[17] indicate $3d^7$ for Fe. Cobalt is taken as $3d^{8.2}$ by interpolation. Using the data in column 5 of Table 1, e.c. for the several intermetallics listed in Table 2 were computed.

Density-of-states data for the $Zr_{1-x}Ti_xFe_2$ and $ZrFe_{2-x}Co_x$ systems[18] are shown in Figs. 4 and 5. Results for the latter system could not be obtained for compositions on the $ZrCo_2$-rich side of the magnetic-moment minimum[12] (lower part of Fig. 5). In the region of the dashed curve in Fig. 5 the temperature dependence of the heat capacity is anomalous. Representative data are given in Fig. 6. The data for $ZrFe_2$ typify a "well behaved" material. Alloys containing 0 to 10 and 50 to 100 mole % $ZrFe_2$ were well behaved. As the $ZrFe_2$ concentration was reduced to 45 mole %, extra heat capacity was noted at the lowest temperatures, and at still lower temperatures the slope of the C_p/T versus T^2 became negative over the temperature range covered (see Fig. 6). For the alloy containing 20 mole % $ZrFe_2$ the γ value is obviously without significance, and the values for alloys such as that with 45 mole % $ZrFe_2$ are of questionable significance. The cause of the anomalous behavior for compositions between 20 and 45 mole % $ZrFe_2$ is not clear. Suggestions[19] involve small ferromagnetic clusters as postulated by Schröder[20] or the hyperfine structure of Co as has been observed for elemental Co.[21] Of these the former seems more likely.

Accepting the e.c.'s listed in Table 2, we see that in the $ZrFe_{2-x}Co_x$ system the e.c. varies from 0.67 to 1.07 and in the $Zr_{1-x}Ti_xFe_2$ system from 0.27 to 1.07. The electronic specific-heat results, and the density of states derived therefrom, may be considered in the light of the Brillouin zone structure for these materials in relationship to the e.c. Relevant information is presented in Table 3. In the $ZrFe_{2-x}Co_x$ it appears that the Fermi surface lies entirely within the B zone. The decline in $N(E)$ with increasing mole fraction of $ZrFe_2$ can be ascribed to filling of the B zone, and in particular to contact at the $\{200\}$ faces.

As regards the $Zr_{1-x}Ti_xFe_2$ system, the sharp decline in $N(E)$ beyond an e.c. of about 0.60 apparently stems from progressive filling of the B zone for the $MgZn_2$ structure, particularly to contact at the $\{1\bar{1}2\}$ faces. The

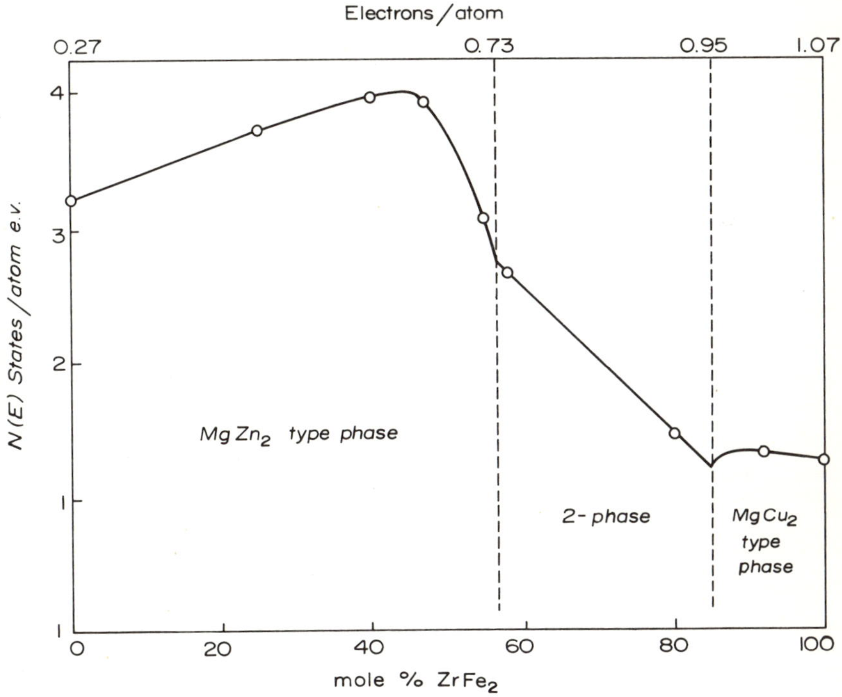

Fig. 4 $N(E)$ versus electron concentration and composition for the $Zr_{1-x}Ti_xFe_2$ system.

destabilizing effect of this interaction causes a transformation into the C_{15} structure at higher e.c.'s.

The analysis of the results for the $ZrFe_{2-x}Co_x$ system suggests that the C_{15} structure would become unstable at slightly higher e.c.'s. In this respect it is significant that $ZrNi_2$ (e.c. = 1.33) does not form, despite a favorable size ratio, whereas YNi_2 (e.c. = 1.00) exists and in the C_{15} structure. ($MgCu_2$, with e.c. = 1.33, does exist but with its Fermi

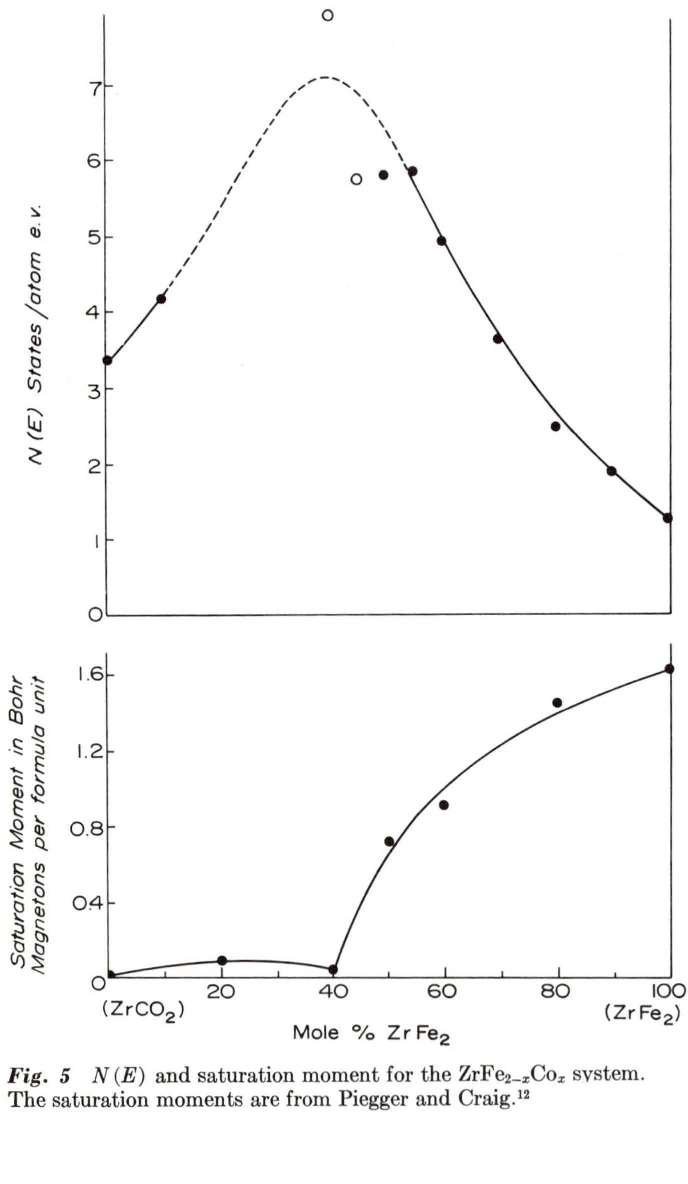

Fig. 5 $N(E)$ and saturation moment for the $ZrFe_{2-x}Co_x$ system. The saturation moments are from Piegger and Craig.[12]

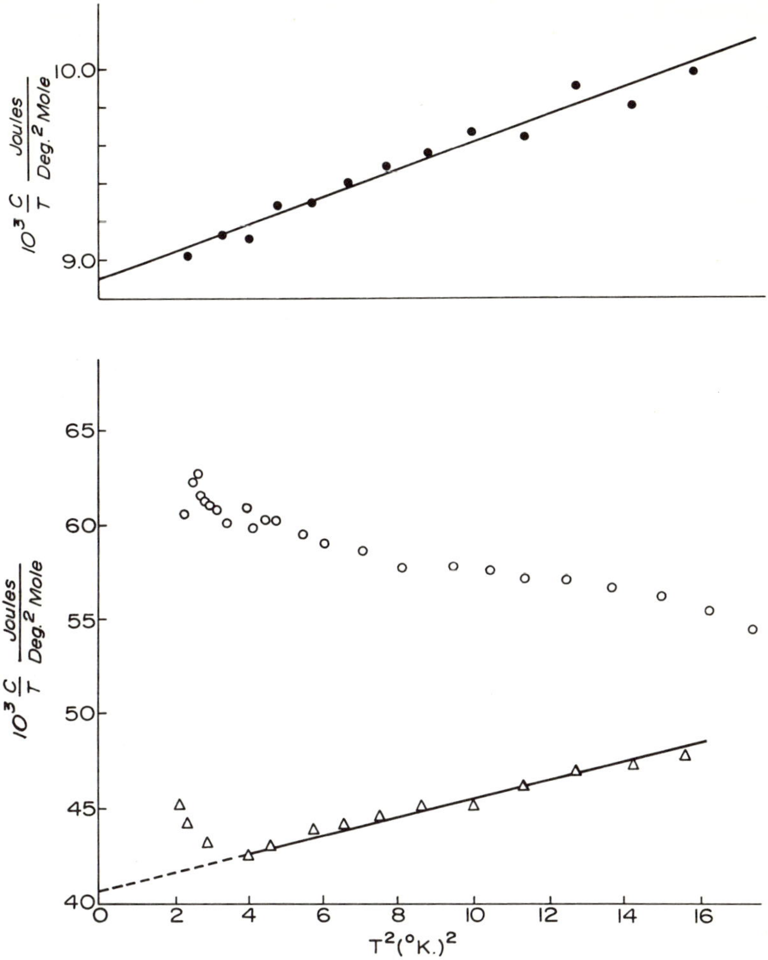

Fig. 6 Heat-capacity data C for $ZrFe_2$, filled circles; $ZrFe_{0.9}Co_{1.1}$, triangles; and $ZrFe_{0.4}Co_{1.6}$, empty circles.

surface extending into the C zone. In this case the energy gap at the {220} faces is sufficiently small to permit overlap into the third zone, whereas in $ZrNi_2$ it evidently is not.)

We have emphasized in the foregoing remarks the importance of FSBZ interactions in the stability of these phases. It should be clear that other factors may also be significantly involved—in particular, the relative electronegativities of the constituent atoms and the possibility of electron transfer. The transfer of electrons to the $3d$ transition metal from its strongly electropositive partner undoubtedly contributes to the stability of $ZrFe_2$, $TiFe_2$, etc. In this respect it is significant that $ZrCu_2$ and $TiCu_2$ do

TABLE 3 ***Electron Concentration in Relation to Brillouin Zone Structure for the*** $MgCu_2$ ***and*** $MgZn_2$ ***Structures***

In electrons per atom

Plane	Zone	Inscribed sphere	Filled zone
$MgCu_2$ structure:			
111	*A*	0.23	0.38
220	*B*	0.99	1.33
311	*C*	1.59	1.83
222		1.81	
$MnZn_2$ structure:			
$1\bar{1}0$		0.19	
110	*A*	1.00	0.67
002		0.23	
$1\bar{1}2$	*B*	0.60	1.33
110		1.00	
112		1.60	
201	*C*	1.68	1.93
004		1.84	
201	*D*	1.68	2.32
004		1.84	

not form despite favorable relative atomic sizes. In these, electron transfer, such as in the Fe and Co compounds, cannot occur. Presumably this together with the unfavorable e.c. is responsible for the nonexistence of these compounds. Electron transfer is of course possible for the corresponding Ni compounds, from which we conclude, in line with the remarks above, that the FSBZ interaction at e.c. = 1.33 is sufficient to render hypothetical $ZrNi_2$ and $TiNi_2$ unstable.

ELECTRONIC SPECIFIC-HEAT STUDIES IN RELATION TO THE STABILITY OF THE DEFECT β-PHASE Ni–Al ALLOYS

These alloys have been considered to be of unusual interest ever since Bradley and Taylor[22] showed that they contain an extensive population of lattice vacancies. This does not occur in Ni-Al or in the alloys rich in Ni. However, in Al-rich alloys the vacancy concentration increases with increasing Al content until about 7% of the lattice sites are empty at the phase boundary. This is clearly not an entropy effect, which leads to 0.001 to 0.1% vacancy population in ordinary solids at temperatures well away from their melting points. The defect Ni-Al system must be energetically favored for the vacancies to be so abundant in it. The characteristics of Ni-Al and Ni-Cu-Al alloys have suggested that e.c. was significantly involved in forming the vacancies. Electronic specific-heat measurements on the β-phase Ni-Al system have been made[23] partly to ascertain the $N(E)$ trend with composition and partly to ascertain whether the origin of the anomalously high-vacancy population could be elucidated.

The composition dependence of $N(E)$ is shown in Fig. 7, together with the Fermi energy relative to that for the alloy containing 44 a/o Al. The relative Fermi energy was obtained by an appropriate integration of the $N(E)$ versus e.c. data, the e.c. being calculated using valencies of 3 and 0 for Al and Ni, respectively. There are two points of interest as regards the results in Fig. 7. First, NiAl has been regarded as an electron compound, with stability originating in FSBZ interactions. It has traditionally been thought that in such materials the Fermi surface is distorted by the proximity of the zone boundary, giving a peak in the $N(E)$ curve. The results show no indication of such a maximum. The second point of interest is that the hypothetical perfect lattice, as regards its Fermi energy, is energetically unfavorable relative to the actual lattice by as much as 0.4 ev. The high population of vacancies indicates, as pointed out above, that the defect lattice has a lower energy than the ideal lattice. The

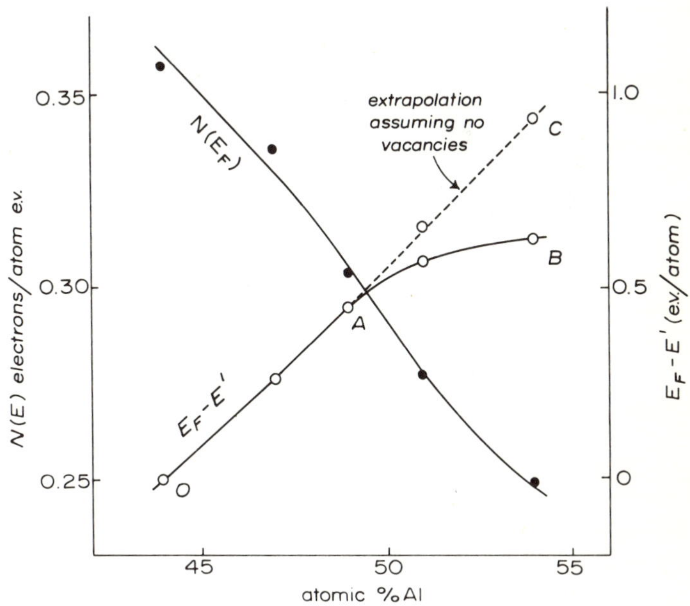

Fig. 7 $N(E)$ and Fermi energy (relative to $Ni_{0.56}Al_{0.44}$) versus compositions for β-phase Ni-Al alloys. The dashed curve shows the Fermi energy for the ideal lattice. The solid curve *OAB* shows the Fermi energy for the vacancy concentration observed by Bradley and Taylor.[22]

measurements of Galli et al.[23] indicate that the Fermi energy is an important, perhaps the principal, contribution to the stability of the defect lattice.

LAVES PHASE CONTAINING LANTHANIDES AND THE $GdCo_{2-x}Al_x$ AND $ErCo_{2-x}Al_x$ SYSTEMS

Laves phases containing the lanthanide elements combined with Fe, Co, and Ni have been extensively studied in this laboratory in recent years.[24] The LnM_2 compounds (Ln = a lanthanide metal or Y) with Fe, Co, and Ni all occur in the C_{15} structure. Magnetic evidence, neutron diffraction work,[25,26] and Mössbauer work on the Fe compounds indicate extensive transfer of electrons to the 3*d* transition metal. We regard this as an important stabilizing factor in these compounds and ascribed the nonexistence of $LnCu_2$ to the absence of this factor.

Fragmentary evidence has indicated that FSBZ interactions may be involved in the stability of the lanthanide Laves phases. Wernick et al.[27] noted that whereas $DyCo_2$ and $DyAl_2$ had the C_{15} structure, DyCoAl was of the C_{14} structure. They noted a similar effect for the corresponding Fe compound. Petzow, Steeb, and their associates in Stuttgart[28] have noted similar effects in U-containing Laves phases, which resemble in some respects their lanthanide counterparts. Insertion of Al undoubtedly raises the electron concentration in these cases, and it appeared that the structure change in these systems might be due to FSBZ interactions and might be similar to the well-investigated effects in the prototype systems. No systematic investigations of the lanthanide systems have been carried out

TABLE 4 ***Stability Ranges and Electron Concentrations in $LnCo_{2-x}Al_x$ Systems***

In electrons per atom

Structure	x in $LnCo_{2-x}Al_x$	Electron concentration at corresponding x values
C_{15}	0–0.5	1.00–1.25*
C_{14}	0.86–1.4	1.48–2.2
C_{15}	1.7–2.0	2.6–3.0

* Computed assuming that Co has absorbed 0.5 electron per atom from the conduction band (see text). Values for $x \geq 1.48$ are computed assuming that Co has absorbed 1 electron per atom.

to date. The work of Wernick et al., while interesting and important, was limited to the three compositions cited.

We have recently begun studies of a series of systems represented as $LnCo_{2-x}Al_x$. Structures are determined for various values of x to ascertain the composition range over which the C_{15} and C_{14} structures exist. Simultaneously magnetic-moment measurements are made to take into account the electrons absorbed by Co. Work to date carried out by Dr. H. Oesterreicher has involved the systems in which Ln is Gd or Er. He has found the C_{14} structure for Gd-Co-Al and Er-Co-Al. Results, as regards location of the phase boundaries, are substantially the same for the two systems, and we provisionally regard all the systems in which Ln is a heavy lanthanide[29] as essentially identical. Phase boundaries as determined from the Gd and Er system are given in Table 4, together with appropriate e.c.'s. The latter merit some comment. Cobalt in several heavy lanthanide-Co compounds has been shown to have a moment of 1 μ_B and is hence regarded as in a $3d^9$ state. Its contribution to the e.c. is hence zero (see Table 1). Dr. Oesterreicher has found zero Co moment in Gd-Co-Al, indicating a $3d^{10}$ configuration. The Co configuration for $GdCo_{1.5}Al_{0.5}$ is as yet unknown, and an intermediate value is assumed, $3d^{9.5}$. If so, Co absorbs 0.5 from the conduction band, and the e.c. for $GdCo_{1.5}Al_{0.5}$ is 1.25. The corresponding values with an assumed $3d^9$ and $3d^{10}$ configuration are 1.5 and 1.0 electrons per atom.

The occurrence of the phase boundaries at the same composition (and hence same e.c.) strongly suggests that the structure changes originate with effects similar to those identified for the prototype structures, i.e., zone filling and FSBZ interactions. On the basis of information given in Table 3 it appears that in $LnCo_2$ the C_{15} structure becomes unstable when there is contact at the $\{220\}$ faces and the B zone is nearing completion. At higher Al concentrations the C_{14} structure forms with the Fermi surface at first in the C zone. This structure persists until the filling of the D zone nears completion; thereupon it reverts to the C_{15} structure. Instability near e.c. = 1.9 might have been anticipated from the filling of the C zone. The fact that the C_{14} structure persists beyond an e.c. of 1.93 suggests that the energy gap at the $\{112\}$ face is small.

The above ideas are quite speculative and obviously need support from additional work before they can be regarded as established. However, it seems clear from work to date on the lanthanide Laves phases as well as on those containing Ti and Zr that e.c. is a factor and FSBZ interactions substantially affect stability here as well as in the prototype systems.

REFERENCES

1. Laves, F., and H. Witte: *Metallwirtschaft* **14**:645(1935); **15**:15, 840 (1936).
2. Raynor, G. V.: *Progr. Metal Phys.* **1**:34 (1949).

3. T. B. Massalski in "Theory of Alloy Phases," p. 63, American Society of Metals, Cleveland, Ohio, 1957.
4. Laves, F.: In *ibid.*, p. 124.
5. Klee, H., and H. Witte: *Z. Physik. Chem.* **202**:352 (1954).
6. Lieser, K. H., and H. Witte: *Z. Physik. Chem.* **202**:321 (1954).
7. Lieser, K. H., and H. Witte: *Z. Elektrochem.* **61**:367 (1957).
8. Bode, H. J.: *Z. Elektrochem.* **61**:376 (1957).
9. Schnabl, H.: *Ber. Bunsenges. Physik. Chem.* **68**:549 (1964).
10. Slick, P. I., C. W. Massena, and R. S. Craig: *J. Chem. Phys.* **43**:2788 (1965).
11. Klar, E., R. M. Mallya, and R. S. Craig: *J. Chem. Phys.*, in press.
12. Piegger, E., and R. S. Craig: *J. Chem. Phys.* **39**:137 (1963).
13. Wallace, W. E.: *J. Chem. Phys.* **41**:3857 (1964).
14. Oesterreicher, J., and W. E. Wallace: To be published.
15. Farrell, J. J., and W. E. Wallace: *Inorg. Chem.* **5**:105 (1966).
16. Wood, J. H.: *Phys. Rev.* **117**:714 (1960).
17. Walker, L. R., G. K. Wertheim, and V. Jaccarino: *Phys. Rev. Letters* **6**:98 (1961).
18. Craig, R. S., C. W. Massena, and R. M. Mallya: To be published.
19. Wallace, W. E., and R. S. Craig: AEC Report, NYO-3454-1, p. 17, May 15, 1965.
20. Schröder, K.: *J. Appl. Phys.* **33**:880 (1960).
21. Heer, C. V., and R. A. Erickson: *Phys. Rev.* **108**:896 (1957).
22. Bradley, A. J., and A. Taylor: *Proc. Roy. Soc.* **A159**:56 (1937). See also H. Lipson and A. Taylor: *ibid.* **A173**:232 (1939).
23. Galli, G., R. M. Mallya, and R. S. Craig: To be published.
24. See Ref. 15 for extensive references to these studies.
25. Moon, R. M., W. C. Koehler, and J. Farrell: *J. Appl. Phys.* **36**:978 (1965).
26. Felcher, G. P., L M. Corliss, and L. M. Hastings: *J. Appl. Phys.* **36**:1001 (1965).
27. Wernick, J. H., S. E. Haszko, and D. Dorsi: *J. Phys. Chem. Solids* **23**:567 (1962).
28. Petzow, G., and O. A. Sampaio: *Z. Metallk.* **56**:503 (1965). See this paper for many references to work of the Stuttgart group.
29. Evidence obtained by Dr. Oesterreicher indicates a different situation in at least some of the $LnCo_{2-x}Al_x$ systems in which Ln is a light lanthanide.

DISCUSSION *on Paper Presented by W. E. Wallace*

FRIEDEL: 1. There is *no* valid relation between the band gap and the X-ray structure factor, if you go out of the nearly free electron into Heine's or Harrison-Pick's type of pseudopotentials.

2. I would expect the *d* band to dominate the values in (Zr, Ti)Fe_2 compounds. Why do you exclude them from your discussion?

3. To what do you attribute the anomalies of γ in Zr(Co, Fe)$_2$ compounds? Have these compounds a magnetic-order temperature in or below that range?

WALLACE: 1. Actually in Bode's calculation the energy gap at the {311} zone boundary was established from the peak in the density-of-states versus electron-concentration curve. The diffraction intensities were only involved in establishing the ratio of the energy gaps of the {311} and {222} faces. The density of states is relatively insensitive to the {222} energy gap, which may explain why Bode's calculation is as valid as it seems to be.

2. Only the *d* electrons associated with Fe are excluded. Those contributed by Ti and Zr are included and are involved in the large density-of-states values.

3. The cause of the anomalies is not entirely clear. If they are of magnetic origin, the effect must be subtle because the Curie temperatures are around 100°K or higher, and magnetic order should be complete. The anomalies may originate with defective portions of the lattice (around dislocations, or the like), where magnetic order is impaired or with the nuclear hyperfine field acting on the Co nucleus.

BECK: Recently A. Jordan at the University of Illinois made low-temperature specific-heat measurements with $Ni_{1-x}Al_{1+x}$ alloys similar to those discussed by Professor Wallace. I can confirm that the results also are quite similar. No indication was found of any effect of the large vacancy concentration reported previously for this phase.

Based on experience with a large number of similar cases, I am strongly inclined to suspect that the "upturn" in the C/T versus T_2 curves at decreasing temperature observed with certain (Zr-Fe-Co) Laves-phase alloys is due to a magnetic contribution to the low-temperature specific heat. This type of anomaly occurs in many binary and ternary systems in which a ferromagnetic alloy is diluted by the addition of a nonferromagnetic component. It always occurs in a limited range of dilution where long-range magnetic order just ceases, but short-range cooperative effects still appear to be present. K. Schröder interpreted such magnetic specific-heat contributions in several cases as results of magnetic clusters.

WALLACE: Possibly the matter should be left at the point of my response to Professor Friedel. However, I shall add a few remarks. If I had to speculate on the origin of the heat-capacity anomalies, I should concur with Professor Beck in ascribing these to magnetic effects. The complexity of the matter is such that the exact process is rather difficult to specify. As an example of this complexity, I cite the following: Despite the fact that Co in $ZrCo_2$ is nonmagnetic and Fe in $ZrFe_z$ is magnetic, Co seems to be magnetic and Fe nonmagnetic in Zr-Fe-Co. This conclusion is reached on the basis of the observations of (1) the existence of a saturation moment (Piegger and Craig, Ref. 12) and (2) unpublished Fe^{57} Mössbauer work on Zr-Fe-Co by myself, which shows no Fe nuclear hyperfine field. Such subtleties as this in the "exchange" of magnetism between Fe and Co could generate extra heat capacity, although the exact process is unclear.

NEVITT: As I understand your analysis of your magnetic data in deriving electron concentration, you have assumed that electrons contributing to the magnetic moment in phases containing Fe are to be excluded from consideration in computing the electron-atom ratio.

The measurements of electronic specific heat, by Professor Beck and his coworkers, on the body-centered cubic solid solution of the first long period suggest that the unpaired *d* electrons that are responsible for the moment contribute to the electronic specific heat. Is it your view that the *d* electrons providing the moment in Fe-containing Laves phases are entirely localized states?

WALLACE: Yes, although the difference is more in degree than in kind. The calculations of Wood showed the Fe wave functions to be more localized for the highest-energy states. Our view is that owing to electron transfer the energies are raised further and become even more localized at thc upper energy limit—so much so that they can be treated as completely localized and excluded from consideration in much the same fashion as the core electrons. The difference between Fe in the Laves phases and in the alloy phases is brought out by Fe^{57} Mössbauer measurements. In the alloy systems, e.g., Fe-Al alloys, the moments and hyperfine fields are not proportional, whereas in the Laves phases they are.

REPORT ON THE DETERMINATION OF THE FERMI SURFACE OF ORDERED β′ PHASES

W. B. Pearson, J.-P. Jan, and Yoshitami Saito

National Research Council, Ottawa, Canada

INTRODUCTION

Many years ago Hume-Rothery and others recognized that the range of stability of the α, β, γ, and ϵ phases formed between B group metals depended on the "electron concentration"—the number of valence or conduction electrons per atom, or precisely, on the number per primitive crystallographic cell. Jones (1934–1937)[1] accounted for this behavior in terms of a falling density of states $N(E)$ with increasing electron concentration, which resulted from prior touching of Fermi surface and Brillouin zone boundaries. This ultimately makes some other structure more stable because of its more favorable $N(E)$ relationship.

According to Jones's principles, the β phases with disordered bcc structure generally become unstable with respect to some other structure when the electron concentration increases to about 1.5 electrons per primitive cell (1 atom) because of falling $N(E)$ resulting from *prior* contact of the Fermi surface with the {110} faces of the dodecahedral first Brillouin zone. The β' phases with the ordered CsCl structure (2 atoms per primitive cell) become unstable when the electron concentration reaches about 3 electrons per cell because of prior touching of the Fermi surface with the {110} faces of the dodecahedral second zone (the first zone is a cube), although the exact electron concentration must depend on temperature and other factors. Although details of Jones's (1937)[1] theory of β-phase stability have been assailed, his main principle of decreasing $N(E)$ has recently received direct experimental confirmation from the determination of the electronic specific heat of β'-CuZn by Veal and Rayne,[2] which indicates a decreasing density of states with increasing electron concentration in the range from 43 to 50 a/o Zn.

We report on the progress of experimental determinations of the Fermi surface of β' phases, undertaken to gain further understanding of problems of phase stability. The possibilities of gaining direct confirmation of Jones's principles are somewhat limited since the *detailed* Fermi surface of alloys and metallic compounds can only be determined at compositions in which each independent crystallographic site is fully and perfectly occupied by one component only. As this condition is relaxed either by imperfect ordering or by nonstoichiometric composition, electron scattering soon becomes too large to permit detailed determination of the Fermi surface by methods such as the de Haas–van Alphen effect, and therefore it can be applied only to ordered β' phases with compositions very close to the equiatomic.

The periodicity in reciprocal field of de Haas–van Alphen oscillations in the magnetic susceptibility gives a measure of extremal cross-sectional areas of the Fermi surface. We have been able to observe these oscillations in stoichiometric-ordered β'-CuZn, β'-AgZn, and β'-PdIn at low temperatures. The last compound is particularly interesting since Hume-Rothery "electron" phases with transition metals as one component occur at the expected electron concentration, provided that it is assumed that the transition metal contributes no electrons to the valence band (is zero-valent). PdIn is reported to occur from about 40 to 53 a/o In; with Pd zero-valent, the electron concentration would be 3 electrons per cell at 50 a/o In, and a complete Fermi surface determination should allow a distinction to be made between say 3.6 and 3 valence electrons per cell.

EXPERIMENTAL DATA

The reduced frequencies of the orbits observed for β'-CuZn and one of the corresponding orbits found for AgZn and PdIn are shown in Fig. 1 as a function of crystal orientation when the magnetic field is scanned in the {110} plane. The figure also gives the nonextended extremal orbits predicted by the free-electron model for the β' phase, assuming 3 electrons per primitive cell (cf. Harrison).[3] This model indicates that there are holes in the corners of the first cubic zone, electrons in the second zone, and very small pieces of Fermi surface in the third and fourth zones, though these latter are not expected to exist in the real crystal because of crystal potential. The expected orbits given in Fig. 1 and indicated diagrammatically in Fig. 2 are thus: band 1, octahedral hole orbits (A_1^R, B_1^R, C_1^R); band 2, hole orbits about ⟨100⟩ (A_2^R), hole orbits about

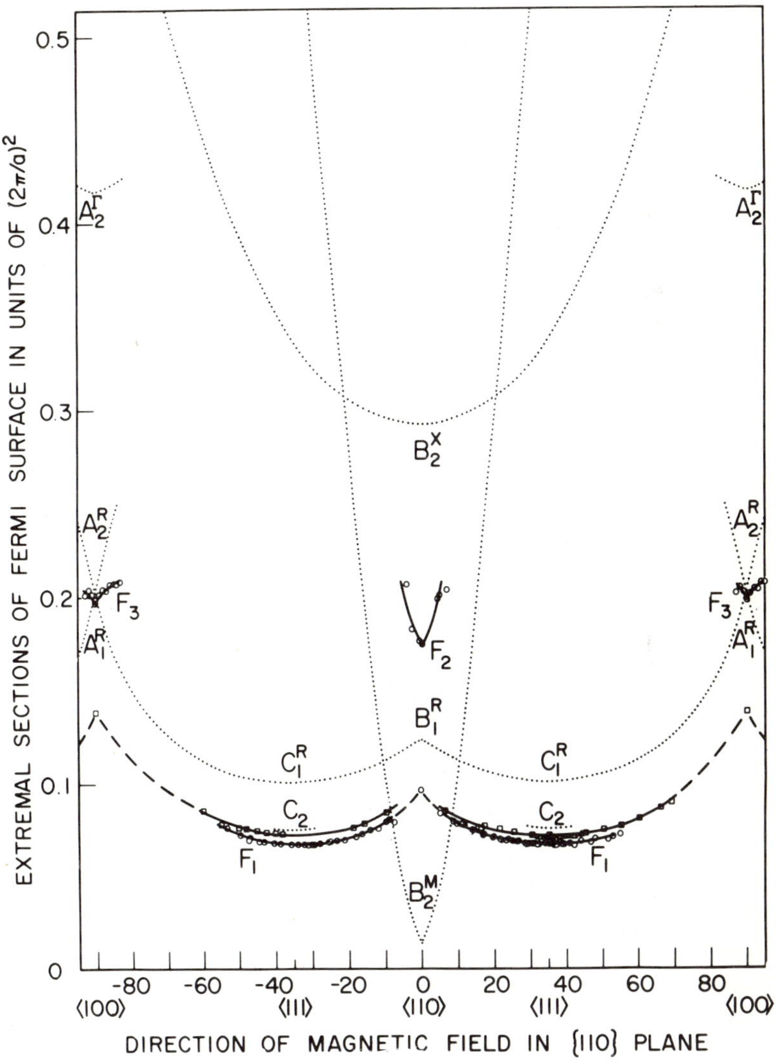

Fig. 1 Extremal sections of Fermi surface in units of $(2\pi/a)^2$ as a function of the direction of the magnetic field scanned in a $\{110\}$ plane. See text for explanation of nomenclature. Circles, observed data for β'-CuZn; squares, β'-PdIn, and triangles, β'-AgZn, observed during F_1 period; dotted line, free-electron predictions of nonextended extremal orbits in bands 1 and 2 for β'-CuZn.

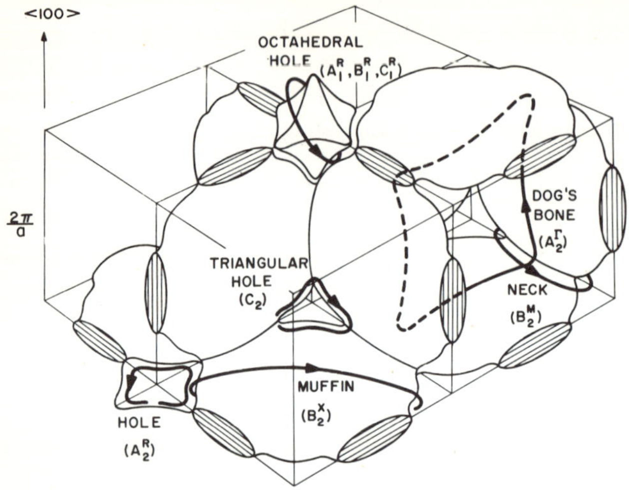

Fig. 2 Diagrammatic sketch of free-electron Fermi surface for β′-CuZn in the extended zone scheme, indicating various extremal orbits.

⟨111⟩ (C_2), electron neck orbits about ⟨110⟩ ($B_2{}^M$), electron muffin orbits about ⟨110⟩ ($B_2{}^X$), and dog's-bone hole orbits about ⟨100⟩ ($A_2{}^\Gamma$).*

The observed orbits are identified with the free-electron predictions by their angular variation. This clearly assigns the CuZn and PdIn orbits labeled F_1 to the holes in the first band, although the magnitude of the observed orbits is actually similar to that expected in the vicinity of C_2. In CuZn, oscillations labeled F_2 have been observed in and close to the ⟨110⟩ direction, and attributed to the neck orbits $B_2{}^M$. The free-electron model cannot possibly be taken as a guide for this particular orbit, since it arises in a region where the free-electron Fermi sphere barely overlaps the {110} faces of the second Brillouin zone, and the lattice potential is expected in these circumstances to modify considerably the free-electron predictions, greatly increasing the diameter of the orbits close to ⟨110⟩. Our identification is supported by the band-structure calculation of Johnson and Amar[4] who predict a neck section of 0.19, which compares favorably with the observed value of 0.17. A further confirmation of the correctness of the identification arises from the possibility of working out a 12-cone model of the Fermi surface, ignoring the energy gaps at the faces of the first (cubic) Brillouin zone. This model accounts very well for the change of electronic specific heat with Zn concentration on a rigid-band model

* In the notation adopted here, A, B, and C refer to ⟨100⟩, ⟨110⟩, and ⟨111⟩ sections, respectively; the index 1 or 2 refers to the band number; and the superscript Γ, M, R, or X indicates that the orbit is centered at the center of the cubic zone, the middle of the cube edge, the corner of the cube, or the middle of a cube face, respectively. C_2 is not centered at a symmetry point and therefore has no superscript.

(Veal and Rayne),[2] if a constant energy gap of 3.5 ev is assumed across the faces of the second (dodecahedral) zone, and with an effective mass of 1.05 in units of free-electron mass. This gap, again, agrees remarkably well with the value calculated by Johnson and Amar.[4] The success of the model indicates that the superstructure ordering probably does not modify appreciably the density of states, and that a discussion of phase stability can probably be carried out in terms of the dodecahedral zone alone.

It should be mentioned that thc first and second bands are degenerate along the edges of the cubic zone, at least when spin-orbit splitting is ignored. As a result, the Fermi surfaces in the first and second band touch at the cube edges, giving rise to complicated multiple orbits and

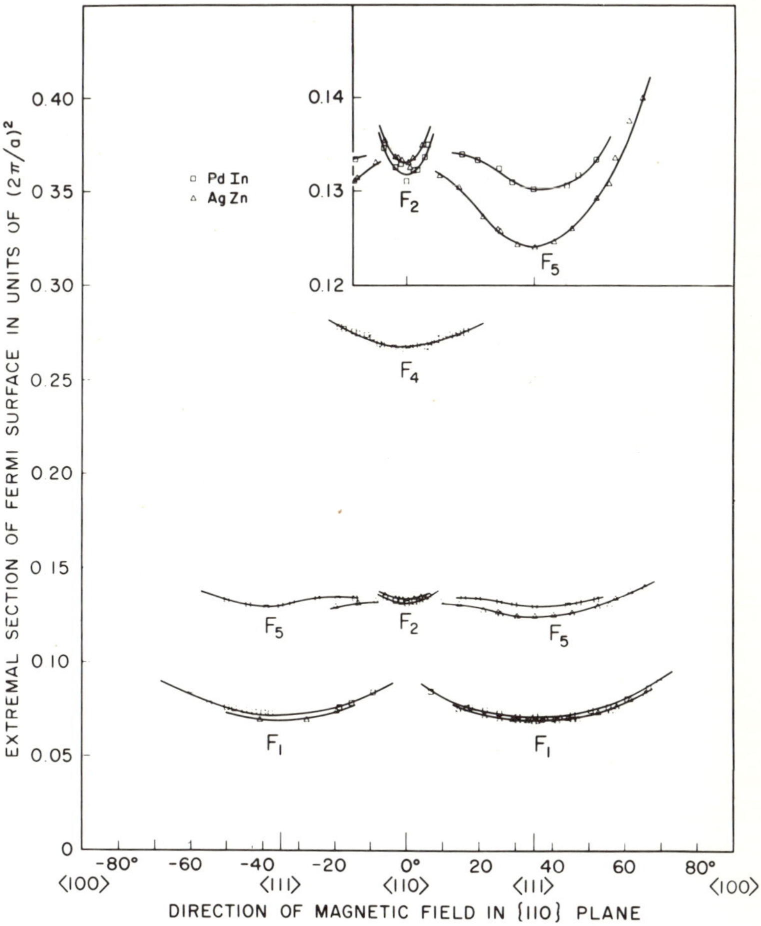

Fig. 3 Extremal sections of the Fermi surface of β'-PdIn and β'-AgZn in units of $(2\pi/a)^2$ as a function of the direction of the magnetic field scanned in a $\{110\}$ plane. Observed data.

even open orbits in some directions, with corresponding complication and blurring of the Landau levels. This could perhaps explain why we cannot observe F_1 for β'-CuZn in the vicinity of $\langle 100 \rangle$, since the actual A_1^R and A_2^R orbits still have four points in common after splitting by the lattice potential. The F_3 oscillations obtained about $\langle 100 \rangle$ are at present attributed to the hole orbits A_2^R in the second band. Our inability to observe the muffin orbit (B_2^X) in CuZn despite residual resistance ratios of the order of 300 is annoying, although the spin-splitting factor $m_c g/m_0$ may well have a value close to unity for B_2^X, giving zero amplitude of the oscillations, since m_c in the free-electron model is $0.5m_0$ for this orbit.

The reduced frequencies of the orbits found for PdIn crystals as a function of orientation when the magnetic field is scanned in the {110} and {100} planes and AgZn is scanned in {110} are shown in Figs. 3 and 4. These

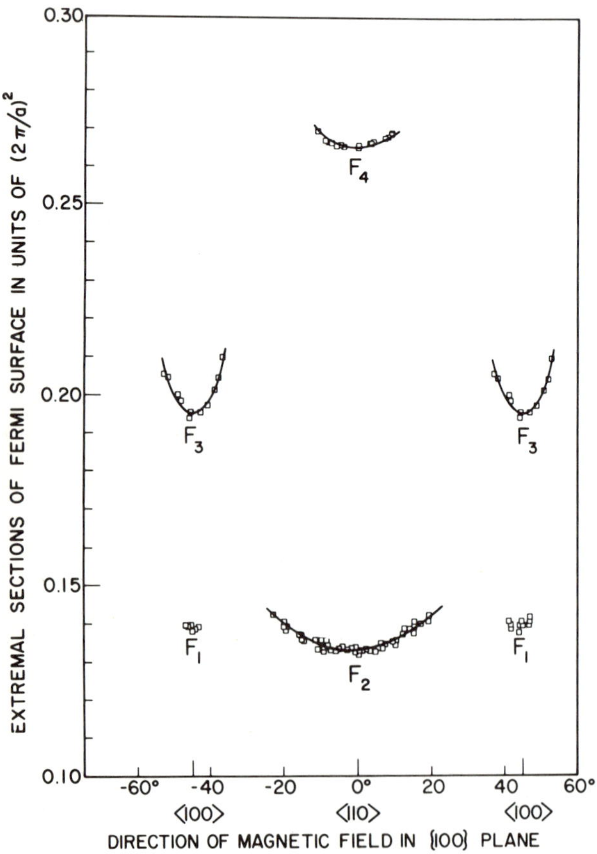

Fig. 4 Extremal sections of the Fermi surface of β'-PdIn in units of $(2\pi/a)^2$ as a function of the direction of the magnetic field scanned in a {100} plane. Observed data.

data were only obtained in the last few days,* and although several of the oscillations (particularly F_5) cannot be identified with certainty at present, the Fermi surfaces of PdIn and AgZn appear to be essentially identical insofar as the same features (F_1, F_2, and F_5) of each have been observed (Fig. 3). In addition, we emphasize the great similarity of F_1 for CuZn, AgZn, and PdIn. The period of 0.13 close to $\langle 110 \rangle$ found for PdIn and AgZn may be due to the neck orbit $B_2{}^M$, and band calculations by Amar et al.[6] predict a neck value of about 0.16 for AgZn. The F_4 oscillations of PdIn appear either to result from the muffins in the second band, which have not been observed for CuZn, or to be the second harmonic of F_2.

Although we have not yet had the opportunity to make use of the results on PdIn, the similarity to AgZn and the apparent absence of data from third-zone "butterflies" and fourth zone "cigars" suggest at present an electron concentration close to 3 per cell, so that Pd does not increase the concentration of electrons in the conduction band beyond that provided by In.

In the case of β'-CuZn the relatively few periods observed compared to those expected has somewhat limited detailed interpretation. Nevertheless we can conclude that the geometry of the observed Fermi surface resembles very much that expected from existing band calculations of Arlinghaus[5] and of Johnson and Amar[4]: that a conduction electron content of 3 per cell is entirely reasonable, and that there exist holes in the corners of the first zone and large electron surfaces in the second zone which make contact with the $\{110\}$ faces. This topology is consistent with the decreasing $N(E)$ with increasing electron concentration observed in specific-heat measurements and required by Jones's theory. We have so far been unsuccessful in the application of pseudopotential theory to interpret the observed Fermi surface of β'-CuZn, and we intend therefore to continue with calculations based on APW methods.

REFERENCES

1. Jones, H.: *Proc. Roy. Soc.* **A144:**225 (1934); **A147:**396 (1934); *Proc. Phys. Soc.* **49:**250 (1937).
2. Veal, B. W., and J. A. Rayne: *Phys. Rev.* **128:**551 (1962).
3. Harrison, W. A.: *Phys. Rev.* **118:**1190 (1960).
4. Johnson, K. H., and H. Amar: *Phys. Rev.* **139:**A760 (1965).
5. Arlinghaus, F.: Quarterly Progress Report 56, p. 4, Apr. 15, 1965 (Solid State and Molecular Theory Group, MIT).
6. Amar, H., and K. H. Johnson: In "Optical Properties and Electronic Structure of Metals and Alloys," p. 586, North Holland, 1966; Amar, H., K. H. Johnson, and K. P. Wang, to be published.

* Although we have been working with PdIn for some 15 months, only very recently have we succeeded in making satisfactory crystals. Data on β'-AgZn were also obtained for the first time a few days ago.

DISCUSSION *on Paper Presented by W. B. Pearson*

LOMER: Arlinghaus gives a good estimate of the Fermi surface. It contains large flat areas parallel to the cube zone faces and about one-third of the zone radius from the boundary. A superlattice vector $t/3$ (i.e., one-third of existing ordered β' reciprocal lattice vector) may couple two such areas across the zone boundary, and a tripling of the cell size in real space would be stabilized. The collapse of the β' phase to form γ may then be explained by this introduction of the (100) reciprocal lattice vector. To get a good value of the necessary matrix element: the phase eliminates two atoms from the block of $3 \times 3 \times 3$ simple cubic cells. It would be interesting to see how many phases adjacent to β' have this threefold periodicity.

MASSALSKI: Judging by the slide which I showed earlier, only the CuZn and AgZn β' phases can be studied in detail by the de Haas–van Alphen method since they do not show a martensitic transformation at the 50-50 composition. Do you know what happens when you move away from the stoichiometric ratio? Do the periods disappear, and is this behavior symmetrical around the 50-50? Have you tried any de Haas–van Alphen experiments with γ-brass?

PEARSON: 1. Naturally as "nonstoichiometry" distributes the other kind of atom on one or other crystallographic site, electron scattering increases, and the oscillations ultimately disappear. For example, at 49 and 51 a/o no oscillations were observed. We should probably be able to observe oscillations due to holes in the first zone within the range 49.9 to 50.1 or 49.8 to 50.2 a/o Zn. This study is on our program, but not yet done: we await crystals.

2. Your slide giving data on the martensitic transformation in β'-AuZn and your discussion indicated that the transition was to a cph structure. Have you in fact identified and confirmed this structural change? Our X-ray rotation photographs at 4.2°K indicated that β'-AuZn transforms as a pseudo single crystal with a cubic cell of side $3a_0$. This might indicate that the cubic cell is the martensitic product, or the X-ray photograph presumably could result from a hexagonal martensitic product coming out along four cubic axes. At any rate the (partially?) transformed AuZn crystal gives many de Haas–van Alphen oscillations.

3. We have prepared several γ-brass crystals, but so far these presumably have rather poor ordering, or else vacancies are present, as the residual resistance ratios are only 3 to 4. As we would expect, these do not give de Haas–van Alphen oscillations.

MASSALSKI: In general, the structures of transformed β' phases tend to be faulted (fcc) cubic, or distorted hexagonal but we have no information on the structural change in β'-AuZn. The metallographic appearance of the transformed structures suggests the possibility that in polycrystalline specimens two transformation structures may be involved.

HEINE: If, as you suggest, a de Haas–van Alphen period is unobservable because $m^* = 0.5$ and $g = 2$, there should be a strong period at twice the area. Do you see this?

PEARSON: No!

ATOMIC-SIZE CRITERIA IN CERTAIN INTERMEDIATE PHASES OF SILVER AND GOLD*

M. V. Nevitt

Metallurgy Division, Argonne National Laboratory, Argonne, Illinois, and the University of Sheffield, England

INTRODUCTION

In a review of the occurrence of transition-metal intermediate phases, prepared 4 years ago,[1] the present author treated atomic size in terms of interatomic distances. In analyzing the influence of atomic size on the occurrence and the unit-cell dimensions of certain phases, use was made of a simple model in which space was occupied by spherical atoms in contact; their diameters were assumed to have values that satisfy the geometric requirements of the crystal structures involved.

For various stoichiometric phases $X_m Y_n$, the distance between nearest-neighbor atoms of the same kind was taken as the atomic diameter characteristic of a particular crystalline environment, d_X or d_Y, while the dis-

* This work was performed under the auspices of the British Science Research Council and the U.S. Atomic Energy Commission.

tance between unlike atoms d_{XY} represented the sum of their respective radii. In the not infrequent instances where more than one equivalent interatomic distance could be identified, the attendant ambiguity in d_X, d_Y, or d_{XY} was resolved arbitrarily by averaging. Values of d_X, d_Y, and d_{XY} were compared with atomic diameters derived from the distances of closest approach in the solid structures of the elements concerned, D_X and D_Y,[2] corrections being applied as necessary to make the diameters correspond to the appropriate coordination numbers.[3]

Estimates of size changes based on differences between D and d can be subject to appreciable random errors when the two quantities have similar values, and can contain systematic errors arising from the arbitrary aspects of selecting interatomic distances and the approximate nature of the correction for different coordination numbers.

There is a fundamental uncertainty, moreover, which existed at the time of the previous survey and which may have assumed larger proportions in the intervening period, concerning the extent to which interatomic distance may be used as a valid criterion of atomic size. With reference to the pure metals, it has been shown[4] that at allotropic transformations atomic volume is conserved to within 1%, although interatomic distances change to a greater extent. Doubt has been expressed as to the validity of atomic diameters derived from distances of closest approach in alloy phases of the noble metals with B subgroup elements,[5] and, in general, experiment and theory appear to be in accord in suggesting that the volume of an atom, rather than the nature or the number of atoms surrounding it, is of first importance in the determination of its energy.[6]

Thus, it seems useful at the present time to reconsider atomic size in certain intermediate phases from the viewpoint of atomic volumes.

For solid solutions the treatment in terms of atomic volume is based on the assumption that the actual volume of a phase is determined by the volume of the atoms composing it through a simple additive relationship, modified to the extent that an atom and the crystalline matrix in which it resides are treated as elastic media.[7] For certain intermediate phases special models have been chosen that can be treated analytically, as for example the hemisubstitutional elastic model for Laves phases.[8]

This paper will not treat the large group of structures considered earlier, or develop detailed models. Its restricted goal is to ascertain for two groups of intermediate phases whether information about the systematics of atomic-size variation derived from interatomic distances is consistent with what may be learned about the effective volumes of the atoms.

By selecting the CsCl (B_2) and $MoSi_2$ (C_{11b}) phases formed by Ag and Au it is possible to observe size effects displayed by essentially the same group of atoms in two different structures, since the major share of both groups of phases involves Sc, Y, and the members of the lanthanide series.

Large numbers of both have been reported recently,[9] and so a more extensive survey can be made than heretofore. A useful aspect of this choice is that Sc, Y, and the lanthanides provide a test of the extent to which interatomic distances can be empirically normalized to provide directly comparable values of D. Among these metals there are several different crystal structures occurring at room temperature which represent variations in stacking sequence. Among the close-packed hexagonal forms the axial ratios (corrected for stacking variations) are mostly nonideal, and differ among themselves by more than 3%. Some uncertainties may exist, therefore, in the atomic diameters assigned to these metals, and it is of interest to see whether discrepancies arise when size effects deduced from D are compared with atomic-volume correlations.

Since these two groups of phases are characterized by large volume-size factors (Table 1), it is possible that their ordered arrangements arise in part as mechanisms to relieve stresses due to misfit. In the CsCl structure this tendency appears to take the form of an effective repulsive interaction between atoms of the same size. In a more easily recognizable manner the $MoSi_2$-type structure at stoichiometry XY_2, where Y is Ag or Au, represents an arrangement that reduces misfit energy. Monatomic layers of the larger atom distributed in a non-close-packed arrangement, i.e., parallel to the basal plane of the ordered body-centered tetragonal cell, are interleaved between similar double layers of the Ag or Au atoms. This stacking scheme should permit a relaxation of stresses across the interfaces between the planes of atoms of dissimilar sizes.[7] The double layers of Ag or Au atoms are dimpled to facilitate the partial "nesting" of the larger atoms in the neighboring layer, as may be seen in the schematic representation of the atomic arrangement on a (110) plane (Fig. 1).

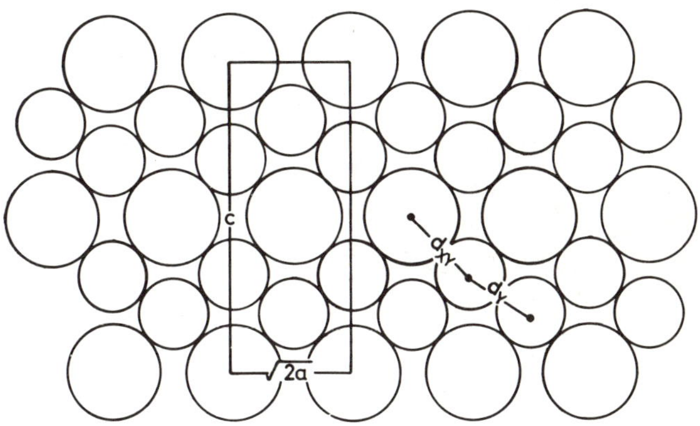

Fig. 1 Arrangement of atoms on (110) in the $MoSi_2$-type structure.

TABLE 1 ***Atomic-size Parameters for Certain Intermediate Phases of Ag and Au***

X	Y	Volume-size factor	CsCl-type phases $D_{Ag} = D_{Au} = 2.80$ A $V_{Ag} = V_{Au} = 17$ A^3				MoSi$_2$-type phases $D_{Ag} = D_{Au} = 2.84$ A $V_{Ag} = V_{Au} = 17$ A^3			
			d_{XY}, A	D_{XY}, A	v, A^3	V, A^3	d, A	$\bar{D}$, A	v, A^3	V, A^3
Zn	Ag	−0.106	2.73	2.75	31.5	32.2				
Li	Ag	0.270	2.74	2.91	31.8	38.6				
Mg	Ag	0.364	2.87	2.95	36.2	40.2				
Sc	Ag	0.470	2.96	2.99	39.6	42.0	2.91	2.98	55.8	59.0
Lu	Ag	0.735	3.06	3.08	44.4	46.5	2.98	3.03	60.0	63.5
Tm	Ag	0.776	3.08	3.09	45.1	47.2	2.99	3.04	60.5	64.2
Er	Ag	0.806	3.10	3.10	45.9	47.7	3.01	3.03	61.7	64.7
Ho	Ag	0.830	3.11	3.11	46.3	48.1	3.02	3.06	62.1	65.1
Dy	Ag	0.858	3.13	3.12	47.1	48.6	3.03	3.06	62.9	65.6
Tb	Ag	0.883	3.14	3.13	47.8	49.0	3.04	3.07	63.6	66.0
Gd	Ag	0.941	3.16	3.14	48.7	50.0	3.06	3.08	64.6	67.0
Sm	Ag	0.948	3.18	3.14	49.8	50.1				
Y	Ag	0.958	3.13	3.14	47.5	50.3	3.03	3.08	62.9	67.3
Nd	Ag	1.01	3.22	3.16	51.5	51.2				
Ce	Ag	1.02	3.25	3.17	52.8	51.4				
Pr	Ag	1.03	3.24	3.18	52.3	51.5				
La	Ag	1.19	3.28	3.22	54.6	54.3				
Mn	Au	−0.282	....	....	....	....	2.77	2.71	45.4	46.2
Zn	Au	−0.106	2.72	2.75	30.7	32.2				
Ti	Au	0.0470	....	....	....	....	2.81	2.85	49.8	51.8
Li	Au	0.271	2.68	2.91	29.8	38.6				
Hf	Au	0.317	....	....	....	....	2.88	2.94	54.0	56.4
Mg	Au	0.364	2.83	2.95	34.8	40.2				
Zr	Au	0.371	....	....	....	....	2.90	2.95	54.7	57.3
Sc	Au	0.470	2.92	2.99	38.3	42.0	2.88	2.98	53.8	59.0
Lu	Au	0.735	3.03	3.08	42.9	46.5				
Tm	Au	0.776	3.04	3.09	43.5	47.2				
Er	Au	0.806	3.06	3.10	44.4	47.7				
Ho	Au	0.830	3.08	3.11	44.8	48.1	3.00	3.06	60.5	65.1
Dy	Au	0.858	3.08	3.12	45.0	48.6	3.00	3.06	61.0	65.6
Tb	Au	0.883	3.10	3.13	45.9	49.0	3.02	3.07	61.8	66.0
Gd	Au	0.941	3.12	3.14	46.6	50.0	3.03	3.08	62.6	67.0
Sm	Au	0.948	3.14	3.14	47.4	50.1				
Y	Au	0.958	3.08	3.14	45.2	50.3	3.00	3.08	61.0	67.3
Nd	Au	1.01	3.17	3.16	49.0	51.2				
Pr	Au	1.03	3.18	3.18	49.8	51.5				

NOTE: Only phases with well-defined stoichiometry and lattice parameters measured at room temperature are included.

CsCl-TYPE PHASES

Figure 2*a* shows the manner in which $d_{X\text{Ag}}$ and $d_{X\text{Au}}$ vary with D_X. The deviations, positive and negative, from the solid line representing $(D_X + D_{\text{Ag}})/2 = (D_X + D_{\text{Au}})/2 = D_{XY}$ were assumed in the earlier review[1] to represent "expansions" and "contractions" relative to the lattice dimensions calculated from the diameters of the atoms in their own pure crystals. When the specific volume (volume per formula weight) computed from d_{XY} is compared with the equivalent one calculated from D_{XY}, positive and negative differences as large as 3.5 and 28%, respectively, are observed. The effective size of the Au atom appears to be consistently smaller than that of Ag.

The corresponding relationship involving atomic volumes is shown in Fig. 2*b*. The measured specific volume v is plotted against V_X, the atomic volume of the X component.[4] The sum of the atomic volumes of the components, $V_X + V_{\text{Ag}} = V_X + V_{\text{Au}} = V$, is given by the solid line. It is clear that both plots reveal the same gross features of the relationship between the sizes of the atoms in the phases and in the corresponding pure metals. However, the absence of an exact correspondence between the signs of $(d - D)$ and $(v - V)$ indicates that only limited reliability can be assigned to "expansions" and "contractions" inferred from atomic diameters. The signs of $(d_{XY} - D_{XY})$ and $(v - V)$ could be brought into better agreement if a larger value were assigned to D_Y.

Although none of the elastic models proposed previously is directly applicable to these ordered phases, the negative values of $(v - V)$, which are seen to predominate in Fig. 2*b*, are not unexpected when considered within the general context of an elastic model. In all the phases except ZnAg and ZnAu the atomic volume of Ag or Au is smaller than the mean atomic volume of the matrix, and the compressibility of the Ag or Au atom is probably smaller than the effective compressibility of the matrix that surrounds it, the latter supposition being based on a comparison of the macroscopic compressibilities of Ag and Au with those of the X components. Under these conditions the specific volume of the phase is likely to be smaller than the sum of the atomic volumes.

The atomic volume and the macroscopic compressibility of Zn are considerably closer numerically to those of Ag and Au than to those of the other X components in the series, and so proportionately smaller deviations from volume additivity should be expected for ZnAg and ZnAu. This is observed in the plot, where the points lying closest to the ordinate are for these phases. The trend toward smaller negative values of $(v - V)$ with larger V_X and the transition to positive deviations shown by the phases of Ag near $V_X = 2V_{\text{Ag}}$ are, on the other hand, not easily rationalized in the same manner. The lanthanide metals with the largest atomic volumes are, in fact, the ones with the highest compressibilities, and so a change in the sign of $(v - V)$ does not seem to be predictable on elastic grounds.

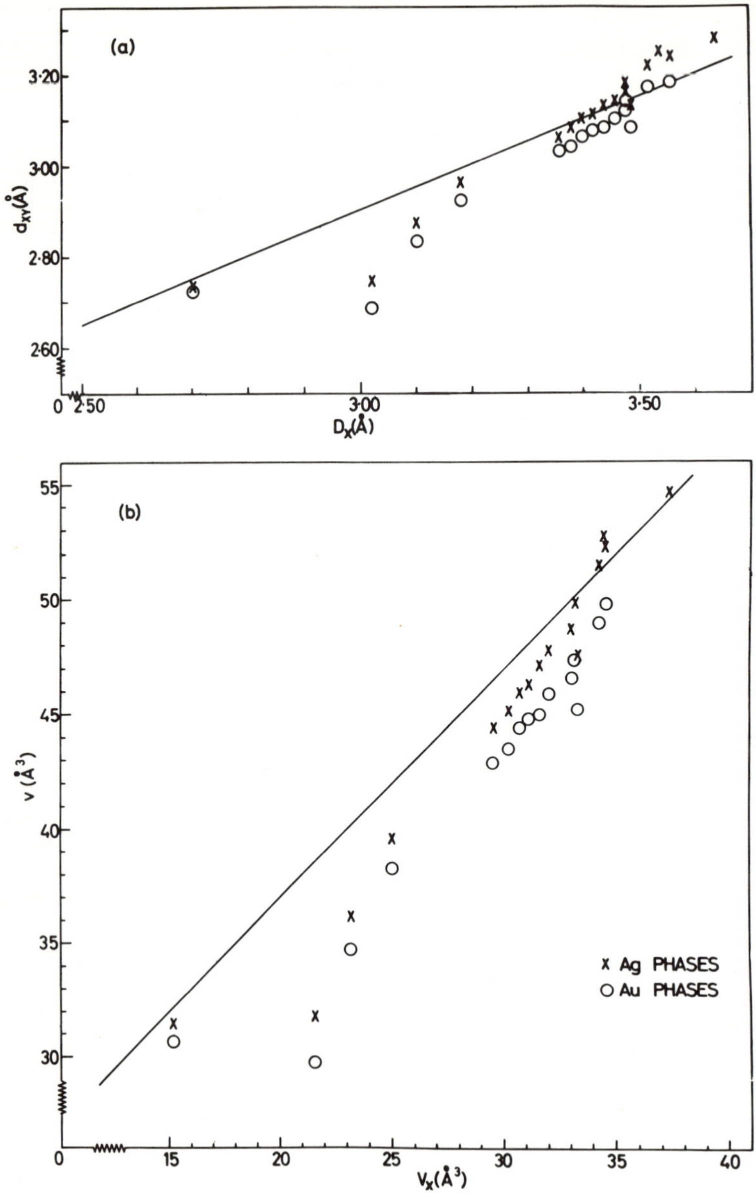

Fig. 2 Variations in interatomic distances and specific volumes for CsCl-type phases of Ag and Au.

$MoSi_2$–TYPE PHASES

The structure of these phases corresponds to space group $I4/mmm(D_{4h}^{17})$, with atoms of X in $2(a)$ at 0,0,0 and ½,½,½, and atoms of Y (Ag or Au in this discussion) in $4(e)$ at $00z$; ½, ½, ½ + z; 0, 0, − z; ½, ½, ½ − z. The value of z is approximately ⅓, but the precise values are not known for most of the phases. Until the z parameters are refined and interatomic distances known more accurately, only $\bar{d}$, which is approximately the average of the distances d_{XY} and d_Y shown in Fig. 1, can be calculated.

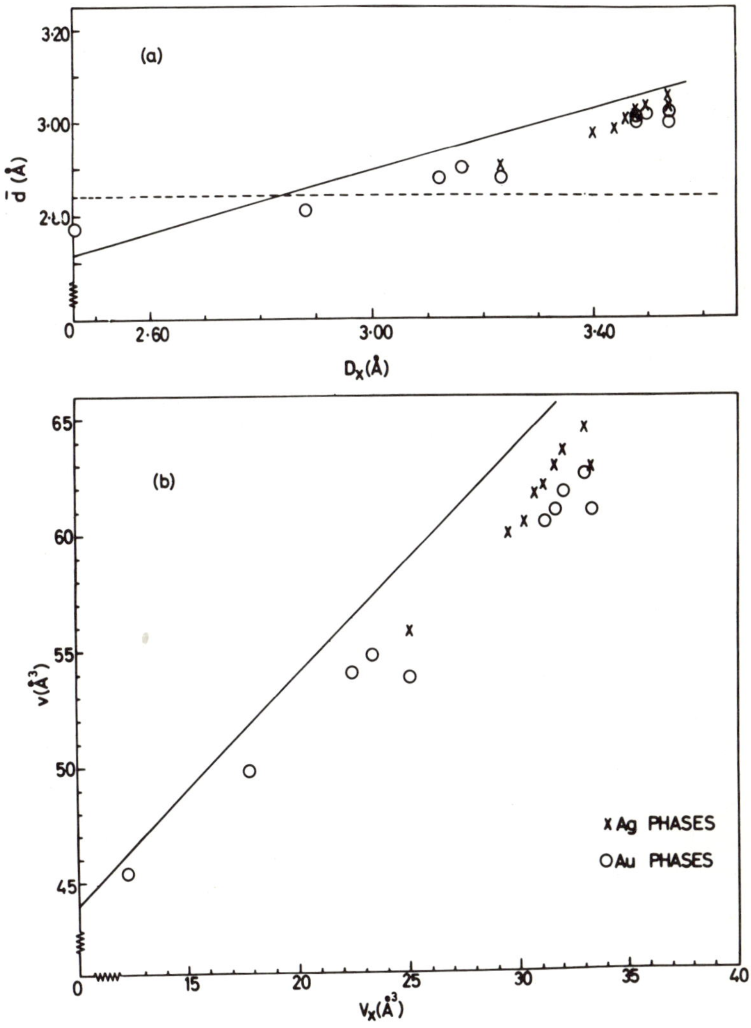

Fig. 3 Variations in interatomic distances and specific volumes for $MoSi_2$-type phases of Ag and Au.

It was demonstrated in the previous review[1] of this structure that there are positive and negative deviations of $\bar{d}$ from D_Y whose magnitudes depend on the value of D_X, and that the deviations vanish for $D_Y \cong D_X$. The effect may be observed in Fig. 3*a* by noting the displacements of the values of $\bar{d}$ from the dashed line representing D_Y for Ag and Au. Considered in terms of atomic diameter, these changes in the interatomic distances were regarded as correlated adjustments in the sizes of X and Y to maintain contacts between like and unlike atoms. Measurements of specific volume do not provide an independent assessment of this effect. On the other hand, the values of $\bar{d}$ can also be compared with $\bar{D} = (D_X + 2D_Y)/3$, given by the solid line; $(\bar{d} - \bar{D})$ has the same significance as $(d_{XY} - D_{XY})$ for the CsCl-type phases, and indicates for all the phases except $MnAu_2$ a decrease in the average atomic diameter relative to the average of the diameters of the atoms in their pure metals. In Fig. 3*b*, where the specific volume is plotted against V_X, a similar contraction is demonstrated by the negative deviation of the measured specific volume from the sum of the atomic volumes given by the solid line. The difficulty in assessing D (and also V) for Mn[4] may be responsible for the disagreement in the signs of $(d - D)$ and $(v - V)$ for $MnAu_2$.

There is again an indication that Au has a smaller effective size than Ag when present in these phases.

SUMMARY

For two groups of intermediate phases involving Ag and Au, an attempt has been made to compare the use of atomic diameters derived from interatomic distances and atomic volumes in establishing relationships between the sizes of the various atoms in the phases and in their pure metals. It has been shown that in these particular phases the two criteria reveal the same general trends concerning size variation, but differences between d and D cannot always be used reliably to infer volumetric differences. Correlated size changes by individual atoms are more easily recognized by considering atomic diameters, whereas a preference for the atomic-volume criterion appears to lie in its potential use in formulating analytical relationships to describe the observed effects, particularly by means of elastic models. Although such analyses have not been made for the two structures considered, it appears that the first-order effect, i.e., predominantly negative deviations from additivity of atomic volumes, is consistent with the relative atomic volumes and compressibilities of the components.

Although Ag and Au have the same atomic diameters and atomic volumes in their pure metals, the effective size of Au is smaller than that of Ag in both the groups of phases considered.

REFERENCES

1. Nevitt, M. V.: In P. A. Beck (ed.), "Electronic Structure and Alloy Chemistry of the Transition Metals," pp. 101–178, Interscience Publishers, Inc., New York, 1963.
2. Teatum, E., K. Gschneidner, Jr., and J. Waber: Los Alamos Scientific Laboratory Report LA 2345, 1960.
3. Laves, F.: In "Theory of Alloy Phases," pp. 124–198, American Society of Metals, Cleveland, Ohio, 1956.
4. Rudman, P. S.: *Trans. Met. Soc. AIME* **233**:864 (1965).
5. Massalski, T. B., and H. W. King: *Progr. Mater. Sci.* **10**:1 (1961).
6. Mott, N. F.: *Repts. Progr. Phys.* **25**:218 (1962).
7. Friedel, J.: *Advan. Phys.* **3**:446 (1954); *Phil. Mag.* **46**:514 (1955).
8. Rudman, P. S.: *Trans. Met. Soc. AIME* **233**:872 (1965).
9. Dwight, A. E.: Fifth Rare Earth Conference, Ames, Iowa, 1965. See also W. B. Pearson: "Handbook of Lattice Spacings and Structures of Metals and Alloys," 2d ed., Pergamon Press, New York, in press.

DISCUSSION *on Paper Presented by M. V. Nevitt*

HEINE: It seems to me that there are situations in which the atomic volume is an appropriate measure of atomic size, and others where a hard-sphere diameter is relevant. In "simple" metals (i.e. nontransition, etc.), the valence electron density is to a zeroth approximation constant, and it is the atomic volume that is important because an electron gas strongly resists compression. The same will apply approximately to alloys of these metals among themselves, but there should be deviations related to electronegativity. It would be useful to have an analysis along those lines for theory to conjecture on. However, in the noble metals and elements immediately to their left, there are large full, or nearly full, *d* shells which behave as hard spheres. The classic picture is that in the pure noble metals the interatomic spacing is determined by the *d* shells, the atomic volume that the *s* electrons would like to take up being somewhat smaller. Thus in alloys where the *d* shells are kept out of contact, the (smaller) atomic volume becomes appropriate, and it would be interesting to see whether one could account for a whole range of data by a unique volume. That for Au should be smaller than that for Ag because of the greater binding of the valence electron.

NEVITT: Dr. Heine's comments are very much appreciated. The suggestion regarding the possible size behavior of the noble metals when their *d* shells are kept out of contact seems to have particular pertinence, offering as it does a possible hint for the observed size disparity between Ag and Au. In this same connection it may be significant that when the data given in Fig. 1 are extrapolated to $V_x = 0$ they yield a value for V (noble metal) of about 10 to 12 A^3, which is smaller than the atomic volume of the noble metal in its own structure. Although it is doubtful that this extrapolated atomic volume for the noble metal should be taken quantitatively, its smaller value is consistent with Dr. Heine's suggestion. It would be of interest to test this hypothesis in other structures in which the noble-metal atom is isolated.

A problem that has annoyed workers in the field of alloy chemistry has been the inability of the noble metals to form Laves phases with the lanthanide metals, although the transition metals immediately to the left do form the phases. As mentioned in the present paper, the $MoSi_2$-type structure appears to be the substitute structure that the noble metals employ in minimizing their misfit energy with the lanthanides at the composition XAg_2 and XAu_2. The layering arrangement of the $MoSi_2$-type structure, while it retains Ag-Ag and Au-Au contacts, may in fact require less compensative size adjustment by the noble-metal atom than the Laves phases. If the "hard sphere" model is particularly appropriate for the noble-metal atoms with filled *d* shells, the preference for the $MoSi_2$-type structure may be understandable.

MASSALSKI: Can you think of any situation in which the phases of the type you have discussed are present in any given binary system at two different compositions? Then one could draw a line through the respective atomic volumes and see where it intersects the binary axes. This would provide an idea of effective atomic volumes that could be used to derive a size factor in terms of atomic volumes.*,†

NEVITT: Dr. Massalski has raised an interesting point, which in fact can be tested. In several systems, notably Zr-Pd, Hf-Pd, Zr-Au, and Hf-Au, the $MoSi_2$-type phase occurs at A_2B and AB_2.

The procedure suggested by Dr. Massalski leads to the following extrapolated volumes and volume-size factors:

System	Extrapolated atomic volume, A^3				Volume-size factor
	Zr	Hf	Pd	Au	
Zr-Pd	23.0	....	13.9	....	0.66
Hf-Pd	....	22.4	13.9	....	0.61
Zr-Au	23.4	....	....	15.7	0.49
Hf-Au	....	22.3	....	15.7	0.42

The extrapolated atomic volumes of Zr and Hf are in agreement with the atomic volumes in their pure metals; Pd and Au have effective volumes that are smaller than the atomic volumes associated with their pure metals. As a result, the volume size factors in the systems Zr-Au and Hf-Au are rather larger than the values given in Table 1 of my paper.

* T. B. Massalski and H. W. King, *Progr. Mater. Sci.* **10**:1 (1961).

† H. W. King, in T. B. Massalski (ed.), "Alloying Behavior and Effects in Concentrated Solid Solutions," Gordon and Breach, Science Publishers, Inc., New York, 1965.

THE ELECTRONIC SPECIFIC HEATS OF ALLOY PHASES BASED ON THE NOBLE METALS*

T. B. Massalski, G. A. Sargent, and L. L. Isaacs
Mellon Institute, Pittsburgh, Pennsylvania

INTRODUCTION

In this paper we wish to review briefly the status of the measurements of the low-temperature specific heats in alloys based on the noble metals. Work in this field is being pursued now in a number of laboratories. We make no attempt here to discuss all the available data, but rather we wish to point out certain observed trends, difficulties, and discrepancies.

We define alloys of the noble metals as the group of phases obtained when Cu, Ag, or Au is alloyed with any of the non-transition elements from groups IIB to VB of the periodic table as follows:

Cu	Zn	Ga	Ge	As
Ag	Cd	In	Sn	Sb
Au	Hg	Tl	Pb	Bi

* The authors are grateful to Prof. S. Noguchi and Dr. H. Montgomery for permission to mention their unpublished data, and to Prof. B. A. Green for exchange of information and discussions. This work has been supported in part by the U.S. Atomic Energy Commission.

In the resulting alloy systems, apart from those with Tl, Pb, and Bi, extensive ranges of solid solubility are observed in both primary solid solutions and intermediate phases, particularly in the range of electron concentration between 1 and 2. The electronic band structure of these alloys is of special interest because for many years they have served as a testing ground for various theories of alloy phases.[1,2]

Measurements of the linear term γT in the specific heat, where γ is the

***TABLE 1 Comparison of Recent Heat-capacity Measurements on* Cu, Ag, *and* Au**

Metal	Authors	Ref.	γ, millijoule/(mole) $(K)^2$	A, millijoule/(mole) $(°K)^4$
Cu	Corak et al.	1	0.688	0.0478
	Rayne	2	0.686	0.0473
	Griffel et al.	3	0.691	0.0486
	Ramanathan and Srinivason	4	0.720	0.0523
	Manchester	5	0.696	0.0478
	Du Chatenier and De Nobel	6	0.721	0.0500
	Veal and Rayne	7	0.692	0.0483
	Kneip et al.	8	0.697	0.0479
	Phillips	9	0.695	0.0485
	Isaacs	10	0.698	0.0485
	Clune and Green	11	0.702	0.0474
	Martin	12	0.696	0.0471
Ag	Keesom and Pearlman	13	0.661	0.1706
	Corak et al.	1	0.610	0.1700
	Rayne	14	0.613	0.1679
	Ramanathan and Srinivason	4	0.616	0.1838
	Du Chatenier and De Nobel	6	0.682	0.1679
	Filby and Martin	15	0.646	0.1679
	Isaacs	10	0.645	0.1622
	Green and Culbert	16	0.646	0.1695
	Green	17	0.656	0.1671
	Martin	12	0.646	0.1702
	Montgomery and Pells	18	0.644	0.1676
Au	Du Chatenier and De Nobel	6	0.740	0.4311
	Ramanathan and Srinivason	19	0.764	0.4391
	Corak et al.	1	0.740	0.4360
	Zimmerman and Crane:			
	Single crystal	20	0.767	0.4390
	Polycrystal	20	0.740	0.4405
	Martin	21	0.729	0.4338
	Isaacs	10	0.729	0.4504

electronic specific-heat coefficient and T the absolute temperature, began in this field in the middle 1950s[3,4] mainly because of the expected relationship between γ and the density of states at the Fermi surface. The determination of the various details of the Fermi surface in the noble metals,[5] and in some alloys,[6] has provided an extra incentive for further work.

MEASUREMENT DIFFICULTIES

The linear term in the specific heat is obtained by extrapolation to the absolute zero of temperature T of the plot of C/T versus T^2, based upon the approximate relationship $C = \gamma T + AT^3$, where C is the specific heat, and γ and A are the usual coefficients relating to the electronic specific heat and the Debye temperature, respectively.

The numerical value of γ in the noble metals and their alloys is unfortunately very small. In the alloys of the transition metals γ values are of the order of 10 units of millijoules/(mole) (°K)^2. In the pure Cu, Ag, and Au they are less than one of this unit, while the changes of γ on alloying are of the order of $1/100$ of this unit. Thus, possible small contributions from various secondary effects may become dominant.

References for Table 1

1. Coark, W. S., M. P. Garfunkel, C. B. Satterthwaite, and A. Wesler: *Phys. Rev.* **98**:1699 (1955).
2. Rayne, J. A.: *Australian J. Phys.* **9**:189 (1956).
3. Griffel, M., R. W. Vest, and J. F. Smith: *J. Chem. Phys.* **27**:1267 (1957).
4. Ramanathan, K. G., and T. M. Srinivasan: *J. Sci. Ind. Res. India* **16B**:277 (1957).
5. Manchester, F. D.: *Can. J. Phys.* **37**:989 (1959).
6. DuChatenier, F. J., and J. De Nobel: *Physica* **28**:181 (1962).
7. Veal, B. W., and J. A. Rayne: *Phys. Rev.* **128**:551 (1962).
8. Kneip, G. D., Jr., J. O. Betterton, Jr., and J. O. Scarbrough: *Phys. Rev.* **130**:1687 (1963).
9. Phillips, N. E.: *Phys. Rev.* **134**:A385 (1964).
10. Isaacs, L. L.: *J. Chem. Phys.* **43**:307 (1965).
11. Clune, L. C., and B. A. Green, Jr.: *Phys. Rev. Letters* **16**:A1 (1966); and private communication.
12. Martin, D. L.: *Phys. Rev.* **141**:576 (1966).
13. Keesom, P. H., and N. Pearlman: *Phys. Rev.* **98**:548 (1955).
14. Rayne, J. A.: *Proc. Phys. Soc.* **B69**:482 (1956).
15. Filby, J. D., and D. L. Martin: *Can. J. Phys.* **40**:791 (1962).
16. Green, B. A., Jr., and H. V. Culbert: *Phys. Rev.* **137**:A1171 (1965).
17. Green, B. A., Jr.: *Phys. Rev. Letters* **16**:A1 (1966); and private communication.
18. Montgomery, H., and G. P. Pells: Conference on Electronic Structures of Alloys, Sheffield, England, 1963; and private communication, 1966.
19. Ramanathan, K. G., and T. M. Srinivasan: *Proc. Ind. Acad. Sci.* **49**:55 (1959).
20. Zimerman, J. E., and L. T. Crane: *Phys. Rev.* **126**:513 (1962).
21. Martin, D. L.: *Phys. Rev. Letters* **12**:723 (1964).

The difficulties associated with the calorimetric technique deserve a brief mention. Thus far, measurements in the alloy phases have been reported only in the He^{IV} range, between approximately 1.5 and 4.2°K. This, naturally, magnifies the extrapolation problems. Discussions of difficulties connected with the calibration of the thermometers, the operation of calorimeters, and the processing of data are available in the literature. Here we wish only to emphasize that different techniques exist. On the whole, a random error of between 0.1 and 0.5% of the measured γ values is claimed

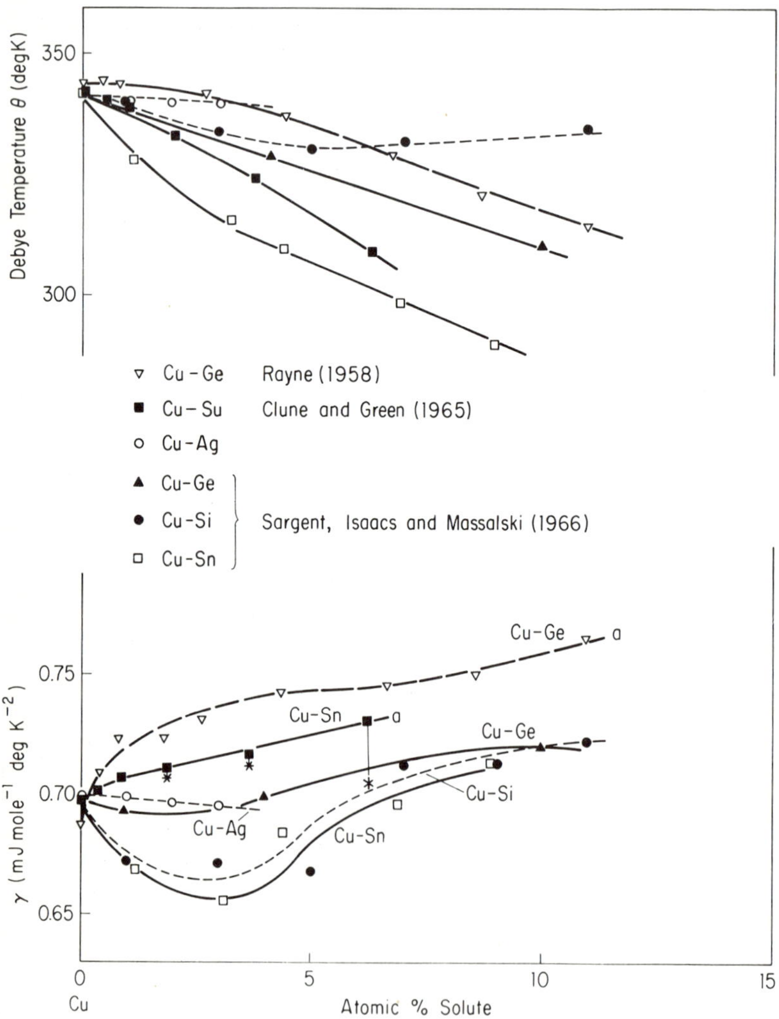

Fig. 1 The variation with composition of the electronic specific-heat coefficient γ and the Debye temperature θ_0 in Cu-based alloys.

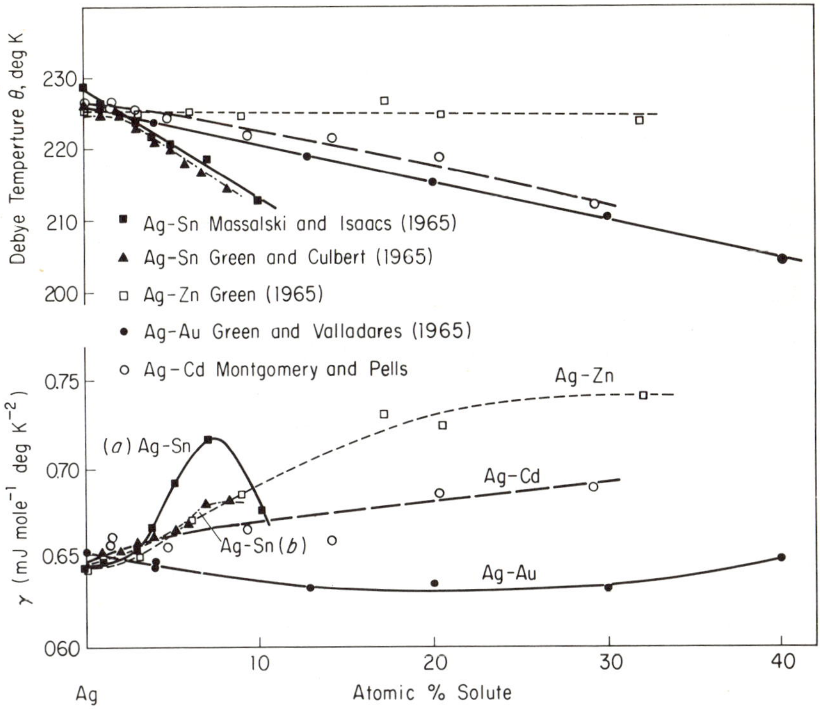

Fig. 2 The variation with composition of the electronic specific heat coefficient γ and the Debye temperature θ_0 in Ag-based alloys.

in the more recent investigations, and a systematic error of between 1 and 2%. Hence the total expected errors are often only slightly smaller than the differences that are sought between successive alloy samples. In practice, judging by the available data, the errors appear to be greater than those stated. A summary of the published values for pure solvents Cu, Ag, and Au is given in Table 1. Even in the pure metals the scatter between different results is quite substantial.

REVIEW OF AVAILABLE DATA

Results known to us on Cu- and Ag-based systems are plotted in Figs. 1 to 5. Both the electronic specific-heat coefficient γ and the Debye temperature θ_0 are included. The general trend in both groups of alloys is for θ_0 to decrease in the α primary solid solutions and to increase slightly in the intermediate phases such as β, γ, or ζ.

In the more concentrated alloys γ values tend to be higher than the values for the pure solvents, except in the γ phase of the Cu-Zn system (see Fig. 4) where γ is below the value for Cu. In dilute alloys, between

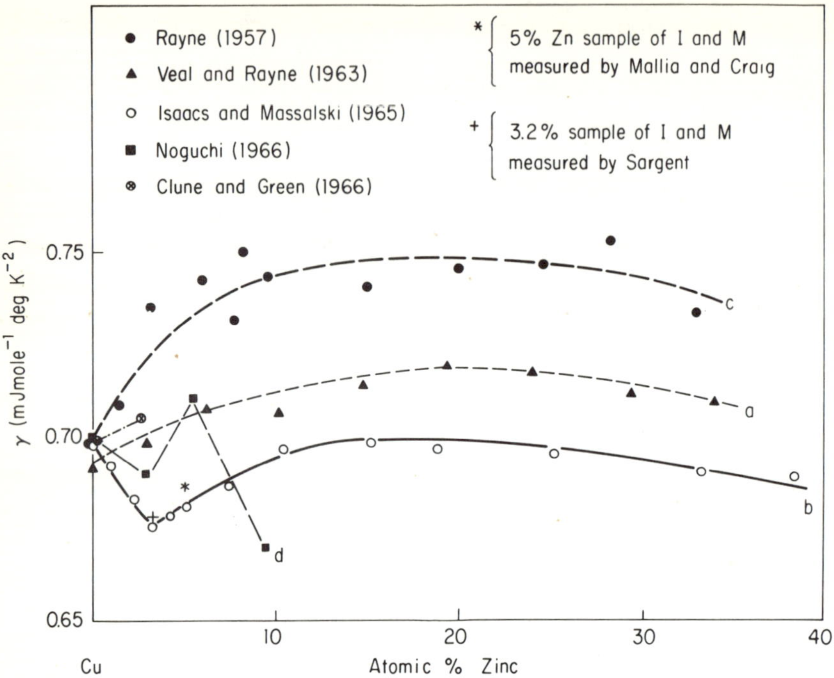

Fig. 3 The variation with composition of the electronic coefficient γ in the α phase of Cu-Zn alloys.

approximately 0 and 3 a/o solute, there is some disagreement about the initial slope of the γ trends. This is best illustrated for the Cu-Zn alloys (Fig. 3). Exceptions to the increasing trend of γ in the primary solid solutions are found in Cu-Ag,[27] Ag-Au,[28] and Cu-Si[18]; also in Cu-Ge and Cu-Sn, but in the latter two systems there is disagreement between different sets of measurements[4,16,18] (see Fig. 1).

Measurements of γ that have been reported for a few inter-noble alloys, namely Ag-Au and Cu-Ag, and for a few intermediate phases (cubic CuZn β- and γ-brasses[8,9] and close-packed hexagonal ζ phases in Ag-Sn,[10] Cu-Ge, and Cu-Si[11]) are also included in the figures. We wish to emphasize that some of the plotted results are by no means final and are considered by the authors as preliminary. In some cases repeated tests are planned, and in others data are not yet available regarding the detailed composition and the impurity content of the samples.

SOME CORRECTIONS APPLICABLE TO DATA

Most samples are machined out of cast ingots and are subsequently homogenized. If the liquidus and solidus curves in the phase diagram are fairly flat, a gradient of composition is established in the ingot during solidification that may be difficult to remove during relatively short homogenization. X-ray lattice parameter tests of areas near the top and bottom of the ingots sometimes indicate differences of composition of up to 1 a/o.

Even more troublesome is the possible presence of paramagnetic impurities of Fe, Co, or Ni. It has been shown that even very small amounts of such impurities contribute to a relatively large increase in the measured γ. Based upon the data obtained with dilute Cu-Fe,[12] Cu-Co,[13] and Cu-Ni[14] alloys, one can estimate that one part per million of Fe, Co, and Ni contributes approximately 0.0008, 0.00005, and 0.00002, respectively, in units of millijoules/(mole) (°K)^2. Hence, although a few parts per million of Co or Ni is relatively unimportant, the Fe impurity must be taken into account. In Fig. 3 for example, curves *a* and *c* are thought to be the result of different impurity levels.[15] The displacement between the two curves indicates an impurity level in curve *c* alloys of some 45 ppm Fe. However, the displacement is evident only in the concentrated alloys. In fact, it is somewhat surprising that frequently considerable differences in γ are observed in the more concentrated alloys whereas γ values reported for the pure starting metals show a good agreement (see Figs. 1 to 3). This

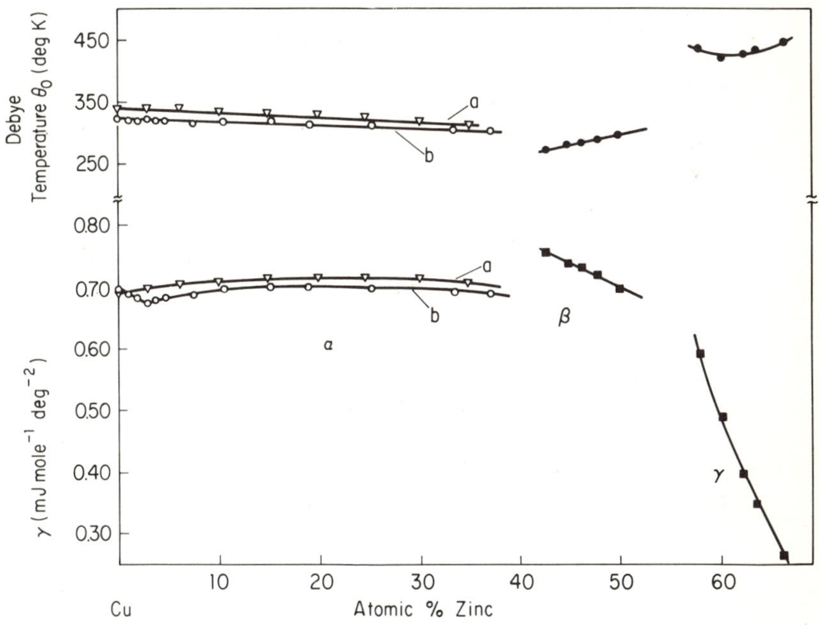

Fig. 4 The variation of the electronic coefficient γ and of the Debye temperature θ_0 in the intermediate phases β and γ of Cu-Zn system, shown in reference to the primary solid solution.

suggests that paramagnetic impurities are most likely introduced with the solute material, or by a direct handling and machining of the cast samples.

In Cu-Zn α alloys only curve c appears to be associated with contaminated alloys. Yet there is substantial difference between the remaining curves a, b, and d (Fig. 3). In the Cu-Sn system[16] the trend in the curve Cu-Sn (a) would be altered into that indicated by asterisks if a correction for the known amounts of Fe is applied. Even then, however, the results differ substantially from the curve marked Cu-Sn. There is also substantial disagreement between curves Cu-Ge (a) and Cu-Ge (see Fig. 1). We conclude that corrections for the presence of paramagnetic impurities, such as Fe, produce substantial modifications in the measured data. However, the difference in the impurity levels alone is unlikely to be responsible for the differences shown in Figs. 1 to 3 between data on the same systems.

Another correction that may be applied to data, if the results are to be considered in terms of the band structure of the solvent metal, is the

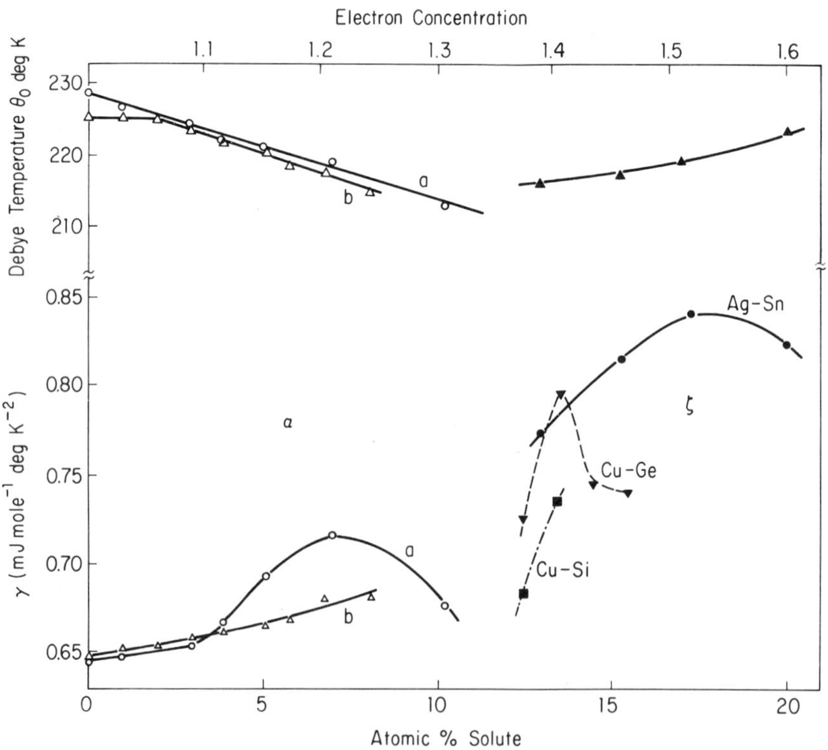

Fig. 5 The variation of the electronic coefficient γ and Debye temperature θ_0 in the close-packed hexagonal intermediate-phase ξ of the Ag-Sn system, shown in relation to similar trends in the primary solid solution. Preliminary results for the ξ phase of Cu-Si and Cu-Ge alloys are shown.

correction for the volume changes, i.e., the correction for lattice expansion or lattice contraction on alloying.[3,7] For free electrons a decrease in the lattice spacing will cause a relative decrease in the density of states.[3] However, Green[7] considers that if there is contact between the Fermi surface and the Brillouin zone, a decrease in the lattice spacings on alloying will produce a positive contribution to γ. According to this interpretation, the relatively large increase of γ values in the Ag-Zn systems (see Fig. 2) is the result of a substantial volume decrease that occurs on alloying. The data plotted in Figs. 1 to 5 have not been corrected for the volume change effects.

Finally there may well be a short-range order effect in the alloys, which will depend upon the heat treatment and preparation. This should certainly be considered, particularly for the concentrated alloys.

DISCUSSION OF DATA

Dilute Alloys

The plots shown in Figs. 1 to 3 clearly reveal the rather unsatisfactory state of affairs. In the Cu-based systems the Cu-Ge (a) curve (Fig. 1) and the Cu-Zn (c) curve (Fig. 3) appear to be associated with fairly contaminated alloys.[15] Hence the γ values are uncertain. Some of the Cu-Sn samples associated with curve a, Fig. 1, also contain significant amounts of Fe, particularly the 6.21 a/o Sn sample.[16] However, even if allowance is made for a positive contribution to γ arising from these impurities, both the results of Veal and Rayne[15] for α-Cu-Zn and those of Clune and Green[16] for α-Cu-Sn suggest a slightly positive trend in γ values on alloying, i.e., a trend above the value of pure Cu. On the other hand, the results obtained thus far by the present authors[17,18] suggest an initially downward trend in α phases of Cu-Zn, Cu-Ge, Cu-Sn, and Cu-Si, followed by an increase to values slightly above that for Cu in solute-rich samples. Thus there is disagreement about the initial slope of γ in these systems, and the reason for this is not at all clear. The preliminary results communicated to us recently by Noguchi[19] did not resolve the problem (see Fig. 3).

In dilute alloys of Ag with B subgroup elements the agreement is better, and a slightly positive initial trend of γ, above the value of pure Ag, is indicated in all alloys. If the γ values are plotted as a function of electron concentration e/a for the systems Ag-Sn[20,21] and Ag-Cd,[22] the initial slopes are nearly the same, indicating that at constant e/a the difference between strongly scattering ions of Sn and somewhat less strongly scattering ions of Cd is not substantial.[20]

From the theoretical point of view, a semirigid-band approach[23] would suggest that, following the onset of contact between the Fermi surface and the Brillouin zone in the pure noble metals, the initial downward trend in the density of states (and hence in γ) is more likely. However, various secondary masking effects have been also proposed; the considerations of possible contributions from the d-band electrons,[15] from the possible contact between the Fermi surface and the square {200} faces of the Brillouin zone,[15,24] and from the electron-phonon interactions[25,26] further complicate the matter. In the latter case there exists a possibility of obtaining a positive contribution to γ if the Fermi surface lies close to a spike in the density of states. This is essentially the situation in all three noble metals where the Fermi level must lie to the right of, and rather close to, the peak associated with the contact between the Fermi surface and the hexagonal {111} zone faces. According to Jones,[25] the spreading out of the spike due to electron-phonon interactions on alloying can produce a "virtual" positive contribution to γ. Dawber and Turner[26] point out, however, that the net change in γ depends very critically on the closeness of the Fermi level to the density-of-states peak. For dilute Cu-Zn alloys they predict an initially negative slope. Thus both the experimental and theoretical proposals for the Cu-Zn alloys are contradictory.

Comparison of data for dilute Ag-Cd[22] and Ag-Sn[20,21] alloys at constant e/a, and of data for inter-noble-metal alloys Cu-Ag[27] and Ag-Au,[28] suggests that the positive "virtual" contribution to γ as discussed by Jones[25] is not as substantial as predicted theoretically. A similar conclusion can be drawn from the data for other Cu-based systems, but here more caution is needed since there is disagreement between experimental results.

Concentrated Alloys

All γ values in concentrated α-phase alloys of Cu and Ag with the B subgroup elements fall slightly above those for the pure elements. This appears to be a general phenomenon. There is no clear indication, however, whether or not there is a decrease in γ just before the termination of the primary solid solubility.

There are a number of theoretical proposals for the case of concentrated alloys: the contribution to an increase in γ because of the interaction between the Fermi surface and the square {200} faces of the Brillouin zone,[24,29] the variable-band model,[30] the contribution from the d-band electrons,[15,29,30] and the possible contribution due to overlap into the next higher zone.[17] We believe that the overlap into higher zones within the range of the α phases seems unlikely, as is also a rigid-band behavior in the alloys. Undoubtedly the band gaps change, most likely decrease, as proposed originally by Cohen and Heine.[30] Optical data for Cu-Zn alloys support this possibility.[31] In addition, the position of the d band in the alloys appears to be crucial,[29,30] as is the possibility of screening of excess charge on solute atoms by the conduction electrons.[32] Since the d band must interact with the conduction band, s-d interactions will affect the shape of the Fermi surface. The top of the d band is likely to be lowered in energy on adding B subgroup elements. However, if bound states are formed as the conduction band is filled, it may be displaced faster downward in energy on alloying, relative to the d band, with the net result that there is little change in the energy at the Fermi level but a relatively large displacement of the conduction band with respect to the d band. Lettington[29] suggests that this effect may make it possible to bring interaction with the $\{200\}$ faces of the zone into the range of the primary solid solution, even though the band gaps decrease on alloying. The relationship between γ and $N(E_F)$ is clearly a complex one, and the measured values of γ in the alloys may reveal but little information about the band structure of the solvent.

In the inter-noble alloys such as Ag-Au or Cu-Ag, the electron concentration does not change. In both systems a downward trend in γ is revealed on alloying, and the possible positive virtual contribution due to electron-phonon enhancement is not apparent. Cu-Ag alloys show a trend that can be extrapolated to the value of γ for pure Ag, whereas Ag-Au alloys indicate a shallow minimum in γ at about 25 a/o Au (see Fig. 2). The general trend with composition in Ag-Au alloys may be compared with the recently established optical measurements of the polar-reflection Faraday effect in this system which indicate that the contact area between the Fermi surface and the Brillouin zone varies linearly and increases slightly between pure Ag and pure Au.[33] Thus the two trends do not agree, again pointing to a complex relationship between γ and the density of states of the solvent metals. Stern[34] in discussing alloys proposes a "charging effect" that is small in Ag-Au alloys with respect to the conduction electrons but large for the d-band electrons. Such charging effects further complicate the understanding of the trends in the electronic specific heats.

INTERMEDIATE PHASES

Figures 4 and 5 show γ and θ_0 measurements in the electron phases β, γ, and ζ in relation to the primary solid solutions. On this scale the differences between different investigations in the primary-solid-solution range are relatively small. The γ values for the β-brass lie above those for the α phase, as predicted by the simple electron theory of Jones.[35] Both in β and γ phases electronic specific-heat coefficients decrease with increasing Zn content, suggesting that contact has occurred between the Fermi surface and the principal faces of the Brillouin zone (or Jones's zone). When the primary solid solution is followed by a ζ phase as in Ag-Sn, Cu-Si, and Cu-Ge, the trend in the electronic specific-heat coefficient falls again above the values for the primary solid solution; the γ values increase further on alloying or pass through a maximum (see Fig. 5). The behavior of the ζ-Ag-Sn alloys has been interpreted[10] in terms of a nearly rigid band picture in which the peak in γ values is due to contact of the Fermi surface with the {10.1} faces of the Brillouin zone, and the initial sharp increase is due to an overlap of electrons into the second zone. However, here again more data are needed before a detailed picture can be established.

REFERENCES

1. Massalski, T. B., and H. W. King: *Progr. Mater. Sci.* **10**:1 (1961).
2. Hume-Rothery, W., and G. V. Raynor: "Structures of Metals and Alloys," 4th ed., Institute of Metals, London, 1962.
3. Rayne, J. A.: *Phys. Rev.* **108**:22 (1927).
4. Rayne, J. A.: *Phys. Rev.* **110**:606 (1958).
5. See for example W. A. Harrison and M. B. Webb (eds.): "The Fermi Surface," John Wiley & Sons, Inc., New York, 1960.
6. See for example T. B. Massalski (ed): "The Alloying Behavior and Effects in Concentrated Solid Solutions," Gordon and Breach, Science Publishers, Inc., New York, 1965.
7. Green, B. A., Jr.: See abstract in *Phys. Rev. Letters* **16**:A1 (1966); also private communication.
8. Veal, B. W., and J. A. Rayne: *Phys. Rev.* **128**:551 (1962).
9. Veal, B. W., and J. A. Rayne: *Phys. Rev.* **132**:1617 (1963).
10. Isaacs, L. L., and T. B. Massalski: *Phys. Rev.* **141**:634 (1966).
11. Sargent, G. A., and T. B. Massalski: To be published.
12. Franck, J. P., F. D. Manchester, and D. Martin: *Proc. Roy. Soc.* **A263**:414 (1961).
13. Crane, L. T., and J. E. Zimmerman: *Phys. Rev.* **123**:1 (1961).
14. Guthrie, G. L., S. A. Friedberg, and J. E. Goldman; *Phys. Rev.* **113**:45 (1959).
15. Veal, B. W., and J. A. Rayne: *Phys. Rev.* **130**:2151 (1963).
16. Clune, L. C., and B. A. Green, Jr.: See abstract in *Phys. Rev. Letters* **16**:A1 (1966); and private communication.
17. Isaacs, L. L., and T. B. Massalski: *Phys. Rev.* **138**:A134 (1965).
18. Sargent, G. A., L. L. Isaacs, and T. B. Massalski: To be published.
19. Noguchi, S.: Private communication, 1966.
20. Green, B. A., Jr., and H. V. Culbert: *Phys. Rev.* **137**:A1168 (1965).
21. Massalski, T. B., and L. L. Isaacs: *Phys. Rev.* **138**:A139 (1965).
22. Montgomery, H., and G. P. Pells: Conference on Electronic Structures of Alloys, Sheffield, England, 1963; and private communication, 1966.
23. Ziman, J. M.: *Advan. Phys.* **10**:1 (1961).
24. Hume-Rothery, W., and D. Roaf: *Phil. Mag.* **6**:55 (1961).
25. Jones, H.: *Phys. Rev.* **134**:A958 (1964).
26. Dawber, P. G., and R. E. Turner: Private communication, to be published.
27. Sargent, G. A., L. L. Isaacs, and T. B. Massalski: *Phys. Rev.* **143**:A420 (1966).
28. Green, B. A., Jr., and A. A. Valladares: *Bull. Am. Phys. Soc.* **10**:127 (1965); and private communication.
29. Lettington, A. H.: *Phil. Mag.* **11**:863 (1965).
30. Cohen, M. H., and V. Heine: *Advan. Phys.* **7**:395 (1958).
31. Biondi, M. A., and J. A. Rayne: *Phys. Rev.* **115**:1522 (1959).
32. Friedel, J.: *Advan. Phys.* **3**:461 (1954).
33. McAlister, A. J., E. A. Stern, and J. C. McGroddy: *Phys. Rev.*, in press, 1966.
34. Stern, E. A.: *Physics* **1**:255 (1965); also sbstract in *Phys. Rev. Letters* **16**:A4 (1966).
35. Jones, H.: *Proc. Phys. Soc.* **49**:250 (1937).

DISCUSSION *on Paper Presented by T. B. Massalski*

FRIEDEL: The rigid band model is founded on the first order perturbation approximation. I want to stress again that this is *not* a good approximation for electronic specific heats as soon as the change of nuclear charge of an impurity is comparable or larger than the number of valence electrons per atom.

This is self-evident in the free electron picture of the matrix. The change of specific heat γ by alloying is then related to the variation of the phase shifts δ_e with energy at the Fermi level E_m.

Take for instancc CuZn. Depending on the exact approximation made, one finds either of the two following pictures, which correspond to whether the perturbing potential accepts a bound state or not:

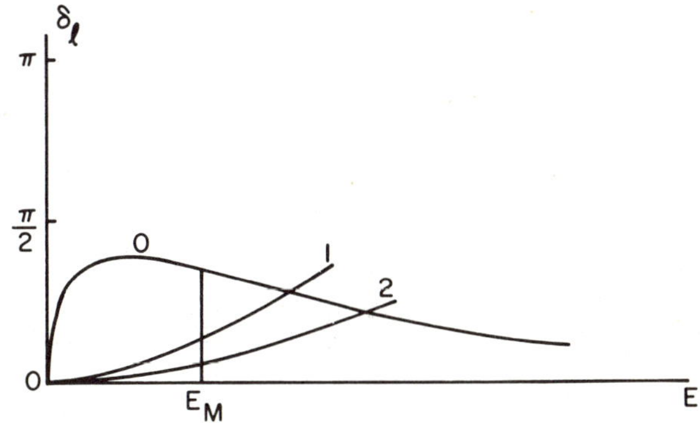

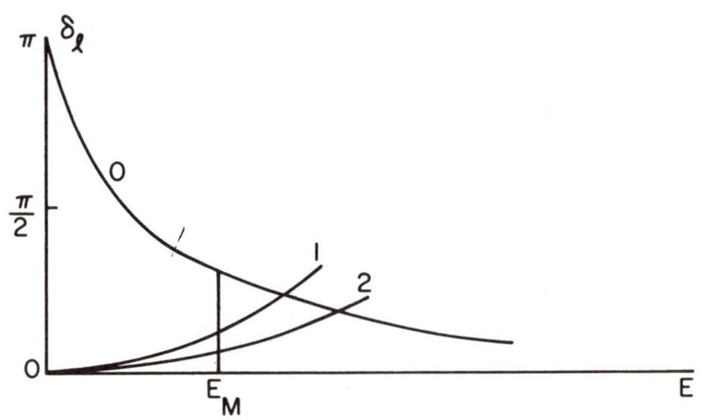

The screening condition requires that

$$\frac{2}{\pi}\sum_{l} (2l+1)\delta_l = 1$$

Since in the upper case the maximum of δ_0 is necessarily below $\pi/2$ (because below the dotted line which corresponds to the apparition of a bound state), and because δ_1 is still small at the maximum of δ_0, the Fermi level is necessarily on the right of this maximum.

One sees that, in either case, $d\delta_0/dE_m$ will be negative, contrary to the Born (low energy) approximation at the back of the rigid-band model. Indeed, I think that the free-electron model for Cu, Ag, Au might well explain the fact that γ does not vary much or indeed decreases when solutes with higher valencies are added.

MASSALSKI: The data that we have presented certainly do not seem to support any rigid-band approximation. Our aim was to point out that the magnitude of the γ coefficients in the alloys of the noble metals is nearly of the same order as the errors involved. Hence some disagreement between experimentalists is likely. Your two pictures for the Cu-Zn alloys are interesting. I can see why the upper case may not change much on alloying and may perhaps decrease initially. How do you explain a possible initial increase?

FRIEDEL: If real, it might be due to a Brillouin zone effect?

GASEOUS INTERMETALLIC MOLECULES

John Drowart

Laboratoire de Chimie Physique Moleculaire
Université Libre de Bruxelles, Brussels, Belgium

INTRODUCTION

Until 1950 the only gaseous diatomic metallic and intermetallic molecules for which the dissociation energies were known were those of the alkaline and of the group VA elements. These molecules were identified and their dissociation energies obtained from their electronic spectra.[1,2]

The existence of a number of other molecules (Zn_2, Cd_2, Hg_2, intergroup IA, Mg–group IA, and group IA–group IIB) was shown by the observation of diffuse absorption spectra.[1]

In the last 12 years, the spectroscopic investigation[5–28] of electronic spectra in thermal emission or in absorption in high-temperature King furnaces containing metals or alloys has led to the identification of some 25 additional diatomic homonuclear and intermetallic molecules. Simultaneously the application[29–55] of mass spectrometry to the thermochemical study of vapors in equilibrium with elements and compounds at high temperature has made it possible to identify independently about 15 of the

TABLE 1 *Diatomic Metallic Molecules*

Molecule	Ground state	ω, cm^{-1}	$x\omega$, cm^{-1}	r, A	D_0°, kcal/mole	Ref.
Li_2	$^1\Sigma$	351.43	2.592	2.672_5	25.0 ± 1.2	1, 2
Na_2	$^1\Sigma$	159.23	0.723	3.078_5	17.3 ± 0.7	1, 2
K_2	$^1\Sigma$	92.64	0.354	3.923	11.8 ± 1.2	1, 2
Rb_2	$^1\Sigma$	57.28	0.96	.....	10.8 ± 1.2	1, 2
Cs_2	$^1\Sigma$	41.99	0.08	.....	10.4 ± 1.0	1, 2
Cu_2	....	266.1	1.025	.....	(48)	6, 7
	($^1\Sigma$)	....	....	(2.68)	45.5 ± 2.2	34–36, 43
	$^1\Sigma$	264.56	....	2.219	....	21
Ag_2	....	192.4	0.643	....	(42)	6, 9
	($^1\Sigma$)	....	....	(2.68)	37.6 ± 2.2	34–36, 43
Au_2	....	190.9	0.420	....	(62)	6, 8
	($^1\Sigma$)	....	....	(2.68)	51.5 ± 2.2	34–36, 43
				2.4719	....	28*a*
Be_2	....	....	....	....	(14)	35
Mg_2	....	....	....	....	(7)	4
Ca_2	....	....	....	....	(5)	3
Zn_2	....	....	....	....	(6)	2
Cd_2	....	....	....	....	2.1 ± 0.5	2
Hg_2	....	....	....	....	1.4 ± 0.07	2
B_2	$^3\Sigma$	1051.3	9.4	1.589	....	1
					65.5 ± 5.5	37, 45
Al_2	($^3\Sigma$)	(400)	....	(2.5)	46	38
	$^3\Sigma$	350.01	2.022	2.466_8	43	11
Ga_2	($^3\Sigma$)	(225)	....	(2.5)	32	38
In_2	($^3\Sigma$)	(135)	....	(3.0)	19.5 ± 2.5	41, 57
	....	(142)	....	....	....	112
Tl_2	....	....	....	....	(14)	35
	....	(136)	....	....	....	11*a*
Sc_2	(g = 5)*	(230)	....	(2.7)	38.0 ± 5	47, 50
Y_2	(g = 5)	(230)	....	(2.8)	37.3 ± 5	47
La_2	(g = 5)	(230)	....	(2.8)	57.6 ± 5	47
Si_2	$^3\Sigma$	506.7	....	2.25	....	5
					74 ± 5	35, 39
		510.98	2.02	2.246	73.9†	10
Ge_2	($^3\Sigma$)	(370)	....	(2.44)	64 ± 5	35, 40
Sn_2	($^3\Sigma$)	(300)	....	(2.80)	45.8 ± 4	35, 43
Pb_2	($^3\Sigma$)	256.5[1]	....	(3.08)	23 ± 5	35
As_2	$^1\Sigma$	429.44	1.120	....	90.8	1, 2
Sb_2	$^1\Sigma$	269.85	0.59	....	70.6 ± 1.5	41
Bi_2	$^1\Sigma$	172.71	0.3227	....	44 ± 2	42, 46

TABLE 1 ***Diatomic Metallic Molecules (continued)***

Molecule	Ground state	ω, cm^{-1}	$x\omega$, cm^{-1}	r, A	D_0°, kcal/mole	Ref.
Ti_2	....	....	....	2.659_4	....	28
				....	<58	49
Cr_2	....	....	....	....	<36 (33)	50
Mn_2	....	....	....	....	18	61
					<21	49
Co_2	(g = 2.45)	(325)	....	(2.30)	39 ± 6	49
Ni_2	(g = 3.16)	(335)	....	(3.16)	54.5 ± 5	48
Pd_2	....	....	....	....	<33	51

* Statistical weight only was estimated.
† From a predissociation at 25,877 cm^{-1} assumed here to be an exact one.

spectroscopically observed and about 15 other such molecules and to measure their dissociation energies with good accuracy.

In this manner the homonuclear diatomic molecules of the principal elements and of a few transition metals are presently known, together with several intergroup IA, IB, and group IA–group IIA, IIIA, and IVA intermetallic molecules.

DISSOCIATION ENERGY OF GASEOUS METALLIC AND INTERMETALLIC MOLECULES

The dissociation energies of the diatomic metallic and intermetallic molecules are summarized in Tables 1 and 2.

For those molecules that have been studied spectroscopically the electronic ground state, the vibration frequency, the anharmonicity, and the interatomic distance are given whenever possible.

The dissociation energy listed for the metallic molecules is mostly taken from the compilations by Herzberg[1] and Gaydon[2] when based on convergence limits or predissociations.

For the intermetallic molecules the values given are, in general, obtained by the Birge Sponer extrapolation of the vibrational levels of the ground state, $D''_0 = \omega^2/4x\omega$. Whereas the uncertainty limits associated with this procedure can be as high as 30%,[2] the comparison with the mass spectrometric results suggests that they are probably smaller, except possibly for the Au–group IIA molecules for which no cross-checks can be made.

For those molecules that have been observed mass spectrometrically the dissociation energy D_0° was obtained from third law calculations, $D_0^\circ = -RT \ln K - T\,\Delta((G^\circ - H_0^\circ)/T)$. Here K is the equilibrium constant

TABLE 2 ***Diatomic Intermetallic Molecules***

Groups	Molecule	Electronic state	ω, cm^{-1}	$x\omega$, cm^{-1}	r, A	D_0'', kcal/mole	D_0°, kcal/mole	Ref.
IA–IA	NaK	$^1\Sigma$	123.29	0.40	....	....	14.3 ± 0.7	1, 2
	NaCs	....	98	....	....	....	....	1
	NaRb	$^1\Sigma$	106.64	0.455	....	....	13.1 ± 1.5	1, 2
	RbCs	....	49.4	....	....	....	....	1
IA–IIA	KHg	($^2\Sigma$)	....	....	....	....	1.08 ± 0.5	60
	CsHg	($^2\Sigma$)	....	....	....	....	1.10 ± 0.05	60
IB–IB	CuAg	....	229.5	1	....	38	....	12
		($^1\Sigma$)	(250)	....	(2.51)	....	40.7 ± 2.2	43
			231.8	0.80	....	48	....	15
	CuAu	....	250	0.7	....	64	....	13, 14
		($^1\Sigma$)	....	....	(2.51)	....	54.5 ± 2.2	43
	AgAu	....	(200)					
		($^1\Sigma$)	(210)	....	(2.68)	....	47.6 ± 2.2	43
IB–IIA	AuBe	$^2\Sigma$	600.63	3.53	2.0605	73	....	22
	AuMg	....	308.2	1.1		61	....	14
		$^2\Sigma$	306.10	....	2.4427	....	....	22
	AuCa	$^2\Sigma$	221.85	1.08	(2.67)	57	....	18
	AuSr	$^2\Sigma$	153.33	0.19	(2.54)	88	....	18
	AuBa	$^2\Sigma$	128.8	0.18	....	66	....	18
IB–IIIA	AuAl	....	332.59	1.14	....	70	....	14
		$^1\Sigma$	333.0	1.16_3	2.338	64	....	20
	AuGa	0+	225.62	0.567	....	46	....	22

IB–IVA	CuSn	($^2\Sigma$)	(215)	....	(2.58)	....	41.4 ± 4	44
	AgSn	($^2\Sigma$)	(190)	....	(2.76)	....	31.6 ± 5	44
	AuSi	$^2\Pi_{1/2}$	391.2	....	....	....	....	22
			390.9	1.35	....	81	....	25
	AuGe		251.7	0.30	....	....	....	22
		($^2\Pi$)	249.64	....	....	....	....	24
	AuSn	($^2\Sigma$)	(200)	....	(2.69)	....	57.5 ± 4	44
		$^2\Pi$	190.4	1.26	....	....	....	19
	AuPb	($^2\Sigma$)	158.6	0.6	....	30	....	23
IB– transition	AuCr	(g = 6)	(250)	....	(2.44)	....	50.4 ± 3.5	51
	AuPd	(g = 2)	(200)	....	(2.60)	....	33.3 ± 5	51
IIB–IIIA	CdIn	$^2\Sigma$	67.3	0.1	....	32	....	27
	ZnIn	$^2\Sigma$	146.7	0.7	....	7.7	....	26
	HgIn	$^2\Sigma$	151	0	....	....	....	27
	HgTl	....	....	....	....	0.7	....	1
IIIA–IVA	BSi	(g = 5)	....	....	....	....	68 ± 6	52
IIIA–VA	InSb	($^3\Sigma$)	(260)	....	(3.15)	....	....	41
IIIB–IIIB	YLa	(g = 5)	(230)	....	(2.8)	....	47.3 ± 5	47
IVA–IVA	GeSi	($^3\Sigma$)	409	....	(2.39)	....	71 ± 5	40
IVA–VA	PbBi	($^2\Sigma$)	....	....	....	....	31 ± 5	53
VA–VA	SbBi	$^1\Sigma$	220.0	0.50	....	(70)	....	1

TABLE 3 Intermetallic Molecules Observed Only

LiK, LiRb, LiCs, KRb, KCs[1]
MgK, MgNa, MgRb, MgCs[1]
CdK, CdNa, CdRb, CdCs[1]
ZnK, ZnCs, ZnRb[1]
HgK, HgNa, HgRb, HgCs[1]
AuIn,[22] AuU[54]
AsSb[55]

of the reaction $AB(g) \rightarrow A(g) + B(g)$, in which the partial pressures are deduced from the measured ion intensities; $(G^\circ - H_0^\circ)/T = -R \ln (Q/N)$ is the free-energy function, Q/N being the partition function calculated from the spectroscopically determined or estimated (in parentheses) statistical weight, vibration frequency, and interatomic distance. The total uncertainty limits, due in essentially equal parts to uncertainties in the free-energy functions when estimated and in the relative ionization cross sections (or partial pressures based thereon) are ±2 to 5 kcal/mole. For further details the reader is referred to general descriptions of the method and results.[56–59]

For KHg and CsHg the dissociation energy is derived from scattering experiments,[60] while for Mn_2 it results from theoretical calculations.[61] The values for Be_2 and Tl_2 are estimated.[35]

Table 3 lists a number of diatomic molecules that have been observed but for which no dissociation energies are available.

Table 4 gives the polyatomic molecules and their atomization energies. The concentration of these molecules in the vapor as well as that of the diatomic metallic or intermetallic molecules is quite low, 10^{-2} to 10^{-5}, and less that of the atom (except in the well-known cases of As_4, Sb_4, and Bi_2).

DISCUSSION

The dissociation energy of the homonuclear metallic molecules (Table 1) decreases regularly with increasing atomic number within each group, except for the two typical transition-metal triads Cu_2, Ag_2, and Au_2; Sc_2, Y_2, and La_2; where the heaviest member is strikingly more stable. Simultaneously typical alternations with the group occur.[34]

When, however, the ratio of the heat of sublimation of the atom to the dissociation energy of the molecule $\alpha = \Delta H(M)/D(M_2)$ is examined,[62] great regularities appear which are no longer restricted to a given group.

The values form three categories in which $0.5 \leq \alpha \leq 1$ for elements with s^2p^3, s^2p^4, and s^2p^5 ground-state electronic configurations; $1.2 \leq \alpha \leq 2.5$ for s^1, s^2p^1, s^2p^2, s^2d^n, $d^{10}s^1$ elements, and $\alpha > 5$ for elements with s^2 and $d^{10}s^2$ ground-state configurations.

TABLE 4 Atomization Energy of Polyatomic Molecules

		In kcal per mole		
Si_3, 173 ± 10[32,40]	Ge_3, 153 ± 10[32,40]	Sn_3, 120 ± 10[30,40]	Ge_2Si, 164 ± 10[40]	$InSb_2$, 102 ± 5[41]
Si_4, 275 ± 10	Ge_4, 240 ± 10	Sn_4, 120 ± 10	Ge_3Si, 255 ± 10	
Si_5	Ge_5	Sn_5		
Si_6	Ge_6			
Si_7	Ge_7			
Si_8	Ge_8			

The first of these categories is the metalloids in which the intermolecular bonding in the crystal lattice is much smaller or comparable to the bonding in the molecule.

The second category includes the metals other than the group IIA and group IIB ones which form the third category. For these atoms with a quasi–rare-gas electronic configuration the bonding in the condensed phase is presumably due to promoted configurations which lie, however, too high to be compensated by the formation of the diatomic molecule, which is hence of the van der Waals type.

On the basis of the regularity in the α values, the suggestion by Pauling[63] that, in the group IA elements, analogous bonds are responsible for the stability of the diatomic molecule and of the metal, in which it is, however, stronger by interaction with the nearest and next nearest neighbors, was extended[62] to the other members of the second category. For most of these bonding would occur through promoted configurations.

This hypothesis is supported by the closely parallel variations in the heats of sublimation and the excitation energy to the same configuration and the spin uncoupling energy[64] or to analogous excited states[62] in the transition[62] and rare-earth metals[65–66] and in the transition-metal[67] or rare-earth[65] oxides. Similar correlations also appear in the structure of the metals and intermetallic compounds in the condensed phase.[68–70]

The particular trends in the dissociation energy of Cu_2, Ag_2, and Au_2; Sc_2, Y_2, and La_2; Cr_2, Mn_2, Co_2, and Ni_2 are then explained[43,44,49] by differences in the excitation energy to promoted configurations or more probably by interaction between electronic states of the same symmetry formed from different atomic states, which are sufficiently close to interact.[71] For instance in Cu, Ag, and Au the energy of the d^9s^2 excited level is 32, 86, and 26 kcal, respectively, above the $d^{10}s^1$ ground state; in Sc, Y, and La the energy of the d^2s state is 33, 31, and 9 kcal above the ds^2 ground state.

The resulting, rather high valencies attributed[62,76] to the group IB elements, in particular in the gaseous molecule, agree with correlations between force constants and bond order[72,73] or dissociation energies and estimated bond lengths.[74] They are now also confirmed by the experimentally measured interatomic distances in Cu_2 ($r_{Cu_2} = 2.220$ A) and Au_2 ($r_{Au_2} = 2.472$ A) compared to those in the group IA molecules ($r_{Li_2} = 2.673$ A, $r_{Na_2} = 3.079$ A, $r_{K_2} = 3.923$ A). These measured r_{Cu_2} and r_{Au_2} values are appreciably smaller than those estimated (see Table 1) from the lattice parameters.[75–77]

Analogous factors probably determine[43,44] the trends in the stability of the gaseous intermetallic elements in relation to that of the corresponding homonuclear molecules and of the condensed phase.

Comparisons are shown in Table 5. Column 2 gives the measured dissociation energy of the intermetallic molecules, column 3 the average of

TABLE 5 ***Comparison of Dissociation Energy of Gaseous Homonuclear and Intermetallic Molecules and Alloys***

Molecule MM'	D, kcal/mole	$\frac{1}{2}[D(M_2) + D(M_2')]$, kcal/mole	Δ, kcal/mole	ΔH mix ($x = 0.5$), kcal/mole
NaK	14.3	14.6	− 0.3	+0.2 (298)*
NaRb	13.1	14.1	−1.0	
KHg	....	9.5	....	−1.0 (600)
CaHg	....	6		
CuAg	40.7	41.5	−0.8	+2.0 (1400)
CuAu	54.5	48.5	+ 6.0	−2.5 (720)
AgAu	47.6	44.5	+ 3.1	−2.2 (800, 1350)
AuBe	73	33	+40	
AuMg	61	30	+31	
AuCd	57	29	+28	
AuAl	64	49	+15	
AuGa	46	42	+ 4	
CuSn	41.4	40.3	+ 1.1	−1 (1423)
AgSn	31.6	36.3	− 4.7	−0.4 (1248)
AuSi	81	64	+17	
AuGe	....	59		
AuSn	58	43	....	−5.2 (873)
AuPb	(30)	38	....	−0.3 (873)
CdIn	(32)	10	+22	+0.7 (723)
ZnIn	8	12	− 4	+1.6 (700)
BSi	68	70	− 2	
YLa	47	47	0	
InSb	35	45	−10	−1.5 (973)
GeSi	71	70	+ 1	
SbBi	(70)	81	(−11)	(+0.6)(1073)

* Numbers in parentheses are °K.

that of the corresponding homonuclear molecules, column 4 the difference Δ, and column 5 the enthalpy of mixing[78] for the alloy ($x = 0.5$).

The intergroups IA (NaK and NaRb), IVA (GeSi), IIIB (YLa), and the group IIIA–IVA (BSi) molecules have dissociation energies close to the average. This is also true for the intergroup IVA (SiC and GeC[39]) gaseous monocarbides which have, perhaps arbitrarily, not been considered as intermetallic molecules.

In the AuM molecules, Δ is quite large, indicating a positive balance between the promotion energy in one or the other atom and the additional bonding gained. It also corresponds to an extra ionic energy and a marked difference in electronegativity χ of the two elements in Pauling's relation:

$$D(MM') = \tfrac{1}{2}(D(M_2) + D(M_2')) + 23(\chi - \chi')^2$$

The differences in electronegativity so obtained for Au, Cu, Sn, Al, and Si agree well with those in the literature: Au = 2.4, Cu = 1.8, Sn = 1.6, Al = 1.5, Si = 1.8.[75] In reverse, the values for Cr and Pd yield, together with D_0°(AuCr) and D_0°(AuPd), dissociation energies $D_0^\circ(Cr_2)$ and $D_0^\circ(Pd_2)$ close to the upper limits resulting from the nonobservation of the corresponding molecules.

The high electronegativity of Au also appears in the dissociation energy of the Au–group IIA molecules where probably D_0''(AuSr) is somewhat high. The trend in and the absolute values of the dissociation energy resemble those that occur in the group IIA monoxides, D_0°(BeO) = 106, D_0°(MgO) = 86, D_0°(CaO) = 99, D_0°(SrO) = 102, D_0°(BaO) = 131 kcal/mole.[80] The last closely follow that expected from an ionic model.[81,82] In this picture Au would by capture of an electron tend toward a quasi–rare-gas configuration $d^{10}s^2$(Hg). A similar qualitative analogy between the dissociation energy of the AuM and MO molecules also appears in the Au–group IIIA and Au–group IVA molecules (D_0°(BO) = 190; D_0°(AlO) = 115; D_0°(GaO) = 100; D_0°(InO) = 76; D_0°(SiO) = 190; D_0°(GeO) = 156; D_0°(SnO) = 126; D_0°(PbO) = 88 kcal/mole.

The small interatomic distances in the AuBe (2.060 A), AuMg (2.443 A), and AuAl (2.338 A) molecules are also to be noted. The last is appreciably smaller than the average for Au_2 (r_{Au_2} = 2.472 A) and Al_2 (r_{Al_2} = 2.467 A).

The trends in the difference between the measured dissociation energy and the average for the homonuclear molecules are, as Table 5 shows, rather parallel to the heat of mixing in the alloy. High positive values of Δ are usually associated with high negative heats of mixing, while negative Δ values are associated with small or positive heats. A general parallelism between the bonds in the alloy and the intermetallic molecule, analogous to that between the bonds in the metal and its gaseous molecules. is thereby suggested.

An exception is InSb. The heat of formation of the compound[79] and the heat of mixing of the alloy at 700°K[78] are quite negative. The dissociation energy of the molecule is, however, much lower than the average for In_2 and Sb_2, which equals the dissociation energy of the isoelectronic Sn_2 molecule. An analogous relative difference in stability occurs in BN compared to C_2 [41] and probably the other group IIIA–group VA molecules. There is a parallel decrease in the atomization energy of the lattice in corresponding compounds.[79]

In other isoelectronic molecules such as CdIn and CuSn, ZnIn and AgSn, HgIn and AuSn, or HgK and AuCa, and HgCs and AuBa, the dissociation energies are completely different. The low values of the dissociation energy in the Cd and Hg molecules are apparently related to the great difference in excitation energy of the s^1p^1 levels in the group IIA and group IIB elements.

The polyatomic molecules Si_3 to Si_7, Ge_3 to Ge_8, Sn_3 to Sn_5, Ge_2Si are, together with SiC_2, Si_2C, Si_2C_2, Si_2C_3, Si_3C,[39] Ge_2C, Ge_3C, and GeC_2,[40] the analogues of the C_3 to C_5 [29,32,84] and presumably larger C species in which the stability alternates from even- to odd-membered molecules. A molecular orbital treatment[83] has shown the odd-membered species to have closed-shell structures and therefore the higher stability. In these polyatomic molecules the approximate constancy of the ratio between atomization energy and heat of sublimation again appears.

In view of the number of Au molecules already known and for which there are spectroscopic data for several electronic states, the study of further AuM species, especially for transition metals, will certainly give a better insight into the factors determining bonding in gaseous intermetallic molecules.

The same applies to molecules that have a sufficiently low number of electrons to make possible theoretical calculations, as already done for Mn_2 [61] and in progress for Sc_2.[85]

The parallelism between the bonding in the condensed phase and in the gaseous species suggests, on the other hand, that gaseous intermetallic molecules may be observed for the phases for which the Engel-Brewer theory[66–70] predicts a high stability.

ACKNOWLEDGMENTS

The author is indebted to L. L. Ames, R. F. Barrow, R. Houdart, C. Moser, and G. Verhaegen for communicating data prior to publication.

REFERENCES

1. Herzberg, G.: "Diatomic Molecules," D. Van Nostrand Co., Inc., Princeton, N.J., 1950.
2. Gaydon, A. G.: "Dissociation Energies," Chapman & Hall, Ltd., London, 1963.
3. Brewer, L.: University of California Radiation Laboratory Report 2854, 1955.
4. Soulen, J. R., P. Shtapitanonda, and J. L. Margrave: *J. Phys. Chem.* **59**:132 (1955).
5. Douglas, A. E.: *Can. J. Phys.* **33**:801 (1955).
6. Ruamps, J.: *Compt. rend.* **239**:1200 (1954).
7. Kleman, B., and S. Lindkvist: *Arkiv Fysik* **8**:333 (1954).
8. Kleman, B., S. Lindkvist, and L. E. Selin: *Arkiv Fysik.* **8**:505 (1954).
9. Kleman, B., and S. Lindkvist: *Arkiv Fysik* **9**:385 (1955).
10. Verma, R. D., and P. A. Warsop: *Can. J. Phys.* **41**:152 (1963).
11. Ginter, D. S., M. L. Ginter, and K. K. Innes: *Astrophys. J.* **139**:365 (1964).
11*a*. Ginter, D. S., M. L. Ginter, and K. K. Innes, *J. Phys. Chem.* **69**:2480 (1965).

12. Ruamps, J.: *Compt. rend.* **238**:1489 (1954).
13. Ruamps, J.: *Colloq. Spectroscop. Intern.* **VI**:329 (1956).
14. Ruamps, J.: *Ann. Phys. Paris* **4**:1111 (1959).
15. Joshi, K. C., and K. Majumdar: *Proc. Phys. Soc.* **78**:197 (1961).
16. Schiltz, J.: *Compt. rend.* **252**:1750 (1961).
17. Schiltz, J.: *Compt. rend.* **253**:1777 (1961).
18. Schiltz, J.: *Ann. Phys. Paris* **8**:67 (1963).
19. Colette, M., and J. Schiltz: *Compt. rend.* **257**:2092 (1963).
20. Barrow, R. F., and D. N. Travis: *Proc. Roy. Soc.* **A273**:133 (1963).
21. Travis, D. N., and R. F. Barrow: *Proc. Chem. Soc.* **64**:(1962).
22. Barrow, R. F., W. J. M. Gissane, and D. N. Travis: *Nature* **201**:603 (1964); *Proc. Roy. Soc.* **A287**:240 (1965).
23. Houdart, R., and P. Colette: *Compt. rend.* **260**:5746 (1965).
24. Houdart, R.: *Compt. rend.* **261**:2609 (1965).
25. Houdart, R.: *Compt. rend.* **262**:550, 1966.
26. Santaram, C., and J. G. Winans: *Phys. Rev.* **136**:A57 (1964).
27. Santaram, C., and J. G. Winans: *J. Mol. Spectr.* **16**:309 (1965).
28. Åslund, N., R. F. Barrow, W. G. Richards, and D. N. Travis: *Arkiv Fysik* **30**:171 (1965).
28*a*. Ames, L. L., R. F. Barrow, and W. J. M. Gissane, to be published.
29. Chupka, W. A., and M. G. Inghram: *J. Chem. Phys.* **21**:371, 1313 (1953); **22**:1472 (1954); *J. Phys. Chem.* **59**:100 (1955).
30. Honig, R. E.: *J. Chem. Phys.* **21**:573 (1953).
31. Honig, R. E.: *J. Chem. Phys.* **22**:126 (1954).
32. Honig, R. E.: *J. Chem. Phys.* **22**:1610 (1954).
33. Koehl, G.: *Z. Naturforsch* **9*a***:913 (1954).
34. Drowart, J., and R. E. Honig: *J. Chem. Phys.* **25**:581 (1956).
35. Drowart, J., and R. E. Honig: *J. Phys. Chem.* **61**:980 (1957).
36. Schissel, P.: *J. Chem. Phys.* **26**:1276 (1957).
37. Chupka, W. A.: Argonne National Laboratory Report 5667, p. 75, 1957.
38. Chupka, W. A., J. Berkowitz, C. F. Giese, and M. G. Inghram: *J. Phys. Chem.* **62**:611 (1958).
39. Drowart, J., G. De Maria, and M. G. Inghram: *J. Chem. Phys.* **29**:1015 (1958).
40. Drowart, J., G. De Maria, A. J. H. Boerboom, and M. G. Inghram: *J. Chem. Phys.* **30**:308 (1959).
41. De Maria, G., J. Drowart, and M. G. Inghram: *J. Chem. Phys.* **31**:1076 (1959).
42. Porter, R. F., and C. W. J. Spencer: *J. Chem. Phys.* **32**:943 (1960).
43. Ackerman, M., F. E. Stafford, and J. Drowart: *J. Chem. Phys.* **33**:1784 (1960).
44. Ackerman, M., J. Drowart, F. E. Stafford, and G. Verhaegen: *J. Chem. Phys.* **36**:1557 (1962).
45. Verhaegen, G., and J. Drowart: *J. Chem. Phys.* **37**:1367 (1962).
46. Aldred, A. J., and J. N. Pratt: *J. Chem. Phys.* **38**:1085 (1963); J. H. Kim and A. Cosgarea, Jr., *J. Chem. Phys.* **44**:806 (1966).
47. Verhaegen, G., S. Smoes, and J. Drowart: *J. Chem. Phys.* **40**:239 (1964).
48. Kant, A.: *J. Chem. Phys.* **41**:1872 (1964).
49. Kant, A., and B. Strauss: *J. Chem. Phys.* **41**:3806 (1964).
50. Verhaegen, G.: PhD thesis, University of Brussels, 1965.
51. Ackerman, M., F. E. Stafford, and G. Verhaegen: *J. Chem. Phys.* **36**:1560 (1962).
52. Verhaegen, G., F. E. Stafford, and J. Drowart: *J. Chem. Phys.* **40**:1622 (1964).
53. Unpublished data, this laboratory.
54. Storms, E. K.: "Thermodynamics," IAEA, Vol. 1, p. 309, Vienna, 1966.
55. Honig, R. E.: *Anal. Chem.* **25**:1530 (1953).

56. Inghram, M. G., and J. Drowart: In "High Temperature Technology," p. 219, McGraw-Hill Book Co., New York, 1960.
57. Drowart, J.: In E. Rutner, P. Goldfinger, and J. P. Hirth (eds.), "Condensation and Evaporation of Solids," p. 255, Gordon and Breach, Science Publishers, Inc., New York, 1964.
58. Goldfinger, P.: In R. I. Reed (ed.), "Mass Spectrometry," p. 265, Academic Press Inc. New York and London, 1965.
59. Drowart, J., and P. Goldfinger: *Angew. Chem.*, in press.
60. Morse, F. H., and R. B. Bernstein: *J. Chem. Phys.* **37**:2019 (1962).
61. Nesbet, K. K.: *Phys. Rev.* **135**:A460 (1964).
62. Verhaegen, G., F. E. Stafford, P. Goldfinger, and M. Ackerman: *Trans. Faraday Soc.* **58**:1926 (1962).
63. Pauling, L.: *Phys. Rev.* **54**:898 (1938).
64. Griffith, J. S.: *J. Inorg. Nucl. Chem.* **3**:15 (1956).
65. White, D., P. M. Walsh, L. L. Ames, and H. W. Goldstein: In "Thermodynamics of Nuclear Materials," p. 417, IAEA, Vienna, 1962.
66. Jørgensen, C. K.: *Mol. Phys.* **7**:417 (1964).
67. Berg, R. A., and O. Sinanoglu, *J. Chem. Phys.* **32**:1082 (1960).
68. Brewer, L.: In P. A. Beck (ed.), "Electronic Structure and Alloy Chemistry of the Transition Elements," p. 221, Interscience Publishers, Inc., New York, 1963.
69. Brewer, L.: In V. F. Zackay (ed.), "High-strength Materials," p. 221, John Wiley & Sons, Inc., New York, 1965.
70. Brewer, L.: This colloquium, p. 39.
71. Mulliken, R. S.: *Phys. Rev.* **50**:1017, 1928 (1936).
72. Guggenheimer, K. M.: *Proc. Phys. Soc.* **58**:456 (1946).
73. Gordy, W.: *J. Chem. Phys.* **14**:305 (1946).
74. Pauling, L.: *J. Phys. Chem.* **58**:662 (1954).
75. Pauling, L.: *J. Am. Chem. Soc.* **69**:542 (1947).
76. Pauling, L.: "The Nature of the Chemical Bond," Cornell University Press, Ithaca, N.Y., 1960.
77. Hume-Rothery, W., and G. V. Raynor: "Structure of Metals and Alloys," 4th ed., Institute of Metals, London, 1962.
78. Hultgren, R., R. L. Orr, P. D. Anderson, and K. K. Kelley: "Selected Values for Thermodynamic Properties of Metals and Alloys," John Wiley & Sons, Inc., New York, 1963.
79. Goldfinger, P.: In R. E. Willardson and H. L. Goering (eds.), "Compound Semiconductors," p. 483, Reinhold Publishing Corporation, New York, 1962; P. Goldfinger and M. Jeunehomme: *Trans. Faraday Soc.*, **59**:2851 (1963).
80. Drowart, J., G. Exteen, and G. Verhaegen: *Trans. Faraday Soc.* **60**:1920 (1964).
81. Brewer, L., and D. F. Mastick: *J. Am. Chem. Soc.* **73**:2045 (1951).
82. Chandrasekharaiah, C. M.: *J. Phys. Chem.* **68**:2020 (1964).
83. Pitzer, K. S., and E. Clementi: *J. Am. Chem. Soc.* **81**:4477 (1959); J. J. Strickler and K. S. Pitzer: In B. Pullman and P. O. Löwdin (eds.), "Molecular Orbitals in Chemistry," Academic Press Inc., New York, 1966.
84. Drowart, J., R. P. Burns, G. De Maria, and M. G. Inghram: *J. Chem. Phys.* **31**:1131 (1959).
85. Verhaegen, G., and C. Moser: Private communication, 1966.

PHASE STABILITY AND CRYSTAL CHEMISTRY OF COMPLEX COMPOUNDS CONTAINING TRANSITION ELEMENTS AND NONMETALS

Hans Nowotny

Institut für Physikalische Chemie, University of Vienna, Vienna, Austria

Friedrich Benesovsky

Metallwerk Plansee AG, Reutte, Austria

INTRODUCTION

This paper describes in brief the present state of the alloy chemistry of transition-element (T) carbides, borides, silicides, and related phases reviewed 4 years ago.[1] For many of the various specific problems, reference is made to the previous paper. Although the main problem of binary transition-element carbides, the bonding in TiC_{1-x} for example, is still unresolved, a systematic extension to more complex phases, such as T^{I}-T^{II}-X (nonmetal), T-M (meta metal)-X, T-X^{I}-X^{II}, appears to reveal many regularities of the electronic state of this class of compounds. This is obviously due to the fact that one more parameter (concentration) is free for adjustment. There are many examples found in alloy chemistry, such as ternary Laves phases, exhibiting a rather perfect correlation between type of crystal structure ($MgZn_2$, $MgCu_2$, $MgNi_2$) and the valence electronic concentration (VEC).

BINARY CARBIDES

Before discussing some new results on complex hydrides, nitrides, carbides, and borides, the so-called "subcarbides" of formula T_2C demand additional exploration. It was previously assumed that the C atoms randomly fill the octahedral voids of a hexagonal close-packed metal lattice, but Parthé and Sadagopan[2] have shown an ordered structure for Mo_2C, the space group being D_{2h}^{14}. In the meantime Rudy et al.[3] found two hexagonal subcarbides, α-Mo_2C and β-Mo_2C, having a slightly different amount of C and a minor difference in the c value. From a thermodynamic calculation on the basis of simple models one can also conclude that complete ordering of the C atoms takes place for low temperatures. A deviation from L'_3 type (random distribution) has been detected by Terao[4] for vapor-deposited Nb_2C; an ordering according to the ϵ-Fe_2N type occurs. Very recently another type of ordering, namely the anti-CdI_2 structure, has been found for Ta_2C by neutron diffraction.[5] It should be mentioned furthermore that, according to Rudy et al.,[6] here again two subcarbides (α-Ta_2C and β-Ta_2C) do exist, derived from differential thermoanalysis experiments.

V_2C

The V subcarbide is reported to be of the L'_3 type,[7] and this was recently confirmed by neutron-diffraction experiments for a sample containing V_2C and minor quantities of VC.[5] In fact, there is agreement for samples containing V_2C + VC insofar as the V_2C pattern can always be indexed with a hexagonal cell required for the L'_3 type. However, for the majority of combinations V_2C + metal or $(V, Cr)_2C$ + metal an orthorhombic splitting was observed. Evaluation of the powder pattern of V_2C for these samples leads to the following data (in angstroms):

$$a = 4.56_7 = c_{\text{hex}}$$

$$b = 5.74_2 = 2a_{\text{hex}}$$

$$c = 5.03_7 \approx \sqrt{3}a_{\text{hex}}$$

For D_{2h}^{14} with 8 V atoms in $8d$, $x = 0.250$, $y = 0.130$, $z = 0.078$; with 4 C atoms in $4c$, $y \approx \frac{3}{8}$. In other words V_2C (orthorhombic) is isotypic with ζ-Fe_2N or Mo_2C.[8] The orthorhombic V_2C exists over a wide range in the ternary system V-Cr-C, including a composition $(V_{\sim0.6}Cr_{\sim0.4})_2C$ at 1350°C. It is of interest to note that increasing Cr/V substitution shifts the ratio

$2c/b$ more toward $\sqrt{3}$ (pseudohexagonal).[9] The formation of the orthorhombic and hexagonal modification of V_2C depends therefore on the mode of formation. As an ad hoc explanation one can consider an epitaxial equilibrium with the (111) plane of the cubic monocarbide VC_{1-x} and (0001) of V_2C being more favorable for the hexagonal type, while the close-packed plane for the cubic body-centered V, namely (110), is pseudohexagonal only. Hexagonal V_2C + metal can nevertheless be obtained by hot-pressing a mixture of $VC_{1-x} + V$. However, an additional anneal (13 hr at 1400 to 1450°C) yields again the orthorhombic V_2C.[8] On the other hand, it is to be assumed that transformation of the orthorhombic V_2C to the hexagonal L'_3 type takes place at higher temperatures. Brauer and Schnell[10] recently described the fairly complicated situation with V_2C by saying that unknown effects, such as impurities or nonequilibria, do not allow one to obtain specific information on the phase boundaries of V_2C. The question of the stability of a cubic CrC or CrC_{1-x} has to be answered negatively. Although VC_{1-x} dissolves up to 40 mole % CrC at 1350°C, and apparently somewhat more in arc-melted samples, there is no evidence for an extension of the $(V, Cr)C_{1-x}$ phase close to Cr carbide. This is in agreement with investigations and thermodynamic evaluations of other Cr-containing carbide systems[11]; the free-enthalpy change $\Delta G[Cr_3C_2 \rightarrow Cr_3C_2(B_1)] = 4070 - 1.00T$ cal/atom Cr shows clearly the instability of the cubic Cr carbide.

VCr_2C

Within the group VA-VIA C combinations, the Cr-containing systems differ not only with respect to crystal structure of the binary Cr carbides—Hägg's principle does not hold—but also by a tendency for the formation of ternary carbides. VCr_2C_2 was found to be rather stable; the crystal structure, determined in the meantime, reveals a combination between trigonal prisms and octahedral groups.*

COMPLEX CARBIDES AND NITRIDES CHARACTERIZED BY T_6X OCTAHEDRA

The existence of the octahedral group $[T_6X]$ was already demonstrated to be a fruitful principle for a systematic search of complex carbides and similar compounds.[12] The condition for the $[T_6X]$-octahedra formation is again closely connected with Hägg's concept. However, generally speaking, the range of the critical ratio r_C/r_T appears to be somewhat expanded. The common features of the whole class with respect to composition may be seen from the following series: T_3MX (perovskite carbide type or filled Cu_3Au type), T_3M_2X (filled β-Mn type), T_2MX (H phases), $T_8M_4X_3$

* P. Ettmayer, G. Vinek, and H. Rassaerts, *Monatsh. Chem.* (in press).

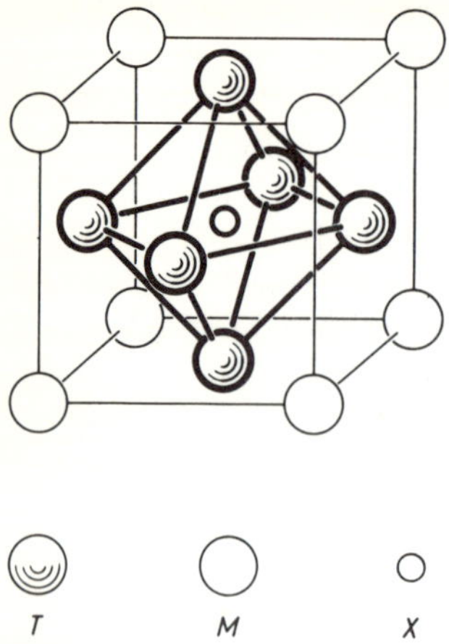

Fig. 1a Perovskite carbide type. [T_6X] octahedra are linked by corners.

(η-carbide type or filled Ti_2Ni type), T_5M_3X (filled Mn_5Si_3 type). However M, standing for meta metals or a b element, can sometimes be substituted for by a second transition element. Furthermore, the amount of X is often less than mentioned above: for example, C-deficient. The characteristic partial lattice of this uniform class is built up by linking of the [T_6X] groups by corners (Fig. 1), by edges (Fig. 2), or by planes (Fig. 3).

COMPLEX CARBIDES WITH A PARTIAL LATTICE LINKED BY CORNERS

With respect to the perovskite carbides, the most regular crystal structure of this class, progress has been made in the last few years based on the fact that transition elements of the IIIA, IVA, and VA groups are also convenient for the formation of T_3MX compounds (Table 1). In other words, there are now representatives from groups IIIA to VIIIA.

From these findings it is presumed that most of the rare-earth elements are able to form perovskite carbides, particularly with IIIB and IVB elements (Al, Ga, In, Tl, Sn, Pb). At least in some cases the corresponding

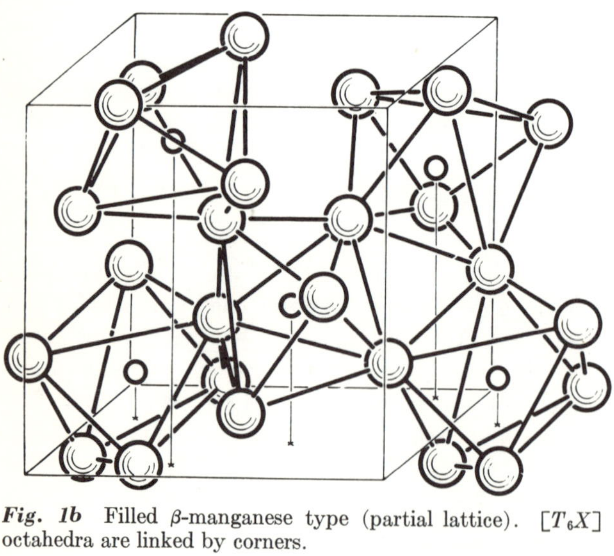

Fig. 1b Filled β-manganese type (partial lattice). [T_6X] octahedra are linked by corners.

TABLE 1 ***Lattice Parameters of New Phases Having Perovskite Carbide Structure***

In angstroms

T = IIIA, IVA, VA

Compound	Parameter	Reference	Compound	Parameter	Reference
Sc_3AlC	4.48	13	$Gd_3AlC_{0.9}$	4.90	13
Y_3AlC	4.89	13	$Dy_3AlC_{0.7}$	4.85	13
Y_3TlC_{1-x}	4.88_9	14	Dy_3GaC_{1-x}	5.07	15
La_3InC_{1-x}	5.14	15	Dy_3InC_{1-x}	4.88	15
Ce_3InC_{1-x}	5.10_3	14	Dy_3SnC_{1-x}	4.85	15
Ce_3TlC_{1-x}	5.09_6	14	$Ho_3AlC_{0.7}$	4.81	13
Ce_3SnC_{1-x}	5.10_1	14	Ti_3AlC	4.15_6	16
Ce_3PbC_{1-x}	5.11_2	14	Ti_3InC	4.19_9	17
Pr_3GaC_{1-x}	5.16	15	Ti_3TlC	4.20_9	17
Pr_3InC_{1-x}	5.09	15	Ti_3InN	4.190	17
Pr_3TlC_{1-x}	5.08	15	Ti_3TlN	4.191	17
Pr_3SnC_{1-x}	5.06	15	$Ti_3Au(N, O)_{1-x}$	4.096	18
Pr_3PbC_{1-x}	5.07	15	$V_3Au(N, O)_{1-x}$	3.964	18
$Nd_3AlC_{0.9}$	4.96	13			

binary phase T_3M does not exist, indicating that these perovskite carbides are essentially ternary phases. Further experimental work is necessary for the precise determination of the composition, particularly with respect to the amount of C. Furthermore, the anomalous behavior of the lattice parameter of the Ga-containing perovskites has to be explored. One could

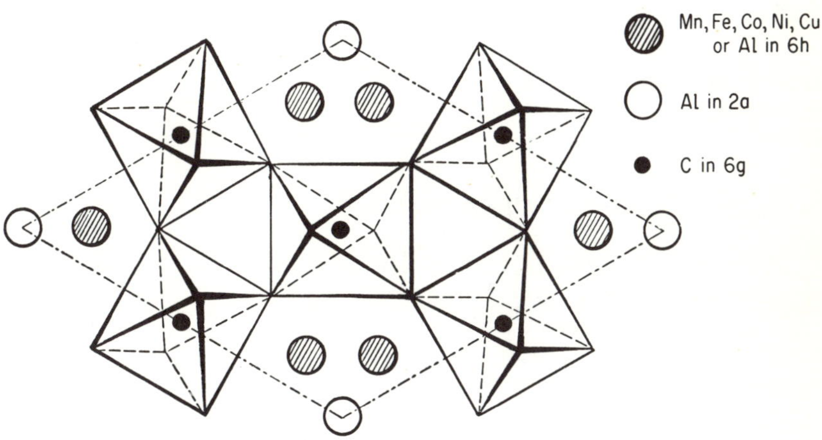

Fig. 1c κ(kappa)-carbide type filled additionally by Al atoms (IIA). $[T_6X]$ octahedra are linked by corners.

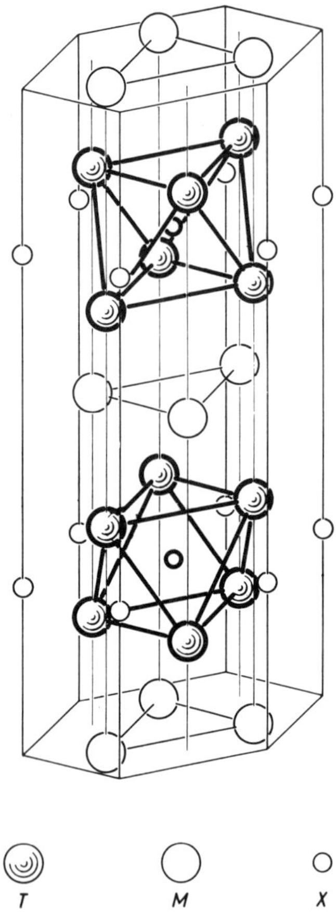

Fig. 2 Structure of *H* phase. $[T_6X]$ octahedra are linked by edges.

add finally the phase $(Cr_{\sim 0.7}Pt_{\sim 0.3})PtC_x$.[19] However, the problem of whether this phase is a true ternary compound or a solid solution which has its origin in the ordered phase $CrPt_3$ of the Cu_3Au type is still unsolved. It is obvious that besides the typical meta metals (IIB, IIIB, and IVB), group IB and VIIIA metals also come into consideration as components. This situation is even more pronounced for complex carbides, nitrides, and oxides with the filled or partially filled β-Mn type (Table 2): for example Nb_3Au_2N, where the Nb and Au atoms possess the 12-fold and 8-fold positions respectively (ordered β-Mn structure).

Au can be substituted randomly by Zn or Ga. The atomic parameters do not differ very much from the precise values for β-Mn; x_{Nb} was found to be 0.206, $x_{Au,Ga}$ = 0.07. In the few cases where comparison can be made, the cell volume of the carbide is always somewhat larger than that of the corresponding nitride. Similar phases were found by Evans and Jack (see K. Schubert[26]) within the systems Mo-{Mn, Fe, Co, Ni}-N, by Kuo[20] for W-Fe-C, W-Mn-Fe-C, $Cr_9W_2Fe_{10}C$, and finally for W_3Re_2C.[21] However nothing is known about the degree of ordering. It is remarkable that Mo may be replaced in Mo_3Al_2C by up to 60 a/o Cr and 30 a/o Mn. A higher amount of Mn leads to formation of the κ phase.

κ (KAPPA) AND RELATED PHASES

Progress within this new group has been achieved while investigating combinations such as Mo-{Mn, Fe, Co, Ni, Cu}-Al-C and W-{Mn, Fe}-Al-C, leading to a set of κ-like phases. A detailed study of the crystal structure of $W_9Co_3C_4$ (κ-carbide) according to Schönberg[22] reveals immediately linked octahedral $[T_6X]$ groups which build up again a partial lattice. It can be seen from Fig. 4 that the linking mode is characterized by common corners of the $[T_6X]$ groups which are much more regular than those of the filled and ordered β-Mn type. The κ-phase $W_9Co_3C_4$

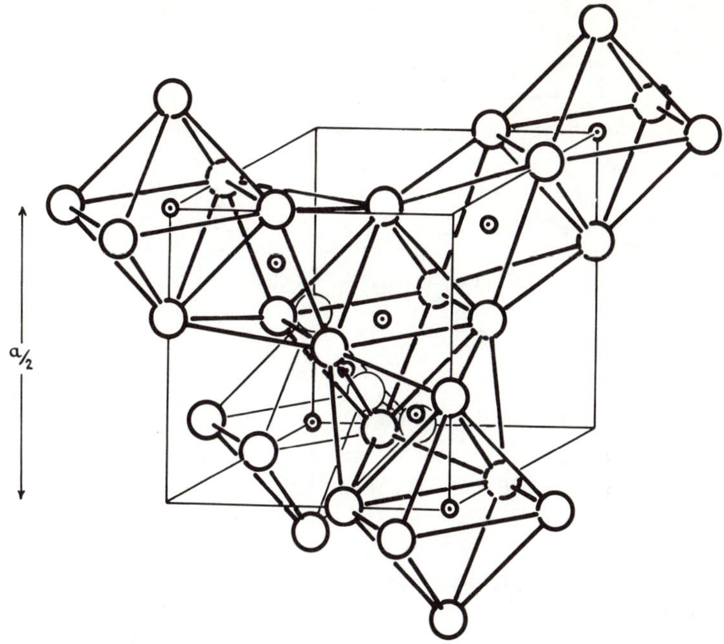

Fig. 3a μ-Carbide type (partial lattice). $[T_6X]$ octahedra are linked by planes.

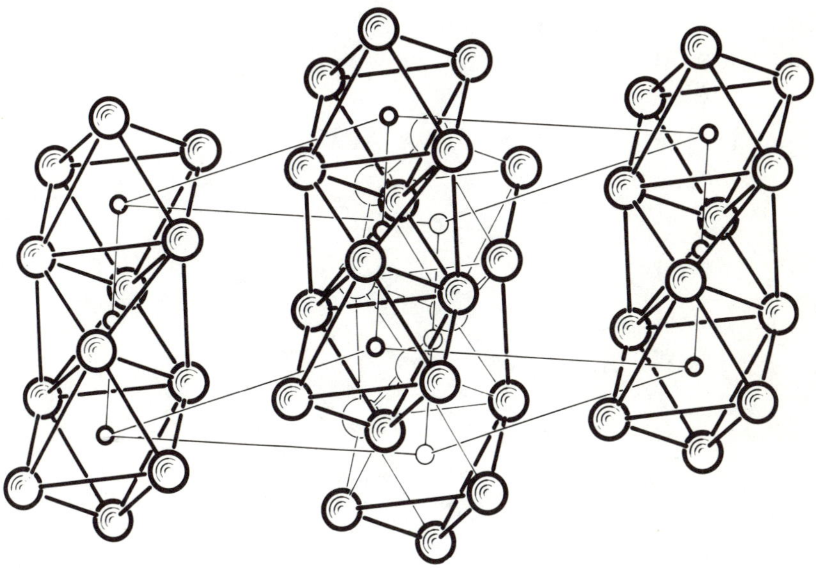

Fig. 3b Filled Mn_5Si_3 type (partial lattice). $[T_6X]$ octahedra are linked by planes.

TABLE 2 ***Lattice Parameters of Complex Phases Having Filled (Ordered) β-Mn Structure***

In angstroms

Compound	*Parameter*	*Compound*	*Parameter*
Nb_3Al_2C	7.07_2–7.07_9*	$V_3(Au_{2/3}Ga_{1/3})_2N_x$	6.726
$Nb_3(Au_{2/3}Ga_{1/3})_2C_x$	7.084	Nb_3Al_2N	7.03_4
Ta_3Al_2C	7.03_8	$Nb_3Au_2N_x$	7.085
Mo_3Al_2C	6.86_0–6.86_6*	$Nb_3(Au_{2/3}Zn_{1/3})_2N_x$	7.052
V_3Zn_2N	6.60_6	$Nb_3(Au_{2/3}Ga_{1/3})_2N_x$	7.048
V_3Ga_2N	6.62_0	$Ta_3(Au_{2/3}Ga_{1/3})_2N_x$	7.034

* Al poor.

was first observed by Rautala and Norton[23] for a slightly different composition. Later on indications were found by Kuo[24] for the existence of the corresponding phases in W-{Mn, Ni}-C systems. Whitehead and Brownlee[25] reported a κ phase of formula $W_{16}Ni_3C_6$. Although the proposed structure provides a partial order only because of the Co-W substitution, the W content appears to be too high in the W-Ni-C κ phase. As was recognized from the formation of η carbides (see later), Al was found to be a suitable B subgroup element for stabilizing complex carbides, and new representatives of κ-like phases could be detected, as is shown in Table 3. A characteristic feature of the κ phases is the ratio c/a equal to unity. The lattice parameters of the Al-containing quaternary phases are slightly higher than those of the ternary, as is expected from the substitution for example of Mn/Al. However, in addition the Al atoms do also fill the 2a

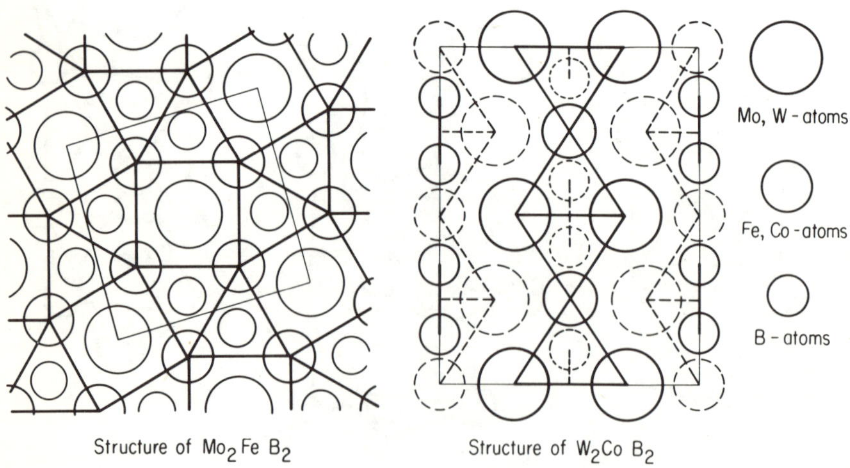

Fig. 4 Left, crystal structure of Mo_2FeB_2. Ordered U_3Si_2 type, projection parallel to [001]. Right, crystal structure of W_2CoB_2. Projection parallel to [001].

TABLE 3 ***Lattice Parameters of κ-Carbides and κ-like Complex Carbides****

In angstroms

System	Parameter		Ratio c/a
	a	c	
$W_{10}Co_3C_4$	7.848	7.848	1.000
$W_9Co_3C_4$	7.826	7.826	1.000
$W_{16}Ni_3C_6$	7.818	7.818	1.000
W-Ni-C	7.848	7.848	1.000
Mo-Mn-Al-C	7.87_6	7.86_7	0.999†
Mo-Fe-Al-C	7.849	7.848	1.000
Mo-Co-Al-C	7.95_0	7.84_3	0.986
Mo-Ni-Al-C	7.89_3	7.85_0	0.995
$Mo_{12}Cu_3Al_{11}C_6$	7.95_2	7.865	0.989
W-Mn-C	7.756	7.756	1.000
W-Mn-Al-C	7.90_3	7.78_7	0.986
W-Fe-Al-C	7.89_5	7.85_7	0.995
Mn_3Al_{10}	7.543	7.898	1.048
Mn_3Al_9Si	7.513	7.745	1.031
Co_2Al_5 ($D8_{11}$ type)	7.656	7.543	0.992

* Included are also the intermetallic compounds (without nonmetal) representing essentially the parent lattice.

† Homogeneous range.

positions, a filling of the metal polyhedra (ikosahedra) which clearly completes the resemblance of the parent lattice of these carbides to the elementary cell of Mn_3Al_{10} or Mn_3Al_9Si.[26]

Thus the atomic positions $6h$, and $6h$ and $12k$ for the metals in the space group D_{6h}^4 are in both cases identical, the parameter values being nearly the same. In contrast to the proposed structure for $W_9Co_3C_4$, it is assumed that the C atoms only fill the $6g$ positions, forming octahedral $[T_6X]$ groups. Inasfar as a considerable number of transition-element atoms are substituted for by Al, one has to expect an even lower degree of filling by C atoms due to the fact that the bonding T-C is appreciably stronger than Al-C bonds.

It is interesting to mention furthermore that a similar relationship between the parent lattice of the η-carbides (W_3Fe_3C) and another aluminide, namely $Cr_4Al_{13}Si$, has been found by Schubert.[26]

H PHASES

About 40 compounds of this type (see Table 4) are already described in detail[12]; a new representative Sc_2InC has been recently reported.[27] Some

TABLE 4 ***Lattice Parameters of H Phases***

In angstroms

H phase	Parameter		Ratio c/a
	a	c	
Ti_2CdC	3.09_9	14.41_4	4.65_3
Ti_2AlC	3.04	13.6_0	4.47
Ti_2GaC	3.06_4	13.30_5	4.34_2
Ti_2InC	3.13_2	14.06	4.19_0
Ti_2AlN	2.99_4	13.61	4.54_4
Ti_2GaN	3.00_4	13.30_2	4.42_8
Ti_2InN	3.07_4	13.97_6	4.54_7
Ti_2GeC	3.07_9	12.9_3	4.19_9
Ti_2SnC	3.18_6	13.63	4.27_8
Zr_2InC	3.34_7	14.90	4.45_5
Zr_2TlC	3.36_3	14.78_9	4.39_8
Zr_2InN	3.27_7	14.83_6	4.52_6
Zr_2TlN	3.30_1	14.71	4.45_5
Zr_2SnC	3.34_7	14.5_9	4.35_9
Zr_2PbC	3.38_4	14.66_6	4.33_3
Hf_2InC	3.30_7	14.73	4.45_5
Hf_2TlC	3.32_2	14.62_5	4.40_3
Hf_2InN	3.23_0	14.74	4.56_4
Hf_2SnN	3.31_2	14.3_9	4.34_4
Hf_2PbC	3.35_8	14.46_5	4.30_8
V_2AlC	2.91_3	13.1_4	4.51_1
V_2GaC	2.93_8	12.8_4	4.37_0
V_2GaN	2.89_2	12.93	4.47_3
V_2GeC	3.00_1	12.25	4.08_3
Nb_2AlC	3.10_3	13.8_3	4.46_4
Nb_2GaC	3.13_1	13.56_5	4.33_2
Nb_2InC	3.17_2	14.37	4.53_0
Nb_2SnC	3.24_5	13.77	4.24_3
Ta_2AlC	3.07_5	13.8_3	4.49_8
Ta_2GaC	3.10_4	13.57	4.37_1
Cr_2AlC	2.86_6	12.8_2	4.48_3
Cr_2GaC	2.88_6	12.61_6	4.37_0
Cr_2GeC	2.95_4	12.08	4.09
Mo_2GaC	3.01_7	13.18	4.36_9

TABLE 5 ***Lattice Parameters of Transition-element Carbohydrides and Subcarbides***

In angstroms

Ti_2CH $a = 3.03_5$ $c = 5.04$ $c/a = 1.66$	V_2C $a = 2.88_5$ $c = 4.56_0$ $c/a = 1.58_0$	$(Cr_{\sim 0.8}W_{\sim 0.2})_2C$ $a = 2.83_5$ $c = 4.45$ $c/a = 1.57$
Zr_2CH* $a = 3.33_1\text{–}3.33_8$ $c = 5.45_5\text{–}5.46_7$ $c/a = 1.63_8$	Nb_2C $a = 3.117$ $c = 4.954$ $c/a = 1.58_8$	Mo_2C $a = 3.002$ $c = 4.724$ $c/a = 1.57_4$
Hf_2CH $a = 3.29_3$ $c = 5.37_5$ $c/a = 1.63_2$	Ta_2C $a = 3.094$ $c = 4.918$ $c/a = 1.589$	W_2C $a = 2.992$ $c = 4.721$ $c/a = 1.578$

* The Zr-hydride γ phase reported by Hägg[31] is obviously identical with the above listed Zr carbohydride.

more H phases with group IIIA elements can be expected which would expand the more metallic branch of the H phases, characterized by an increasing c/a ratio. The linking of $[T_6X]$ octahedras by edges is already provided for by the subcarbide cell, as was discussed previously.[28]

A similar arrangement occurs for the so-called carbohydrides, which were detected independently by Rexer and Peterson.[29,30] Carbohydrides of the approximate formula T_2CH exhibit a hexagonal close-packed parent lattice, the C atoms filling octahedral voids while the H atoms are preferentially accommodated in tetrahedral voids. One can compare the group IVA carbohydrides with the sucarbides of group VA and VIA metals, disregarding the order. As is shown in Table 5, the c/a ratio decreases continuously from left to right. Ti, Zr, and Hf carbohydrides are 10 kcal more stable than the corresponding hydrides. Large amounts of H (up to 20 a/o H) can be dissolved in group IVA monocarbides if a C deficiency is present: e.g., $ZrC_{\sim 0.6}H_{\sim 0.4}$.[30]

COMPLEX η-CARBIDES

Within the scheme established by Kuo many η phases of formula $T_3^{I}T_3^{II}C$ have been reported,[1] among these the compounds Nb_3Cr_3C or Ta_3Cr_3C. However, in a more recent study it was not possible to corroborate these results.[32] Starting from very pure components no formation could be observed even after a long anneal at 1350°C, a mixture of NbC_{1-x} solid solution + $NbCr_2$ (Laves phase) being present. In hot-pressed samples, containing some N and O, the Nb_3Cr_3C phase was occasionally found.

TABLE 6 ***Lattice Parameters of η-Carbides and Similar Phases Derived from the Parent Lattice*** **Ti_2Ni**

In angstroms

Phase	*Parameter*	*Phase*	*Parameter*
$Ti_4Zn_2C_x$	11.55_8	$Ti_4Zn_2N_x$	11.49_6
$Zr_4Zn_2C_x$	12.16_0	$Zr_4Zn_2N_x$	12.13_1
$Hf_4Zn_2C_x$	12.04_4	$Hf_4Zn_2N_x$	11.96_6
$Nb_4Zn_2C_{x(x \gtrsim 3/2)}$	11.74	$Nb_4Zn_2N_x$	11.54_7
Ta_3VAl_2C	11.6_7	$(Nb_3Zn_3N_x)$	
Nb_3CrAl_2C	11.7_1	Zr_3Cu_2Al	N- or O-containing
Ta_3CrAl_2C	11.6_0	Zr_3Ag_2Al	N- or O-containing
Nb_3MnAl_2C	11.6_9	Hf_3Cu_2Al	N- or O-containing
Ta_3MnAl_2C	11.6_1	Hf_3Ag_2Al	N- or O-containing
Nb_3FeAl_2C	11.6_3	Nb_3Cu_2Al	N- or O-containing
Ta_3FeAl_2C	11.5_7	Zr-Cu-Zn	N- or O-containing
Ta_3CoAl_2C	11.56	Zr-Ag-Zn	N- or O-containing
Nb_3Ni_2AlC	11.5_5	Zr-Au-Zn	N- or O-containing
Ta_3Ni_2AlC	11.4_9	Hf-Cu-Zn	N- or O-containing
Ta_3CuAl_2C	11.6_2	Hf-Au-Zn	N- or O-containing

Therefore the compound Nb_3Cr_3C appears to be additionally stabilized by another nonmetal or another metal, as was confirmed by a systematic investigation within the quaternary combinations: {Nb, Ta}-{Cr, Mn, Fe, Co, Ni, Cu, Zn}-Al-C.[33] From these results (Table 6) it turns out that Al is an extremely good η-carbide stabilizer. As can also be seen from Table 6, there are numerous phases containing group IB and IIB elements. Some of these η phases are stabilized by a very minor amount of O or N.

By the general formula $(T_6T'_2)M_4X_3$ the diamondoid linking of $[T_6X]$ groups does allow filling of the $8a$ and $16c$ positions. However, there is a variation with respect to the mode of filling. Thus, Leciejewicz[34] has shown that, for W_6Fe_6C, only filling of the $8a$ position takes place. On the other hand, $Nb_8Zn_4C_3$ seems to be completely filled by C atoms. A remarkable variation does also occur with respect to the metal positions because of the substitution between Al and the smaller metal atoms. The precise arrangement (degree of metal ordering) is not known for each of the listed compounds, and occasionally two closely related η-carbide phases are present. It was found that Nb_3CrAl_2C is ferrimagnetic at low temperatures while Nb_3Cr_2AlC is not.[35] One must also mention that Nb_3Fe_2, reported in the literature[36] as a compound having Ti_2Ni type (parent lattice of the η-carbides), obviously does not exist; however, addition of C and Al yields easily the Nb-Fe-Al-C η-carbide. Presumably Nb_3Fe_2 (a = 11.26) contained a small amount of stabilizing elements.

THE FILLED Mn_5Si_3 TYPE

No detailed discussion is planned concerning this type; reference may be made to a very recent paper on this subject by Parthé, Jeitschko, and Sadagopan.[37] However, it is of interest to note that B can also be accommodated within the metal octahedra, as is shown schematically in Fig. 3. V_5Ge_3B, $Nb_5Ga_3B_x$, Nb_5Ge_3B, Ta_5Al_3B, $Ta_5Ga_3B_x$, and $Ta_5Ge_3B_x$ are typical examples.[38] Besides these and one perovskite boride detected by Stadelmaier[39] there are no complex borides that exhibit $[T_6B]$ octahedra. The reason for this can be recognized in the departure from the critical ratio r_B/r_T. The Mn_5Si_3 structure is an exceptional case insofar as linking of the octahedra takes place in one direction only. One can therefore understand the filling with even larger atoms, thus forming the Ti_5Ga_4 structure.

COMPLEX BORIDES

Shifting to higher values of r_X/r_T, the octahedral surrounding of the nonmetal is generally replaced by the trigonal prism or quadratic antiprism. This is well known for Cr_3C_2, $Cr_{23}C_6$, WC, Fe_3C, and many borides.

Complex borides of formula $T_2^{I}T^{II}B_2$ have been detected within the systems: {Mo, W}-{Fe, Co, Ni}-B, Mo_2FeB_2 crystallizing with an ordered U_3Si_2 type, while the others belong to a new type (W_2CoB_2 structure), as is seen in Fig. 4. One can recognize immediately the trigonal prism in both cases and the pair formation of the B atoms. The space group is D_{2h}^{25}, and the atomic positions are:

$$\begin{array}{ll} 4\text{ W in }4f & x = 0.205 \\ 2\text{ Co in }2a & \\ 4\text{ B in }4h & y \approx 0.30 \end{array}$$

The lattice parameters are listed in Table 7. A ternary phase Mo_2NiB_2 has already been reported by Steinitz and Binder.[40]

Complex borides of $Cr_{23}C_6$ type were already known such as $Fe_{23}C_3B_3$ or $T_{23-x}M_xB_6$ ($x = 2$ to 3.5) due to Stadelmaier et al.[41] It is less surprising that a great many elements can tolerate M because of the relatively small amount involved; in fact, 23 new complex borides of this type have been observed [42] (Table 8).

The ordering of the larger metal atom in the 8c position, such as is the case for $Mo_2Fe_{21}C_6$,[26] has been confirmed for $W_2Co_{21}B_6$ and $Hf_2Co_{21}B_6$. Similarly, ordering in $Hf_3Ni_{20}B_6$ is very likely for the large atoms in 8c and 4a, at least completely random distribution: e.g., Hf-Ni has to be excluded.

TABLE 7 ***Lattice Parameters of Complex Borides***

In angstroms

Compound	Parameter		
	a	*b*	*c*
Mo_2FeB_2	5.782	. . . .	3.148
Mo_2CoB_2	7.087	4.564	3.164
Mo_2NiB_2	7.075	4.557	3.179
W_2FeB_2	7.124	4.610	3.148
W_2CoB_2	7.075	4.561	3.177
W_2NiB_2	7.077	4.559	3.186

As can be seen from Table 8, there are τ phases existing with a metal ratio quite different from 21:2 or 20:3. The wide region of the Re-Fe-B τ phase is of special interest, no or almost no ordering being found. The same is true for the other Re-containing phases. From this observation many more τ phases are to be expected. Only recently Stadelmaier[43] emphasized the fact that a departure from the ratio boron/metal = 6:23 to lower values can occur. This connects with the valence electron concentration, the highest B deficiency being found for $Ni_{21}Sn_2B_5$. It is obvious that these complex compounds are governed by relationships similar to those of the previously mentioned class.

TABLE 8 ***Lattice Parameters of So-called*** τ ***Phases***[43] ***Having*** $Cr_{23}C_6$ ***Type***

In angstroms

Phase	*Parameter*	*Phase*	*Parameter*
$Co_{21}Sc_2B_6$	10.54_6	$(Ni_{21}U_2B_6)$*	10.65_2
$Co_{21}U_2B_6$	10.62_0	$Ni_{20}Ti_3B_6$	10.57_7
$Co_{21}Ti_3B_6$	10.54_2	$Ni_{20}Zr_3B_6$	10.59_4
$Co_{21}Zr_2B_6$	10.58_2	$Ni_{20}Hf_3B_6$	10.64_9
$Co_{21}Hf_2B_6$	10.57_4	$Ni_{21}V_2B_6$	10.48_5
$Co_{21}V_2B_6$	10.48_6	$Ni_{21}Mn_2B_6$	10.49_5
$(Co_{21}Nb_2B_6)$*	10.52_0	$Ni_{21}Sb_2B_6$	10.59_8
$Co_{21}Mo_2B_6$	10.50_5	$Mn_{11}Re_{21}B_6$	11.19_3
$Co_{21}W_2B_6$	10.50_6	$(Re_{0.9-0.4}Fe_{0.1-0.6})_{23}B_6$	11.3–11.0
$Co_{21}In_2B_6$	10.58_0	$(Re, Co)_{23}B_6$	11.0_4
$Co_{21}Sb_2B_6$	10.59_5	$Re_{12}Ni_{11}B_6$	11.25
$Ni_{21}Sc_2B_6$	10.56 (10.60)		

* Deviation from the above composition.

NEW *G* PHASES

There are some good reasons which justify a discussion of the problem of *G* phases in connection with the preceding structural types. Many complex transition-element silicides and germanides were found to have the $Mg_6Cu_{16}Si_7$ crystal structure, where the atoms occupy equipoint positions of the Th_6Mn_{23} type. Beck et al.[44] suggested that these ternary compounds be called *G* phases. Further examples were detected by Gladishevskii et al.[45] From a structural inspection one can recognize in compounds of this type rather large voids at ¼, ¼, ¼, etc., in the (110) plane which offer space enough for small nonmetal atoms. However, no systematic work has been done so far.

New representatives of *G* phases have been found, as was expected from the work of Beck et al.[44] for Hf-containing silicides and germanides; it was discovered, however, that *G* phases also occur with complex beryllides, zincides, and aluminides (Table 9). In contrast to the silicides or germanides the composition, and therefore atomic distribution, of two components is different.

According to the formula $Ta_6Cu_8Be_{(1+6+8)}$ Be atoms occupy one-half of the usual group VIIIA or IB metal positions ($32f$, $x_{Be} = 0.17$), the atomic parameters for Cu atoms in $32f$ being $x_{Cu} = 0.375$; for the Ta atoms $x_{Ta} = 0.193$. The tendency for substitution Fe/Be or Cu/Be is well known from the Laves-phase beryllides, such as $(Fe_{1/2}Be_{1/2})Be_2$. With respect to the complex aluminides it should be mentioned that the *G* phase $(Hf_6Ni_7Al_{16})$ does obviously correspond to the ternary phase $TiNiAl_2$ ($a = 11.90$ A) found by Raman and Schubert.[46] The replacement of Ni atoms by Al atoms and in turn Si atoms by Ni atoms indicates that the stability of this structural type is mainly influenced by the geometrical factor.

TABLE 9 ***Lattice Parameters of G Phases***

In angstroms

G phase	Parameter	*G* phase	Parameter	*G* phase	Parameter
$Zr_6Pd_8Be_{15}$	11.25	$Zr_6Co_8Be_{15}$	11.10	$(Zr_6Cu_7Zn_{16})$	12.10
$Hf_6Pd_8Be_{15}$	11.20	(HfCoBe)*	11.00	$(Hf_6Cu_7Zn_{16})$	12.01
$Ti_6Cu_8Be_{15}$	10.78_4	$(Zr_6Ni_8Be_{15})$	11.06	$(Zr_6Ni_7Al_{16})$	12.08
$Zr_6Cu_8Be_{15}$	11.18_5	$Hf_6Ni_8Be_{15}$	10.99	$(Hf_6Ni_7Al_{16})$	12.00
$Nb_6Cu_8Be_{15}$	10.83_3	$Nb_6Ni_8Be_{15}$	10.73	$Hf_6Fe_7Si_{16}$	11.49_0
$Ta_6Cu_8Be_{15}$	10.77_5	$Ta_6Ni_8Be_{15}$	10.68	$Hf_6Co_7Si_{16}$	11.37_8
Ta-Cu-Be(Be-rich)	10.72_5			$Hf_6Co_7Ge_{16}$	11.56_2
				$Hf_6Ni_7Ge_{16}$	11.56_1

* Departure from the composition.

CONCLUSION

The most interesting aspect of the uniform class of complex phases with $[T_6X]$ octahedra can be seen in the general mode of filling voids of a T-atom environment in a more or less ordered parent lattice T_aM_b by nonmetals X. The importance of the construction element $[T_6X]$ is clearly shown by the formation of a partial lattice, e.g., $\{T_3X\}$, $\{T_2X\}$, or extended structural elements, which can be compared with the corresponding binary T_2X or TX phases. This view is greatly supported by the fact that the T-X bonding in the complex phases is basically the same as in the binary compounds. From a few examples in Table 10 concerning interatomic distances T–X for complex and simple carbides and nitrides one can recognize the small differences between the corresponding phases. Nevertheless there is a tendency for contraction of the T–X distance within the complex phase. The cell volume of the complex nitrides is always somewhat smaller than that of the corresponding carbides, in complete agreement with the behavior of the respective binary carbides and nitrides. Similarly, the degree of filling varies for complex phases with $[T_6X]$ octahedra, as is the case for binary T-X compounds. The corresponding features are even more pronounced. Thus some perovskite carbides, as well as η-carbides and η-oxides, both cubic, exhibit large domains of homogeneity similar to the cubic monocarbides or monoxides, while the H phases strongly connected with the hexagonal close-packed subcarbides mainly exist only within a small range. The high stability of the H phase, e.g., Ti_2AlC, can be deduced immediately from the fairly high heat of formation of TiC and TiAl according to the reaction $TiC + TiAl = Ti_2AlC$ The most stable complex carbides and nitrides are again those containing a transition element of the IVA and VA groups, as is true for the binary phases. The role of the meta metal is not fully understood; from the great variety one may think of a rather neutral component because of the lack of M-X bonds. One can presume, however, that the meta metal atoms act as an ordering component, increasing the stability of the parent lattice. The shorter distances T-X agree with this concept or, in other

TABLE 10 ***Interatomic Distances T–X in Complex Carbides Compared with Those in the Corresponding Binary Phases***

In angstroms

Compound	*T–X*	*Compound*	*T–X*
TiC	2.16 (2.14 for C defect)	Mo_2C	2.16
Ti_2AlC	2.12	Mo_3Al_2C	2.14
Ti_3AlC	2.08	$Mo_{12}Cu_3Al_{11}C_6$	2.08
Ti_2ZnC_x	2.14	$Mo_{4.8}Si_3C_{0.6}$	2.15

words, an extension of the critical radii ratio r_X/r_T from 0.59 (Hägg) to 0.62 occurs (0.65 according to Stadelmaier[47] for perovskite carbide phase.) Although for some of the perovskite phases a relationship between the number of d electrons of the T element and the amount of filling by the nonmetal appears to exist, there is no indication for the validity of such a regularity for the other types. One can speculate on whether or not the nonmetal is a donor filling the d band of the T element, and if so to what extent an electron transfer takes place within these complex carbides and nitrides.

REFERENCES

1. Nowotny, H.: in P. A. Beck (ed.), "Electronic Structure and Alloy Chemistry of the Transition Elements," John Wiley & Sons, Inc., New York and London, 1963.
2. Parthé, E., and V. Sadagopan: *Acta Cryst.* **16**:202 (1963).
3. Rudy, E., St. Windisch, and Y. A. Chang: Technical Report AFML-TR-65-2, part 1, vol. 1, 1965.
4. Terao, N.: *J. Appl. Phys. Japan* **3**:104 (1964).
5. Bowman, A. L., T. C. Wallace, J. L. Yarnell, R. G. Wenzel, and E. K. Storms: *Acta Cryst.* **19**:6 (1965).
6. See H. D. Heetderks, E. Rudy, and T. Eckert: Technical Report AFML-TR-65-2, parts 2 and 3, vol. 1, 1965.
7. See E. K. Storms and R. J. McNeal, *J. Phys. Chem.* **66**:1401 (1962).
8. Yvon, E., W. Rieger, and H. Nowotny, *Monatsh. Chem.* **97**:689 (1966).
9. Rassaerts, H., R. Kieffer, and H. Nowotny: *Monatsh. Chem.* **96**:1536 (1965).
10. Brauer, G., and W. D. Schnell: *J. Less-Common Metals* **7**:23 (1964).
11. Rudy, E., and Y. A. Chang: In F. Benesovsky (ed.), *Plansee Proc. 5th Seminar*, 1965.
12. Nowotny, H., W. Jeitschko, and F. Benesovsky: Symposium sur la Métallurgie des Poudres, Paris, 1964.
13. Rosen, S., and P. G. Sprang, 13th Annual Conference on Application of X-ray Analysis, Denver, Colo., 1965.
14. Jeitschko, W., H. Nowotny, and F. Benesovsky: *Monatsh. Chem.* **95**:1040 (1964).
15. Haschke, H., H. Nowotny, and F. Benesovsky: *Monatsh. Chem.* **97**:716 (1966).
16. Jeitschko, W., H. Nowotny, and F. Benesovsky: *Monatsh. Chem.* **95**:319 (1964).
17. Jeitschko, W., H. Nowotny, and F. Benesovsky: *Monatsh. Chem.* **95**:436 (1964).
18. Philipsborn, H. v.: Dissertation, Eidgenöessische Technische Hochschule, Zürich, 1964.
19. Ganglberger, E., H. Nowotny, and F. Benesovsky: *Monatsh. Chem.* **96**:1781 (1965).
20. Kuo, K.: *Trans AIME J. Met.* **8**:97 (1956).
21. Kuzma, Yu. B., V. I. Lakh, V. Ya. Markiv, B. I. Stadnik, and E. I. Gladishevskii: *Poroshkovaya Met. Akad. Nauk Ukr. SSR* **3**:40 (1963).
22. Schönberg, N.: *Acta Met.* **2**:837 (1954).
23. Rautala, P., and J. T. Norton: *J. Metals* **4**:1045 (1952).
24. Kuo, K.: see discussion *J. Metals* **5**:744 (1953).
25. Whitehead, K., and L. D. Brownlee, *Planseeber. Pulvermet.* **4**:62 (1956).
26. See K. Schubert, "Kristallstrukturen mehrkomponentiger Phasen," Springer-Verlag OHG, Berlin, Göttingen, Heidelberg, 1964.
27. Toth, L. E., W. Jeitschko, and C. M. Yen: *J. Less-Common Metals* **10**:29 (1966).
28. Nowotny, H., F. Benesovsky, and W. Jeitschko: *Planseeber. Pulvermet.* **2**:31 (1964).

29. Rexer, J., and D. T. Peterson: In Waber, Chiotti, and Miner (eds.): International Symposium on Compounds of Interest in Nuclear Reactor Technology, AIME IMD Special Report 13, 1964. Bittner, H.: *Monatsh. Chem.* **95**:1514 (1964). Goretzki, H., H. Bittner, and H. Nowotny: *Monatsh. Chem.* **95**:1521 (1964).
30. Goretzki, H., E. Ganglberger, H. Nowotny, and H. Bittner: *Monatsh. Chem.* **96**: 1563 (1965).
31. Hägg, G.: *Z. Physik. Chem. Leipzig,* **B11**:433 (1931).
32. Rassaerts, H., F. Benesovsky, and H. Nowotny: *Planseeber. Pulvermet.* **13**:199 (1965).
33. Reiffenstein, E., H. Nowotny, and F. Benesovsky: *Monatsh. Chem.* **96**:1658 (1965).
34. Leciejewicz, J.: *J. Less-Common Metals* **7**:318 (1964).
35. Reiffenstein, E.: Thesis, University of Vienna, 1966.
36. Goldschmidt, H. J.: *Research* **10**:289 (1953).
37. Parthé, E., W. Jeitschko, and V. Sadagopan: *Acta Cryst.* **19**:1031 (1965).
38. Rieger, W., H. Nowotny, and F. Benesovsky: *Monatsh. Chem.* **96**:98 (1965).
39. Stadelmaier, H. H.: *Z. Metallk.* **55**:378 (1964).
40. Steinitz, R., and I. Binder: *Powder Met. Bull.* **6**:123 (1953).
41. Stadelmaier, H. H., G. Hofer, M. Kotyk, J. D. Schöbel, and B. B. Helms: *Metall* **18**:460, 1065, 1285 (1964); **19**:121 (1965).
42. Ganglberger, E., H. Nowotny, and F. Benesovsky: *Monatsh. Chem.* **96**:1144 (1965); *ibid.* **101**:494 (1966).
43. Stadelmaier, H. H.: In International Symposium on Compounds of Interest in Nuclear Reactor Technology, *loc. cit.*
44. Spiegel, F. X., D. Bardos, and P. A. Beck: *Trans. Met. Soc. AIME* **227**:575 (1963).
45. Gladishevskii, E. I., W. Ja. Markiv, and Iu. B. Kuzma, *Dopovidi Akad. Nauk Ukr.* **4**:481 (1962).
46. Raman, A., and K. Schubert, *Z. Metallk.* **56**:99 (1965).
47. Stadelmaier, H. H.: *Z. Metallk.* **52**:758 (1961).

Agenda Discussion:
INTERMETALLIC COMPOUNDS

chairman **P. A. Beck**

Department of Metallurgy and Materials Research Laboratory, University of Illinois, Urbana, Illinois

secretary **H. M. Davis**

U.S. Army Research Office, Durham, North Carolina

Beck introduced the session by calling for a discussion on the significance of the electronegativity factor, and on the possibility of its quantitative measurement.

Hume-Rothery noted that from the very early days of alloy chemistry it was known that structures of alloys were influenced by the alloying of an electropositive metal with an electronegative one, resulting in what was then called the electrochemical effect. The question is whether it could be estimated quantitatively. In an extreme case, such as sodium chloride, one has ionic bonding, with the sodium losing the electron and the chlorine gaining it. With an alloy such as Au-Mg, one gets the same sort of thing happening to some extent. When one tries to estimate it quantitatively, the standard electrode potentials are not what one wants, since they represent equilibria between the metal and the ions in aqueous solution. On the whole, a not too unsatisfactory quantity seemed to be the Pauling electronegativities, but it is unfortunate that the Pauling electronegativities were rather indefinite. The idea of Pauling was that the normal covalent bond would be such that the energy of the A-B linking was the arithmetic mean:

$$E(A\text{-}B) = \tfrac{1}{2}[E(A\text{-}A) + E(B\text{-}B)]$$

Pauling found that if an electropositive element is alloyed with an electronegative one, the bonding is stronger than that suggested by this mean energy, and he regarded the excess, Δ, as coming from ionic interaction. Then he found that the Δ values themselves could not be expressed additively, but that one could calculate the difference by what he called the "electronegativities," such that, if one were dealing with two kinds of atoms in a particular linkage, and one called X_A and X_B the electronegativities, then, approximately,

$$X_A - X_B \propto \sqrt{\Delta}$$

These are the true Pauling electronegativities, but when they are applied to the theory of metals, it is found that in the general case one cannot get the normal covalent bond between two metal atoms although there are a few cases, such as the alkali diatomic molecules, where this is possible.

In order to get over this difficulty, the definition was changed. If M is a metal and forms a halide MCl_x, then the electronegativity is defined as the work required to separate one chlorine atom from MCl_x. This definition of electronegativity is thus based on the heats of formation of the halides. Hume-Rothery considers this definition as the least unsatisfactory. One of his colleagues at Oxford thought that electronegativity was a useful concept "provided you do not define it precisely."

Hume-Rothery remarked that many people tried to relate the shapes of equilibrium diagrams to electronegativity differences, but he wondered whether this was justifiable or not. He also noted that the electronegativity varies with the valency so that in the transition metals, where a series of valency states exist, each valency state has its own electronegativity, and, in general, the electronegativity increases with the valency. It had been suggested that one might get over this difficulty by using the work function of the metal itself, although in that case one would be dealing with a surface property and not with a volume property. Hume-Rothery concluded that what he really wanted to know was whether it would be possible to find a more useful and more precise way of defining electronegativity.

Nowotny discussed electronegativity according to Mulliken in terms of the reactions

$$A + B \rightarrow A^+ + B^- + I_A - E_B \quad (1)$$

$$A + B \rightarrow A^- + B^+ + I_B - E_A \quad (2)$$

where A and B are the two components, I_A and I_B the ionization potentials of these components, and E_A and E_B their electron affinities. The factor $(I + E)$ is then a measure of the electronegativity. If $I_A + E_A = I_B + E_B$, the electronegativity difference is zero. He noted that it was not correct to assume that the electronegativity for a certain element was always the same, since different ionic states have different electronegativities.

He felt that the concept of electronegativity helped greatly in the evaluation of bonding problems, even in metal chemistry.

Heine wondered what "electronegativity" corresponded to in band theory. In the non-transition metals, the following quantities, measured on an absolute energy scale relative to the vacuum, appear relevant to the cohesion of a pure metal: firstly, the Fermi level, equal to minus the work function ϕ when corrected for the surface dipole; secondly, the bottom of the band E_0; and thirdly, perhaps the mean energy in the band, equal to $E_0 + \frac{3}{5}\epsilon_F$ where ϵ_F is the Fermi bandwidth for a free-electron gas (Fig. 1). If we made an alloy out of unit cells of the pure metals and then removed imaginary barriers between them, the interaction would be presumably related to the difference between some such parameters for the two metals. Heine asked whether there was an empirical correlation between any one of these parameters and the observed electronegativity effect. Hume-Rothery observed that for practical purposes one would have to know the values of these parameters for all elements. Heine replied that one could calculate these quantities for simple metals from the measured cohesive energies, using the correlation energy of a free-electron gas (which is now sufficiently well established). He thought the noble metals could be similarly handled. If from these simple metals one could obtain some significant correlation, one could probably extend the appropriate parameter to transition metals. Beck inquired whether the calculation for a free-electron parabolic band could be extended to the case of the d band. Heine replied that for the d band, insofar as one was talking about the s electrons, one could also supply the data, and one could extend it to the

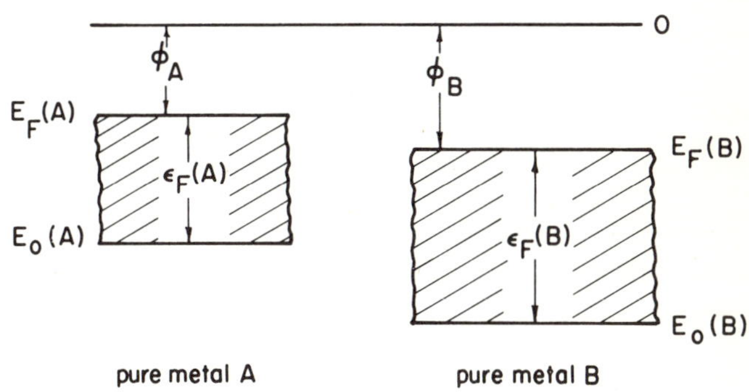

Figure 1

d bands themselves. Hume-Rothery asked how one could know the Fermi energy of "funny metals like osmium." Heine did not know about "funny metals like osmium" but, for the noble metals and for elements to the right of them in the periodic table, one could estimate the Fermi energy from the observed cohesive energies by putting in the energy of the free-electron gas. From the fundamental band-structure calculations which have been done, together with the cohesive energies, one could also make quite a good guess as to the Fermi energies for the transition metals.

Drowart remarked that one can now derive the difference in the electronegativity of two metal atoms more directly than via halides. This is, for example, the case for Au and Cu, for which the dissociation energies of the gaseous Au_2, Cu_2, and AuCu molecules are known. The electronegativity differences so obtained agree with those derived from the halides or from thermochemical data for condensed phases. With regard to Nowotny's comments, Mulliken defined the electronegativity as $X_A = (I_A + E_A)/2$ where I is the ionization potential and E the electron affinity. Since trends in the ionization potentials and the energies of excited states of the atoms and also in the work functions of the corresponding metals are analogous, these different quantities and the electronegativities should give similar correlations. Friedel observed that for a substitutional impurity (neglecting size effects) the electronegativity should tell one the amount of the excess charge taken by the impurity atomic cell when the (neutral) impurity is introduced. He stressed that there was no simple criterion for this. If, following Heine, one pictures an impurity B atom as a piece of B metal stuck in the A matrix, either one of two conditions would lead to electron transfer from B to A:

1. The Fermi energy E_M of B higher than that of A
2. The energy E_C at the bottom of the band of the B metal lower than that of A (because of a scattering effect)

Clearly, a difference of work function and thus of the Fermi energies is not enough. The differences in Fermi energy and in energy at the bottom of the conduction band work in opposite ways; and a criterion related to the average energy $E_F = E_C + \frac{3}{5}(E_M - E_C)$ is clearly no good either. Heine repeated that both quantities could be calculated from the pseudopotentials which had been published for 25 elements, and one could try empirically to see which factor was more relevant.

Lomer remarked that electronegativity, from the bond-energy point of view, was one thing and the transfer of charge another. If we were to discuss whether charge was transferred to an impurity atom cell or out of it, we must know where the boundary of the cell was. The classic difficulty of how to treat the group III to V compounds; that is whether they are like covalently bonded compounds, in which the pentavalent element has given an electron to the trivalent, or whether the bonding was more ionic, getting toward the iron sulfide structure, was simply a question of language and where to draw the cell boundaries. We have to be very careful indeed in trying to set up criteria which have to do with the electron distribution, when attempting to define a term which has to do with the energy of the system, and yet not trying to look in detail at the wave function. Seeger replied that in favorable cases there was a possibility of deciding this question by experiment. An example is a case like sulfur, where there are all sorts of ionic states in molecules. One measures the X-ray levels and compares them with the X-ray levels that one finds in various solids. Then, one could decide what the change of the electron density in the immediate neighborhood of the sulfur atom is. In such a case one might indeed be able to decide what charge transfer occurred in a solid compound.

Nevitt discussed the validity of the isomer shift (from the Mössbauer effect) as an assessment of charge transfer. In the CsCl phases of gold with the lanthanides, his co-workers and he found that the isomer shift depends on the electronegativity difference of the partner.[1] In fact, the sense and the size of the isomer shift were in the right direction to be accounted for by the electronegativity effect. Friedel remarked that the isomer shift only measured the *s* part of the electron-wave functions, while the charge transfer could also involve *p* or higher orbitals; so one could not measure the total. One should be somewhat careful in analyzing the experimental results. In a postmeeting written discussion, Nevitt remarked that the isomer shift is influenced by the *p* electrons via shielding effects; so even if some electrons have *p* character, the isomer shift measures a net change. This net change may be proportional to the total charge transfer, especially if the ratio of *s*- to *p*-like character remains constant.

Kubaschewski said that the quantity of interest was the *lattice energy*. The lattice energy depended largely on distance, and the ionic sizes are approximately constant. We could try to calculate that energy, but why not derive it from empirical data? What was then surprising were the additional effects one then found. For instance, if the electronegativities were used to calculate the heats of formation of compounds, such as HCa_3Sb_2, then the results were correct within 2 kcal. However, the electronegativities for Sb, Bi, etc., have been derived from heats of formation of halides, in which Sb and Bi are cations, and yet they still give the right answers when they are anions. What Kubaschewski required was some similar simple relationship for the covalent bond. He added that these electronegativities were used in all sorts of combinations, and they always give at least the right direction, even when the differences are very small.

Beck then introduced a new topic: the type of bonding as a function of structure. Kubaschewski asked: What is bond type? Various people seemed to hold different views. It appears that in NiAs phases the c/a ratio decreases as the metallic nature increases. When one compares this with the heats of formation, there is a general trend to less negative heats of formation, ending with a pure NiAs type, the $InCu_2$, which appears to be rather metallic. The CsCl type is another structure, which seems to cover a very wide range of metallic bonding. The tolerance for changing bond type is rather large, varying from the ionic CsCl to the metallic β phases. In measuring heats of formation in ternary systems, Kubaschewski selected the system Ni-Ti-Al in the hope that it would prove simple, since the structure of both phases NiAl and TiAl is closely related to CsCl. Nevertheless, there is no complete solid solution, apparently because of a change in bond mechanism.

The final point Kubaschewski made concerned the use of the size factor where metallic bonding was involved. The major structural types, such as the Laves phases, the $CaZn_5$, and the $CaZn_{13}$ types, can be plotted as in Fig. 2 which shows the ratio of the atomic radii versus the ratio of the heats of sublimation. The higher the heat of sublimation of the larger metal, the more likely are deviations from the ideality. If the assumption is made that, in the first approximation, the bonding in the metallic alloys is additive, then one can work out curves at which the stability would become zero. Kubaschewski pointed out that this was a gross simplification which neglected the electrochemical factor, and, indeed, phases that do not fit under the curves have quite frequently strong differences in electronegativity. Massalski pointed out that in zinc one had three choices for the size of the zinc atom—the closest distance of approach, the longest distance of approach, or a value calculated from the volume.

The discussion then turned to bonding in Laves phases, opened by Laves, who spoke about $MgCu_2$, using a hard sphere model. Taking the distance between Mg and Cu atoms as 1.0, the distance between copper atoms is 0.86, and the distance between magnesium atoms is 1.05. Therefore the quotient

$$Q = \frac{d_A + d_B}{2d_{AB}} = 0.94$$

where d_A is the distance between two A atoms, etc. On the average, therefore, different atoms are further apart than like atoms, and this is a unique characteristic of this type of crystal. On the other hand, in the ideal case the radius ratio should be $\sqrt{3/2} = 1.23$. Although the ratio of the Goldschmidt radii of the components varies over rather wide limits, Laves thinks that in forming these phases the atoms adjust to the ideal radius ratio by electron transfer. In those cases where one cannot expect electron transfer, one has nearly the exact ratio, e.g., 1.235 for KNa_2. In his opinion in this case the good fit is the reason the compound exists.

The volume for K in the bcc structure is 74.9 A^3; the volume for two Na atoms in the elementary bcc structure is 78.4; and the sum of the volumes is 153.3, compared with the volume for the molecule KNa_2 in the AX_2 structure of 148.5. One has a small contraction, but by forming this AX_2 structure the space-filling principle is followed. Even with predominantly covalent bonding some electron transfer may take place so that the atoms adjust their size. From a geometrical point of view this would give good space filling. For example, in the compound $LaNi_2$ where the radius ratio of the elements is very large, of the order of 1.6, the lanthanum transfers electrons to the nickel so that the lanthanum becomes positive, the nickel negative, and the radius ratio approaches the ideal value. Where the radius ratio is small, for example in $FeBe_2$, one might expect the reverse electron transfer. Laves stated that the stability of the $MgCu_2$-type

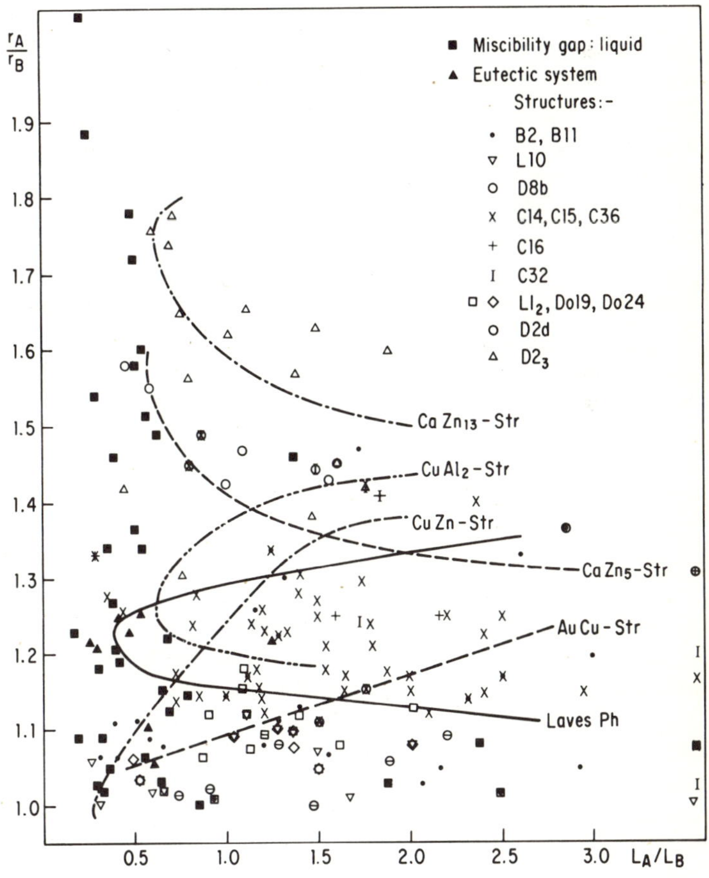

Fig. 2 The ratio of the atomic radii r_A/r_B versus the ratio of the heats of sublimation L_A/L_B for the major structure types.

structure is due to the good space filling, and the atoms may be positive or negative, with covalent or heteropolar bonds, just to obtain the optimum space filling. Beck inquired whether this meant that the bond energy came from the requirement of space filling, but Laves said it comes from the kind of bonds that are actually present: for example, the bonding energy in KNa_2 is low, but in $LaNi_2$ the bonding energy is quite high, suggesting heteropolar bonding as well as metallic bonding. Further, in these phases, when the melting point is very high, there is a strong contraction; on the other hand, where there is no contraction compared to the elementary components, the melting point is low. For example, in the system K-Na there is no large bonding energy because the radius ratio is right. Beck remarked that, according to this concept, one would expect the largest heteropolar bonding energies where the size misfit is largest, and Laves agreed.

Brewer urged that one should first study the stability of the liquid. The next step is to consider the various possible ways of ordering the atom in the liquid, and which of these would give the most stability. He felt that Laves phases achieve an unusual stability because of their high density of packing. Brewer had predicted the conditions for the occurrence of Laves phases in relation to the location of the components in the periodic table. For the alkalies, they are not found to the left of the eleventh group. For the alkaline earths, they are not found to the left of the ninth group. For the third group, as for the lanthanides, they are not found to the left of the seventh group; for Ti, Nb, and Ta they are found for only Cr, Mn, Fe, and Co (the first element being A, the second B in AB_2); and finally for Zr and Hf, for the sixth to ninth groups. These include all known cases and also many possible phases whose existence has been demonstrated, since he made the predictions based on the Hildebrand regular-solution treatment and the internal-pressure treatment. If one considers only the first 18 elements of the first six groups of the periodic table (all of which have the bcc structure at high temperatures) and mixes the solids or the liquids, one can use the regular-solution equation to represent not only the liquid miscibility but also the bcc solubility, just by taking the energies of vaporization divided by the molar volume. If one does that, one finds that the alkali metals cannot be mixed in the liquid with any of these metals until one gets far to the right side of the periodic table. It turns out that the internal pressure is the critical factor in determining the range of occurrence of Laves phases. When one goes from the alkali metals to the alkaline earths, the internal pressures rise, and one has a better chance of having a liquid with a higher stability, which with additional stabilization will go to a Laves phase. This is the indicated trend of internal pressure, and by the time one gets to Hf and Zr one has a high enough internal pressure so that they form miscible liquids with all the elements to the right of the periodic table. The sizes are also appropriate; so one can form a wide range of Laves phases. Ti, Nb, and Ta have appropriate internal pressures, but the sizes are too small. These considerations were sufficient to account for all the observed Laves phases. Beck asked whether the same considerations as to the liquid applied also in relation to other phases, and Brewer replied that they do indeed apply to every one. His first step in every problem of this type is to consider the bonding in the liquid. Some of the previous discussion which has emphasized what goes on in the ordered solid is really the tail wagging the dog. The main bonding effect occurs without the regular arrangement of atoms in the solid.

Nowotny gave an example of complete miscibility between an alkali metal halide and the alkali metal in the liquid state and stressed that this was not a unique example.

Enderby stated that it is very dangerous to consider the liquid bonding as an indication of the type of bonding in the solid. Liquid CdSb, for example, is essentially metallic, whereas in the solid it is nearly covalently bonded. If one gets large volume changes, one must be very careful.

Massalski returned to the discussion of size factor. If one wants a rough and ready discussion of sizes in solid solution, as in the Hume-Rothery criterion, all one needs to do is to consider the atom size in the pure metal. But as soon as one goes to alloys, the "size" of atoms changes. In Laves phases one has a criterion of space filling and radius ratio to work back from, but the only thing one can do to estimate the size of the atoms in, for example, the γ phase is to extrapolate the lattice-parameter trend to the pure metal. If one has a compound, one cannot do this, because the range of stability is too narrow, but one does know that atom sizes may be different in different phases of a single binary system, and perhaps the atomic volumes may be extrapolated through a number of successive compounds. Pearson commented that at a recent symposium in New York (March 1, Met. Soc. AIME) K. A. Gschneidner reviewed work on a calculation of deviations from Vegard's law. As he remembered it, taking some conservative arbitrary criterion of "agreement" and studying more than 250 systems, the percentage of successful agreement was about $37 \pm 3\%$, whether an electronic or elastic model or a combination of both was used. This analysis is interesting since the relatively poor agreement, which is roughly independent of the model of attack, seems to indicate one of two things. Either it is trivial evidence of the fact that the precise definition of the size of an atom is impossible, or it suggests that we have not yet found the main principle governing deviations from Vegard's law, and the different models used are both giving secondary corrections, perhaps dependent on the unknown effect.

Brewer again returned to the change of structure on going from liquid to solid. In some cases there is only one electronic structure available.

Thus in niobium, d^5s is the only configuration possible, and there is thus only one structure possible. In other cases, for example liquid manganese, three electronic configurations are possible. With regard to size, he agreed that size can vary, but stressed that there is a limit to the variation of size one can have. Laves compounds are not electron compounds, but the question of which Laves phase one will get, once the existence of a Laves phase is predicted on size grounds, is determined by electronic considerations.

Friedel, contrary to Brewer, sees no need to assume that the number of d electrons in transitional metals is the same as in the free atoms. For instance, Friedel thinks that one can have more electrons per atom in metallic nickel or copper than in the free atoms. The origin of such an $s \rightarrow d$ promotion in the metal is, he believes, to be found in the fact that, in the metal, the s electrons are very different from what they are in atoms: They overlap very much and give rise to a very broad conduction band where there is plenty of kinetic energy available. Heine said that there is a great difference between a metal and a nonmetal, in that a metal could conduct electricity: All chemical effects, and in particular differences in energy between different term values in an atom, are determined ultimately by the Coulomb interaction between the electrons and the atomic potential; and in a metal this is different—it is screened, compared to what it is in a free atom. If one is going to talk about different configurations, one has to think of the energy as being screened, and the whole energy scale considerably collapsed.

Beck then introduced the next topic, the stability of nickel arsenide structures. Pearson commented that the normal metal or metalloid atoms in this structure lie in a close-packed array, and the transition-metal atoms lie in the octahedral holes of this array (Fig. 3). The "anion" octahedra about the transition-metal atom share faces along the c axis; so one has immediately the possibility of strong T-T interaction across the shared faces of the octahedra (T = transition-metal atom). Secondly, one has the possibility of a strong T-T interaction in the basal plane, but not as strong as the other, because the coordination octahedra only share edges. Thirdly, there is the possibility of an interstitial getting into the trigonal bipyramidal hole: the center is midway between the two tetrahedral holes in this array, and in the "filled-up" nickel arsenide structure (as

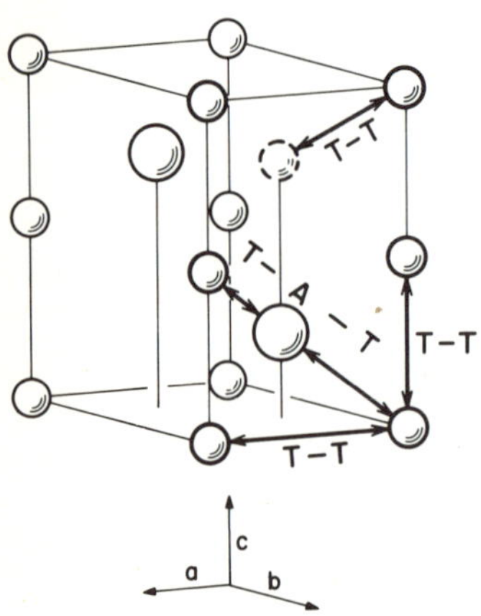

Fig. 3 The nickel-arsenide structure.

one goes to the transition-metal-rich side) this site is occupied. That is, in general, the closest interaction for the transition-metal atoms. In addition, there is the possibility of indirect T-A-T interaction, which may be moderately strong.

There are three types of compounds with the NiAs structure. All the phases with this structure contain a transition metal, except the one with the composition AuSn. First, there are those with a large c/a; e.g., TiS with $c/a = 1.94$; there are those with nearly ideal ($c/a = 1.63$) axial ratio; and then there are the typically metallically bonded ones with c/a as low as 1.26; for example AuSn. Because of the large electronegativity difference, one would expect TiS to have the NaCl structure, but the Ti^{++} radius (calculated from TiO) means that the ratio r_C/r_A is less than 0.414, which is the critical radius at which the anions in the rock-salt structure come into contact. Anion repulsion consequently renders the NaCl structure unstable, and the NiAs structure is formed; but the c/a becomes very large to separate the cations as far as possible. MnTe is an example of the second group: here directional bonding seems to be important. The anion achieves a filled subshell by a process of electron sharing, and the compound is a semiconductor. In the semiconducting group, the nearest T-T approach is 2.89 A in FeS and 3.5 to 4.0 A in MnTe. This is getting close to the limiting distance of interaction beyond which one is in the shell model, as in FeS. Lastly, in the metallic group one has strong T-T interaction. In the metallic compounds, reduction in axial ratio and/or filling of the interstitial site results in strong T-T interactions leading to band formation. In reply to a query by Beck, Pearson remarked that Zeeman had calculated that the optimum Madelung constant for ionic binding in this structure is obtained for a c/a of 1.77.

Beck then invited Nevitt to discuss bonding in Cr_3Si structures. Nevitt stated that there are some 40 or more compounds of this structure. The major component in every case is a transition metal having an unfilled d shell; the second element may be (for example) a transition metal from the cobalt or nickel groups, or gold, or from groups IIIB, IVB, or VB. Several compounds of this type have high superconducting transition temperatures, and the resulting studies have produced a lot of additional information. Recently APW calculations have been made for the structure.

Essentially, the structure is cubic with the B atom at the cube corner and the cube center, while the transition-metal atoms form three mutually orthogonal sets of chains. There are strong indications of bonding along these chains. V_3Co and V_3Ni are completely miscible, but neither shows significant miscibility with V_3Si for example; thus, there appear to be different kinds of bonding, depending on whether the second atom is a transition metal or not. There are strong indications of directional bonding: recently, the superconducting transition was studied as a function of an applied strain tensor, along the cube edges in V_3Si. Both the normal state resistivity and the transition are very sensitive to this. The recent APW calculations have shown that when the minor component is not a transition metal, as for example V_3Ga, the Fermi surface contains very little $4p$ character. When the partner is cobalt, as for example in V_3Co, then there is a fairly strong interaction between the $3d$ states of cobalt and the $3d$ states of vanadium.

Friedel reported a theoretical analysis by Labbé,[2] in Orsay, of the low-temperature martensitic transformation that occurs in some of the compounds just discussed, especially V_3Si. Experimentally the transformation temperature T_t is about 10° above the superconducting temperature T_c. The lattice distortion ϵ was discovered by Barrett et al. It is a small compression of two families of linear chains with a corresponding extension of the third family, so as to keep the atomic volume constant. This shear decreases sharply just below T_t so that, if the transition is of the first order, as suggested by the hysteresis phenomenon, the latent heat is small (Fig. 4).

Labbé's model originates from the chainlike structure of the lattice. If one treats the narrow d band of the transitional atoms separately, one can certainly neglect the overlaps between atoms on the same chain in a tight binding approximation. One then knows that $E(k)$ has a sinusoidal form; the small slopes at both ends of the band lead to large values of the density of states, becoming actually infinite at the edges in this simple model. The different d orbitals actually lead to three families of such d bands with essentially the same average energy but different widths. Each family is at least triply degenerate in the cubic phase, from the equivalence of chains parallel to the three sides of the cube. There is furthermore a broad (sp-like) band due to the $4s$ states of the transitional atoms and the valence electrons of the B atoms (Si in V_3Si).

There is a strong suggestion, from the very high electronic specific heat and from the sharp decrease of magnetic susceptibility with temperature, that the Fermi level E_M falls within a sharp peak of density of states such as those pictured in Fig. 5.

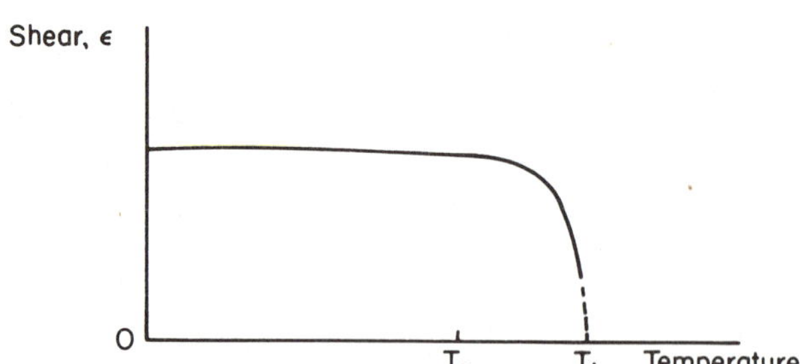

Fig. 4 Shear strain in martensitic transformation of V_3Si.

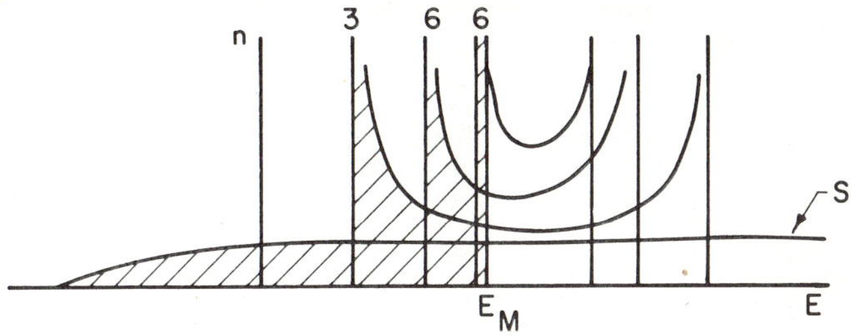

Fig. 5 Labbé's model for the cubic phase.

One can then wonder whether one could not have an electron-induced lattice distortion of a kind similar to the Jahn-Teller effect for bound electronic states in insulators. Indeed, it is easy to see that the strain corresponding to the martensitic transformation will widen the d bands corresponding to two families of chains, and narrow the bands corresponding to the third family (Fig. 6).

Then, if the Fermi level is near enough to a band edge, the few electrons (or positive holes) in the corresponding band will all go into the lower part of the tetragonally split band, thus lowering its energy.

Without going into details, we can say that this type of model requires a certain number of parameters. Most of them are known at least in order of magnitude: the d-band width, its variation with distortion, the part of the shear modulus due to other than d electrons. The only really free parameter is the (small) number of electrons in the nearly empty d sub-band (or the small number of positive holes in a nearly full d sub-band). Fixing this number to about 0.06 per atom and taking for the other parameters reasonable values, Labbé was able to explain the following properties:

1. Low-temperature lattice distortion $\epsilon(0)$
2. Temperature variation of the distortion, with a strong reduction near the transition temperature and an order-of-magnitude estimate of this temperature

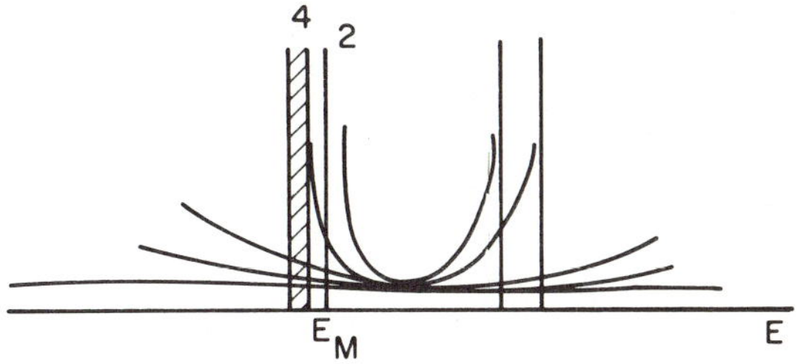

Fig. 6 The central portion of Fig. 5 in the tetragonal distortion.

3. Values and temperature variation in both phases of the shear modulus corresponding to the tetragonal distortion (especially its strong temperature variation in the cubic phase)
4. Values of the different electronic specific heats measured both at Bell Labs and at Orsay in the cubic and tetragonal phases
5. The order of magnitude and the temperature variation of the magnetic susceptibility

Finally, Labbé has shown that this type of band Jahn-Teller effect by *homogeneous* lattice distortion is more stable than that introduced by a *periodic* lattice distortion (i.e., frozen-in elastic wave), when as here the Fermi level falls within a sharp peak in the density of states. A periodic distortion with a wavelength equal to that of the Fermi d electrons would have been stable otherwise.

Brewer commented that two aspects of this structure had so far not been dealt with. The first is that transition elements of the first long period can form much stronger bonding in this type of structure, and the most stable structure has a combination of an element from the first long period together with a "heavy" transition element such as platinum. Secondly, he commented about electronic configuration. Variations from stoichiometry are accommodated in the main by vacancies on the two sublattices; there is little evidence of A atoms on B sites, or vice versa.

One may write the sequence of phases versus the average number of s, p electrons beyond the approximately five d electrons per atom in alloys of the later transition metals as

e/a	Phase
1.0–1.5	bcc
1.1–1.7	Cr_3Si
1.2–1.9	σ
1.4–2.0	α-Mn
1.8–2.0	β-Mn
1.8–2.0	hcp

The selection where ranges overlap is based on size-factor considerations.

Beck then invited Nowotny to introduce the subject of stabilization of intermetallic phases by interstitials. Nowotny stated that in the system Zr-Si there are numerous phases, and one observes that a new ternary phase is stabilized by 0.3 a/o carbon (Fig. 7). Hence, very minor amounts of interstitial can stabilize new phases. Nevitt[3] showed some years ago that in systems of titanium with transition metals a phase of composition Ti_2X was stable where X is Ni or Co. These phases absorbed significant amounts of oxygen. In the system Ti-Fe, Ti_2Fe is not stable, but a phase with a similar structure is stable when approximately 5 a/o of oxygen is present. In the system Ti-Mn, the phase is stabilized by 12 to 13 a/o oxygen. In all four systems, the maximum solubility is about 14 to 15 a/o oxygen (Fig. 8).

Another problem, not so interesting for the physicist but of great interest to the chemist, is the continuous change in bonding. Figure 9 shows the c/a ratio versus the group number for the H phase. One can see that in these typical metallic carbides there is a continuous range from the more metallically bonded, such as Ti_2CdC down to even Ti_2SC, corresponding to a considerable amount of ionic bonding. The similarity between these complex carbides and the simple carbides can be seen by comparing the transition-metal–carbon distances:

TiC	2.16	Mo_2C	2.16
Ti_2AlC	2.12	Mo_3Al_2C	2.14
Ti_3AlC	2.08	$Mo_{12}Cu_2Al_{11}C_6$	2.08
$Ti_4Zn_2C_x$	2.14	$Mo_{4.8}Si_3C_{0.6}$	2.15

In the complex carbides the distances are thus a little smaller than in the simple binary carbide phases.

In the tantalum carbide TaC_{1-x} the highest stable composition (smallest x) is paramagnetic; however, with carbon deficiency one gets a diamagnetic value (Fig. 10). He believes that this is evidence of very strong electron-pair bonding between the metal atom and carbon as the most important contribution to the bonding system. In reply to a question by Beck, he preferred not to describe the bonding as covalent.

Friedel reported on a systematic study of the ferromagnetic properties of transition-metal borides, done by Mrs. Cadeville and Dr. Meyer[4] in Strasbourg and also in part at the Swedish Atomic Energy Laboratory.

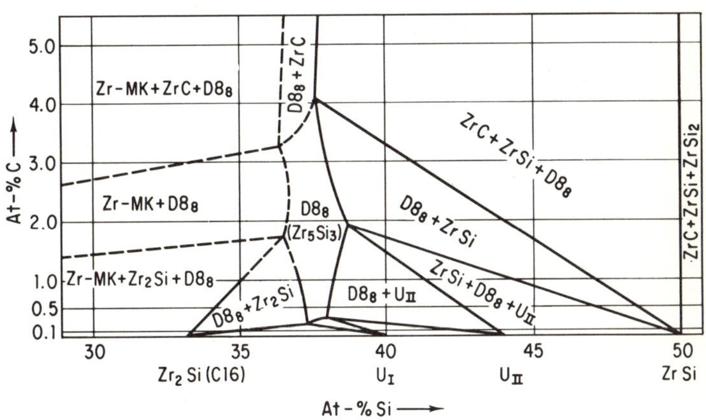

Fig. 7 Part of the Zr-C-Si ternary diagram.

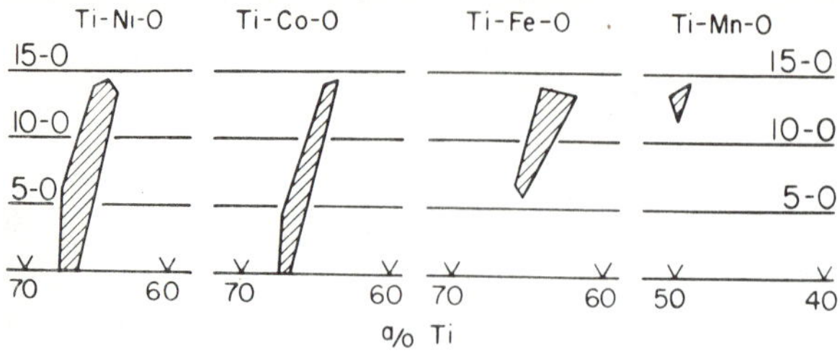

Fig. 8 Ranges of concentration of O in Ti_2Ni-type phases in the ternary systems Ti-Ni-O, Ti-Co-O, Ti-Fe-O, and Ti-Mn-O.

He thought that they threw an interesting light on the electronic structure of these compounds, when compared with the properties of ferromagnetic transitional alloys.

For binary transitional alloys, the Slater-Pauling curve gives the average magnetic moment per atom μ (in Bohr magnetons) as a function of the average atomic number (Fig. 11, curve 1).

The straight line with 45° negative slope on the right was explained by Mott[5] as due to the progressive filling of an incompletely filled half of the d band, with spin opposite to that of the magnetic moment (Fig. 12). The straight line with 45° positive slope on the left, for Fe-based alloys, is attributed to the removal of electrons from the other half of the d band, with spin parallel to the magnetic moment. Finally, the strong deviations

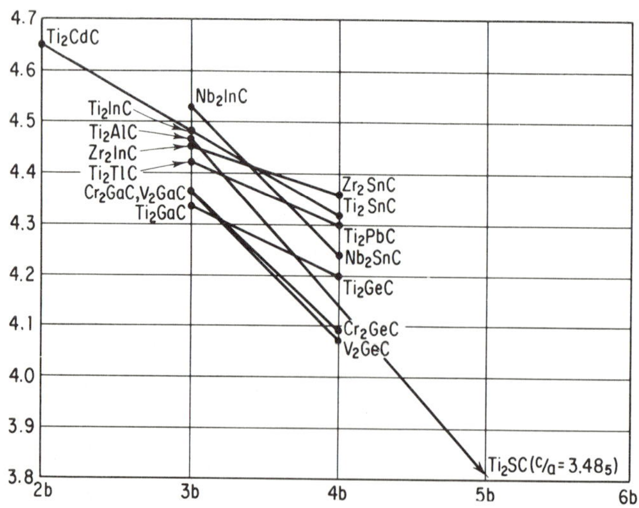

Fig. 9 The axial ratio c/a of the H phases X_2YC versus the group number of the Y component.

that occur for nickel- and cobalt-based alloys with elements far on the left in the periodic table are attributed to the emptying of a spin-up d state, pushed above the Fermi level by the strongly repulsive perturbation due to the impurity (Friedel,[6] fig. 3) (our Fig. 13).

The magnetic moments for borides of the $B(TT')_2$, $B(TT')$, and $B_2(TT')$ types are given by the other curves in Fig. 11; these are very similar to the right-hand side of the Slater-Pauling plot for transition-metal alloys (TT'), if the average atomic number plotted for each boride specimen is that for the transitional elements alone. In Fig. 11, curves 2 and 3 give schematically the results for the BT_2 and BT types of borides, respectively. Furthermore, as in the transition-metal alloys, the data for the compounds which deviate from the 45° line of the Slater-Pauling curve define slopes that can be numerically explained by the model of Fig. 13. In the compounds, as in the alloys, these deviations correspond to strong residual resistivities, which can be numerically explained as a resonance

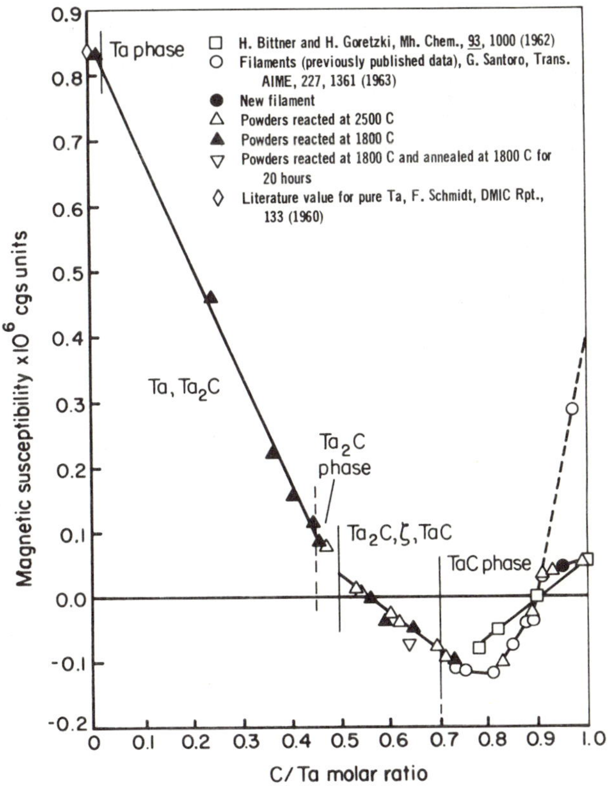

Fig. 10 Magnetic susceptibility versus composition in the Ta-TaC system.

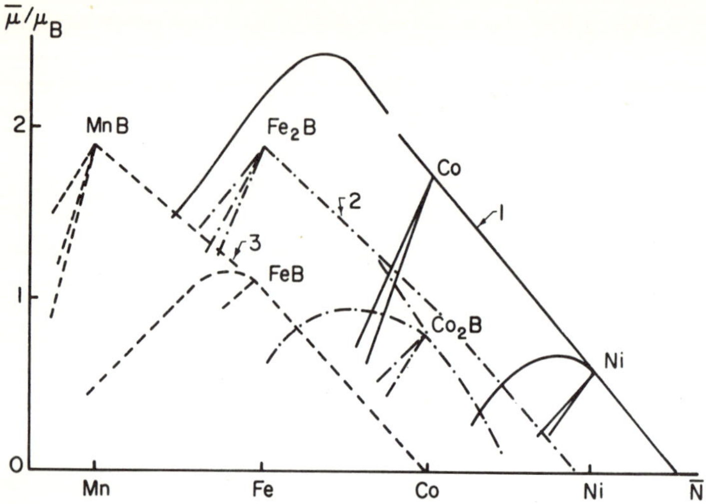

Fig. 11 Average magnetic moment per transitional atom versus average atomic number of transitional elements. Curve 1: alloys of transitional metals $T_xT'_{1-x}$; curve 2: alloys of transitional borides $T_{2x}T'_{2-2x}B$; curve 3: alloys of transitional borides $T_xT'_{1-x}B$.

scattering of the conductive (s) electrons by the (virtual) bound $3d^5$ state which is being emptied.

In conclusion, the evidence strongly suggests that these borides have a partially filled d band with properties very similar to those of the transitional metals and alloys, and that each boron atom fills this band with about 1.5 electrons. This does not necessarily mean an electron transfer from boron to transition-metal atoms: it could simply be that the presence of the boron stabilizes the d band with respect to the s band on the transition-metal atoms, thus producing an $s \rightarrow d$ electron transfer within the transition-metal atoms.

Heine remarked that he would just like to mention again the data of Geballe et al. on NbN which they interpret as indicating that the Fermi level lies in a pure s band with the d band presumably being pushed up above E_F. If the two p electrons of N lie below the bottom of any of the Nb bands and overlap slightly onto Nb in a bonding sense, then the d band

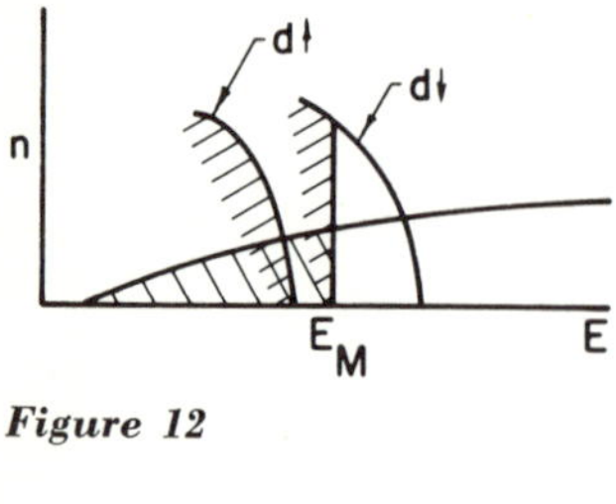

Figure 12

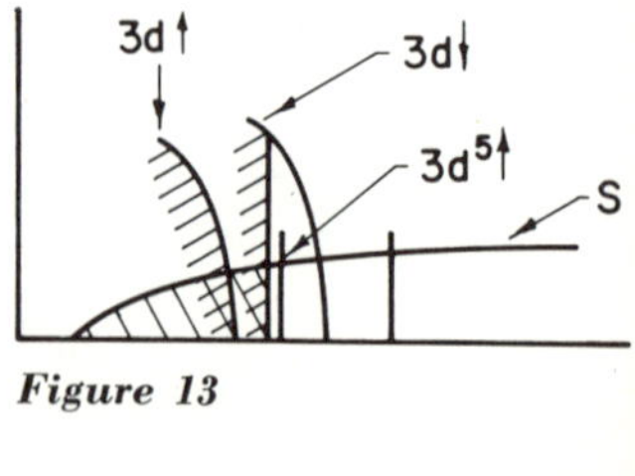

Figure 13

of Nb is somewhat raised by antibonding. This mechanism presumably could apply only with atoms at the beginning of the transition series, for if the d band were nearly full, as in Ni, one would lose as much through the antibonding as one gains through the bonding. This picture would suggest that in phases containing two transition-metal atoms such as Ni and Mo, the N (or C) would prefer to sit between the Mo rather than the Ni. Perhaps the lesson to draw from the conflict of different interpretations is that one needs the most careful study by all possible means of a few typical cases, to establish what the bands really do.

Friedel remarked that concerning the electronic structure of transitional carbides and nitrides of the TC kind, there is a controversy raging, and one should be careful in one's statements. It is true that there is a minimum of electronic specific heat γ near TiC in the periodic table (Costa, Colloque CNRS sur les composés semimétalliques, 1965) and that γ is in VN, NbN similarly small. But some people think that the band filled on the left of Ti has sp character, and the one on the right has d character, as Heine stated. Some other people think that they are two halves of a d band, as for Cr. Other people think that both bands are of mixed d sp character. He personally tended to favor the second picture, at least for carbides.

Lomer said that in transition metals it is believed that part of the paramagnetism is due to Pauli-type spin reorientation, and part due to induced orbital effects of the Kubo–Van Vleck kind. If a substance like NbN is in fact diamagnetic, this would imply not only a low density of states, but also a major increase in separation of the occupied and unoccupied bands to suppress the induced orbital contribution. Thus if compounds of elements as far up the transition period as Nb or Cr are not paramagnetic, the nonmetallic elements must be taken to have caused a very large splitting of the bands.

Nowotny produced a slide which showed NbN to be paramagnetic. Lomer felt that this was strong evidence of occupied d states, since the density of states is low. Nowotny agreed, and remarked that zirconium and hafnium carbides were diamagnetic. The paramagnetism of titanium carbide approaches zero as the composition approaches stoichiometry. One could speculate that with $4 + 4$ electrons one gets a minimum. The same is true of tantalum carbide; the minimum is shifted toward an overall electron concentration of 8. The same is also true for niobium and vanadium carbides, but the latter does not exist at the "stoichiometric" composition. Nowotny felt that "valency" in this connection is difficult to define. Brewer remarked that if an electron-transport experiment is performed on oxygen, nitrogen, or carbon in zirconium, the interstitial goes to the cathode. However, several discussers pointed out that this type of experiment does not measure the "charge" on the interstitial.

REFERENCES

1. Kimball, C. W., G. M. Kalvius, and M. V. Nevitt: *Bull. Am. Phys. Soc.* **11**:267 (1966).
2. Labbé, J.: *J. Phys.*, in press, 1966.
3. Nevitt, M. V.: In P. A. Beck (ed.), "Electron Structure and Alloy Chemistry of the Transition Metals," pp. 101-178, Interscience Publishers, Inc., New York, 1963.
4. Cadeville, M. C., and J. P. Meyer: *Compt. Rend.* **255**:3391 (1962); M. C. Cadeville: Thèse (Strasbourg), 1965; M. C. Cadeville and E. Daniel: *J. Phys.*, in press.
5. Mott, N. F., and H. Jones: "Metals and Alloys," Oxford University Press, New York, 1936.
6. Friedel, J.: *Nuovo Cimento* **S2**:287 (1958); E. Daniel: *J. Phys. Chem. Solids* **23**:975 (1962).

Part Four

INTERSTITIAL PHASES AND SOLUTIONS

STRAIN-ORDERING SYSTEMS OF THE VANADIUM-OXYGEN TYPE

J. L. Meijering
Inorganic Chemistry Laboratory, Technical University, Delft, Holland

INTRODUCTION

In a body-centered cubic lattice the octahedral interstitial sites—and also the tetrahedral sites—have tetragonal symmetry. They are divided into three sets: x, y, and z (Snoek[1]), depending on the direction of the shorter axis. The distribution of atoms like C, N, or O over the x, y, and z sites can be made unequal by strain, giving rise to the Snoek damping effect. In the absence of external strain the equilibrium occupancies of the three sorts of sites can be unequal if the concentration of the interstitial atoms is large enough to cause "strain order."[2,3] Then a martensite structure (body-centered tetragonal with $c/a > 1$) can become more stable than the bcc structure. In the system V-O, for instance, such a phase is thermodynamically stable at about 20 a/o O.

However, tetragonal phases with $c/a < 1$ and body-centered orthorhombic structures have also been found, e.g., in Ta-O and Ta-H, respectively. Therefore it seems worthwhile to extend Zener's strain-interaction calculations toward larger concentrations of the interstitial atoms. We use a regular enthalpy approximation

$$H = p(x^2 + y^2 + z^2) + 2a(xy + yz + zx) \tag{1}$$

where, for instance, x is the fraction of x sites occupied by interstitial atoms, while p and a are interaction parameters. The concentration

$$c = x + y + z \tag{2}$$

is expressed in, say, atoms O per atom V. It is more convenient to transform (1) into

$$H = ac^2 - b(x^2 + y^2 + z^2) \tag{3}$$

where $b = a - p$ is positive owing to the strain interaction, while a may be positive or negative. H and the other thermodynamic functions are taken for one mole of V.

The simplest approximation for the entropy would be the Gibbs expression

$$S = -R[x \ln x + (1 - x) \ln (1 - x) + y \ln y + (1 - y) \ln (1 - y) + z \ln z + (1 - z) \ln (1 - z)]$$

We would then have a completely long-range order treatment, analogous to Shockley's treatment[4] of ordering in systems of the Cu-Au type using four sets of sites on the fcc lattice. However, in our case it is both desirable and feasible to take account of an important aspect of short-range order. If an atom is situated in a z site, the occupation of other z sites *in general* is enhanced. But for two neighboring z sites ("above" and "below" the first one) the opposite is true,[2] because the two V atoms in between are pushed apart. Therefore the model was modified by prohibiting occupancy of two consecutive x sites along an x direction, etc. Because of the unidimensional nature of the restriction the entropy is easily* calculated to be

$$\begin{aligned} S = -R[&x \ln x + (1 - 2x) \ln (1 - 2x) - (1 - x) \ln (1 - x) \\ &+ y \ln y + (1 - 2y) \ln (1 - 2y) - (1 - y) \ln (1 - y) \\ &+ z \ln z + (1 - 2z) \ln (1 - 2z) - (1 - z) \ln (1 - z)] \end{aligned} \tag{4}$$

Now x, y, and z can vary only between 0 and ½, and c between 0 and 1½. The system becomes asymmetric.

The free enthalpy can be written in the form

$$G = H - TS = ac^2 + G_x + G_y + G_z \tag{5}$$

with

$$G_u = -bu^2 + RT[u \ln u + (1 - 2u) \ln (1 - 2u) - (1 - u) \ln (1 - u)] \tag{6}$$

* By calculating the mixing entropy in rows of—for instance—Nx vacant sites (each attached to a preceding occupied site) and $N(1 - 2x)$ other vacant sites.

DISTRIBUTION OF THE INTERSTITIAL ATOMS OVER THE DIFFERENT SITES

First, only unary behavior is considered: the distribution of the interstitial atoms over the sites is varied to give a minimum G at a given c and T, but no separation into phases of different concentration is allowed yet. Then the term ac^2 in Eq. (5) can be ignored, and one may calculate a graph of unary transformations in a Tc diagram with T proportional to b.

In equilibrium $\delta G/\delta x = \delta G/\delta y = 0$. This amounts to

$$-\frac{2bx}{RT} + \ln\frac{x(1-x)}{(1-2x)^2} \equiv f(x) = f(y) = f(z) \tag{7}$$

Figure 1 gives the function $\ln u + \ln(1-u) - 2\ln(1-2u)$. It has a point of inflection E at $u = \frac{1}{2}(1 - \sqrt{3}/3)$. Imagine a series of parallel

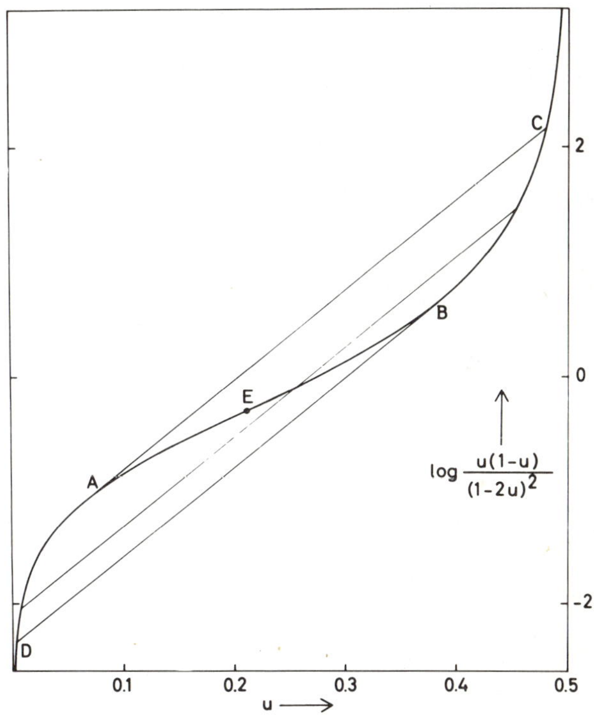

Fig. 1 Auxiliary curve for computing the distribution of interstitial atoms over x, y, and z sites. The curve is obtained by plotting log(tan 2α) versus $(1 - \cos\alpha)/2$ and has a point of inflection at E. The slope of the parallel lines corresponds to $RT = 0.11b$, as an example. Between the tangents AC and BD each line cuts the curve in three points, yielding a set of three possible values for x, y, and z.

lines drawn with slope $2b/RT$. For each line the u coordinates of its intersection points with the curve are values of x, y, and z satisfying Eqs. (7). For $RT = b/3\sqrt{3}$ the slope is equal to that of the curve in its inflection point. Thus for higher T each line cuts the curve in one point only. Then x, y, and z must be equal to one another everywhere, and the solid solution is cubic.

But for lower temperatures there are sets of lines intersecting the curve in three points. By using all three u values of one line an orthorhombic solution is obtained. One may also use, however, only two or one of the u values and so produce tetragonal and cubic solutions. It can be shown that combinations where the middle value of u is used more than once represent unstable equilibria. There remain two cubic, four tetragonal, and one orthorhombic solution for each line intersecting the curve in three points, all at different concentrations c. At a given temperature the infinite set of parallel lines (also above AC and below DB in Fig. 1) gives one solution at each concentration, apart from two rather narrow regions where cubic and tetragonal solutions overlap. Calculation of G yields the transformation points cubic-tetragonal.

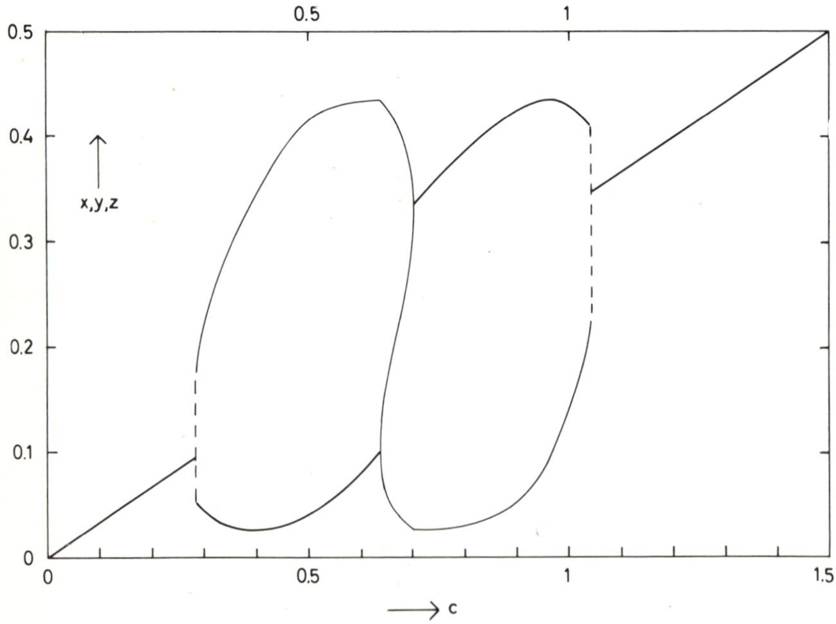

Fig. 2 Distribution curves at $T = b/3R \ln 10$, which is 75.2% of the top temperature in Fig. 5. x, y, and z are the fractions of the different sites occupied by interstitial atoms and add up to the concentration c. Thick curves (lower left and upper right) indicate two coinciding occupation curves, and on the two parts of the thick diagonal line x, y, and z are all equal.

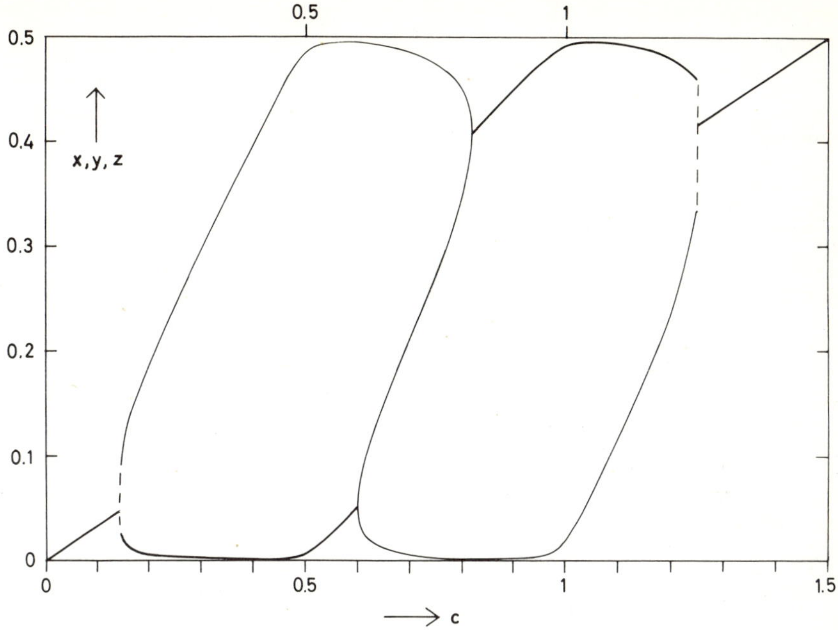

Fig. 3 Distribution curves at $T = b/5R \ln 10$, which is 45.1% of the top temperature in Fig. 5. The two minimum occupation numbers are 0.0006.

The distribution of the interstitial atoms over the x, y, and z sites is given as a function of concentration for two temperatures in Figs. 2 and 3.

When c becomes large enough, the strain interaction causes preferential occupation of one of the three sorts of sites, say z. With c increasing beyond the transformation cubic-tetragonal, the preferred sites become still more attractive—because of the cooperative nature of the strain interaction—and even capture atoms from the x and y sites. Their identical occupations go through a minimum and rise again thanks to the entropy. After having become large enough, they part at the second-order transformation tetragonal-orthorhombic. In the orthorhombic region x, y, and z lie on a continuous S-formed curve. If we take $z > y > x$, then the y sites are more favorable than the x sites with respect to energy, and more favorable than the z sites with respect to entropy. The transformations at higher c are the counterparts of those mentioned earlier.

The maxima of the double- and single-occupation curves have equal height, and the same applies to their minima. However, the maximum and minimum of the single curve—which are at the transformations tetragonal-orthorhombic—are discontinuous, with du/dc zero toward the tetragonal sides only. Their concentrations are $2u_A + u_C$ and $2u_B + u_D$ (cf. Fig. 1). The concentrations at the minimum and maximum of the

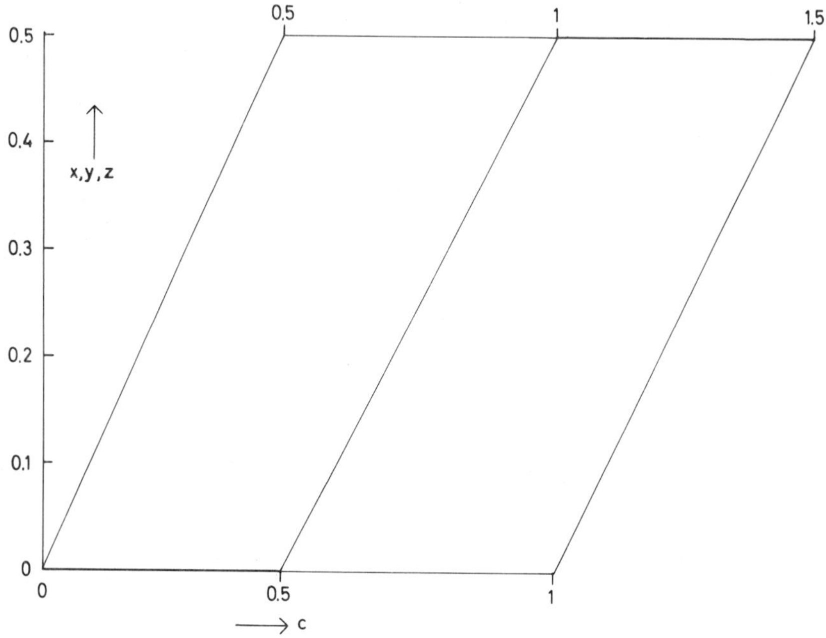

Fig. 4 Distribution lines at $T = 0$. Between $c = 0$ and 0.5 two occupation numbers are zero, and between $c = 1$ and 1.5 two are equal to 0.5.

double-occupation curve are $2u_D + u_B$ and $2u_C + u_A$, respectively. The values of u_A and u_B are roots of the equation $u(1 - u)(1 - 2u) = RT/2b$.

At $T = 0$ (see Fig. 4) the entropy plays no role, and matters are quite simple. One has to make $x^2 + y^2 + z^2$ as large as possible with $x + y + z$ constant. Thus, when interstitial atoms enter, they occupy, say, z sites only until these are full.* So from $c = 0$ to $\frac{1}{2}$ the alloy is tetragonal with $c/a > 1$. Between $c = \frac{1}{2}$ and 1 a second sort of sites is being filled; then x, y, and z are unequal, and the alloy is orthorhombic. For c between 1 and $1\frac{1}{2}$ two sorts of sites are full, and the alloy is again tetragonal, but now with $c/a < 1$.

At T above $b/3R\sqrt{3}$ the alloy is cubic everywhere and the triple-occupation line runs uninterrupted from $c = 0$ to $c = 1\frac{1}{2}$.

Figures 2, 3, and 4 are rather analogous to Shockley's[4] fig. 10.

THE UNARY TRANSFORMATION DIAGRAM

In Fig. 5 the Tc diagram is given. The transformation tetragonal-cubic is first order everywhere, apart from the top point where it is second order. The transformations orthorhombic-tetragonal are all second order.

* Because of the restriction introduced this is really half full, the z sites on a row being alternately occupied and empty.

The two boundaries of the orthorhombic region end in an oblique cusp in the top point at $c = \frac{1}{2}(3 - \sqrt{3}) = 0.634$ and $T = b/3R\sqrt{3} = 0.1925b/R$, which will be called T_t. Their final slope $dT/dc = -4\sqrt{3}T_t$, but the curvatures are infinite there. The final width of the orthorhombic concentration domain is proportional to the $\frac{3}{2}$ power of $(T_t - T)$.

The left-hand boundary of the orthorhombic region is retrograde. Between $c = 0.634$ and 0.640 the alloy at rising T shows the transformations orthorhombic $\rightarrow$ tetragonal $(c/a > 1) \rightarrow$ orthorhombic $\rightarrow$ tetragonal $(c/a < 1) \rightarrow$ cubic.

At low temperatures both boundaries have a faint inflection point, and their slopes at $T = 0$ are $3\sqrt{3}T_t$ and minus half this value.

The transformation curve tetragonal-cubic is continuous at the top, with $d^2T/dc^2 = 20T_t/9$. The slopes at $T = 0$ are $b/2R \ln 2 = 3\sqrt{3}T_t/2 \ln 2$ and minus half this value. The former value of dT/dc (at $c = 0$) is essentially in accord with the result of Zener[2] for dilute solutions. There is a small numerical difference because the root of an equation was graphically computed to be 0.34, while it is exactly $\frac{1}{3}$.

That the initial slopes at the right are half those at the left is due to the fact that introduction of an extra vacant site in a row of alternate occupied

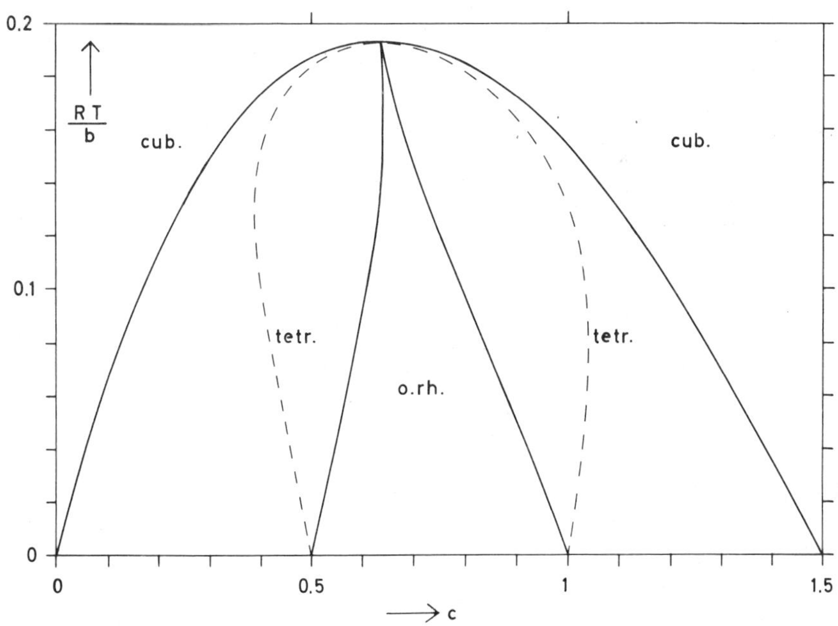

Fig. 5 Transformation diagram. Orthorhombic region (orh) between two tetragonal regions, with $c/a > 1$ on the left and $c/a < 1$ on the right. The dashed curve in those two regions gives the concentrations where two of the occupation numbers are minimum or maximum (cf. Fig. 2).

and vacant sites entails an entropy effect twice that caused by introduction of an occupied site in a vacant row. Indeed, in the former case the removal of an interstitial atom creates a succession of three vacant sites; this "double fault" can dissociate, however, into two separate faults, each consisting of two successive vacant sites.

The transformation entropy tetragonal-cubic becomes zero at $T = 0$ and $T = T_t$; in between it passes through two maxima, about $0.07R$ at the left and about $0.15R$ at the right.

Some specific heats due to the rearrangement of the interstitial atoms were also calculated. At the top the extra c_p of the orthorhombic phase is $(72/67)R\sqrt{3}$, and that of the tetragonal phase $R\sqrt{3}$. It is remarkable that for the latter phase (with $c = 0.634$) $dc_p/dT = -\infty$ at $T = T_t$. It appears that c_p goes through a maximum before the tetragonal phase becomes unstable at decreasing temperature.

The entropies and specific heats mentioned are for 1 mole of V.

SOME GENERALIZATIONS

If it is assumed that an interstitial atom on a z site prevents the occupation of *two* or more sites in each direction of a z row, the entropy can be calculated easily also. Presumably the type of the transformation diagram will not change much, although the concentrations decrease.

A more interesting question is whether the second-order nature of the transformation orthorhombic-tetragonal is peculiar to the regular model used. The simple interchange of two axes after having become equal (cf. the S-formed curve in the orthorhombic region of Fig. 3) appears quite normal. It *would* be possible, however, for such a transformation to be first order if the interaction energy deviates rather strongly from the regular approximation. A ferromagnetic Curie point can be first order[5] instead of second order as required by the Weiss approximation.

It is instructive to compare such transformations to miscibility gaps in binary and ternary systems. The normal Curie point corresponds with the critical point of a normal binary miscibility gap. Suppose, however, that the mixing enthalpy is given by the symmetric function $ax(1 - x) \times [1 - px(1 - x)]$, with positive a and $p > \frac{1}{2}$, x being an atomic fraction. Then, with Gibbs's mixing entropy, one finds a eutectoid three-phase equilibrium where the gap divides into two gaps which vanish each in a critical point at rising T. The nonvariant three-phase transformation corresponds with a first-order Curie point.

The first-order transformation cubic-tetragonal in the strain-ordering systems corresponds with the eutectoid four-phase equilibrium in a ternary regular system with equal positive interaction energies of the three component pairs.[6]

BINARY PHASE DIAGRAMS

When diffusion of the interstitial atoms is fast enough to allow phase separation, we have binary behavior. The term ac^2 in Eqs. (3) and (5) must now be taken into account. Figure 6 gives G as a function of c when $a = 0$, and $T = b/5R \ln 10$ as in Fig. 3. A quadruple tangent can be drawn to the lower side of the curve. The four tangent points are at equal distances, and this also applies to the tangent points of the—physically unimportant—triple tangent which can be drawn to the upper side. The curve as a whole is, however, not periodic.

The kinks at the unary first-order transformation points cubic-tetragonal are not distinct on this scale and are therefore marked by arrows. The second-order transformation points tetragonal-orthorhombic are indicated by dots. Here the curvature is discontinuous, d^2G/dc^2 being zero at the tetragonal side and negative at the orthorhombic side.

The concentrations of the four tangent points as a function of temperature yield the phase diagram of Fig. 7; the homogeneity regions of the two tetragonal phases have zero width.

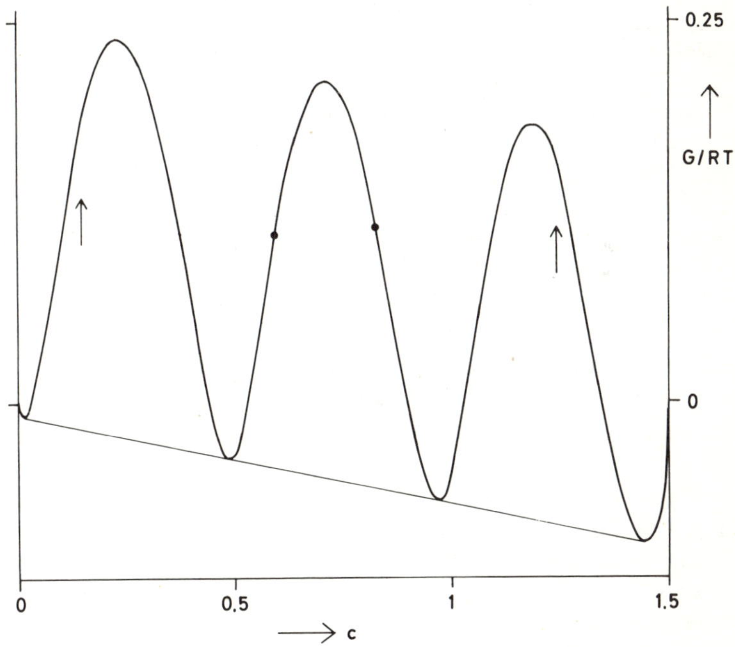

Fig. 6 $G/RT + 2.5c$ as a function of c, for $a = 0$ and $RT = b/5 \ln 10 = 0.0869b$. The linear term was added to bring the ends of the curve to the same height. Between the two dots the homogeneous alloy is orthorhombic, between the dots and the arrows tetragonal. But only at two concentrations can the noncubic alloys remain homogeneous. The quadruple tangent corresponds to the lower three tie lines in Fig. 7.

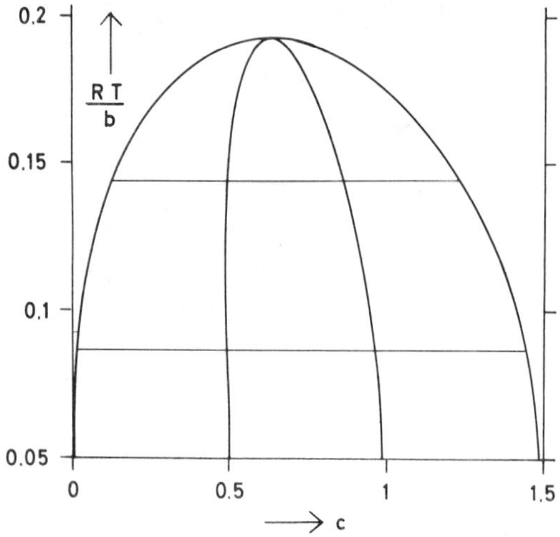

Fig. 7 Binary phase diagram for $a = 0$. The inner two-phase region is tetragonal $(c/a > 1)$ + tetragonal $(c/a < 1)$. The two outer two-phase regions are cubic + tetragonal. Tie lines have only been drawn at $T = b/3R \ln 10$ and $b/5R \ln 10$, the temperatures of Fig. 2 and Fig. 3.

If we now let the parameter a become *negative*, the term ac^2 bends the middle part of the Gc curve upward. Consequently only a double tangent cubic + cubic remains stable. The binary diagram then shows only a miscibility gap in the cubic domain, with a critical point at the same concentration but at a higher temperature than the top point in Figs. 5 and 7. The phase diagram computed for $a = -0.2b$ is given in Fig. 8.

If, on the other hand, we let a become *positive*, the middle part of the Gc curve bends downward, and the quadruple tangent in Fig. 6 breaks up into three double tangents. Both homogeneous tetragonal regions now acquire finite width when T is between zero and T_t. The top point remains the same as that in Fig. 7, but the two-phase regions are narrower. At $T = b/3R \ln 10$ (compare the upper tie lines in Fig. 7) the three tie lines are between $c = 0.22$ and 0.40; 0.58 and 0.755; and 0.96 and 1.12 when $a = 0.2b$.

The orthorhombic phase becomes stable with respect to phase separation when a/b exceeds $^{40}\!/_{67} = 0.597$, its first appearance being at the top. The whole orthorhombic domain of Fig. 5 has become stable only when a/b attains 1. The second-order transformation curves tetragonal-orthorhombic remain single curves in the binary diagrams, like Curie lines. It is convenient, therefore, to lump the tetragonal and orthorhombic domains together to a "noncubic" domain. This domain shows a miscibility gap closing in a critical point below T_t for a/b between 0.597 and 1. For

$a = 0.618b$ the phase diagram is given in Fig. 9. The parameter ratio a/b was chosen at 0.618 in order to bring the critical temperature at $b/3R \ln 10$, where x, y, and z had been computed (cf. Fig. 2) and d^2G/dc^2 was accurately known.

The value of d^2G/dc^2 is smallest in the orthorhombic region, and thus it was expected that the critical point would be in this region. However, the minimum value of d^2G/dc^2—at least at this temperature—proved to be at the right-hand-side boundary of the orthorhombic region, where d^2G/dc^2 jumps upward. This means that the miscibility gap ends in a "sharp" critical point.[7,8]

For $a > b$ there is no miscibility gap any more in the noncubic domain. The phase diagram now is very much like Fig. 5. The second-order transformation curves tetragonal-orthorhombic are completely stable, but the first-order transformation curves are replaced by two-phase regions cubic + tetragonal, astraddle the unary curves.

For $a = 2b$ the maximum width Δc of the two-phase region appears to be somewhat greater than 0.01 at the right-hand side and somewhat smaller than 0.01 at the left-hand side. These widths decrease on increasing a/b.

EXPERIMENTAL SYSTEMS

Well-known body-centered tetragonal phases with $c/a > 1$ are of course the martensites in the systems Fe-C and Fe-N. In V-N[9] and Ta-N[10] such phases have also been reported. Apparently body-centered tetragonal

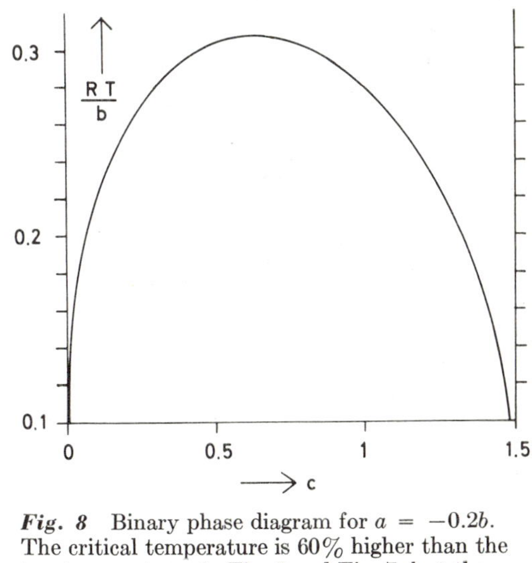

Fig. 8 Binary phase diagram for $a = -0.2b$. The critical temperature is 60% higher than the top temperatures in Fig. 5 and Fig. 7, but the concentration is the same.

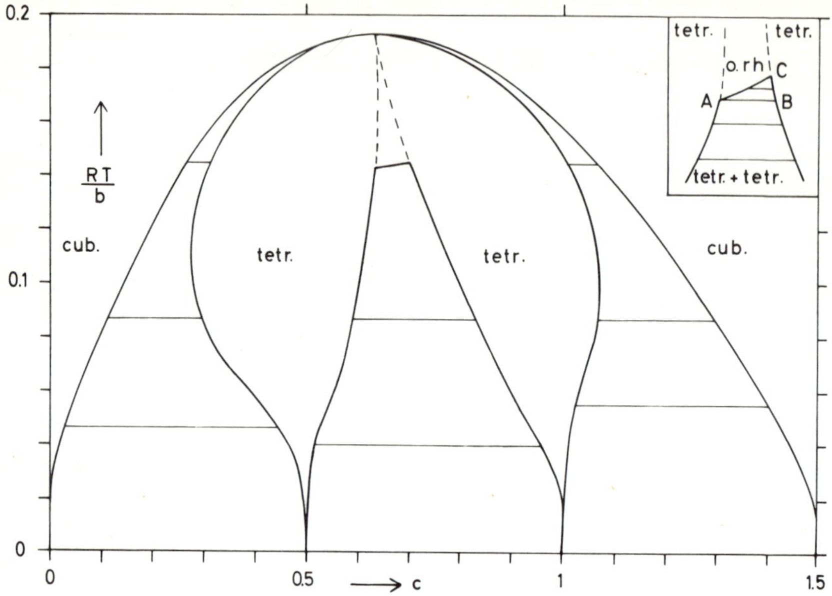

Fig. 9 Binary phase diagram for $a = 0.618b$. In the lower temperature region (say at $RT/b = 0.05$) the boundary concentrations of the miscibility gap in the tetragonal domain may well be 0.01 or 0.02 in error. The dashed second-order transformation curves enclosing the orthorhombic domain are the same as in Fig. 5, but only the top parts are stable. The *insert* in the upper right-hand corner shows qualitatively the situation around the "sharp" critical point at C. The transformation curve orthorhombic-tetragonal ($c/a < 1$) is tangent to one boundary of the miscibility gap, in C. The other boundary shows a pronounced kink at A, where it is nontangent to the transformation curve orthorhombic-tetragonal ($c/a > 1$). Between the tie line AB and C the two-phase region is orthorhombic + tetragonal ($c/a < 1$).

phases with c/a smaller than one and body-centered orthorhombic phases are confined to systems of V, Nb, and Ta with O and H (see Table 1).

In the system V-O the tetragonal phase with $c/a > 1$ is stable, and in V-H probably also. Hexagonal and NaCl structures also appear, for example, in V-O. They oppose apparently the occurrence of body-centered noncubic phases in the systems of V, Nb, and Ta with N and C. This does not mean that theory and experiments are at variance. Even the fact that in Nb-O, Ta-O, and Nb-H the body-centered tetragonal phase with $c/a > 1$ has not been found is not necessarily in conflict with the strain-order theory, as the other body-centered noncubic phases do not seem to have been obtained in (metastable) equilibrium with the bcc phase.

However, in Nb-O and Ta-O the tetragonal phase (with $c/a < 1$)

appears to have a smaller O concentration than the orthorhombic phase, which is unexpected from a theoretical point of view. One of the shortcomings of the theory is the neglect of superstructure formation.

There is another difficulty with the orthorhombic phases in Nb-H and Ta-H. Libowitz[13] describes these as having unit-cell axes differing by a few percent only, as one would expect, from unequal occupations of x, y, and z sites. But the unit cells of these phases are mostly given with a and b being about $\sqrt{2}$ times as large; two of the three original cube edges have equal length, but they are not exactly perpendicular to each other.[14] Perhaps occupation of tetrahedral and octahedral sites by H is responsible for this anomaly.

Finally we mention an interesting system of a bcc non-transition metal with H. In the system Ba-H[15] the melting point of Ba is raised by H. The solid solubility—in equilibrium with the saline hydride BaH_2—increases with temperature and attains the composition $BaH_{1.35}$ where apparently the structure is still body-centered cubic. This is presumably due to the large lattice parameter of Ba.

TABLE 1 ***Occurrence of Body-centered Tetragonal and Orthorhombic Phases in the Binary Systems of* V, Nb, *and* Ta, *with* O *and* H**

Host metal	Interstitial element					
	Oxygen			Hydrogen		
	Tetragonal ($c/a>1$)	Ortho-rhombic	Tetragonal ($c/a<1$)	Tetragonal ($c/a>1$)	Ortho-rhombic	Tetragonal ($c/a<1$)
V	1	2	2	3	—	—
Nb	—	4	5	—	3	—
Ta	—	5, 6	5	3	3	—

References for Table 1

1. Seybolt, A. U., and H. T. Sumsion: *Trans. AIME* **197**:292 (1953).
2. Westman, S.: *Acta Chem. Scand.* **17**:749 (1963).
3. Libowitz, G. G.: "The Solid-state Chemistry of Binary Metal Hydrides," p. 36, W. A. Benjamin, Inc., New York, 1965.
4. Steeb, S., and J. Renner: *Z. Metallk.* **56**:531 (1965).
5. Brauer, G., and H. Müller: *Mém. Chim. Minérale*, p. 63, 16th Congress of IUPAC, Paris, 1957.
6. Schönberg, N.: *Acta Chem. Scand.* **8**:240 (1954).

REFERENCES

1. Snoek, J. L.: *Physica* **8**:711 (1941).
2. Zener, C.: *Trans. AIME* **167**:550 (1946).
3. Zener, C.: "Elasticity and Anelasticity of Metals," University of Chicago Press, Chicago, 1948.
4. Shockley, W.: *J. Chem. Phys.* **6**:130 (1938).
5. Bean, C. P., and D. S. Rodbell: *Phys. Rev.* **126**:104 (1962).
6. Meijering, J. L.: *Philips Res. Repts.* **6**:183 (1951).
7. Landau, L. D.: *Phys. Z. Sowjet.* **11**:26 (1937).
8. Meijering, J. L.: *Philips Res. Repts.* **18**:318 (1963).
9. Rostoker, W., and A. Yamamoto: *Trans. ASM* **46**:1136 (1954).
10. Gebhardt, E., H. D. Seghezzi, and W. Dürrschnabel: *Plansee Proc. 3rd Seminar*, p. 291, 1958.
11. Seybolt, A. U., and H. T. Sumsion: *Trans AIME* **197**:292 (1953).
12. Westman, S.: *Acta Chem. Scand.* **17**:749 (1963).
13. Libowitz, G. G.: "The Solid-state Chemistry of Binary Metal Hydrides," p. 36, W. A. Benjamin, Inc., New York, 1965.
14. Brauer, G., and R. Hermann: *Z. Anorg. Chem.* **274**:11 (1953).
15. Peterson, D. T., and M. Indig: *J. Am. Chem. Soc.* **82**:5645 (1960).

DISCUSSION *on the Paper Presented by J. L. Meijering*

FRIEDEL: In the low concentration range, I wonder whether the uniform mode of coupling between impurities that you describe is not superseded by the kind of "frozen randomness" observed by Känzig for dilute O molecules in ionic solids and explained by Brout along lines similar to those used for magnetic coupling of dilute CuMn alloys. This is to be expected from the fact that the coupling between two interstitials is complex in nature, so that if they are especially near to each other in the (random) solid solution, their relative orientation has very little chance to be along the uniform mode you assume.

There are several features that would distinguish this mode of coupling:

1. A broader peak in specific heat around the coupling temperature, with a high temperature tail in c^2T^{-2}.

2. A coupling temperature increasing proportionately with concentration, but with a slope not necessarily equal to that for the uniform mode at higher concentrations.

3. A broadening of the internal friction frequency (proportional, I think, to concentration).

4. A remanent distortion definitely less than the permanent distortion produced by an applied stress.

MEIJERING: The use of one long-range interaction parameter is crude, even with the exclusion of occupation of consecutive sites on a row. However, the introduction of more parameters with different ratios for different systems would prohibit the calculation of an overall diagram.

DEHLINGER: Perhaps one could clarify the picture by using elastic dipoles in analogy to magnetic dipoles. They form stresses both by themselves and by their interaction.

MEIJERING: I think Professor Zener could say something on that.

ZENER: I concur with Professor Dehlinger's opinion that the elastic interaction between interstitial atoms in bcc metals must be treated as the interaction of magnetic dipoles. Such a treatment might invalidate many of the conclusions of Professor Meijering, which conclusions were based upon the concept of isotropic lattice distortion.

LOMER: The behavior of the two tetragonal phases and the orthorhombic phase is very similar to AuMn magnetic effects. The magnetic superlattice produces axial distortions of the lattice analogous to the ordered interstitial, and the entropy of reorienting spin may not be so different from that for reordering the C interstitials. Do the phase boundaries in fact take exactly the form sketched by Professor Meijering?

MEIJERING: I don't know exactly the phase boundaries in the Au-Mn system.

RUDMAN: A tetragonally distorted CsCl structure occurs in some transition-metal alloys. It has just occurred to me that the origin of this tetragonality might be due to an analogue in substitutional solutions of the effect that Dr. Meijering has described in interstitial solutions: in deviating from stoichiometry, the extra atoms may align in some preferential crystallographic direction.

BECK: Tetragonally distorted CsCl-type phases were found in the V-Ru, Nb-Ru, and Ta-Ru systems at the equiatomic composition.* Evidence was obtained by A. Dwight for the martensitic nature of the tetragonal phase, suggesting strongly that at high temperatures the ordinary cubic CsCl phase is stable, and that the tetragonal distortion takes place on cooling to room temperature. At compositions somewhat off the equiatomic, the cubic CsCl phase is retained on cooling. In some cases a second distortion also occurs in addition to the one mentioned above, so that a slightly orthorhombic structure results. A. Dwight found recently that in ternary alloys the boundaries separating the compositions of the tetragonal structure from those of the cubic and the orthorhombically distorted structures extend in a direction of constant electron concentration.

RUDMAN: In view of Professor Beck's answer that the tetragonality occurs primarily at stoichiometry, my proposed mechanism does not appear to be operative.

BREWER: Raub and Fritzsche† and Raub, Beeshow, and Fritzsche‡ have shown, as I recall, that both bcc Nb and Ta become tetragonally distorted upon addition of Ru before the stoichiometric CsCl phase is reached.

* Greenfield and Beck, *Trans. AIME*, 1956.

† E. Raub and W. Fritzsche, *Z. Metallk.* **54**:21, 317 (1963).

‡ E. Raub, H. Beeshow, and W. Fritzsche, *Z. Metallk.* **54**:451 (1963).

THE SOLUTION OF INTERSTITIAL ELEMENTS IN THE BODY-CENTERED CUBIC EARLY TRANSITION METALS*

A. D. McQuillan

Department of Physical Metallurgy, University of Birmingham, Birmingham, England

The primary purpose of the protracted program of research of which the work described in this paper forms a part was to elucidate the reasons for the formation of extensive ranges of solid solubility of interstitial nonmetallic elements in the body-centered cubic early transition metals. The ability of these metals to dissolve large quantities of H and comparatively large amounts of O, N, and C is a characteristic of the metals which marks them off from all other metallic elements. The problem has historically received much attention both experimentally and theoretically since its inception by Sieverts in the 1930s. A number of theoretical treatments of interstitial solutions have been published, notably by Lacher,[1] Scheil,[2] Speiser and Spretnak,[3] Moon,[4] and Hoch.[5] All treatments previously

* The author would like to acknowledge the considerable assistance obtained from Miss V. W. Russell and A. D. Wallbank who obtained the experimental results which form the background of the paper.

published have been of a fairly generalized nature, and all have contained simplifying assumptions. The greater part of the present treatment is based on the special case of H solutions in the body-centered cubic form of the metals.

Briefly the reasons for choosing H were that it has a simple electronic structure, that it forms the more extensive ranges of interstitial solid solutions in the particular metals, and that the thermodynamic functions of both the gas and the solutions are such that they can be readily elucidated experimentally by measurement of the H equilibrium pressure of the metallic solutions. It is not appropriate to consider here either the experimental methods or results; these will be published elsewhere. They involve the measurement of equilibrium pressures over a wide pressure range extending from 10^{-4} torr upward, and they include the determination of many physical properties of the solutions. A brief description of the treatment of the experimental results relating to equilibrium H pressures will, however, be given because this will have some bearing on the theoretical interpretation of the results.

THERMODYNAMIC ANALYSIS OF EXPERIMENTAL RESULTS

Experimental determinations yield information on the relationships between concentration of dissolved H, temperature, and the pressure of gaseous H in equilibrium with dissolved H. The work is always repeated using deuterium in place of H.

From the known thermodynamic properties of gaseous H and deuterium, it is possible to obtain the partial free energy of the dissolved gases based on a common origin which is taken to be a hypothetical gas of molecular H at absolute zero for both H and deuterium solutions. Thus a direct comparison of the results with H and deuterium can be made. The basic relationships between the thermodynamic quantities are

$$\mu_{\mathrm{H}} = \frac{1}{2}\mu_0 + \frac{RT}{2}\ln\frac{p}{p_0} = \bar{H}_{\mathrm{H}} - T\bar{S}_{\mathrm{H}}$$

where μ_{H} is the partial free energy of H (or deuterium) in solution in the metal, μ_0 is the standard free energy of H at pressure p_0, and p is the equilibrium pressure of H at temperature T. $\bar{H}_{\mathrm{H}}$ and $\bar{S}_{\mathrm{H}}$ are respectively the partial enthalpy and entropy of the dissolved H. Given the quantity μ_{H} derived from experimental results, $\bar{H}_{\mathrm{H}}$ and $\bar{S}_{\mathrm{H}}$ can readily be obtained. It is in the analysis of these quantities that assumptions have to be made.

In very dilute solutions containing less than 1 a/o H or deuterium, it has been assumed that the following statements are appropriate.

1. The molecules of the interstitial element dissociate on entering into solution in the metal.

2. The interstitial elements occupy the tetrahedral interstitial sites in the body-centered cubic metal, of which there are six per metal atom.

3. The interstitial elements are randomly distributed among the available sites in the metal. Later it will be necessary to define the number of sites available.

4. The expansion of the metal with changes in temperature and H concentration is negligibly small.

5. The contribution to the partial free energy of the thermal vibration of the interstitial atoms can be obtained by treating the very light atoms of H and deuterium as a set of three-dimensional Einstein oscillators. At our present state of knowledge it is not possible to include terms which relate to the thermal vibration of the metal atoms.

6. The behavior of H and deuterium solutions differs only in that the two types of atom have different masses and hence different vibrational frequencies.

Using the assumptions given above, the partial free energy of dissolved interstitial element per gram atom as a function of temperature T and composition x expressed in terms of the ratio of interstitial to metal atoms is

$$\mu = E + \tfrac{3}{2}Nh\nu + 3RT \ln\left[1 - \exp\left(-\frac{h\nu}{kT}\right)\right] + RT \ln \frac{x}{6}$$

In this equation ν is the vibrational frequency of the interstitial atoms in a potential well of depth E, N is Avogadro's number, and R is the gas

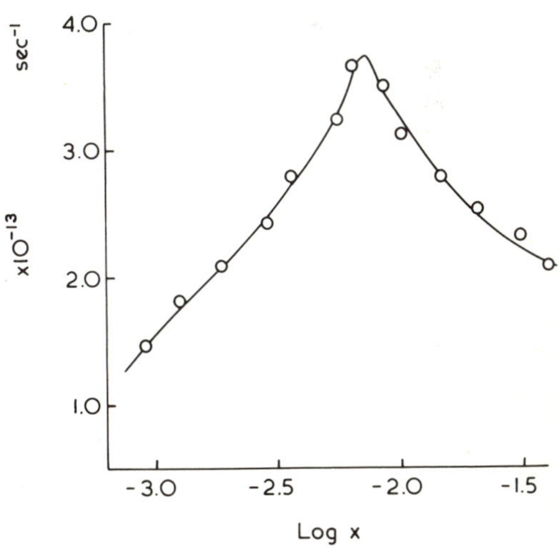

Fig. 1 The vibration frequency ν of H in Ti at 917.3°C as a function of the H concentration x, where x is expressed in atomic ratio of H atoms per metal atom.

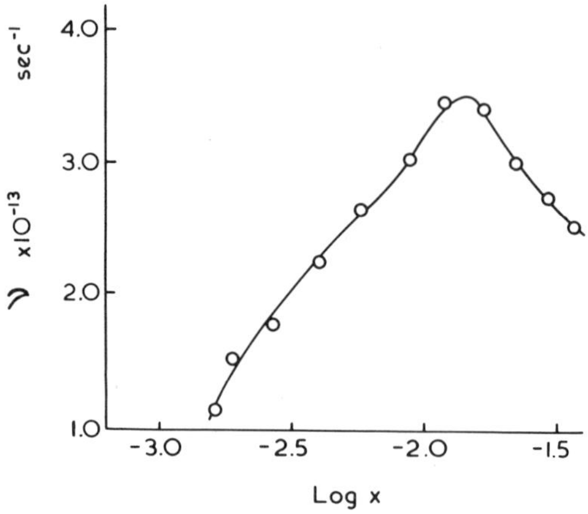

Fig. 2 The vibration frequency ν of H in Zr at 917.3°C as a function of H composition x.

constant. In previous treatments of the problem of interstitial solutions it has been assumed that E and ν were not functions of x or T. Our experimental results indicate quite clearly that this assumption is erroneous.

The vibrational frequency ν can be obtained by comparing the experimentally derived partial free energies of H and deuterium at equivalent concentrations and temperatures, on the assumption that the terms E and $RT \ln (x/6)$ in the above equation do not change. Once the vibrational frequencies of H and deuterium, which because of the different masses will be in the ratio $\sqrt{2}:1$, are known, E may be calculated. As would be expected the greater the depth of the potential well in which the interstitial atom finds itself, the greater is the vibrational frequency. The frequencies

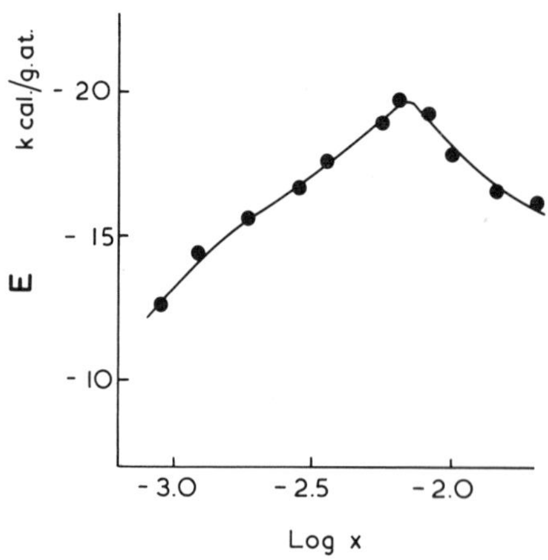

Fig. 3 The bonding energy E of H in Ti at 917.3°C as a function of H concentration x.

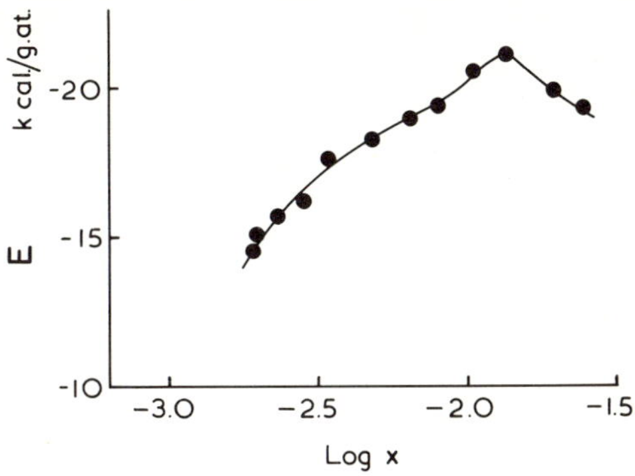

Fig. 4 The bonding energy E of H in Zr at 917.3°C as a function of H concentration x.

are found to lie in the range 1 to 4 × 10^{-13} sec^{-1}, whereas the bonding energy E, which is the difference between the bonding energy of an H atom in a molecule of H at 0°K, and its energy in the dissolved state, is of the order of 10 kcal/g atom. The experimental findings will be discussed first in terms of H solutions in β-Ti and β-Zr, and second in terms of H-Nb solutions where the absence of a phase transformation to a hexagonal close-packed structure allows results to be obtained over a much wider range of H concentrations and temperatures.

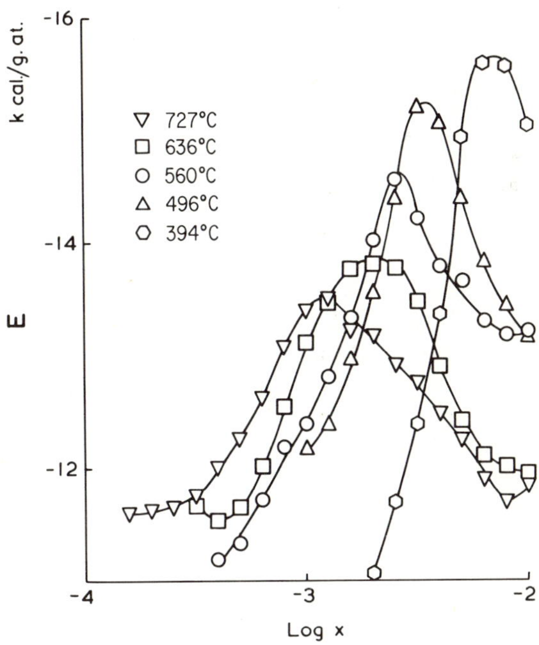

Fig. 5 The bonding energy E of H in Nb at various temperatures as a function of H concentration x.

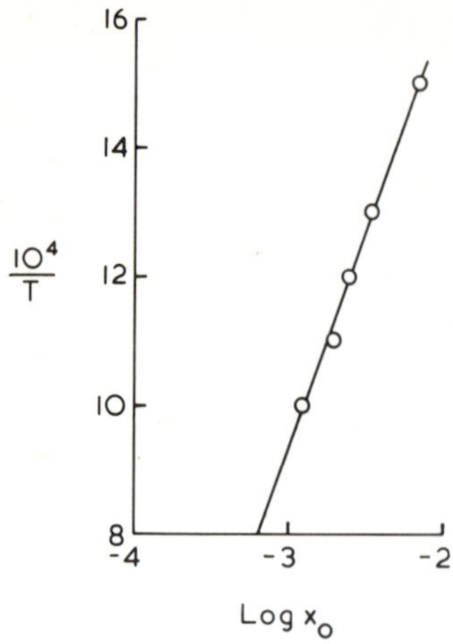

Fig. 6 Curve of log x_0 against $1/T$ (T in °K) where x_0 is the composition at which the isothermal curves in Fig. 5 exhibit a maximum value of E.

Ti and Zr Solutions

The existence of the phase transformation in both metals restricts investigation of dilute H solutions to the temperature range above about 900°C, and experimental difficulties prohibit determination above 1000°C. In the following, isothermal curves at 917.3°C will be presented and discussed, and we shall be concerned only with concentration-dependent effects. Figures 1 and 2 show the change of vibrational frequency of dissolved H as a function of H concentration in the body-centered cubic forms of Ti and Zr, and Figs. 3 and 4 the bonding energies of H in the metals obtained by subtracting the partial configurational entropy and partial vibrational free energy of H from the experimentally derived partial free energy of H. All curves rise to a peak in the neighborhood of $x = 0.01$, and also there is a notable decrease in E at lower H concentrations, with no tendency for E to attain a constant value with decreasing x. Any theory of the bonding forces between dissolved H atoms and the metals must be capable of explaining this unexpected effect.

Nb Solutions

In the Nb-H system it is possible to obtain experimental results over a range of temperatures from 1000 to 250°C, and hence the effect of temperature on the position of the maximum, which also appears in the curves of vibrational frequency and bonding energy as a function of H concentration in Nb, can be followed. As shown in Fig. 5, the maximum occurs at higher H concentrations, the lower the temperature, and furthermore Fig. 6 would seem to indicate that we are dealing with a thermally activated process.

THE BONDING ENERGY OF DISSOLVED H

We will consider first the way in which the bonding energy of H varies with concentration at high temperatures in very dilute solutions of H in Ti and Zr before reaching the critical composition x_0 at which the maximum value of E occurs. The first few H atoms would appear to be but weakly bonded to the metal. The quantum-mechanical problem has been briefly treated by Isenberg[6] in general, and he concludes that H should always be completely ionized when in solution in a metal. In terms of this assumption, the dissolution of an H atom in a metal can be envisaged as the ionization of a free atom and the introduction of the positive ion into an interstitial site, followed by the introduction of the electron into the conduction band of the metal. The presence of interstitial positive charges in the metal will affect the periodicity of the lattice, but in the absence of pronounced Brillouin zone effects, this should, besides causing minor distortion in the shape of the density-of-states curve for conduction electrons, have the effect of decreasing the energy of the conduction electrons as a direct result of the added positive charge. The subsequent introduction of electrons to restore electrical neutrality will increase the mean energy of the electrons by an amount that will depend on the density of states at the Fermi level. A high density of states at the Fermi level should, therefore, lead to strong bonding of dissolved H, and a low density to the opposite effect.

The density of states at the Fermi level of the body-centered cubic early transition metals as a function of electron-atom ratio, based on the results of electron specific heat (Cheng et al.,[7] Morin and Maita[8]), magnetic susceptibility (Jones and McQuillan[9]), and superconducting transition temperature (Hume and Blaugher[10]), is known, and has the form shown in Fig. 7. Both Ti and Zr, for which the e/a ratio is 4, have a low density of states, and hence it would be expected that the bonding energy of H would be low. The density of states, however, rises rapidly with increasing

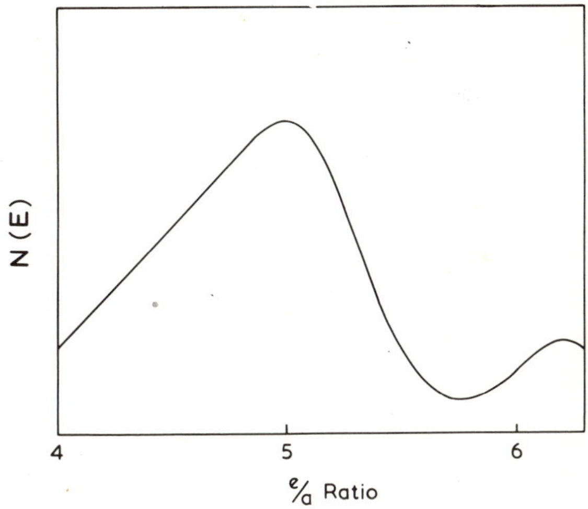

Fig. 7 Generalized form of the density of states at the Fermi level $N(F)$ as a function of the electron-atom ratio e/a for the body-centered cubic early transition metals.

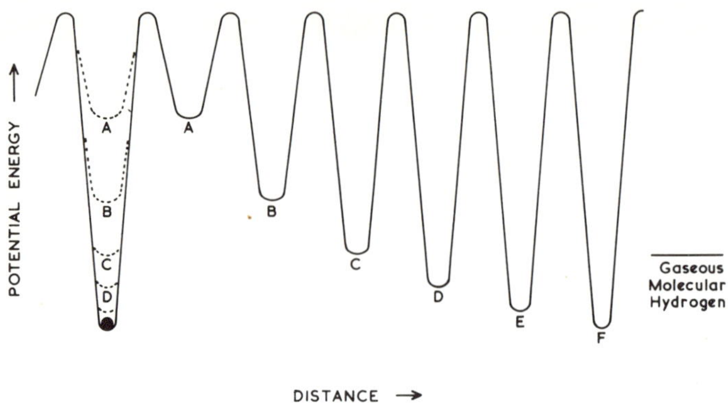

Fig. 8 Diagrammatic representation of the potential energy along a series of interstices adjacent to an interstice occupied by an H atom.

e/a ratio, and since we are envisaging H contributing electrons to the conduction band, the bonding energy of H would increase with increasing H concentration. This is in accord with the experimental results for Ti and Zr solutions, but does not explain the existence of the maximum in the curve of H bonding energy as a function of concentration. In the present picture a maximum could only arise if there were to be a strong repulsive interaction between H atoms which makes its appearance when the concentration of H reaches a magnitude where H atoms begin to approach within a few atomic spacings of one another. At a composition of 1 a/o, the mean separation of H atoms is about five atomic spacings, and hence there is a finite possibility of fairly large numbers of H atoms approaching one another to within closer distances. Evidence that such a repulsive interaction exists can be obtained from the fact that although there are six tetrahedral interstitial sites per metal atom, no H solution in the body-centered cubic form of early transition metals has ever been found to exceed the composition equivalent to M_5H_6, where M represents the metal. From this it would appear that each H atom effectively occupies five interstitial sites. Since each tetrahedral interstice in a body-centered cubic structure is surrounded by four other sites at a distance of $\alpha/2\sqrt{2}$, where α is the lattice parameter of the metal, this assumption is not unreasonable, and indeed has been used in the analysis of Speiser and Spretnak[3] and all subsequent analyses.

A calculation based on this simple assumption indicates quite clearly that the maximum in the Ex curves occurs at too low a value to be explained in this way. A more realistic picture is obtained if it is assumed that the energy of a pair of H atoms as they approach each other in the metal varies in the manner shown in Fig. 8. The broken lines within the potential well on the extreme left represent the successive changes in shape and depth of this well when a second H atom approaches from the right. The energy of an H atom in gaseous molecular H is given, for comparison, on the right-hand side of the diagram.

Moon[8] has considered this problem in a simplified form, and we have attempted a more detailed calculation of the configurational entropy of dissolved H atoms based on the actual spatial arrangement of tetrahedral sites in the body-centered cubic lattice shown in Fig. 9. It is not possible to give details of the calculation here, but the lines along which it was developed will be indicated. It was first assumed that the repulsive interaction decreased rapidly to negligibly small values as the separation of H atoms increased, and hence it was assumed that an H atom could not approach another H atom within a certain distance; this distance will include θ sites. If there are m metal atoms, the first H atom can be arranged in $6m$ ways; the second in $(6m - \theta)$ ways; but the third can be arranged in less than $(6m - 2\theta)$ ways since in certain configurations of the two previous H atoms the regions of forbidden sites around the two atoms overlap and there will be less than 2θ forbidden sites. Thus the number of ways of arranging the third atom can be expressed as $(6m - 2\theta')$ where θ' is very slightly smaller than θ. As more H atoms are considered, the effective θ' will decrease continuously, and very rapidly since not only will the number of closely separated pairs of H atoms increase but also triplets at various mutual separations will arise. At this stage the calculation becomes impossibly complex even for a computer program. In dilute solutions the problem is just tractable provided θ is known. To obtain a value for θ a reexamination of the thermodynamic experimental results was undertaken by abandoning the assumption that the H solutions were perfectly random, and obtaining the vibrational frequencies of the H and deuterium atoms from the temperature variation of the partial entropy of H and deuterium,

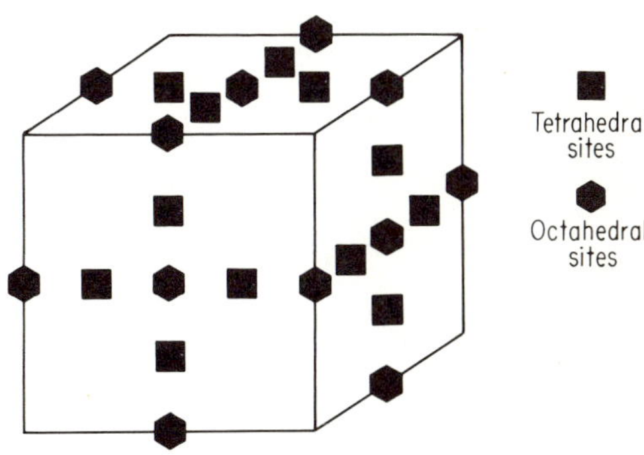

Fig. 9 Unit cell of a body-centered cubic lattice showing the positions on the surface of the cell of the interstitial sites. There are no sites within the body of the cell.

which should only depend on the change of the partial vibrational entropy $\bar{S}_v$ of the form

$$\bar{S}_v = 3R\left[\ln\left(1 - \exp\left(-h\nu/kT\right)\right) - \frac{h\nu}{kT}\frac{\exp\left(-h\nu/kT\right)}{1 - \exp\left(-h\nu/kT\right)}\right]$$

When this was done, it was apparent that the solutions departed from ideal-solution behavior at the concentration x_0 for which E was a maximum. This is illustrated in Figs. 10 and 11. In Fig. 10 the frequencies ν_H of H and $\sqrt{2}\nu_D$ of deuterium as a function of concentration in Ti are seen to be equal at all compositions up to x_0 but diverge thereafter. The bonding energy E as a function of concentration in Fig. 11 has the identical values for H and deuterium at all concentrations even when x exceeds x_0. The same effects were observed in Zr. The divergence of ν_H and $\sqrt{2}\nu_D$ for $x > x_0$ is to be expected if repulsive interactions between dissolved atoms are occurring, since all interstitial atoms will not have the same vibration frequency because they will not all have the same bonding energy. Those close together will have lower bonding energies and hence lower vibration frequencies, and the collection of atoms can no longer be considered as having a single vibrational frequency. The analysis for $x > x_0$ yields only an effective frequency, but even so this gives rise to the same values of E for H and deuterium.

The values of x_0 differ between Ti and Zr solutions, x_0 being smaller for Ti. This can be attributed to the lattice parameter of Zr being greater than that of Ti. Two H atoms in equivalent relative sites in the two

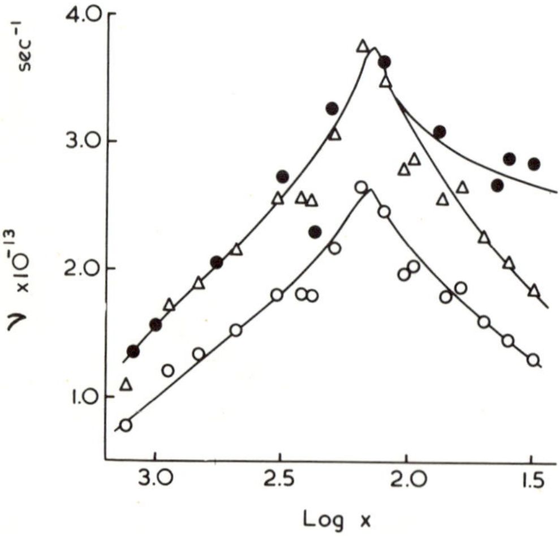

Fig. 10 Vibrational frequency ν_H of H ● and ν_D of deuterium ○ as a function of composition when in solution in Ti. The values of $\sqrt{2}\nu_D$ are denoted by △.

Fig. 11 Bonding energy E of hydrogen ● and deuterium ○ as a function of composition when in solution in Ti. To obtain this curve we used the values of ν given in Fig. 10 together with a correction dependent on the effect of departure from ideal solution on the partial configurational entropy when $x_0 > x$ have been used.

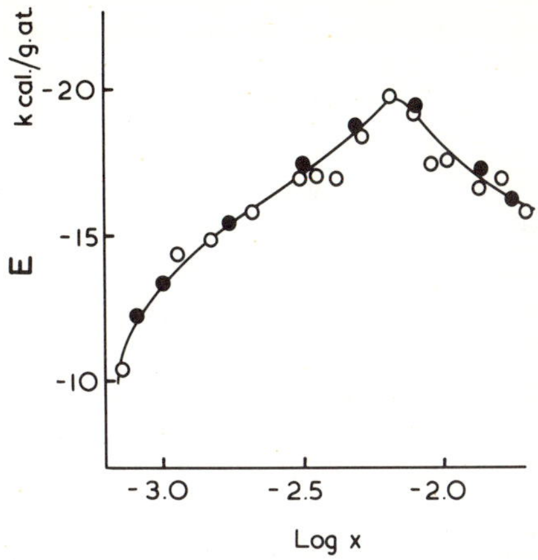

metals will be physically further apart in Zr, and hence the repulsive interaction will be smaller in magnitude in Zr than Ti. Once it is known that repulsive interactions become effective at x_0, the number of sites occupied by an H atom can be deduced. The number turns out to be 19, and the closest separation of H or deuterium atoms in Ti and Zr can be seen by examination of the spatial arrangements of interstitial sites around any given site to be $(\sqrt{10}/4)\alpha$, where α is the lattice parameter of the metal lattice. There are eight such sites around any given site, and these, as well as all sites at closer separations, lie so that a line joining the central site to any of the sites within a distance of $(\sqrt{10}/4)\alpha$ does not pass through a metal atom, considering the metal atom as a sphere of the accepted atomic size. All sites at greater separations have metal atoms lying between their line of centers.

There are two possible mechanisms for the repulsive interaction of adjacent H atoms. It is possible that the presence of an H atom distorts the spatial arrangement of the surrounding metal atoms, which in turn brings about a decrease in the bonding energy of H atoms in adjacent sites. The apparent existence of a rule that H atoms cannot occupy pairs of sites where the line of centers does not pass through a metal atom suggests strongly, however, that the repulsive interaction is more directly electronic in origin than the displacement of metal atoms. This is especially so in H solutions where the interstitial atom is very small, if indeed it can be said to have any meaningful size. In the case of larger interstitial atoms such as O or C, the local distortion of the metal may well play an important role.

In the simple model of an H solution in which the electron enters the conduction band, the charge on the positively charged H ion must be electrically shielded by a local increase in the spatial density of electrons. The local increase in amplitude of the wave function of conduction electrons caused by the excess positive charge is then responsible for the screening of this charge. It is also possible to consider the positively charged interstitial ion as giving rise to a virtual bound state. In the absence of any detailed knowledge of the electronic structure of the body-centered cubic forms of Ti and Zr, speculation on these two possibilities might be considered unproductive, were it not that the maximum in the Ex curve in dilute solutions of H in Nb depends markedly on temperature. If the origin of the repulsive interaction is caused by the interaction of quasi-localized electrons from adjacent interstitial atoms, the temperature dependence of this quantity would seem to indicate that the effective number of sites occupied by an H atom decreases with decreasing temperature; i.e. the repulsive interaction becomes shorter in range at lower temperatures, and hence the onset of the maximum is delayed until higher concentrations.

A mechanism whereby this effect could be achieved can be visualized in terms of a series of localized electron states in the neighborhood of an interstitial H atom lying energetically below the Fermi energy of the metal. Increasing temperature would then be expected to lead to the excitation of the bound electron to higher-energy states, where, being less closely bound to the interstitial atom, it would extend the range of the repulsive interaction between adjacent H atoms.

This model of the state of a bound electron around an H ion in Nb is supported by two other aspects of the behavior of the H solutions in Nb. The first is that although Nb metal has a high density of states at the Fermi level, the bonding energy of H at very low H concentrations is not markedly higher than in Ti and Zr. Furthermore, since Nb has an electron-atom ratio which places it at the maximum of the density of states against the e/a ratio curve shown in Fig. 7, it does not exhibit a constant or slowly decreasing H bonding energy with increasing H concentrations, as would be expected if the electron from the H atom entered the conduction band. Second, there is a miscibility gap in the Nb-H system below 300°C, and at room temperature this has widened until there is only a negligibly small primary solubility of H in Nb which can coexist with a body-centered tetragonal concentrated solution containing about 40 a/o H. If at low temperatures there are localized states associated with interstitial H ions, then at high H concentrations when the H atoms are close together, it would be possible for these states to form a band of electron states lying below the Fermi level which could further enhance the stability of H solutions. Unfortunately the results on the Nb-H system have only just been obtained, and in the absence of time for more detailed calculations it would not be productive to speculate further on the feasibility of this model. The existence of pronounced localized states associated with interstitial H atoms cannot be established experimentally at room temperature since dilute H solutions in the body-centered cubic metals are not stable at room temperature.

THE MAGNETIC SUSCEPTIBILITY OF Zr DIHYDRIDE

The only H solutions stable at low temperatures in the early transition metals contain large concentrations of H. In an attempt to find evidence in support of the idea of a band of semilocalized electrons associated with interstitial H atoms, the magnetic susceptibility of various saturated hydrides was measured. The only one to exhibit anomalous magnetic behavior was Zr dihydride. This has a face-centered cubic structure for the metal atoms, and because of this there are only two tetrahedral sites per metal atom. These are more widely spaced than the equivalent sites in a body-centered cubic structure, and they are, of course, filled in the dihydride.

The magnetic susceptibility of Zr hydride is shown in Fig. 12. Below 300°C there is a marked increase in susceptibility, which appears to reach a saturated value just below room temperature. This low-temperature susceptibility is an order of magnitude greater than the high-temperature value. The susceptibility is not dependent on the magnetic field strength, nor is the curve in any way related to a Curie-Weiss curve. Decrease in H concentration does not change the shape of the curve but only the extent of the excess susceptibility at low temperatures. The enormous

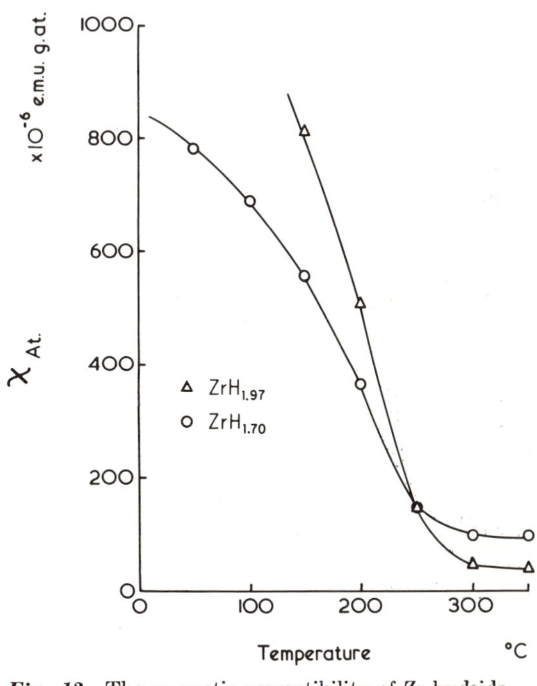

Fig. 12 The magnetic susceptibility of Zr hydride as a function of temperature.

increase in paramagnetic susceptibility of the hydride phase could possibly be explained in terms of the formation of an electron band based on an ordered array of H atoms in a face-centered cubic lattice at low temperatures, in such a way that the energy band happened to lie at the Fermi level of the conduction band of the metal. The greatly enhanced density of states at the Fermi level would then be responsible for the high susceptibility at low temperatures.

INVESTIGATION OF Ti–O–H SOLUTIONS

So far in this paper, attention has been concentrated solely on H solutions in the early transition metals, and it could be argued that such solutions are hardly typical of interstitial solutions in the metals, since H would, as postulated here, be likely to form a positive ion, whereas O, N, and C would tend to form negative ions. It is possible to dissolve H in Ti which already contains dissolved O, and measurement of H equilibrium pressures indicates quite clearly that the pressures are greater at equivalent temperatures and H concentrations in O-containing metal than in the pure metal. This immediately suggests that there is no association between dissolved H and O, but rather that the contrary applies.

To investigate this problem, a series of Ti-O alloys were prepared containing a few atomic percent of O. These were used to obtain equilibrium H pressure data for dilute solutions of H of exactly the same nature as described earlier in this paper. It was found that the bonding energy of H at any particular H concentration had not changed because of the presence of O. The H solutions at high dilution were still ideal, but the partial free energy of H was greater at each concentration of H by an amount that increased with the O content of the metal, and that was greater the higher the temperature of the metal. It was apparent that the only term in the expression for the partial free energy of H that could give rise to this effect was the configurational entropy term $RT \ln (x/6)$.

The results were analyzed in terms of a repulsive interaction between dissolved O and H atoms, where it was assumed that an O atom in a large octahedral site keeps H from entering a certain number of ϕ sites adjacent to it. This has the effect of reducing the number of tetrahedral sites available to the H atoms by the product of ϕ and the number of O atoms. A very good agreement between the experimental results was obtained if ϕ was assumed to have the numerical value 11.8. Now each octahedral site in a body-centered cubic structure, as may be seen from Fig. 9, has four tetrahedral sites at a distance of $\alpha/4$, and eight further sites at $(\sqrt{5}/4)\alpha$. Thus there are 12 tetrahedral sites close to an octahedral site, and since the experimentally derived value of ϕ was 11.8, it seems reasonable to assume that these are the 12 sites that cannot be occupied by H if an O atom is occupying the central octahedral site. Once more the spatial relationship between the octahedral site and the forbidden tetrahedral sites is such that the line of centers does not pass through a metal atom. It would appear, therefore, that the repulsive interaction between O and H obeys the same law as between two H atoms. Thus if it is assumed that H forms a positive ion on entering into solution in body-centered cubic Ti, it must also follow that O does likewise.

The idea that O in solution in Ti could form a positive ion is somewhat surprising. However, an O atom which has lost an electron will have a somewhat disklike shape, which is ideal for fitting into the octahedral site of a body-centered cubic metal. This site has four coplanar metal atoms equispaced from the center of the site, and two others in the perpendicular direction which are considerably closer. It would not be expected that this type of bonding would lead to a high bonding energy for O in the metal, and indeed the solubility of O in body-centered cubic Ti is somewhat restricted, except at very high temperatures where the solution gains extra stability from the configuration entropy.

CONCLUSION

The major conclusion to be drawn from the present paper is that, with interstitial solutions of H in the body-centered cubic form of the early transition metals, it is extremely dangerous to assume that one is dealing with ideal solutions having a constant bonding energy in dilute solutions containing less than 1 a/o H. This may well also be true for other interstitial solid solutions.

REFERENCES

1. Lacher, J. R.: *Proc. Roy. Soc.* **A161**:525 (1937).
2. Scheil, E.: *Arch. Eisenhuettenw.* **22**:37 (1951).
3. Speiser, R., and J. W. Spretnak: *Trans. ASM* **47**:493 (1955).
4. Moon, K. A.: *Trans. AIME* **227**:116 (1963).
5. Hoch, M.: *Trans. AIME* **230**:238 (1964).
6. Isenberg, I.: *Phys. Rev.* **79**:736 (1950).
7. Cheng, C. H., C. T. Wei, and P. A. Beck: *Phys. Rev.* **120**:426 (1960).
8. Morin, F. J., and J. P. Maita: *Phys. Rev.* **129**:1115 (1963).
9. Jones, D. W., and A. D. McQuillan: *J. Phys. Chem. Solids* **23**:1441 (1962).
10. Hume, J. K., and R. D. Blaugher: *Phys. Rev.* **123**:1569 (1961).

DISCUSSION *on the Paper Presented by A. D. McQuillan*

LOMER: Presumably any modification of the thermal entropy of the matrix is included in your definition of the "vibrational frequency" of the H atoms. They could be separated by neutron inelastic scattering experiments.

The structural distribution of H atoms over octahedral to tetrahedral sites should also be possible by neutron work.

McQUILLAN: I agree entirely with your suggestions, and I would be very happy if we could collaborate in carrying out such experiments.

McLELLAN: In calculating ν from the entropies you have to neglect the effect of H on the frequency spectrum of the solvent metal lattice. This may be a small effect but since ν is given by

$$\nu = \frac{kT}{h} \exp\left(1 - \frac{S}{3k}\right)$$

for a classical assembly of Einstein oscillators, a small error in S could have a large effect on your ν values. Do you have any comment on this?

McQUILLAN: The reason why I have some faith in the corrections of my values for the vibrational frequency of the H in solution in Ti and Zr is that the same values of E as a function of concentration are obtained when the thermal entropy is computed independently from the experimental data from H and deuterium solutions. This is shown in Fig. 11.

MEIJERING: With regard to the possibility of having a low entropy without long-range order I would like to point out that apparently glasses may obey Nernst's theorem. This can be understood with the following model. Suppose one presses a stack of close-packed spheres on a floor which is not flat but irregularly wrinkled. If the number of spheres is $n \times n \times n$, the entropy will be of the order n^2k and per mole it goes to zero for large n.

WALLACE: Our work with H dissolved in bcc Ta has always been interpreted as indicating attractive interactions between two solute particles. You postulate repulsive interactions in bcc Ti and Zr. It is well known that dilute solutions of H in Ti and Zr, upon cooling, exhibit phase segregation in which an H-rich phase precipitates out. How do you reconcile this with an assumed repulsive potential?

McQUILLAN: If there is attractive interaction between H atoms in solution in Ta, then, since there are six tetrahedral interstices per metal atom, I fail to see why the saturation composition of such solutions does not correspond to the formula TaH_6. I believe that the saturation value of H in the early transition metals is determined by the filling of the d sub-band at an electron-per-atom ratio of about 5.8 where the bonding energy becomes small and is no longer able to compensate for the repulsive interaction between dissolved H atoms.

ORIANI: A comment on the increasing attractive "electronic" interaction with increasing H concentration: the explanation offered by the author is the opposite of that generally expected, and offered for example by Carl Wagner, that H solubility should decrease as the Fermi energy rises.

A comment on the author's treatment of the partial molal entropy of vibration as assigned completely to the H ions as Einstein oscillators: remembering that the partial molal volume of H in these solutions is about 1 to 5 cm^2/g atom, the immediate neighborhood of the solute must be very distorted. Therefore, it would seem that the metal ions near the solute must also have their vibrational characteristics modified.

McQUILLAN: I would be glad if Dr. Oriani could explain our results in terms of the Wagner theory. So far I have been unable to do this.

I am very aware that I have not included the change of vibrational characteristics of the metal atoms in the immediate vicinity of a dissolved H atom. I know no method whereby this can be done. It must be remembered, however, that the metal is at temperatures well in excess of its Debye temperature, and hence the partial thermal entropy of the metal might be expected to be rather insensitive to further distortion over and above the large thermal disorder of the metal atoms.

THE THERMODYNAMICS OF DILUTE INTERSTITIAL SOLID SOLUTIONS*

Rex B. McLellan
Department of Mechanical and Aerospace Engineering and Materials Science, Rice University, Houston, Texas

INTRODUCTION

In analyzing experimental thermodynamic data for solid solutions it is convenient to classify the solutions into two categories: In one category we deal with solutions in which the interaction between solute atoms manifests itself in a nonrandom solute atom distribution. The second category is a special case of the first, where the solution is dilute enough to enable solute-solute interactions to be ignored and the solute atom distribution is random. In the general category of solutions the partial energy of solution $\bar{E}_u$ is a function of composition and the partial entropy $\bar{S}_u$ is nonideal. In the second class of solution, which may be termed quasi-regular, $\bar{E}_u$ is not dependent on the solute concentration, and although the entropy may be nonideal, the configurational contribution to the partial entropy is ideal and thus easily calculated. From an experimental

* This work was supported by NASA grant NsG-6-59.

point of view we can say that a solution is *quasi-regular* when it is dilute enough for the solute to obey Henry's law.

Most of the statistical theories of solutions assume that the state sum Ω_i due to the internal degrees of freedom (terms due to vibrational, magnetic, or electronic specific-heat changes) is separable from the configurational state sum Ω_c so that the state sum Ω for the whole solution can be written as

$$\Omega = \Omega_i \sum_c g(E_c) \exp\left(\frac{-E_c}{kT}\right) \tag{1}$$

where $g(E_c)$ denotes the number of nondegenerate states in the configuration of energy E_c. The evaluation of the combinatorial factor $g(E_c)$ in Eq. (1) is simple for a chain of atoms, and the calculation for a two-dimensional lattice (Ising model) can be carried out when the problem is formulated in an appropriate matrix notation.[1] However, the exact evaluation of $g(E_c)$ for a three-dimensional lattice meets with great mathematical difficulty and has not yet been accomplished. Much of statistical solution theory consists of attempts to estimate the thermodynamic parameters of solutions by making approximations for $g(E_c)$ in Eq. (1).

The simplest approximation is the zero-order Bragg and Williams[2] approximation in which $g(E_c)$ is replaced by a term corresponding to random mixing in spite of the fact that the energetics of the solution are nonideal. A better approximation is the quasi-chemical model[3] in which the energy of solution is characterized by pairwise interaction energies which are assumed to be independent of composition. The number of configurations corresponding to a given energy is found by counting the possible orientations of each pair in the lattice. If the degree of short-range order in a solution is determined, the excess thermodynamic functions can be calculated from the quasi-chemical model. For a few metallic solid solutions (concentrated Cu-Au alloys) the model is remarkably successful but in general agreement is poor.[4] It is to be noted that both the Bragg-Williams and the quasi-chemical models do not take non-configurational entropy changes into account.

Because of the incomplete state of the statistical interaction theories the interpretation of experimental thermodynamic data on concentrated solutions is subject to more limitations than the analysis of data for quasi-regular solutions. In these latter solutions the combinatorial factor $g(E_c)$ reduces to its value in an ideal solution, and the solid-solution crystal is characterized by a single interaction energy. Thus the state sum and the chemical potential of a solute atom can be written down explicitly. The chemical potential $\mu_u{}^s$ of a solute atom in a quasi-regular solution is given by[5]

$$\mu_u{}^s = \bar{E}_u - T\bar{S}_u{}^v + kT \ln \frac{\theta/\beta}{1 - \theta/\beta} \tag{2}$$

where θ is the ratio of the numbers of solute atoms to solvent atoms, β is the number of interstices of a given kind per lattice atom, and $\bar{E}_u$ is the energy required to insert a solute atom into the solution measured with respect to the energy of an atom at rest in a vacuum. The term $\bar{S}_u{}^v$ is the partial excess entropy of a solute atom,

$$\bar{S}_u{}^v = \lim_{C_u \to 0} \left(\frac{\partial S}{\partial N_u} \right)_v + k \ln \frac{\theta/\beta}{1 - \theta/\beta} \tag{3}$$

where S is the total entropy of the solution. Throughout v is used to denote solvent and u to denote solute.

Equation (2) can be used in the analysis of solubility data for quasi-regular solutions, and the parameters found are important in the development of the theory of interstitial solid solutions. The energy $\bar{E}_u$ can be used to check fundamental quantum-mechanical calculations for the case when the interaction energy is not complicated by solute-solute terms. Furthermore, since the partial configurational entropy

$$-k \ln \frac{\theta/\beta}{1 - \theta/\beta}$$

is known explicitly, analysis of the solubility data yields $\bar{S}_u{}^v$, the non-configurational partial entropy. For solutions where no magnetic or electronic specific-heat changes occur, this excess partial entropy can be interpreted as a vibrational entropy, owing to the perturbation of the normal modes of the solvent lattice which occurs when a solute atom is inserted. The parameters $\bar{E}_u$ and $\bar{S}_u{}^v$ are also useful in the development of theories for concentrated solutions since the partial energies and entropies in concentrated solutions must approach $\bar{E}_u$ and $\bar{S}_u{}^v$ in the limit as the solute concentration approaches low values.

This report will present a review of some of the models used to represent quasi-regular interstitial solid solutions, together with a discussion of the applications of the models to the analysis of data for the variation of the solute solubility with temperature. Instead of attempting a general approach the discussion will be limited to a consideration of two groups of interstitial solutions—N and C dissolved in Fe, and H alloys.

The solutions and models discussed are in thermodynamic equilibrium with a second phase. The solvent lattice is "perfect" in that it does not contain defects such as vacancies or dislocations at a concentration which would provide low-energy sites where solute atoms could locate and give rise to a measurable deviation from the normal lattice solubility.

Fe-BASED INTERSTITIAL SOLID SOLUTIONS

The Fe-C system is one of the most intensively studied binary systems. Many data exist on the equilibrium of dilute solid solutions of C in both

α- and γ-Fe with a second phase, both when the second phase is a solid or liquid (equilibrium diagram work) and when the second phase is gaseous. If the thermodynamic functions of the C atoms in the second phase are known, then experimental data for the variation of the equilibrium with temperature can be analyzed to give the thermodynamic parameters of the C atoms dissolved in the solid.

Smith[6] and Dünwald and Wagner[7] have measured the equilibrium between ferrite and austenite and mixtures of $CO + CO_2$ and $H_2 + CH_4$ at several temperatures, and more recently Ellis, Davidson, and Bodsworth[8] determined the equilibrium between austenite and a gas containing CH_4, H_2, CO_2, CO, and H_2O at two temperatures. In the experiments of Smith and Dünwald and Wagner the variation in the ratios $r_1 = P_{CH_4}/P_{H_2}^2$ and $r_2 = P_{CO}^2/P_{CO_2}$ with the C content in ferrite and austenite was measured. From a knowledge of the equilibrium constants for the reactions

$$2H_2(g) + C(gr) \rightarrow CH_4(g)$$
$$CO_2(g) + C(gr) \rightarrow 2CO(g) \qquad (4)$$

the activity a_c of the C in solution relative to graphite (gr) can be calculated from the measurements of r_1 and r_2. It was found in both sets of experiments that the C atoms in ferrite (at 750 and 800°C) obey Henry's law up to the limit where the $\alpha \rightarrow \gamma$ transition occurs. The austenite equilibrium was measured at 1000, 1070 and 1200°C, and it is found that the solute obeys Henry's law up to about $1 \rightarrow 2$ a/o, but at higher C contents there is a marked departure and the C activity increases with C content more rapidly than Henry's law would predict.

Let us first give a brief discussion of the models which have been considered to explain the deviation from Henry's law in the more concentrated austenite, and then proceed to a discussion of the very dilute solutions.

Darken and Smith[9] considered a simple statistical model in which a C atom in its octahedral site in the fcc lattice has either one or no neighboring C atoms in the 12 nearest-neighbor octahedral sites. This model thus contains two characteristic interaction energies since it is assumed that the number of C atoms having more than one nearest-neighbor C atom is negligible and a solute atom does not interact with another solute atom situated at greater distance than the first shell of interstitial sites. Darken and Smith showed that this model gives reasonable agreement with the measured variation in the activity of C with composition. It is interesting to note that Darken and Smith concluded from the analysis of the data in the light of their model that there was a slight repulsive force between C atoms in austenite which reduces the concenrtation of C-C pairs below that which corresponds to random mixing.

The exclusion of solute atoms from nearest-neighbor sites in an interstitial solution has been a basic postulate in other models used to explain the C-activity data in austenite. This "blocking" was formulated by Speiser and Spretnak,[10] who considered that an integral number z of sites could be blocked by a solute atom. The molar configurational entropy of such a solution in terms of c_u, the atom fraction of solute, is given by the following equation[5] (which assumes that there is no overlap of the sites prohibited by nearest-neighbor solute atoms).

$$S_z^c = -\frac{R}{z}\left\{c_u z \ln \frac{c_u}{\beta} - \beta(1 - c_u) \ln (1 - c_u) + \beta\left[1 - \frac{c_u(\beta + z)}{\beta}\right] \ln \left[1 - \frac{c_u(\beta + z)}{\beta}\right]\right\} \tag{5}$$

For a perfectly random solution (no blocking) $z = 1$ and putting $z = 1$ in Eq. (5) gives the configurational entropy of a random interstitial solution S^c.

$$S^c = -R\left\{c_u \ln \frac{c_u}{\beta} - \beta(1 - c_u) \ln (1 - c_u) + \beta\left[1 - \frac{c_u(\beta + 1)}{\beta}\right] \ln \left[1 - \frac{c_u(\beta + 1)}{\beta}\right]\right\} \tag{6}$$

A plot of these two functions for $z = 5$ is shown in Fig. 1. The entropy curve for the blocked solution does not differ sensibly from that of the random solution until c_u is about 2.0 a/o.

It has been customary to describe the thermodynamics of interstitial solutions in terms of the variation in activity a_c with solute content and, in parallel to the treatment of substitutional solutions, to define a_c by an equation of the type

$$\mu_u^s = \overset{\circ}{\mu}_u + kT \ln a_u \tag{7}$$

using the pure solute as a reference material (graphite in the case of Fe-C solutions). However, it can be seen from Eq. (2) that for an ideal interstitial solution

$$\mu_u^s = \overset{\circ}{\mu}_u + kT \ln \frac{c_u}{1 - 2c_u} \tag{8}$$

where, for simplicity we have put $\beta = 1$. Thus even for an ideal interstitial solution the activity defined in Eq. (7) is not equal to the atom fraction of solute, and a_u is not proportional to c_u in a quasi-regular solution. Figure 1 shows the activity vs. composition curve for an ideal solution, and it can be seen that at solute compositions greater than about 2 a/o there is a sensible departure from linearity. In a rigorous discussion of interstitial solid solutions Henry's law should be restated or the activity should

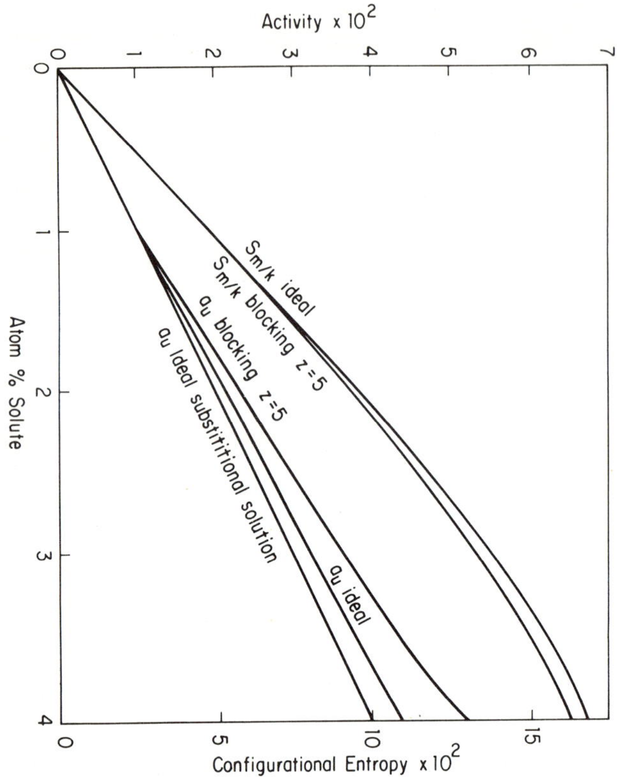

Fig. 1 Configurational entropy $\times 10^2$. Curves of the configurational entropy and solute activity for blocked and ideal interstitial solutions.

be defined by an equation of the form

$$\mu_u{}^s = \overset{\circ}{\mu}_u + kT \ln \frac{a'_u}{1 - 2a'_u} \tag{9}$$

It can be shown[5] that the chemical potential of a blocked solution is given by

$$\mu_u{}^{s'} = \bar{E}_u - T\bar{S}_u{}^v + kT \ln \frac{\theta/\beta}{1 - \theta z/\beta} \tag{10}$$

Thus if $\bar{E}_u$ and $\bar{S}_u{}^v$ are independent of composition, then the composition dependence of the activity at a given temperature is determined by the expression

$$\frac{\theta/\beta}{1 - \theta z/\beta} = \frac{c_u}{1 - c_u(z + 1)} \tag{11}$$

where $\beta = 1$. A plot of this function for $z = 5$ is shown in Fig. 1. It can be seen that there is a strong departure of this function from plot for an ideal solution. The departure becomes experimentally significant between $c_u = 0.01$ and 0.015.

Many models have been set up to explain the thermodynamic behavior of C in γ-Fe in which it is assumed that although blocking is present and the configurational entropy is nonrandom, the partial energy and partial nonconfigurational entropy are independent of solute content [as in Eq. (10)]. Scheil[11,12] and Kaufman, Radcliffe, and Cohen[13] have analyzed the data of Smith and deduced that $z = 5$ from the solute activity data. On the other hand, Schwarzmann and Temkin[14] deduced that a C atom blocks only three other interstitial sites so that $z = 4$. Ellis, Davidson, and Bodsworth[8] determined the variation in a_c with C content in austenite at 925 and 1050°C. At both temperatures a_c began to deviate from Henry's law at between $c_u = 0.03$ and 0.04. The authors concluded from the analysis of their data that $z = 4$.

Let us now turn to a consideration of the Fe-C solutions which are dilute enough to obey Henry's law. McLellan[15] has analyzed the data of Smith and Dünwald and Wagner for the equilibrium of CO-CO_2 and CH_4-H_2 mixtures with ferrite and austenite, utilizing only the data for equilibriums in the range where the C activity (as measured by r_1 and r_2) varies linearly with C content. Suitable equilibrium determinations were made at 750 and 800°C for ferrite and 940, 1000, 1070, and 1200°C for austenite. In this work it was assumed that the solid solutions were quasi-regular so that the chemical potential of a solute atom in solution is given by Eq. (2). This chemical potential was then equated to that of C atoms in the gas, enabling a solubility equation to be deduced. The theoretical solubility equation was used to extract the relevant thermodynamic parameters of the solid solutions from the experimental equilibrium data.

The solubility equation deduced for the CH_4-H_2 equilibria is

$$\frac{\theta/\beta}{1-\theta/\beta} = \frac{r_1 k T^3 \phi}{Z^v_{\mathrm{CH_4}}} \exp\left[-\frac{\bar{E}_u - (\mathring{E}_{\mathrm{CH_4}} - 2\mathring{E}_{\mathrm{H_2}})}{kT}\right] \exp\frac{\bar{S}_u{}^v}{k} \tag{12}$$

where

$$\phi = \frac{\{[(2\pi 2 m_{\mathrm{H}} k)^{3/2}/h^3](8\pi^2 I_{\mathrm{H_2}} k/2h^2)\}^2}{[(2\pi m_{\mathrm{CH_4}} k)^{3/2}/h^3][8\pi^2(2\pi I_{\mathrm{CH_4}} k)^{3/2}/12h^3]} \tag{12a}$$

and for the CO_2-CO equilibria

$$\frac{\theta/\beta}{1-\theta/\beta} = \frac{r_2 \Psi}{kT^{7/4}} \frac{Z^v_{\mathrm{CO_2}}}{(Z^v_{\mathrm{CO}})^2} \exp\left[-\frac{\bar{E}_u + (\mathring{E}_{\mathrm{CO_2}} - 2\mathring{E}_{\mathrm{CO}})}{kT}\right] \exp\frac{\bar{S}_u{}^v}{k} \tag{13}$$

where

$$\Psi = \frac{[(2\pi m_{CO_2}k)^{3/2}/h^3](8\pi^2 I_{CO_2}k/2h^2)}{\{[(2\pi m_{CO}k)^{3/2}/h^3](8\pi^2 I_{CO}k/h^2)\}^2} \tag{13a}$$

In these equations Z denotes the vibrational state sum of the molecules in the gas, I their moments of inertia, m their mass, and $-E°$ is the dissociation energy at 0°K. These solubility equations take the complex temperature dependence of the chemical potential of solute atoms in the gas explicitly into account and enable reliable values of $\bar{E}_u$ and $\bar{S}_u{}^v$ to be obtained from the experimentally determined variation in θ with r_1 and r_2 at different temperatures.

A similar analysis has been applied to the solubility data for nitrogen in ferrite and austenite in equilibrium with N_2 gas.[5] The solubility of N in an interstitial solution in equilibrium with N_2 molecules is

$$\frac{\theta/\beta}{1 - \theta/\beta} = \frac{P_{N_2}{}^{1/2}\lambda}{T^{7/4}(Z_{N_2}^v)^{1/2}} \exp\left(-\frac{\bar{E}_u - \frac{1}{2}E_{N_2}^\circ}{kT}\right) \exp\frac{\bar{S}_u{}^v}{k} \tag{14}$$

where

$$\lambda = \left[\frac{h^3}{(2\pi 2m_N k)^{3/2}} \frac{2h^2}{8\pi^2 I_{N_2}k} \frac{1}{k}\right]^{1/2} \tag{14a}$$

where P_{N_2} is the partial pressure of N_2 molecules. This equation was used to estimate the values of $\bar{E}_u$ and $\bar{S}_u{}^v$ from the experimental solubility data.

The thermodynamic data obtained from these analyses are given in Table 1. The equilibrium data of Ellis, Davidson, and Bodsworth[8] have not been analyzed by this technique but estimates of $\bar{E}_u$ and $\bar{S}_u{}^v$ can be obtained from their published activity data from the equation

$$\frac{c_u}{1 - 2c_u} = \exp\left(-\frac{\Delta\bar{H}_u}{kT}\right) \exp\frac{\bar{S}_u{}^v - S_u^\circ}{k} \tag{15}$$

where $\Delta\bar{H}_u \equiv \bar{H}_u - H_u^\circ$ is the relative partial molar enthalpy of a solute atom, and S_u° is the standard entropy of the pure solute (graphite). The value of $\Delta\bar{H}_u$ deduced obtained from the data is 11.57 kcal/mole, and this is presumably a mean value for the two temperatures at which the activity was measured. Using enthalpy tables[16] to obtain H_u°, the value of $\bar{E}_u$ is -154.5 kcal/mole, in good agreement with the values given in Table 1. The value of $\bar{S}_u{}^v$ estimated from $(\bar{S}_u{}^v - S_u^\circ)$ and tables of standard entropies[16] is $5.77k$.

The advantage of deducing the thermodynamic parameters of a solution from solubility data with a closed solubility equation is that it obviates the use of a reference state, and the partial entropy can be obtained without recourse to extra data on the equilibrium of the gas phase with graphite.

TABLE 1 ***Thermodynamic Data for Fe-C Solutions***

Solution	$\bar{E}_u$, kcal/mole	$\bar{S}_u{}^v/k$
C–α-Fe	−147.2	5.20
C–γ-Fe	−155.2	5.75
N–α-Fe	−74.3	7.20
N–γ-Fe	−81.9	6.20

Moreover, the analysis is carried out in terms of equations which are easily understood physically.

The data in Table 1 were obtained using the assumption that solute atoms occupy only one kind of interstitial site. Recently[17] it has been shown that if the C atoms in α-Fe can occupy both the octahedral and tetrahedral sites, the deviation from linearity at high temperatures of an Arrhenius plot of the diffusivity of C through α-Fe could be explained. The thermodynamic functions of a solution with dual-site occupancy will differ from those of a quasi-regular solution in which only one kind of site is occupied. The differences in partial energy, entropy, and Gibbs potential between a dual-occupancy solution and a single-occupancy solution of the same composition are given by

$$\Delta\tilde{E}_u = \Delta E(1 - \delta) \tag{16}$$

$$\Delta\tilde{S}_u = \frac{\Delta E}{T}(1 - \delta) - k \ln \delta \tag{17}$$

$$\Delta\tilde{\mu}_u = kT \ln \delta \tag{18}$$

where δ is the fraction of solute atoms located at octahedral sites, and ΔE is the energy difference between an octahedral site and a tetrahedral site. The analysis of the diffusion data using this model indicated that at 800°C only one solute atom in 1,000 is located at a tetrahedral site, and thus no effect of their presence on the measured thermodynamic parameters should be seen. They have a measurable effect on the diffusivity at high temperatures since the diffusion rate via tetrahedral sites is relatively large.

The information in Table 1 represents reliable data for the thermodynamics of quasi-regular Fe-C and Fe-N solutions. A brief discussion of the partial energies and entropies has been given,[15] and it was shown that for both N and C solutes electronic specific-heat and magnetic spin changes were probably negligible and $\bar{S}_u{}^v$ arises from vibrational changes.

H-METAL SOLUTIONS

In this section the experimental solubility data for H-metal solutions are presented, previous models for H alloys are briefly discussed, and finally the data are analyzed by assuming a quasi-regular solution model for the solid.

The Experimental Data

The solubility of H in metals has been most extensively measured for the metals in the first and second long periods of the periodic table. The variation of the H solubility with temperature at 1 atm pressure for metals in the first long period is given in Fig. 2, that for metals in the second long period is in Fig. 3, and data for other metals are given in Fig. 4. The sources of the data are indicated in Table 2. The reference works of Smith[18] and Smithells[19] have been indicated repeatedly for metals where they give solubilities recalculated from original data; however, these data points were not included in the computer analysis of the solubility data (next section).

The data in Figs. 2, 3, and 4 cover a wide range of solubilities; at 400°C the solubility of H in Th is greater than that in Al by a factor of 10^8. For metals that dissolve only small amounts of H, changes in lattice parameter with H content are probably too small to be measurable. For metals that dissolve large amounts of H some data are available on changes of lattice parameter and the positions occupied by the H atoms in the metal lattice. Shull, Rundle, and Woolan[20] have shown that for Zr and Th there is little change in lattice parameter until appreciable amounts of H are absorbed, and then there is a change in the lattice structure when the compounds ZrH and ThH are formed. They showed that these compounds are body-centered tetragonal and, by neutron diffraction, that the H atoms are situated at the center of tetrahedrons of metal atoms. Woolan, Cable, and Koehler[21] showed by neutron diffraction that the H atoms occupy octahedral sites in fcc Ni.

This work deals only with H-metal solutions in equilibrium with gaseous H. It is, however, possible to surcharge a metal by making it the cathode in an electrolytic cell, especially if the metal surface is coated with certain catalysts. Occlusion of 10 times the amount of H dissolved in the metal in equilibrium with gaseous H at 1 atm pressure can occur, and this may damage the structure of the metal and even produce large blisters on the surface. The surcharged H alloys are not stable, and the H is occluded in rifts; the pockets of H responsible for the formation of blisters undoubtedly contain molecular H.

Although the occlusive capacity of a metal in cathodic surcharging can be greatly increased by cold working, the equilibrium solubility is not

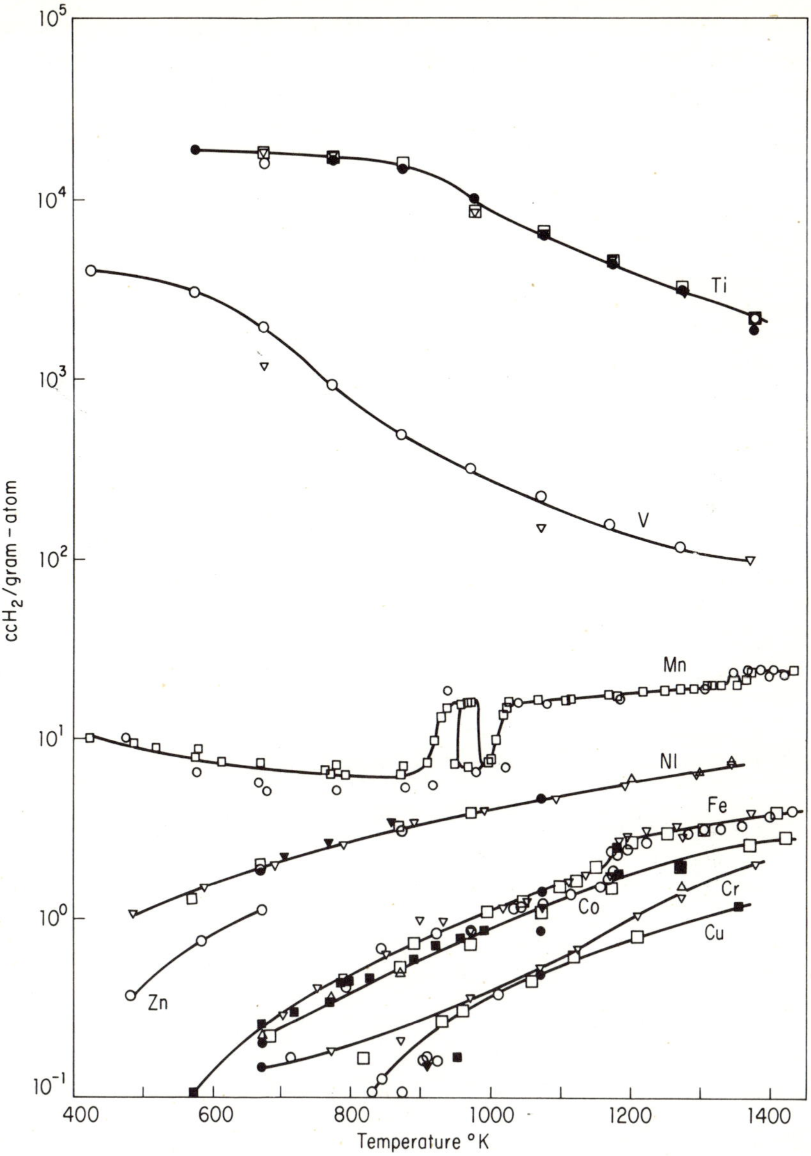

Fig. 2 Variation of H solubility with temperature. (For sources of data see Table 2.)

changed thus. Carmichael, Hornady, Morris, and Parlee[22] have shown that α-Fe has the same H solubility in the well-annealed state as it has after being cold-worked by 75% and annealed for a short time at 430°C. At this annealing temperature the greater part of the cold-worked hardness remains, and there is no microscopically visible recrystallization.

It may have been supposed that the degree of cold working would affect

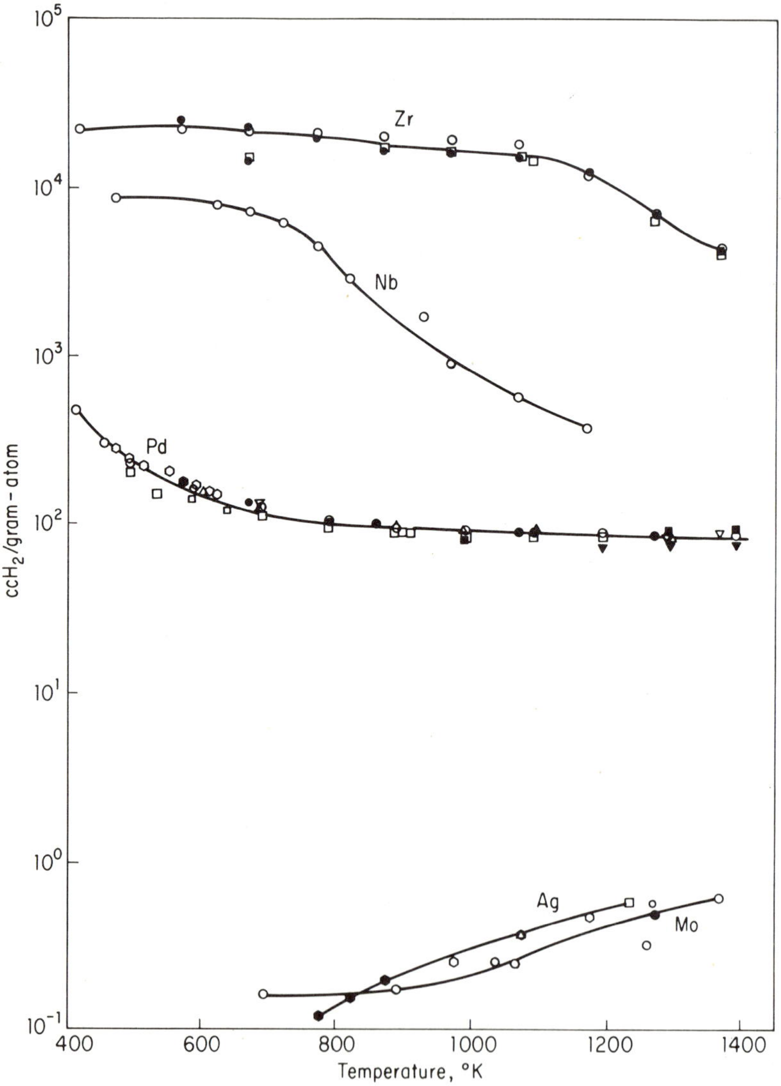

Fig. 3 Variation of H solubility with temperature. (For sources of data see Table 2.)

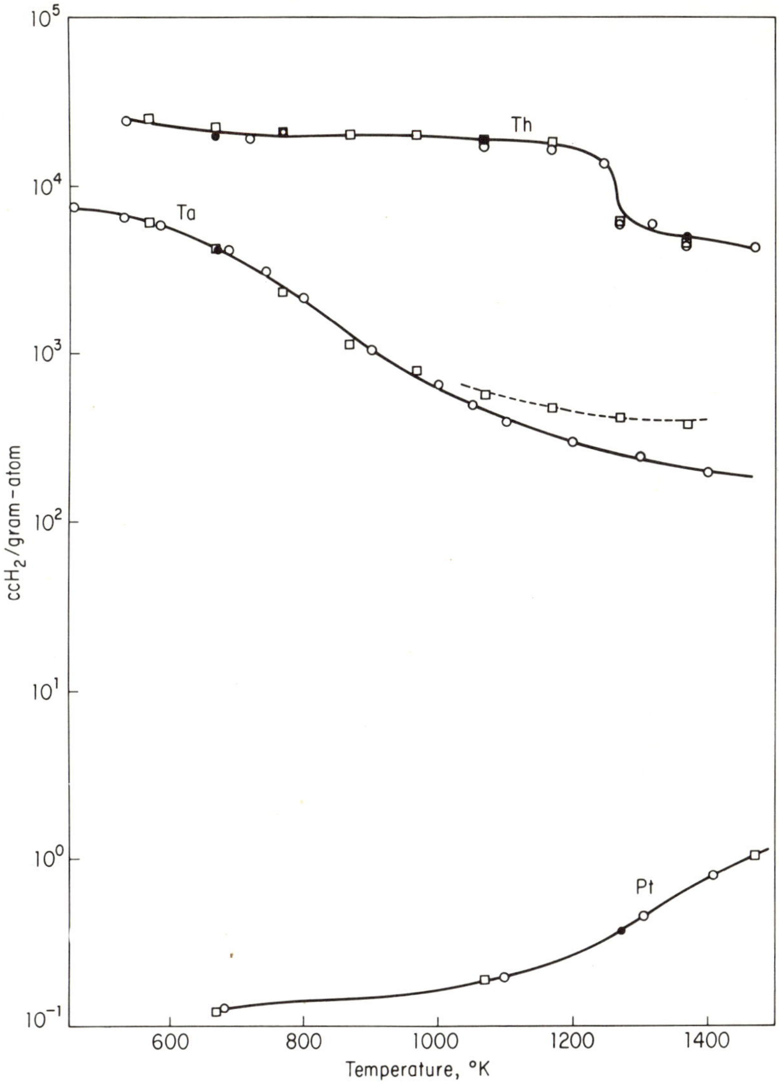

Fig. 4 Variation of H solubility with temperature. (For sources of data see Table 2.)

the H solubility by increasing the number of dislocations in the metal and thus providing low-energy sites for the H atoms. However, it can easily be shown that this effect is not important at temperatures above which the H solubility of most metals can be measured. For bcc α-Fe the dislocations lie along (110) planes, and according to the measurements of Nutting and Brandon,[23] there are 10^{10} dislocations cutting an area of 1 cm^2 in heavily

TABLE 2 *Sources of the H Solubility Data for Figs. 2 to 4*

Al	○	C. E. Ransley and H. Neufeld, *J. Inst. Metals* **74**:599 (1948).
	□	W. Eichenauer and A. Pebler, *Z. Metallk.* **48**:373 (1957).
	●	L. L. Bircumshaw, *Trans. Faraday Soc.* **31**:1439 (1935).
Ti	●	A. Sieverts, *Z. Metallk.* **21**:37 (1929).
	○	D. P. Smith, "Hydrogen in Metals," University of Chicago Press, Chicago, 1948—quoted from Sieverts.
	▽	L. Kirschfeld and A. Sieverts, *Z. Physik. Chem.* **145A**:227 (1929).
	□	A. Sieverts, H. Huber, and L. Kirschfeld, *Chem. Ber.* **59**:2891 (1926).
V	△	D. P. Smith, *loc. cit.*
	○	L. Kirschfeld and A. Sieverts, *Z. Elektrochem.* **36**:123 (1930).
Cr	●	D. P. Smith, *loc. cit.*
	○	C. J. Smithells, "Metals Reference Book," 2d ed., pp. 535–536, Interscience Publishers, Inc., New York, 1955.
	▽	L. Luckemeyer-Hasse and H. Schenk, *Arch. Eisenhuettenw.* **6**:209(1932).
Mn	○	A. Sieverts and H. Moritz, *Z. Physik. Chem.* **180A**:249 (1937).
	□	E. V. Potter and H. C. Lukens, *Trans. AIME* **171**:401 (1947).
	●	C. J. Smithells, *loc. cit.*
Fe	●	D. P. Smith, *loc. cit.*
	○	E. Martin, *Arch. Eisenhuettenw.* **3**:407 (1929).
	□	A. Sieverts, *Z. Physik. Chem.* **77**:591 (1911).
	■	W. Eichenauer, H. Kunzig, and A. Pebler, *Z. Metallk.* **49**:220 (1958).
	▲	C. J. Smithells, *loc. cit.*
	△	A. Sieverts and W. Krumbhaar, *Chem. Ber.* **43**:893 (1910).
	▽	A. Sieverts, G. Zapf, and H. Moritz, *Z. Physik. Chem.* **183A**:19 (1938).
Co	●	D. P. Smith, *loc. cit.*
	■	C. J. Smithells, *loc. cit.*
	○	P. Beckmann, Dissertation, Leipzig—quoted by A. Sieverts and H. Hagen, *Z. Physik. Chem.* **169**:237 (1934).
	□	A. Sieverts and H. Hagen, *Z. Physik. Chem.* **169**:237 (1934).
	△	W. Siegelin, K. H. Lieser, and H. Witte, *Z. Elektrochem.* **61**:359 (1957).
	▽	A. Sieverts, *Z. Metallk.* **21**:37 (1929).
Ni	●	D. P. Smith, *loc. cit.*
	■	C. J. Smithells, *loc. cit.*
	○	K. H. Lieser and G. Rinck, *Z. Elektrochem.* **61**:357 (1957).
	▼	W. Eichenauer, W. Loser, and H. Witte, *Z. Metallk.* **56**:287 (1965).
	□	K. H. Lieser and H. Witte, *Z. Physik. Chem.* **202**:321 (1954).
	△	A. Sieverts and W. Krumbhaar, *Chem. Ber.* **43**:893 (1910).
	▽	A. Sieverts, *Z. Metallk.* **21**:37 (1929).
Cu	●	C. J. Smithells, *loc. cit.*
	□	W. Himmler, *Z. Physik. Chem.* **195**:244 (1950).
	▼	W. Eichenauer, W. Löser, and H. Witte: *Z. Metallk.* **56**:287 (1965).
	○	W. Eichenauer and A. Pebler, *Z. Metallk.* **48**:373 (1957).
	■	K. H. Lieser and H. Witte, *Z. Physik. Chem.* **202**:321 (1954).

TABLE 2 *Sources of the H Solubility Data for Figs. 2 to 4—(Continued)*

Zn	○	K. Iwase, *Sci. Rept. Tokohu Univ. First Ser.* **15**:531 (1926).
Zr	●	C. J. Smithells, "Gases and Metals," p. 161, Chapman & Hall, Ltd., London, 1937.
	○	M. N. A. Hall, S. L. H. Martin, and A. L. G. Rees, *Trans. Faraday Soc.* **41**:306 (1945).
	□	A. Sieverts and H. Moritz, *Z. Anorg. Allgem. Chem.* **247**:124 (1941).
	■	D. P. Smith, *loc. cit.*
Nb	○	A. Sieverts and H. Moritz, *Z. Anorg. Allgem. Chem.* **247**:124 (1941).
Mo	■	C. J. Smithells, "Metals Reference Book," 2d ed., pp. 535–536, Interscience Publishers, Inc., New York, 1955.
	○	A. Sieverts and K. Bruning, *Arch. Eisenhuettenw.* **7**:641 (1934).
Pd	●	A. Sieverts and G. Zapf, *Z. Physik. Chem.* **174A**:359 (1935).
	○	A. Sieverts, *Z. Physik. Chem.* **88**:103 (1914).
	□	J. Hagenacker, Dissertation, Leipzig, 1909—quoted by A. Sieverts, *Z. Physik. Chem.* **88**:103 (1914).
	■	F. Müller, Dissertation, Leipzig, 1911—quoted by A. Sieverts, *Z. Physik. Chem.* **88**:103 (1914).
	△	E. Jurisch, Dissertation, Leipzig, 1912—quoted by A. Sieverts, *Z. Physik. Chem.* **88**:103 (1914).
	▽	E. Jurisch, Dissertation, Leipzig, 1912, and H. Oehme, Dissertation, Leipzig, 1912—quoted by A. Sieverts, *Z. Physik. Chem.* **88**:103 (1914).
	⬡	A. Sieverts and W. Danz, *Z. Physik. Chem.* **34B**:158 (1936).
	▲	A. Sieverts and W. Krumbhaar, *Chem. Ber.* **43**:893 (1910).
Ag	⬡	E. W. R. Steacie and F. M. G. Johnson, *Proc. Roy. Soc.* **A117**:662 (1928).
	●	W. Siegelin, K. H. Lieser, and H. Witte, *Z. Elektrochem.* **61**:359 (1957).
	□	C. J. Smithells, *loc. cit.*
	△	D. P. Smith, *loc. cit.*
Ta	○	A. Sieverts and E. Bergner, *Chem. Ber.* **44**:2394 (1911).
	●	D. P. Smith, *loc. cit.*
	□	A. Sieverts, *Z. Metallk.* **21**:37 (1929).
Pt	○	A. Sieverts and E. Jurisch, *Chem. Ber.* **45**:221 (1912).
	□	D. P. Smith, *loc. cit.*
	●	C. J. Smithells, *loc. cit.*
Th	○	A. Sieverts and H. Moritz, *Z. Anorg. Allgem. Chem.* **247**:124 (1941).
	●	D. P. Smith, *loc. cit.*
	□	A. Sieverts, *Z. Metallk.* **21**:37 (1929).

cold-worked iron. This infers that the number of dislocation sites per square centimeter is 5×10^{17}. At 600°C Fe dissolves 40×10^{17} H atoms per cm^3; thus even at low temperatures in heavily cold-worked metal there are many more H atoms than dislocation sites.

Models for H-Metal Solutions

A model for H-metal solutions was proposed by Fowler and Smithells[24] for those metals whose H solubility is small (Cu, Ni, Fe, Co), in which the H atoms dissolve as protons and have translational freedom in a large fraction f of the volume of the metal. The chemical potential $\mu_H{}^S$ of a proton in the solid can be then written in the form

$$\mu_H{}^S = E_0 + kT \ln \frac{h^3}{(2\pi mkT)^{3/2}} + kT \ln \frac{C_H}{f} \tag{19}$$

where E_0 is the energy of an H atom at rest in the solution with respect to an H atom at rest in a vacuum, and C_H is the number of H atoms in the solid per unit volume. Equating this $\mu_H{}^S$ to the chemical potential $\mu_H{}^S$ in the gas yields the solubility equation,

$$\frac{\rho\theta T^{1/4}}{MP_{H_2}^{1/2}} = \frac{1}{N_0}\frac{(2\pi mk)^{3/2}}{h^3}\left[\frac{h^3}{(2\pi 2mk)^{3/2}}\frac{2h^2}{8\pi^2 I_{H_2}k}\frac{1}{k}\right]^{1/2} f \exp\left(-\frac{E_0 - \frac{1}{2}E^\circ_{H_2}}{kT}\right) \tag{20}$$

where ρ is the density of the metal and M its atomic weight.

There is, however, a serious objection to the translational model since it predicts diffusion coefficients for the diffusion of H through the metal lattice which are much higher than the measured values. In this model the diffusion of H through the metal is analogous to the Knudsen flow of gas molecules at low pressure through a tube whose radius g is small compared to the mean free path of the gas molecules. The diffusion coefficient for such a flow is proportional to $g\bar{v}$, where $\bar{v}$ is the mean velocity of the molecules,[25] so that

$$D = K'g\left(\frac{RT}{M_H}\right)^{1/2} \tag{21}$$

where M_H is the molecular weight of the gas, and the coefficient K' is close to unity. Putting $g = 10^{-8}$ cm, $D = 1.52 \times 10^{-3}$ cm^2/sec at 500°K. The ratio of this calculated D to the smallest observed D (for α-Fe) at 500°K is 2.7×10^1.

For copper ($f = 0.8$) Eq. (20) gives reasonably good agreement with the measured solubility data, but for other metals the agreement is poor.

Fowler and Smithells[24] also deduced a solubility equation for the hydride-forming metals by assuming that the H atoms mix randomly in the metal lattice, do not perturb its vibrational spectrum, and do not interact. It was assumed also that the vibrations of the dissolved H atoms were not excited in the temperature range considered. It was shown that despite these radical assumptions the general shape of the solubility isobars could be explained by the model.

Lacher[26] gave a theoretical treatment of Pd-H solutions in which the partial energy $\bar{E}_u$ is proportional to the fraction of filled sites but the configurational entropy is ideal. The model was qualitatively successful in predicting a critical temperature for concentrated Pd-H alloys below which two solid phases exist.

Eichenauer[27] has presented a theoretical discussion of the solubility and diffusivity of H in metals where the solubility is low. He calculated the vibrational frequency of an H atom in solution from the solubility data, using a quasi-regular solution model for the solid. It was shown that these calculated frequencies were in reasonably good agreement with those calculated from the activation energy for the diffusion of H atoms through the metal lattice, using a diffusion model due to Wert and Zener[28] in which the H atoms vibrate in a parabolic potential well in their lattice sites and the potential energy between adjacent sites varies sinusoidally.

The Quasi-regular Model Applied to H-Metal Solutions

In this work it is assumed that the H atoms in solution are bound at interstitial sites and execute many vibrations before jumping into an adjacent site; in other words, the activation energy for diffusion is appreciably greater than the mean vibrational energy of an H atom. A graphical representation of the model is given in Fig. 5. The dissociation energy of an H_2 molecule is $-E^{\circ}_{H_2}$. The difference in energy between an H atom in the gas and an H atom in the solid is $\bar{E}_u$ and the enthalpy of solution of H_2 molecules per atom is

$$\Delta H = \bar{E}_u - \tfrac{1}{2}E^{\circ}_{H_2} - \tfrac{7}{4}kT \tag{22}$$

It can be seen that previous classification of H-metal solutions into endothermic and exothermic occluders does not reflect a fundamental difference in the solutions since H atoms are always absorbed with the evolution of energy $\bar{E}_u$ and the sign of ΔH depends on whether $\bar{E}_u$ is greater or less than $\tfrac{1}{2}E^{\circ}_{H_2}$.

The treatment of the H-metal solutions is parallel to that of N in Fe: The chemical potential of an H atom in solution is given by Eq. (2), and the solubility equation is given by replacing N with H in Eqs. (14) and (15). For the temperature range involved $Z^{v}_{H_2}$ is essentially unity. It was assumed that the octahedral interstices were occupied so that $\beta = 1$ for fcc lattices and $\beta = 3$ for bcc lattices and for the complex α- and γ-Mn structures the value of β has been taken as 3.

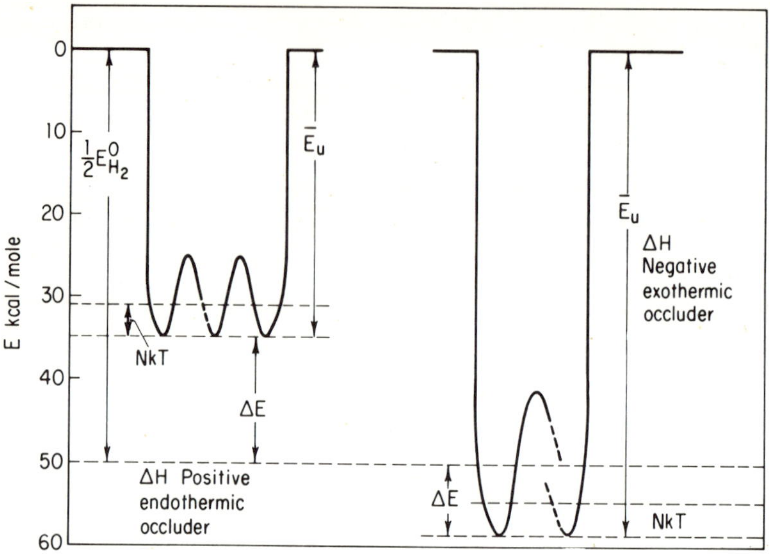

Fig. 5 Bound-oscillator model of H-metal solutions.

Plots of ln $[(\theta/\beta)/(1 - \theta/\beta)][T^{7/4}/P_{H_2}^{1/2}]$ vs. $1/T$ will have a slope given by $-(\bar{E}_u - \frac{1}{2}E^{\circ}_{H_2})/k$ and an intercept given by ln $\lambda + \bar{S}_u{}^v/k$. These plots are shown in Figs. 6, 7, and 8, corresponding to the solubility data given in Figs. 2, 3, and 4. It is to be noted that with the exceptions of Ni and Pt the plots are good straight lines. A computer was used to find the best values of the slopes and intercepts of the plots of Figs. 6, 7, and 8, using the method of least squares, and the values of the partial energy $\bar{E}_u$ and excess partial entropy $\bar{S}_u{}^v$ of H atoms in the solutions are presented in Table 3. For the solvents Th, Zr, and Mo the temperature range covered by the data in the one-phase alloys is so small that no reliance can be placed on the values of $\bar{E}_u$ and $\bar{S}_u{}^v$.

Thermodynamic Parameters of H-Metal Solutions

Table 3 is set out with the metals arranged in the form of the periodic table. Immediately below each solvent metal is given the value of $\bar{S}_u{}^v/k$ and then the value of $-\bar{E}_u$ in kcal per mole.

It can be seen that in all the periods the energy of solution increases (becomes less negative) from left to right, the only exception being Cr in the first long period where the energy of solution of an H atom is anomalously high. In view of the curvature no data have been given for Ni, but the best value for $\bar{E}_u$ is 45.0 kcal/mole.

In the second long period the $\bar{E}_u$ values increase regularly through Zr → Nb → Mo; there is also a regular increase in the first long period

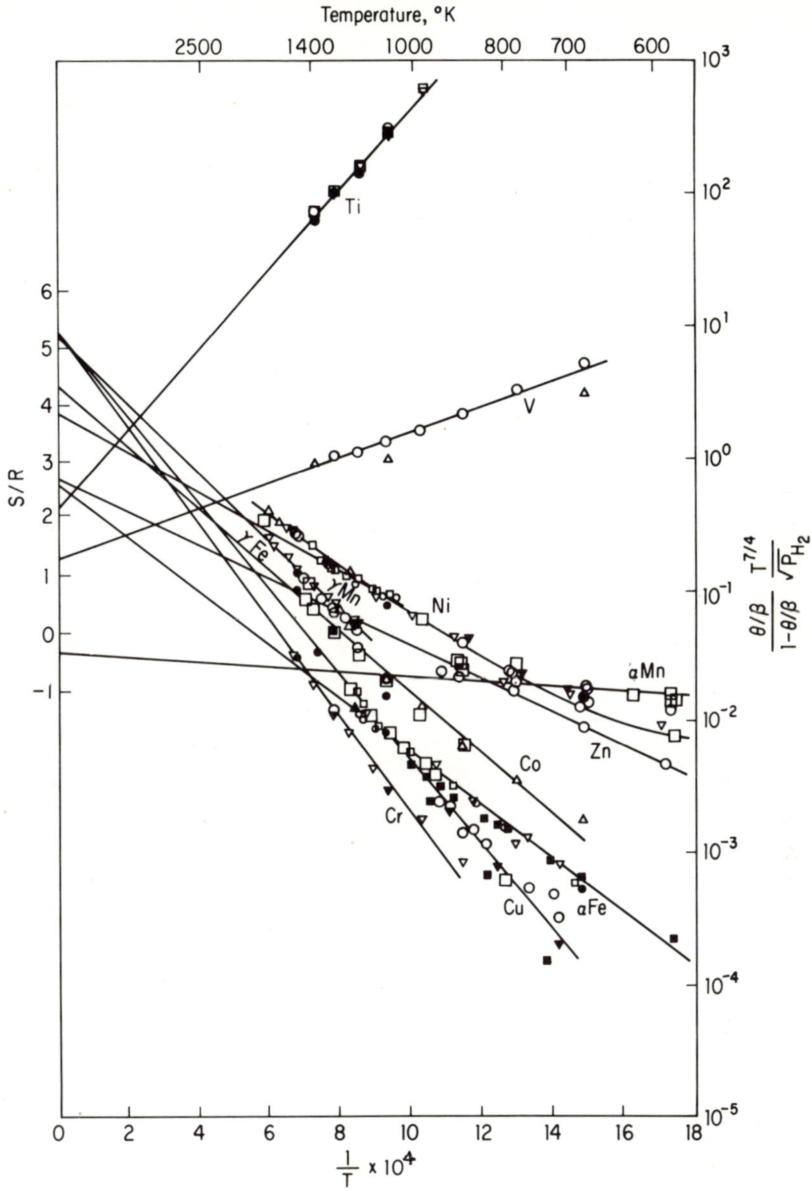

Fig. 6 Thermodynamic analysis of the solubility data of Fig. 2, using the solubility equation (14).

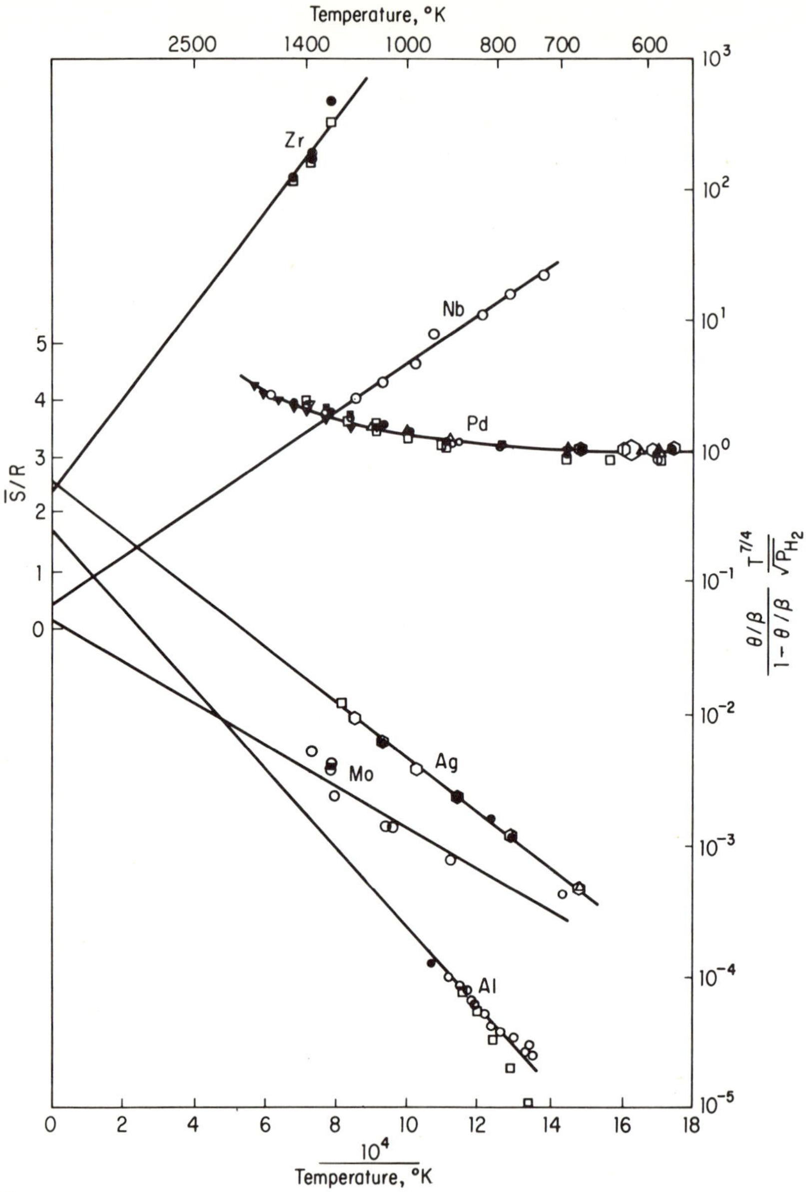

Fig. 7 Thermodynamic analysis of the solubility data of Fig. 3, using the solubility equation (14).

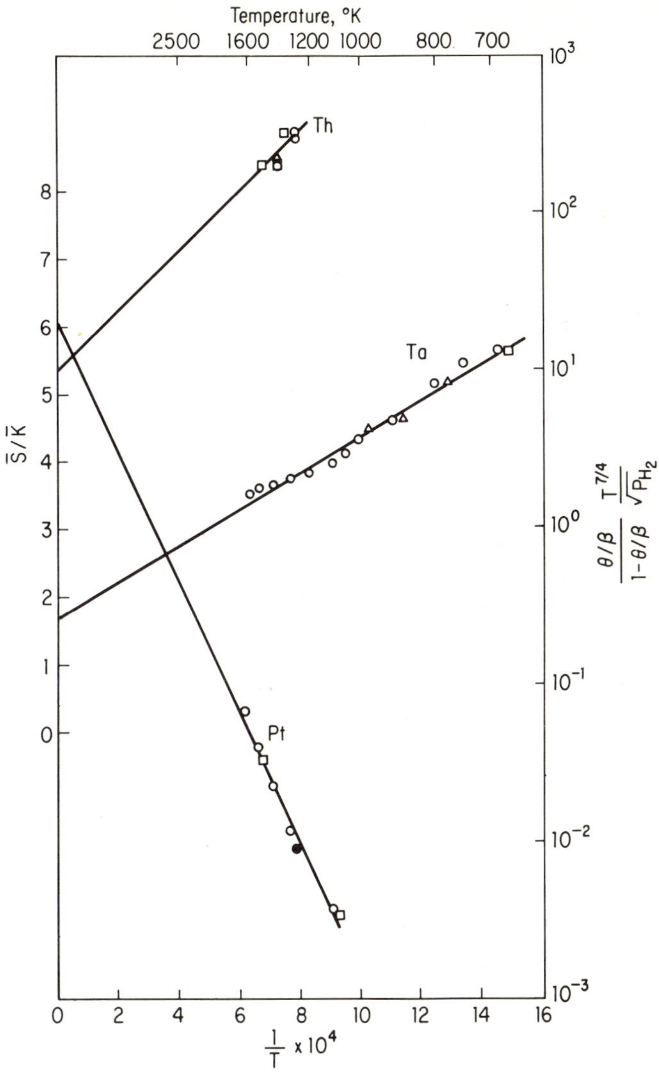

Fig. 8 Thermodynamic analysis of the solubility data of Fig. 4, using the solubility equation (14).

until Cr, and then $\bar{E}_u$ changes only slightly from γ-Mn to Ni. This constancy of $\bar{E}_u$ may be a reflection of the mechanism of solution. Franck[29] suggested that if H dissolves in metals in the form of a proton, then the heat of solution can be written as

$$\bar{E}_u = I_u - W_F + \bar{E}_p \tag{23}$$

where I_u is the ionization energy of H, $\bar{E}_p$ is the heat of solution of a proton, and W_F is the thermionic work function of the metal solvent. If it is assumed that among these transition metals $\bar{E}_p$ does not vary much, then $\bar{E}_u + W_F$ should be constant. For Co, Ni, and α-Fe the values of $\bar{E}_u + W_F$ are 60.55, 61.26, and 60.76 kcal/mole, respectively, but for Cr, $\bar{E}_u + W_F$

TABLE 3 ***Values of $\bar{S}_u^v/k$ and $\bar{E}_u$***

											Al*
											1.69
											37.8
Ti†	V‡	Cr‡	α-Mn	γ-Mn	α-Fe‡	γ-Fe*	Co*	Ni*	Cu*	Zn†	Ga
2.10	1.31	5.30	0.39	3.80	2.53	5.17	4.26		5.21	2.64	
65.7	55.9	32.9	50.9	45.0	42.5	39.8	41.1		36.9	45.9	
Zr†	Nb‡	Mo‡						Pd*	Ag*		
0.24	0.46	0.14							2.45		
73.5	59.8	44.6							42.3		
					Th*						
					5.35						
					60.1						
	Ta‡							Pt*			
	1.74							6.06			
	56.8							32.9			

* Face-centered cubic.
† Close-packed hexagonal.
‡ Body-centered cubic.

is 73.13 kcal/mole. Although there is good evidence from measurements of magnetic susceptibilities in the Pd-H system that H is ionized in Pd with the extra electrons from the H entering the holes in the d band, the data for W_F are too scant to allow any further perusal of the ideas underlying Eq. (23).

There seems to be little correlation between the position of the solvent elements in the periodic table and the value of $\bar{S}_u{}^v/k$. In general, the solutions with the lowest (most negative) heats of solution have smaller partial excess entropies, which would be expected since the more tightly bound protons would vibrate with a higher frequency and give rise to a smaller vibrational entropy. If it is assumed that $\bar{S}_u{}^v$ arises only from uncoupled vibrations of the solute atoms, their Einstein frequency ν can be calculated from the equation

$$\bar{S}_u{}^v = 3k\left\{\frac{h\nu/kT}{\exp\,(h\nu/kT) - 1} - \ln\left[1 - \exp - \frac{h\nu}{kT}\right]\right\} \tag{24}$$

Values of ν calculated from this formula for $T = 1000°K$ enable values for the activation energy E_D for the diffusion of H through the solvent metals to be estimated from the model of Wert and Zener.[28] The values of E_D are all of the right order of magnitude but show no systematic agreement with experimental values of E_D. Such a simple interpretation of the $\bar{S}_u{}^v$ data is almost certainly invalidated by the neglect of the lattice contribution to the excess entropy, the neglect of the change in electronic specific heat, and the uncoupling of electron spins which may occur when the extra electron enters the Fermi energy levels of the solvent metal.

CONCLUSIONS

The work presented in this report involves the analysis of data for the equilibrium between gaseous phases and very dilute interstitial solid solutions in order to derive the useful thermodynamic parameters of solute atoms which are essentially isolated in a solvent matrix. The method developed eliminates the necessity for using extraneous data for equilibria with the phase defining a standard state, and, thereby, the results can be presented as partial quantities insteady of the more usual relative partial quantities.

The partial-energy values calculated by this method have proved useful in discussing problems connected with the solubility of Fe_3C in ferrite[15] and the diffusion of C through ferrite.[17] The partial energy of solution of H in metals has been shown to depend in a fairly regular fashion on the position of the solvent metal in the periodic table.

The excess entropies deduced for Ag-O and Fe-N solutions[5] have been interpreted as vibrational entropies, and their magnitude can be explained[5] by a simple model in which the entropy change due to the shear and dilational strains concomitant with the insertion of a solute atom in the matrix is taken into account.[30] The $\bar{S}_u^v$ data for H alloys are not easy to interpret since it is probable that electronic specific-heat changes make a large contribution to the excess entropy.

REFERENCES

1. Huang, K.: "Statistical Mechanics," Chap. 17, John Wiley & Sons, Inc., New York, 1963.
2. Bragg, W. L., and E. J. Williams: *Proc. Roy. Soc.* **A145**:699 (1934).
3. Guggenheim, E. A.: "Mixtures," Clarendon Press, Oxford, 1952.
4. Oriani, R. A.: *J. Phys. Chem. Solids* **2**:327 (1957).
5. McLellan, R. B.: *Trans. AIME* **230**:1468 (1964).
6. Smith, R. P.: *J. Am. Chem. Soc.* **68**:1163 (1946).
7. Dünwald, H., and C. Wagner: *Z. Anorg. Allgem. Chem.* **199**:321 (1931).
8. Ellis, T., Davidson, I. M., and C. Bodsworth: *J. Iron. Steel Inst.* **220**:582 (1963).
9. Darken, L., and R. P. Smith: Appendix to Ref. 6.
10. Speiser, R., and J. W. Spretnak: *Trans. AIME* **47**:493 (1955).
11. Scheil, E.: *Arch. Eisenhuettenw.* **22**:37 (1951).
12. Scheil, E.: *Arch. Eisenhuettenw.* **30**:315 (1959).
13. Kaufman, L., S. V. Radcliffe, and M. Cohen: In V. F. Zackay and H. I. Aaronson, "Decomposition of Austenite by Diffusion Controlled Processes," p. 313, Interscience Publishers, Inc., New York, 1963.
14. Temkin, M., and L. Schwarzmann: *Zh. Fiz. Khim.* **23**:755 (1949).
15. McLellan, R. B.: *Trans. AIME* **233**:1664 (1965).
16. Rossini, F. D., D. D. Wagman, W. H. Evans, S. Levine, and I. Jaffe: NBS Circular 500, part 1, 1961.
17. McLellan, R. B., M. L. Rudee, and T. Ishibachi: *Trans. AIME* **233**:1938 (1965).
18. Smith, D. P.: "Hydrogen in Metals," University of Chicago Press, Chicago, 1948.
19. Smithells, C. J.: "Metals Reference Book," 2d ed., Interscience Publishers, New York, 1955.
20. Rundle, R. E., C. G. Shull, and E. O. Woolan: *Acta Cryst.* **5**:22 (1952).
21. Woolan, E. O., J. W. Cable, and W. C. Koehler: *J. Phys. Chem. Solids* **24**:1141 (1963).
22. Carmichael, D. C., J. R. Hornady, A. E. Morris, and N. A. Parlee: *Trans. AIME* **218**:862 (1960).
23. Nutting, J., and D. G. Brandon: *J. Iron Steel Inst.* **196**:160 (1960).
24. Fowler, R., and C. J. Smithells: *Proc. Roy. Soc.* **A160**:37 (1937).
25. Present, R. D.: "Kinetic Theory of Gases," p. 60, McGraw-Hill Book Company, New York, 1958.
26. Lacher, J. R.: *Proc. Roy. Soc.* **A161**:525 (1937).
27. Eichenauer, W.: *Mem. Sci. Rev. Met.* **152**:943 (1960).
28. Wert, C., and C. Zener: *Phys. Rev.* **76**:1169 (1949).
29. Franck, J.: *Göttinger Ber.* **2**:293 (1933).
30. McLellan, R. B., and R. Shuttleworth: *J. Phys. Chem. Solids.* **24**:453 (1963).

DISCUSSION *on Paper Presented by Rex B. McLellan*

SEEGER: In my experience it is rather difficult to analyze small deviations from an Arrhenius law for diffusion coefficients. In your model for C diffusion, in addition to the parameters obtainable from the low-temperature data, you have to determine six quantities: (1) free energy characterizing occupancy of T sites, and (2) free energies characterizing jumps out of T sites. Can you specify the error in the experimentally determined quantities?

McLELLAN: I agree completely that it is difficult to analyze the small deviation from the Arrhenius plot when so many constants are involved in the equation that is to be fitted to the data. However, I believe that the Monte Carlo technique has given the best values of the constants in the theoretical diffusivity equation. These constants are physically reasonable, and I take this to mean that the model is compatible with the experimental diffusivity data.

The only experimentally determined quantities are the diffusivity data, and since data from many authors are used, some of whom made no statement of their probable errors, it is impossible to give a figure for the errors involved.

SEEGER: Do you have any comments on Homan's interpretation of the deviation from the Arrhenius law?

McLELLAN: I believe the model of Homan is not compatible with the work of Damask and co-workers on the annealing of irradiation-damaged Fe-C alloys. This work seems to indicate that the C-vacancy complexes have virtually disappeared at a temperature much lower than that at which the diffusion anomaly is seen.

THE ROLE OF THE DEFECT INTERACTION ENERGY ON THE STABILITY OF INTERSTITIAL PHASES*

Michael Hoch

Department of Materials Science, University of Cincinnati, Cincinnati, Ohio

INTRODUCTION

In earlier papers,[1,2] equations were deduced from statistical considerations to determine the activities of components in compounds that deviate from stoichiometry to a great extent. These equations were used, together with thermodynamic data obtained by using solid-state electromotive-force measurements, to determine the vacancy interaction energies in FeO,[1] NbO,[1] TiO,[1] and VO.[4] An attempt was also made to relate those interaction energies to one another.[3]

In this paper, a similar derivation for the interaction energies will be carried out for the refractory metal carbides and nitrides with the NaCl-type structure.

For substitutional solid solutions, the chemical energy of interaction has been evaluated by Blandin and Déplanté[5] from the electronic structure of

* This research has been supported in part by the Air Force Materials Laboratory, Research and Technology Division, Air Force Systems Command, U.S. Air Force.

the defect. For the interstitial solutions where the presence of vacant sites complicates the problem, this cannot be done yet. However, the various energy terms can be calculated from available experimental data; by comparing the thus obtained values of different systems, some regularities in those energy terms can be deduced.

THEORY

The refractory metals of groups IV and V form carbides and nitrides with the NaCl-type structure; these compounds possess a large homogeneity range. The metal sites are all filled, whereas some of the nonmetal sites are empty. Depending on the fraction of the nonmetal sites that are empty, the composition of the material can vary. The formula of those materials, therefore, has to be written AB_{1-x} where A is the metal atom, B is the nonmetal atom (C and N), and x is the fraction of nonmetal atom sites that are empty. From statistical considerations, Hoch[2] has deduced that for this type of nonstoichiometry the following equations apply:

$$RT \ln a_1 \frac{1}{x} = E_1 + \frac{E_{11}}{2} - (1 - x)^2 \frac{E_{22}}{2} - RT \ln K_1(T) \quad (1)$$

$$RT \ln a_2 \frac{x}{1 - x} = E_2 + E_{12} + E_{22} - xE_{22} - RT \ln K_2(T) \quad (2)$$

Subscript 1 refers to the metal and subscript 2 to the nonmetal; a_1 and a_2 are the activities of the components; E_1 and E_2 are the energy required to transfer one mole of material from standard state to the NaCl-type structure; E_{11}, E_{22}, and E_{12} are the interaction energies among the various types of atoms; and $R \ln K_1$ and $R \ln K_2$ are the differences in vibrational entropy between the material in the standard state and in the NaCl-type structure.

The model used in deriving Eqs. (1) and (2) is equivalent to the regular[6] solution approximation for substitutional solutions. Because in the present case a fixed frame is available (the metal sites are always filled the same way), this approximation should represent the experimental data better than in the substitutional case. In the derivation of Eqs. (1) and (2) the small amount of Schottky defects present in the metal and nonmetal sublattice has been ignored, and also the possible Frenkel disorder.

Inspection of Eq. (1) shows that $E_1 + E_{11}/2$ is the energy difference between pure metal in an fcc (face-centered cubic) structure having the lattice parameter of the compound investigated (all interstitial sites empty, $x = 1$) and in the standard state. From Eq. (2) it follows that $E_2 + E_{12} + E_{22}$ is the energy difference between the nonmetal in the stoichiometric compound ($x = 0$) and in the standard state.

E_{22} is the interaction energy between nonmetal atoms in the NaCl-type structure and also the interaction energy between nonmetal vacancies.[2] E_{22} is composed, as pointed out,[2] of two terms: E_{22}(strain) and E_{22}(chemical). The strain term is positive, whereas the chemical term can be positive or negative depending on E_{12}, the bonding between the metal atom and the nonmetal atom. Both components of E_{22} are dependent on the distance between the two defects. To be able to make meaningful comparisons, systems with similar lattice parameters (and distances between defects) must be used. Zr and Hf atoms, on the one hand, and Nb and Ta atoms, on the other, have the same radii and are also in the same group of the periodic table. Their carbides and nitrides can be used to study the variation of E_{22} and the other energy term. Other compounds with metals of the first long period of the periodic table are also included for further comparison.

In several of the systems investigated, not enough data were available to evaluate all the energy terms. In those cases, it was assumed that the energy difference between fcc metal and metal in the standard state $E_1 + E_{11}/2 = (1 \pm 1)$ kcal/mole. This quantity must be positive but cannot be too large as the group IV metals can be present in an hcp (hexagonal close-packed) structure.

EXPERIMENTAL RESULTS

The experimental results of various authors were reevaluated in terms of Eqs. (1) and (2). The standard state was taken as bcc (body-centered cubic) metal, pure graphite, and N_2 at 1 atm pressure. The properties of the elements in the standard state were taken from Elliott and Gleiser.[7] The entropy of formation of the nitrides was also taken from Elliott and Gleiser[7]; the entropy of formation of the carbides was taken from Elliott and Gleiser[7] and from Rudy and Chang.[8] For the nitrides, the sum of the partial entropies of metal and nonmetal was taken as equal to the entropy of formation. For the carbides, it was assumed that the partial entropy of C was the entropy of formation of the carbide. This introduces a small error into the computations.

Group IV Carbides

ZrC. Many pressure measurements of C and Zr in ZrC_{1-x} are available; however, the data are not very consistent. From the vapor-pressure measurements of Pollock[9] for the Zr activity in ZrC_{1-x} equilibrium with graphite,

$$RT \ln a_{Zr} = -40.5 \pm 0.8 \text{ kcal/mole at } 2700°\text{K} \tag{3}$$

TABLE 1 ***Activity of Zr and C in ZrC_{1-x}***

Author	x	Temperature range, °K	$-RT \ln a_C[x/(1-x)]$, kcal/mole	$-RT \ln a_{Zr}(1/x)$, kcal/mole
Pollock[9]	0.1	2641–2747	28.72 ± 0.4	11.1 ± 0.4
Coffman[10]	0.1	2674–2818	28.7 ± 1.0	10.9 ± 2.3
Blackburn[12]	0.03	2717–2859	33.3 ± 1.2	10.8 ± 0.9

From the resonance-line absorption measurements of Coffman, Kibler, Lyon, and Acchione,[10]

$$RT \ln a_{Zr} = -40.5 \text{ kcal/mole at } 2747°\text{K} \tag{4}$$

Pollock[9] and Coffman et al.[10] studied the vaporization of ZrC in vacuum. When stoichiometric ZrC was first vaporized, the lattice parameter increased, and with further vaporization, remained constant. From the lattice parameter–composition curve given by Sara,[11] it can be concluded that the constant-boiling composition they observed is at $ZrC_{0.90}$.

Blackburn[12] measured the Zr and C pressure on starting materials of composition $ZrC_{0.97}$ and $ZrC_{0.67}$, respectively. His data on the $ZrC_{0.67}$ sample show a trend in pressure probably due to rapid composition change, and are discarded. Thus the data in Table 1 are available for evaluating the energy terms in Eqs. (1) and (2). From these data and the additional data given earlier,

$$RT \ln a_{Zr} \frac{1}{x} = (1 \pm 1) - (1 - x)^2(14 \pm 1.5) \text{ kcal/mole} \tag{5}$$

$$RT \ln a_C \frac{x}{1-x} = -(35.2 \pm 4) - (28 \pm 3)x + 2.7 \times 10^{-3}T \text{ kcal/mole} \tag{6}$$

HfC. There are two measurements on the vaporization of HfC, one by Blackburn[12] ("preliminary data") and one by Coffman et al.[10] Blackburn[12] used a sample with initial composition of $HfC_{0.97}$. Coffman[10] noted an increase in lattice parameter after the first vaporization similar to ZrC. Assuming that the lattice parameter–composition curve for HfC is similar to that for ZrC, then the constant-boiling composition is below stoichiometry. This is also borne out by the fact that the C/Hf pressure ratio is lower in the data of Coffman[10] than in the data of Blackburn.[12] Both authors measured the pressure of pure Hf; each author's own data were used to evaluate the Hf activity in HfC. Combining both authors' data, the following equations are obtained:

$$RT \ln a_{\mathrm{Hf}} \frac{1}{x} = (1 \pm 1) - (1 - x)^2(7.5 \pm 4) \text{ kcal/mole} \tag{7}$$

$$RT \ln a_{\mathrm{C}} \frac{x}{1-x} = -(47 \pm 10) + 15(\pm 8)x + 2 \times 10^{-3}T \text{ kcal/mole} \tag{8}$$

These equations place the constant-boiling composition between $x = 0.03$ and $x = 0.1$.

Group IV Nitrides

TiN. McClaine and Coppel[13] measured the N and Ti activities in TiN by measuring the N_2 pressure above TiN_{1-x} as a function of composition, $0.8 < x < 1.0$, and temperature, $1500 < T < 1950°K$. Three measurements of the Ti pressure were also carried out. From their data, one obtains for the N_2 pressure

$$RT \ln p_{\mathrm{N}_2}{}^{1/2} \frac{x}{1-x} = -(57.5 \mp 9.0) - (78.2 \pm 9.1)x + 18.6 \times 10^{-3}T \text{ kcal/mole} \tag{9}$$

$E_1 + E_{11}/2$ can be evaluated from the Ti pressure data of McClaine and Coppel[13] and also from the data of Linevsky.[14] Linevsky[14] measured by a line-absorption method the Ti pressure above TiN as a function of the N_2 pressure. From Eq. (9) the composition of the solid corresponding to the

TABLE 2 ***Evaluation of $E_1 + E_{11}/2$ in TiN***

Author	x	N_2 pressure, atm	Temperature, °K	$E_1 + E_{11}/2$, kcal/mole
McClaine and Coppel[13]	0.133	. . . .	1885	20.0
	0.133	. . . .	1896	19.5
	0.185	. . . .	1930	16.0
Linevsky[14]	. . . .	0.066	2120	16.6
		0.203	2120	16.7
		0.335	2120	17.0
		0.440	2120	17.1
		0.066	2088	16.0
		0.066	2140	16.1
				Avg. 17.2 ± 4

N_2 pressure can be calculated; from this and the Ti pressure, $E_1 + E_{11}/2$ is evaluated. The values of $E_1 + E_{11}/2$ are summarized in Table 2. The agreement between the two authors is very good. The Ti activity can thus be expressed as

$$RT \ln a_{\mathrm{Ti}} \frac{1}{x} = (17.2 \pm 4) - (1 - x)^2(39.1 \pm 4.5) + 3.3 \times 10^{-3}T \text{ kcal/mole} \quad (10)$$

The large value of $E_1 + E_{11}/2$ is surprising, indicating an extremely large energy to transform bcc Ti to an expanded fcc Ti.

ZrN. Kibler, Lyon, Linevsky, and Desantis[15] measured the N_2 pressure above ZrN as a function of composition, $0.03 < x < 0.28$, and temperature, $1596 < T < 2360°K$, respectively. Their data (series I, III) can be represented as

$$RT \ln p_{\mathrm{N}_2}{}^{1/2} \frac{x}{1 - x} = -72.1 - 19.0x + 17.4 \times 10^{-3}T \pm 2.0 \text{ kcal/mole} \quad (11)$$

The data in Eq. (11) on ZrN_{1-x} can be combined with the results of Linevsky[14] to obtain the activity of Zr as a function of composition.

Linevsky[14] also measured the Zr pressure as a function of the N_2 pressure by a line-absorption method. He used the experimental temperature of 2522°K and three N_2 pressures, 0.132, 0.263, and 0.395 atm. From Eq. (11) the composition of ZrN_{1-x} corresponding to those three N_2 pressures can be computed, and, together with the Zr pressure, $E_1 + E_{11}/2$, evaluated. One obtains $E_1 + E_{11}/2 = 0.85 \pm 0.40$ kcal/mole. Thus, the activity of Zr in ZrN_{1-x} can be expressed as follows:

$$RT \ln a_{\mathrm{Zr}} \frac{1}{x} = 0.85 - (1 - x)^2 \times 9.5 + 4.1 \times 10^{-3}T \pm 2.0 \text{ kcal/mole} \quad (12)$$

HfN. Kibler, Lyon, Linevsky, and Desantis[15] measured the N_2 pressure above HfN as a function of composition, $0 < x < 0.18$, and temperature, $1490°K < T < 2356°K$. Their data can be represented by

$$RT \ln p_{\mathrm{N}_2}{}^{1/2} \frac{x}{1 - x} = -(72.40 \pm 1.6) + (15.8 \pm 0.7) \times 10^{-3}T \text{ kcal/mole} \quad (13)$$

Group V Carbides

VC. Fujishiro and Gokcen[16] measured the pressure of V above VC_{1-x} in equilibrium with graphite in the temperature range 2350 to 2550°K. The sample has a composition of $VC_{0.88}$ when in equilibrium with graphite. From these data and the assumptions mentioned earlier,

$$RT \ln a_V \frac{1}{x} = (1 \pm 1) - (1 - x)^2(12.5 \pm 2.3) \text{ kcal/mole} \tag{14}$$

$$RT \ln a_C \frac{x}{1 - x} = -(10.6 \pm 4.6) - (25 \pm 4.6)x + 1.7 \times 10^{-3}T \text{ kcal/mole} \tag{15}$$

NbC. Fries[17] studied the vaporization of NbC_{1-x} by the Langmuir method at 2940°K and also obtained, at certain compositions and temperatures, the partial pressures of Nb and C. In order to use his results where only weight-loss ratios but no sample areas or times are given, Eqs. (1) and (2) have to be combined and transformed. For simplicity, let $y = 1 - x$. By definition, the activity of each component is

$$RT \ln a = RT \ln p - RT \ln p^\circ \tag{16}$$

where p is the pressure of the component in the system and p° is the pressure of the material in its standard state. The vaporization studies by Fries[17] were carried out using the Langmuir method; the pressure of each component can be expressed as

$$RT \ln p = RT \ln \left(w\sqrt{\frac{2\pi RT}{M}} \right) \tag{17}$$

where w is the rate of evaporation in grams per square centimeter per second, and M is the molecular weight of the species. Subtracting Eq. (2) from Eq. (1) and introducing Eqs. (16) and (17) gives

$$RT \ln \frac{w_{Nb}}{w_C} \frac{y}{(1 - y)^2} = E_1 + \frac{E_{11}}{2} - E_2 - E_{12} - E_{22}\left(y + \frac{y^2}{2}\right) + RT \ln \overset{\circ}{p}_{Nb} - RT \ln \overset{\circ}{p}_C + RT \ln K_2(T) - RT \ln K_1(T) + RT \ln \sqrt{\frac{M_{Nb}}{M_C}} \tag{18}$$

Equation (18) now is applied to the data of Fries,[17] to his table 2 and also to those data in his table 1 which contain the molar ratio of C to Nb in the gas phase.

From Elliott and Gleiser[7] and Rudy and Chang,[8]

$$RT \ln \overset{\circ}{p}_{Nb} - RT \ln \overset{\circ}{p}_C + RT \ln K_2(T) - RT \ln K_1(T) + RT \ln \sqrt{\frac{M_{Nb}}{M_C}}$$

is evaluated as -1.25 kcal/mole at 2940°K and at 2679°K. Thus, if

$$RT \ln \frac{w_{\mathrm{Nb}}}{w_{\mathrm{C}}} \frac{y}{(1-y)^2} \tag{19}$$

is plotted versus

$$y + \frac{y^2}{2} \tag{20}$$

a straight line results. Fries's[17] data are plotted in Fig. 1. The points for low values of y at 2940°K and the point for a high value of y at 2679°K fall on a straight line, whereas the points for high values of y at 2940°K fall much higher than this straight line. If we inspect the ordinate (19), the number calculated is too high because either w_{Nb} is high or w_{C} is small. Fries[17] pointed out that in his samples he had a concentration gradient, the C content being low at the surface. This indicates that the rate of C evaporation is too low at high values of y at 2940°K. At 2679°K, apparently the C concentration is not low at the surface. This change of C concentration with temperature during evaporation can be expected. The activation energy for C diffusion is much lower than the activation energy for C evaporation from the system. Thus, as the temperature is raised, the rate of C evaporation increases more rapidly than the rate of C diffusion. At the low temperature, the C diffusion is sufficient to maintain

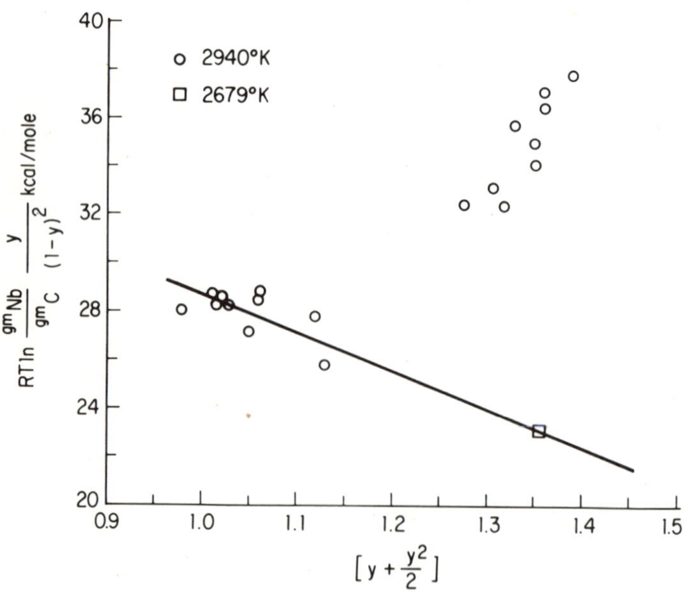

Fig. 1 Activity vs. composition plot for NbC, to evaluate E_{22}.

uniform sample concentration; at the high temperature, it is not. The values of the ordinate (19) at low values of y at 2940°K are correct because they are close to the constant-boiling composition ($y = 0.75$), and, therefore, no large C diffusion is required to maintain uniformity of the sample. From Eq. (18) and the straight line in Fig. 1,

$$E_{22} = 15.2 \pm 2.5 \text{ kcal/mole} \tag{21}$$

Using the only point where Fries[17] gives rate-of-evaporation data and which does not show a concentration gradient in the sample due to evaporation ($y = 0.926$, $T = 2679$°K), the other constants in Eqs. (1) and (2) can be evaluated.

$$RT \ln a_{\mathrm{Nb}} \frac{1}{x} = -(0.05 \pm 1.0) - (1 - x)^2(7.6 \pm 1.3) \text{ kcal/mole} \tag{22}$$

$$RT \ln a_{\mathrm{C}} \frac{x}{1 - x} = -(28.6 \pm 2.5) - (15.2 \pm 2.5)x + 0.4 \times 10^{-3}T \text{ kcal/mole} \tag{23}$$

TaC. Coffman et al.[18] studied the evaporation of TaC_{1-x} as a function of temperature. They determined the composition before and after the experiment by chemical analysis. Assuming that the average composition can be taken as the composition of the run, from the run at 2633°K (where errors due to diffusion are the smallest) one obtains at $x = 0.09$

$$RT \ln \frac{a_{\mathrm{Ta}}}{a_{\mathrm{C}}} \frac{1 - x}{x^2} = 25.6 \text{ kcal/mole} \tag{24}$$

Juenke and Sjodahl[19] measured the C activity in TaC_{1-x} by equilibrating the TaC_{1-x} samples with H_2-methane mixtures at various temperatures. The composition range covered was $0.9 < x < 1$; the temperature range was $2100 < T < 2600$°K. The composition of the solid phase was obtained by determining its lattice parameter. The composition of the gas, determined by chemical analysis, gave the C activity. Juenke and Sjodahl's data give

$$RT \ln a_{\mathrm{C}} \frac{x}{1 - x} = -27.0 \pm 1.3 \text{ kcal/mole} \tag{25}$$

With the other assumptions,

$$RT \ln a_{\mathrm{Ta}} \frac{1}{x} = 1 \pm 1 \text{ kcal/mole} \tag{26}$$

From these two equations,

$$RT \ln \frac{a_{\mathrm{Ta}}}{a_{\mathrm{C}}} \frac{1-x}{x^2} = 28.0 \pm 1.6 \text{ kcal/mole} \tag{27}$$

which is in good agreement with the values of Coffman et al.[18]

DISCUSSION

Table 3 contains all the energy terms that were calculated. $E_1 + E_{11}/2$, the energy difference between fcc metal with the lattice parameter of the interstitial phase and the bcc standard state, is about 1 kcal/mole, in agreement with the assumptions made earlier. The exception is TiN where the value of $E_1 + E_{11}/2$ is very large and no explanation can be given. Because of this, the results on VC are in doubt (V is also in the first long period); a calculation using the large value of $E_1 + E_{11}/2$ found for TiN gives the data in the brackets. E_{22} then increases, but $E_2 + E_{12} + E_{22}$ does not change much.

The values of $E_2 + E_{12} + E_{22}$ (the energy difference between the nonmetal in its standard state and at the stoichiometric composition in the interstitial phase) are equal in the pairs of compounds that are in the same group and have the same lattice parameter, that is, ZrC-HfC (here the errors are quite large), ZrN-HfN, and NbC-TaC. Thus, the bonding of the nonmetal at stoichiometry is the same.

E_{22} is always positive, and in the pairs of compounds increases as the atomic weight of the metal increases. It can be assumed that the strain term is the same in the pair; therefore, the decrease in E_{22} must be due to a better screening by the heavier element.

TABLE 3 ***Energy Terms in Nitrides and Carbides***

Material	$E_1 + E_{11}/2$, kcal/mole	E_{22}, kcal/mole	$E_2 + E_{12} + E_{22}$, kcal/mole
ZrC	....	28.0 ± 3.0	−35.2 ± 4.0
HfC	....	15.0 ± 8.0	−47 ± 10.0
TiN	17.2 ± 4	78.2 ± 9.0	−57.5 ± 9.0
ZrN	0.85 ± 0.4	19.0	−72.1 ± 2.0
HfN	....	0	−72.4 ± 1.6
VC	....	25 ± 4.6	−10.6 ± 4.6
		(67 ± 8)	(−4.1 ± 8)
NbC	−0.05 ± 1.0	15.2 ± 2.5	−28.6 ± 2.5
TaC	....	0	−27.0 ± 1.3

From the equality of $E_2 + E_{12} + E_{22}$, it follows that when E_{22} decreases, E_{12} becomes less negative (as E_2 will not change much), implying a weaker bond between metal and nonmetal atom. E_{22} is much larger in the first long period than in the second and third periods; thus in the compounds formed by metals of the first long period, the bonding between metal and nonmetal is probably more ionic.

Inspection of Eq. (2) and study of the variation of the activity of nonmetal with composition show that a large positive value of E_{22} changes the activity very rapidly with composition and increases the activity to unity at large values of x, i.e., far from stoichiometry. Thus, the preparation of a "stoichiometric" phase becomes easier as one goes down in a group. Also, the stability is less composition-sensitive with the heavier elements.

REFERENCES

1. Hoch, M., A. S. Iyer, and J. Nelken: *J. Phys. Chem. Solids* **23**:1463 (1962).
2. Hoch, M.: *Trans. Met. Soc. AIME* **230**:138 (1964).
3. Hoch, M.: *J. Phys. Soc. Japan* **18**(suppl. II):147 (1963).
4. Hoch, M., and D. Ramakrishnan: *J. Phys. Chem. Solids* **25**:869 (1964).
5. Blandin, A., and J. L. Déplanté: Size Effect and Chemical Interactions, in J. Friedel and A. Guinier (eds.), "Metallic Solid Solutions," W. A. Benjamin, Inc., New York, 1963.
6. Guggenheim, E. H.: "Mixtures," Clarendon Press, Oxford, 1952.
7. Elliott, J. F., and M. Gleiser: "Thermochemistry for Steelmaking," Addison-Wesley Publishing Company, Inc., Reading, Mass., 1960.
8. Rudy, E., and V. A. Chang: Second Progress Report, Contract AF 33(615)-1249, Oct. 1, 1964.
9. Pollock, B. D.: *J. Phys. Chem.* **65**:731 (1961).
10. Coffman, J. A., G. M. Kibler, T. F. Lyon, and B. D. Acchione: Carbonization of Plastics and Refractory Materials Research, Technical Report WADD-TR-60-646, part 2, January, 1963.
11. Sara, R. V.: *J. Am. Ceram. Soc.* **48**:243 (1965).
12. Blackburn, P. E.: Thermodynamic and Kinetic Studies for a Refractory Materials Program, Technical Report ASD-TDR-62-204, part 3, April, 1964.
13. McClaine, L. A., and C. P. Coppel: Equilibrium Studies of Refractory Nitrides: Details of Apparatus and Studies of the TiN System, Technical Report AFML-TR-65-299, part 1, September, 1965.
14. Linevsky, M. J.: Contract AF 33(615)-1150, General Electric Co. Report, Philadelphia, Pa., Nov. 30, 1964.
15. Kibler, G. M., T. F. Lyon, M. J. Linevsky, and V. J. Desantis: Carbonization of Plastics and Refractory Materials Research, Technical Report WADD-TR-60-646, part 3, vol. 2, April, 1964.
16. Fujishiro, S., and N. A. Gokcen: *J. Electrochem.* **104**:835 (1962).
17. Fries, R. J.: *J. Chem. Phys.* **37**:320 (1962).
18. Coffman, J. A., G. M. Kibler, T. R. Riethof, and A. A. Watts: Carbonization of Plastics and Refractory Materials Research, Technical Report WADD-TR-60-646, part 1, February, 1961.
19. Juenke, E. F., and L. H. Sjodahl: Private communication.

Agenda Discussion:
INTERSTITIAL PHASES AND SOLUTIONS

chairman **R. A. Oriani**
Fundamental Research Laboratory, U.S. Steel Corporation, Monroeville, Pennsylvania

secretary **A. L. Bement**
Battelle Memorial Institute, Northwest Research Laboratory, Richland, Washington

The topics that were suggested by the chairman for discussion during this session are:

1. What is the state of ionicity of an interstitial solute and what are the physical consequences?
2. What long-range effects are produced by interstitial solutes both in the electronic structure and in the atomic distribution?
3. To what extent can elasticity theory be used to describe the interactions between interstitial solutes and between solutes and lattice defects?

In order to initiate the discussion, the chairman remarked that it is very important to keep in mind, as had already been emphasized in McLellan's paper, that in a statistical treatment of interstitial solutions the fact that there are a fixed number of interstitial sites that may be occupied by the interstitial atoms makes it mandatory that a Langmuir-isotherm type of description be used. This means that one can expect that in the absence of solute-solute interactions, a straight-line relationship will not be found between activity and concentration. Rather, the analogue to Henry's law is that the activity of the interstitial solute is proportional to $\theta/(1 - \theta)$, where θ is the fraction of interstitial sites that are occupied.

In general, however, one should expect interactions between interstitial solutes. For example, Powers and Doyle[1] measured an attractive interaction between dissolved oxygen atoms in niobium of about 2 kcal/g atom, in good agreement with Fisher's[2] elasticity-theory calculations which took account of the dipole character of the elastic distortion.

Interaction of the interstitial solute with other solute species or with structural imperfections of the lattice may also be important. This morning we heard of the possibility that some interstitial sites may be blocked from occupancy by solute atoms in adjacent sites. Figure 1, from the work of R. P. Smith,[3] demonstrates a rather striking example of this effect and shows isoactivity lines for carbon in fcc iron solid solution for the Fe-Si-C system. Darken[4] has pointed out that when Si replaces Fe substitutionally, it is surrounded by six interstitial sites as nearest neighbors. If one assumes that each of these six nearest-neighbor sites is blocked against occupancy by carbon, then 1 a/o of silicon would block 6% of the sites, or 1 wt % Si would block 12% of the sites. Therefore, 8⅓ wt % Si should result in zero carbon solubility. It is interesting that if one extrapolates the initial portion of any of the isoactivity lines of Fig. 1, the horizontal axis cuts at about 8 wt % Si in agreement with this simple argument.

Interaction of interstitial solutes with vacancies, internal surfaces, or dislocations may have important thermodynamic consequences when the intrinsic solubility of the interstitial component is very low because additional solute content held at sites affected by such structural singularities may be comparable to, or larger than, the solute content at normal lattice sites. Wriedt and Darken[5] have measured the additional nitrogen taken up by cold-worked mild steel and have shown that the excess nitrogen is held at essentially two kinds of extraordinary sites which have been deduced to be associated with edge dislocations and with screw dislocations, respectively. The free energy $\Delta F°$ for transfer from normal sites to sites at edge dislocations was found to be $-20.5 + 0.006T$ kcal/g atom, and the corresponding quantity for transfer to sites associated with screw dislocations was found to be $-2.6 - 0.01T$ kcal/g atom.

Kinetic phenomena may also be affected by attractive interactions between interstitial solutes and structural imperfections. For example, the annealing out of neutron irradiation damage in α-Fe-C alloys shows the effect[6] of an attractive interaction between lattice vacancies and interstitial carbon, and indeed a binding energy of 0.41 ev has been deduced from the annealing experiments.

The diffusivity of carbon in α-Fe has been shown by Smith[7] not to follow an Arrhenius plot, and Homan[8] has invoked the possibility that with increasing temperature, more and more of the carbon flux is carried by carbon-vacancy complexes. Similar reasoning has been applied[9] to deduce a plausible explanation of the marked deviation from a linear Arrhenius plot for the self-diffusion in various bcc metals. In these cases, the thought is that adventitious interstitial solutes, such as oxygen and nitrogen, markedly increase the population of lattice vacancies through an attractive interaction, leading to an anomalous temperature dependence of the self-diffusion.

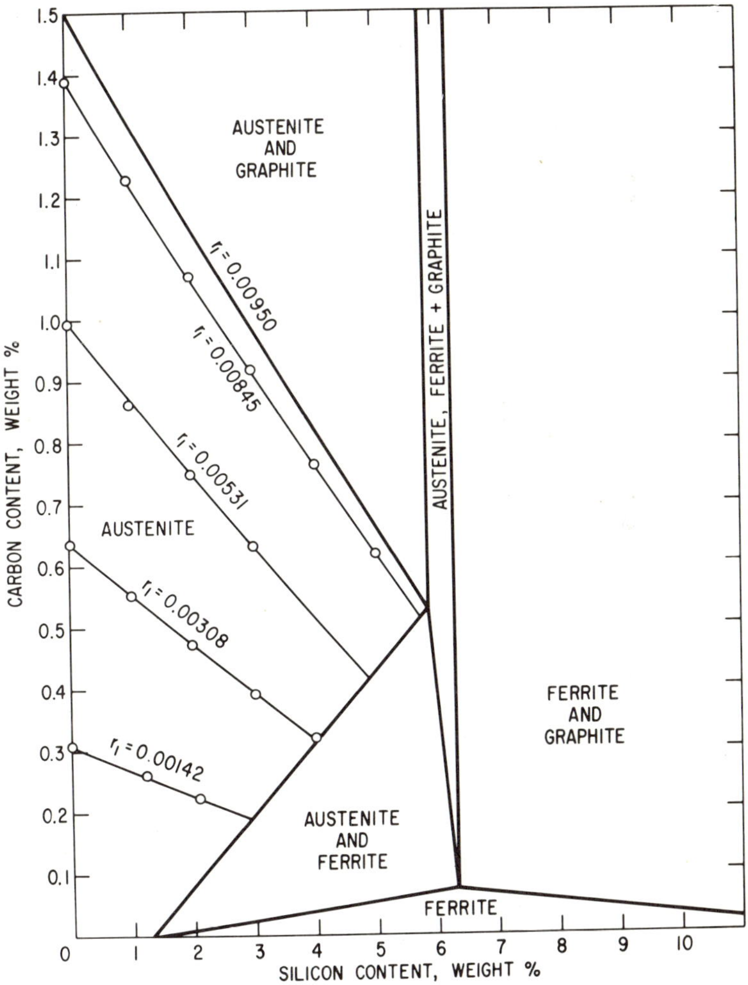

Fig. 1 Isoactivity lines for carbon in austenite at 1000°C. (*After R. P. Smith.*) The parameter r signifies the ratio $p_{CH_4}/p_{H_2}{}^2$.

The chairman proceeded to the question of the state of ionization of the interstitial solute. He remarked that it would be valuable to put a stop to the notion that electromigration experiments can shed light on the state of ionicity of solutes in metals. Table 1 summarizes recent data on the effective charge of interstitial solutes derived from experiments on electromigration. It was pointed out that the very large values for N in α-Fe and C in γ-Fe, as well as the change of effective charge with isotopic mass, are completely incomprehensible from the point of view that these effective charges represent the state of ionicity of the interstitial solute as it rests in the lattice. It was stressed that these numbers are related to the very complex problem of momentum transfer between current carriers and those solute atoms which are in the act of changing sites. Therefore, electromigration depends upon a kinetic property which is only distantly related to the electronic state of the solute at rest.

Another concept that the chairman wished to put to rest is that hydrogen dissolved in a metal does not distort the lattice in any essential way, the usual reason given being that the hydrogen exists merely as a proton in the lattice. The question of what size one should attribute to a solute in a metal is a difficult one even conceptually, but the partial molal volume, or the amount by which the volume of a very large amount of solution changes by the introduction of one gram atom of solute, should offer an important starting point in such considerations. Table 2 assembles data for the partial molal volume of hydrogen in various bcc metals, as well as for nitrogen and carbon for comparison. One sees that the disturbance produced by the hydrogen can be very large, and that it varies from one lattice to another.

TABLE 1 ***Effective Charges of Interstitial Solutes in Metals***

Unit is the electronic charge

Solute	Solid solvent			
	α-Fe	Ni	γ-Fe	Co
C	~+4*	+1.5*	~+4† +9 to +11*	+2.6*
N	+8‡			
H	+0.23§	+0.50§		
D	+0.38§	+0.61§		

* I. I. Kovenskii, *Soviet Phys.—Solid State* **5**:1036 (1963).

† P. Dayal and L. S. Darken, *Trans. Met. Soc. AIME* **188**:1156 (1950).

‡ W. J. McVeagh and J. C. M. Li, Unpublished measurements.

§ R. A. Oriani and O. D. Gonzalez. To be published.

TABLE 2 ***Partial Molal Volume of Some Interstitial Solutes***

Solute	Solvent lattice	$\bar{V}_i$, cm^3/g atom	Reference
H	Pd	1.73	W. T. Lindsay, Jr., Abstracts of American Chemical Society National Meeting, Pittsburgh, March, 1966
H	β-Ti	4.1	Calculated from data of A. D. McQuillan, C. E. Ellis, N. Pessall, A. P. Bennett, J.
H	β-Zr	5.3	Basterfield, and A. D. Wallbank, report from the University of Birmingham, 1964
H	Ta	9.4	Calculated from data in M. Hansen, "Constitution of Binary Alloys," 2d ed., McGraw-Hill Book Company, New York, 1958
H	α-Fe	2.0	R. A. Oriani, *Trans. Met. Soc. AIME*, to be published.
N	α-Fe	5.6	H. A. Wriedt and L. Zwell, *Trans. Met. Soc. AIME* **224**:1242 (1962).
C	α-Fe	~6	H. W. Wagenblast. Private communication

To conclude his remarks, the chairman referred to the effect of internally generated elastic stress upon the solubility of compounds one element of which dissolves interstitially in the adjoining solid solution. Suppose, for example, that a carbide is allowed to precipitate and grow at a temperature at which the carbon can diffuse easily in the matrix phase, but the principal component of the matrix cannot. The precipitation will then generate a stress system in the lattice which will be felt by a spherical precipitate particle as a hydrostatic pressure. If the particles are sufficiently small, dislocation loops will not be generated to relieve the stresses; at larger sizes, dislocation loops will partially relieve the stresses. Under these conditions one can have equilibrium with respect to the mobile component,[10] though not with respect to the immobile component, and the equilibrium solubility will be increased by the stresses. This effect can be sizable and is probably responsible for differences among reported low-temperature solubilities of compounds.

With these remarks the chairman turned to the proposed topics for general discussion and asked for contributions in the area of the state of ionicity of interstitial solutes in compounds and solutions. Nowotny pointed out that measurements of magnetic susceptibility give definite indication that in palladium, hydrogen is present as a proton. The same seems to be true for the hydrides and solutions of hydrogen in the rare earths, titanium, and zirconium from band-theoretical and geometrical considerations. However, the very large bonding energies in TiH_2 and

UH_3, in which the bonding energy is larger than -30 kcal/mole, raises the question of the possibility of bonding as in the saltlike hydride of lithium. Nowotny opined that among the hydrides there exists the possibility of a continuous gradation between H^- and H^+ as one goes from the alkali metals toward the transition metals.

In the case of oxides, Nowotny remarked that there have been various recent calculations based mainly upon the orthogonalized-plane-wave and the augmented-plane wave methods. For example, Bilz[11] has shown that TiO is ionic, and similar conclusions have been reached by Ern and Switendick.[12] However, TiO contains about 15% vacancies in both sublattices. There is stronger evidence that NbO is also ionic and that the vacancy concentration in both sublattices reaches 25%, and in addition that the vacancies are completely ordered. This ordering is explained today in a qualitative manner by introducing the so-called cation-cation bonds, as did Van Arkel.[13] This concept is similar to that discussed by Pearson in a former session, and is identical with the considerations of Laves on the NiAs type of phase which involve the linking of metal atoms into chains.

Combining the ideas of Van Arkel with those of the physicists, Nowotny arrives at the following picture for NbO. Consider two interpenetrating simple cubic lattices of niobium ions and of oxygen ions. The latter will then form an octahedral array. Of the five electrons of the Nb, two are transferred to the oxygen. Of the three remaining electrons, one will be shifted into a conduction band, and the other two will form d_{xy} and d_{yz} orbitals, giving rise to additional bonding between Nb and $Nb.^{++}$

Nowotny then referred to the nitrides and carbides, in which, according to papers by Bilz, Costa, Ern and Switendick, and by Lye at a meeting in Paris last September, the bonding mechanism is similar to that in NbO. In TiC, for instance, there is an ionic component with electron transfer from Ti to C according to Ern and Switendick.

Friedel remarked that Ern's computations are by the augmented-plane-wave method, and that it is difficult in such a method to decide how much electron transfer actually occurs, or even the direction in which it takes place. Friedel further stated that, since TiO has some electrical conductivity, it is difficult to imagine the possibility of a large degree of electron transfer, though this is quite possible for a poor conductor. With this observation both Nowotny and Brewer disagreed. Brewer pointed out that in going from Ce_3S_4 to Ce_2S_3, which has cerium vacancies, one goes from a metallic conductor to a typical refractory compound. Furthermore there is no appreciable change in lattice parameter despite the change in vacancy concentration. It would seem that we have here contributions from both metallic and ionic bonding. Brewer further observed that the adjoining phase, CeS, actually looks like copper and has good electrical conductivity; since there are still unpaired electrons, the bonding is even more metallic. ThS has the same crystal structure as CeS, but has a higher melting point. All the electrons in ThS are used in bonding, but it has an even higher metallic conductivity than CeS. This suggests that there exists a gamut of possibilities in this class of compounds.

Pearson suggested that one might describe TiO in terms of an outer, filled valence band corresponding to the ionic state $Ti^{++}O^{=}$ with the two Ti *d* electrons directed forward, and overlapping with those on neighboring Ti atoms, giving a *d* band. Although the electrical conductivity (and carrier mobility) is relatively low, its temperature coefficient is metallic, suggesting that one observes pure *d*-band conductivity; however, the bond would be basically ionic otherwise. The main question seems to be whether or not it is possible to sustain a filled ionic band together with the *d* band, if the two bands are not overlapping and if the lower one is the ionic band. Friedel said that he has no objections to this kind of description. He is prepared to believe that in TiO the oxygen atoms contain filled shells and that the *d* electrons of the titanium overlap to give a metallic-conduction band, and that according to Nowotny this would correspond to an ionic compound.

Nowotny reminded us that the former concept of the so-called interstitial compound is that the smaller atoms give up electrons to form strong metal-metal bonds. However, the idea has developed that there may be strong bonding between the metallic and the nonmetallic atoms, especially as the distances involved are smaller. In complex carbides and nitrides these distances are even shorter, and one finds strong bonds between the metallic and the nonmetallic atoms and weaker metal-to-metal bonds.

In relation to the conduction mechanism in TiO, Hoch reported on the work of Morin on VO_2, which ranges from semiconducting to metallic in its properties. Morin explains the conduction as being due to overlap of the 3*d* orbitals of the metallic atoms and also to overlap of these with the 2*p* orbitals of the oxygen. The effect of a change in the V/O ratio is to produce a change in the number of electrons in the 3*d* band. Morin would expect similar behavior in the oxides of scandium and titanium.

Laves stated that in $TiO_{0.63}$ the oxygen occupies the hexagonal lattice in octahedral positions, and wondered whether the oxygen gives up electrons to the titanium or takes up electrons. Nowotny alluded to Speiser's conclusion, that there is insufficient space in the octahedral interstices surrounded by Ti atoms for oxygen anions.

There is evidence, according to Wallace, that in the lanthanide hydrides the hydrogen absorbs electrons from the conduction band to form hydride ions. These hydrides are orthorhombic and isostructural with barium and strontium hydrides, which are established ionic compounds. With increasing hydrogen content, the electrical conductivity of the lanthanide hydrides falls by six or seven orders of magnitude, and the temperature coefficient of electrical resistivity becomes opposite to that for metals. In addition, gadolinium, which has a magnetic transition at about 300°K, has the magnetic transition temperature lowered by hydrogenation to 2°K. These facts are taken as evidence that hydrogen is present as an anion in the lanthanide hydrides.

McQuillan contrasted the behavior of the hydrides of the rare earths and actinides with that of the hydrogen solutions in titanium and zirconium. In the latter, the electrical resistance first rises with increasing hydrogen content up to about 20 a/o H, but it drops rapidly with further increase of hydrogen concentration. Indeed, solutions containing 55 a/o hydrogen are much more metallic than is the pure metal. Also, hydrogen additions increase the magnetic susceptibility of these metals. These facts are consistent with the idea that the electrons from the hydrogen atoms enter the conduction band to produce a higher density of states at the Fermi level.

However, Lomer issued a warning on taking the disappearance of metallic conductivity as evidence of ionicity. He pointed out that extensive calculations would be necessary in order to show, for example, that gadolinium hydride when treated in a molecular-orbital fashion would not simply show a large gap. Lomer also stated that, if there is a Fermi surface in a substance, electrostatic screening of an impurity will be complete; hence, in agreement with Oriani with respect to the interpretation of electromigration experiments, one cannot conceive of any component in a metallic matrix carrying a net charge with it. One can say that the question of ionicity of a solute species is simply one of how big is the screening radius, and where one chooses to draw a boundary. In addition, Lomer remarked that one of the few comparatively clear criteria for ionicity involves the infrared activity in the vibrational spectra.

In reply to a question from Oriani concerning the physical implications of the distinction between a bound electronic state about an impurity and screening by the Fermi electrons, Friedel replied that in conductors one can talk of bound states only in the one-electron approximation, and that with many-electron corrections the meaning of a bound state loses its sharpness. Furthermore, it is easily shown that, as far as the electrons at the Fermi surface are concerned, there is no great difference between screening with and without bound states in fairly free electron cases such as CuZn, Ga, Ge. Only in special cases such as the NiCr, V, Ti discussed previously is the concept of an impurity bound state of interest for electrons at the Fermi level. Even then, it is strongly broadened by resonance into a virtual bound level.

Blandin addressed himself to the second topic suggested by the chairman for discussion, the long-range electronic disturbances generated by an interstitial solute. For the case of hydrogen in palladium there are two problems: what happens in the *d* band and in the *s* band, and the nature of the screening of the proton by the conduction electrons. Since the proton vibrates within its interstitial site, consideration must be given to the screening as a function of position. Calculations of this type have been made by Fert, and these agree with his measurements of the variation of the Knight shift in Pd with hydrogen concentration. The Knight shift measures the density of states at the Fermi level at the position of the nucleus, and in the simple case of Pd-H one can measure this quantity as a function of hydrogen concentration.

A curve of the Knight shift of V^{51} as a function of hydrogen concentration in the solid-solution range was shown by Oriani.[14] The Knight shift increases essentially parabolically with increasing concentration up to 40 a/o. In answer to the query as to the significance of these data, Blandin pointed out that there is the difficulty that in vanadium the Knight shift not only represents the contribution of the *s* electrons but also reflects the effect of spin polarization inside the vanadium core; hence only a semiquantitative interpretation of the Knight shift in transition metals is possible. Nevitt added that the Knight shift K in vanadium might be expressed* by the relation

$$K = A\chi_s + B\chi_{\text{orb}} - C\chi_d(T)$$

where χ_s is the *s*-electron susceptibility, χ_{orb} is proportional to the Van Vleck high-frequency susceptibility, and $\chi_d(T)$ is the *d*-electron susceptibility. Now the $N(E)$ curve for the d band of the first long period has a negative slope at the value of the Fermi level in vanadium. Hence, χ_d will decrease as the band is filled up by adding hydrogen, so that K will increase.

Friedel reported on work by Rossier[15] at Orsay on NMR investigation in nonstoichiometric vanadium carbide, VC_{1-x}. Because of the quadrupole moment of the vanadium nucleus, the signal from a vanadium which has a carbon vacancy adjacent to it will be broadened because of the loss of cubic symmetry. Indeed, the nuclear magnetic resonance from the vanadium can be analyzed as a set of lines of different Knight shifts and of different widths. The unbroadened line is attributed by Rossier to vanadium nuclei which do not have carbon vacancies nearby; the other lines to vanadium nuclei with one, two, or three carbon vacancies as nearest neighbors. There are two interesting points deducible from these data. The variation in the intensity of the lines with carbon concentration is not what would be expected from a statistical distribution of vacancies, but can be analyzed in terms of a strong short-range order. The second point has to do with the two contributions to the Knight shift. One contribution arises from neighboring vacancies; it is proportional to their number about any one V atom and is responsible for the splitting of the lines. The

* This relationship was proposed for V_3X compounds by A. M. Clogston, A. C. Gossard, V. Jaccarino, and Y. Yafet, *Phys. Rev. Letters* **9**:262 (1962).

second contribution is due to the more distant vacancies and is responsible for the general shift of the lines with carbon concentration. The two contributions have the same sign but different amplitudes. The rigid-band model, which assumes that all the V nuclei experience the same field, is therefore not a good approximation.

Nowotny finds support for the concept of long-range effects of vacancies in carbon-deficient carbides in the formation of carbohydrides.* Titanium and zirconium carbides take up hydrogen in solid solution with a consequent ordering of the carbon atoms. For example, if $ZrC_{0.6}$ is allowed to absorb hydrogen, neutron diffraction gives clear evidence that a carbon superstructure develops at a composition $ZrC_{0.6}H_{0.7}$. However, if the hydrogen is subsequently removed, all evidence of superstructure disappears. Nowotny feels that this superstructure is the first approach to a connection between zirconium carbide and zirconium hydride.

The case of Ta_2H was discussed by Wallace, who pointed out that although only one-twelfth of the available interstitial sites are occupied by hydrogen, there is an interaction among the hydrogens since at low temperatures the hydrogens become ordered, as neutron diffraction and thermodynamic measurements have shown. The heat-capacity curve of Ta with 10 a/o hydrogen shows a sharp λ peak at 30°C and two other λ peaks at higher temperatures. Neutron diffraction shows that at low temperatures there are groups of four hydrogen atoms clustered together. At the first λ peak, order is partially destroyed, and thermodynamic analysis shows that the configurational entropy is then about 10% of that for random distribution. Perhaps groups of two atoms persist above the first λ anomaly and perhaps the higher λ points represent a breakup of short-range order, though the association of a λ anomaly with the disappearance of short-range order has never been demonstrated.

Lomer suggested that the successive λ peaks described by Wallace may be analogous to the successive magnetic phase changes observed in the rare-earth metals. Suppose there are bits or pieces of Fermi surface left over; then one can introduce a superlattice vector of some kind equal to the separation of two parallel bits of Fermi surface. This would be a self-stabilizing perturbation on the initially nearly random distribution of interstitials. Now, suppose that one has a system of two different wave vectors $\mathbf{q}$ and $\mathbf{q}'$, each one associated with different interaction energies. Then, upon cooling down from an initially random distribution of interstitials, it becomes energetically profitable to order the interstitials into a planar arrangement with an interplanar spacing corresponding to wave vector $\mathbf{q}$. At a lower temperature, however, another interplanar spacing corresponding to wave vector $\mathbf{q}'$ becomes energetically favored. The result

* See H. Goretzki, E. Ganglberger, H. Nowotny, and H. Bittner *Monatsh. Chem.* **96**:1563 (1965).

would be two distinct phase changes which would be difficult to analyze from neutron-diffraction experiments on powders. Rare-earth metals undergo three or four magnetic phase changes by shifts of this kind.

McQuillan discussed the arrangement of octahedral and tetrahedral sites in a bcc lattice. He pointed out that ordering of interstitial solutes does not occur because of a mutual attraction, but rather because of repulsive interactions which allow only specific placements of the interstitials in the three-dimensional array. This specificity will not permit long-range order, but the arrangements represent much more order than that characterized by short-range order. McQuillan feels that Wallace's higher-temperature λ peaks correspond to changes among such arrangements.

Beck wondered whether McQuillan would accept a description in terms of long-range order of very small domain size. Although McQuillan was willing to accept this description, Frank demurred, pointing out that this would not explain the observed λ peaks since for such peaks one must have a large number of cooperative atomic changes.

Meijering contributed the observation that it is possible to have a low configurational entropy without having long-range order; this is exemplified by glasses at low temperatures. One can have an irregular arrangement of atoms in a close-packed stack of layers such that the complete configuration is determined by the atom positions in one layer. In this way one can have low entropy with absence of superstructure. Thus, in McQuillan's or Wallace's cases one may have specific arrangements but not periodic positioning of the hydrogen atoms.

The chairman asked for discussion on the third topic on the agenda, the role of elasticity theory in the elucidation of the interaction between interstitial solutes and between solutes and lattice defects. Zener remarked that in a bcc lattice the elastic interactions of interstitial solute atoms are not isotropic. Hence, for two nearest-neighbor sites about one occupied site there will be repulsive interactions. The net result is the formation of layered structures.

Friedel added that a random distribution of impurities, each of which has a dipolar field, produces at any one site a force field which has an equal probability of being of either sign. Therefore, there is no tendency for alignment of the moments in any one direction. This problem has been put on a quantitative basis by Brout in order to analyze Känzig's observations of diatomic oxygen molecules in ionic solids. Friedel agreed with Beck's observation that one should expect a low-temperature specific-heat contribution from the interaction of stress dipoles, by analogy with the interaction among magnetic dipoles in a nonmagnetic matrix.

Seeger discussed the validity of the nearest-neighbor approximation for the interaction between a solute atom and a vacancy.[16] In this approximation, G_b, the free energy of binding between a solute atom and a vacancy, is finite when these are nearest neighbors and zero otherwise. The free energy of binding has an enthalpy and an entropy term: $G_b = H_b - TS_b$ and both contributions may be taken to be temperature-independent. In terms of nearest-neighbor bonds the free energy and energy of binding can be expressed as $G_b = G_{AB} - G_{BB}$ and $H_b = H_{AB} - H_{AA}$ respectively, where A is the solvent species and B the solute species. By considering a cycle involving removal to infinity of A and B atoms from the solution and from the pure crystals as well as the dissolution of B into A, one finds that $H_b = (H_A{}^c - H_B{}^c - H^s)/z$, in which $H_A{}^c$ and $H_B{}^c$ are the cohesive energies of the pure components, and H^s is the heat of solution of B into the A matrix of coordination number z. The S_b term can be similarly evaluated as $S_b = -\Delta S - (S_A - S_B)/z$, where S_A and S_B are the standard entropies of the pure components, and ΔS is the excess entropy of solution.

Deductions from such a model can be compared with the experimental measurements of Simmons and Balluffi for the Al-Ag and Al-Mg systems. Whereas the nearest-neighbor model seems appropriate for the Al-Ag system the nature of the disagreement in the case of the Al-Mg system led Seeger to deduce that one has to consider interactions beyond the nearest neighbors, and perhaps that close agreement between calculated and experimental values in other systems may be fortuitous.

Work by Déplanté[17] on interaction between impurities and vacancies was recalled by Friedel. Déplanté's model pictures the matrix as a free-electron gas, with the solute atom and the vacancy as small point charges, the vacancy having the excess charge of -1. A computation by perturbation methods of the energy of interaction of these two point charges gives the binding energy. This should give the major contribution to the difference in the activation energy U_D of diffusion of the impurity and the matrix; the other term, due to the change in the energy of motion of the vacancy in the neighborhood of the solute atom, is here neglected.

For normal monovalent matrices such as copper, the right portion of the diagrams of Fig. 2 can be explained within an order of magnitude by an attraction between the vacancy and a solute atom of higher valency than the matrix atom. Screening of the excess charge of the solute is not complete immediately about the atom itself, resulting in an electrostatic binding between the impurity and the vacancy. However, this simple picture does not work so well for transition-metal impurities; for these, Déplanté has used the virtual bound level concept. That is, the d shell of the impurity is within the same energy range as that of the conduction electrons, and hence resonance between the d-shell electrons and the conduction electrons occurs. This resonance leads to long-range effects such as phase shifts and residual resistivity. From the phase shifts one can compute the long-range perturbing potential due to the solute atom, thus the binding energy between the vacancy and transition-metal impurities: one explains in that way the complex shapes of that portion of the diagrams.

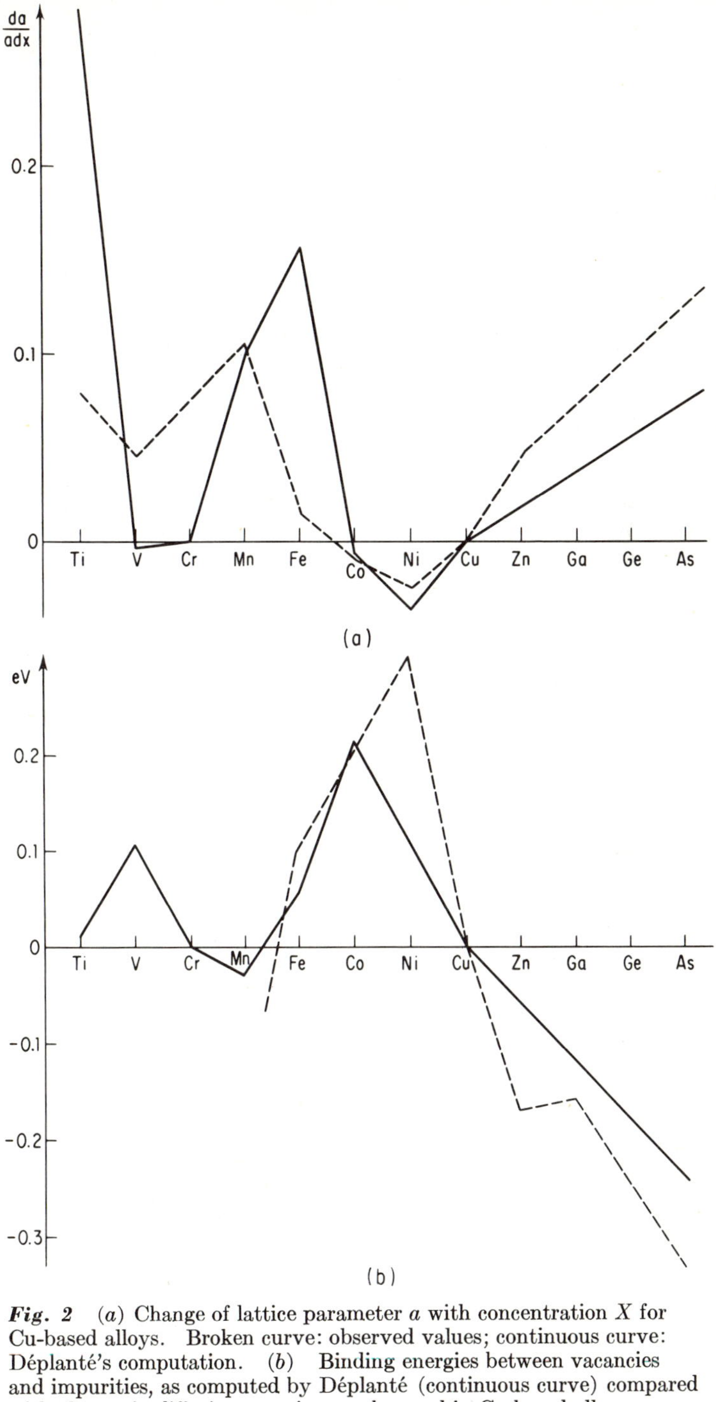

Fig. 2 (*a*) Change of lattice parameter *a* with concentration *X* for Cu-based alloys. Broken curve: observed values; continuous curve: Déplanté's computation. (*b*) Binding energies between vacancies and impurities, as computed by Déplanté (continuous curve) compared with change in diffusion energies, as observed in Cu-based alloys (broken curve).

The same kind of model explains the "size effect," i.e., the change in lattice parameter due to impurity atoms (of Fig. 2). The dilatation represented by the change in lattice parameter occurs because of the repulsion of neighboring ionic charges by the incompletely screened higher-valency impurities.

One may also calculate the volume effect of the vacancy itself. The presence of a vacancy should produce a negative size effect for monovalent metals such as Cu, Ag, and Au, and a positive size effect for Al, as is observed.

Blandin added that in calculations of interactions between vacancies and impurities in alloys of copper and of aluminum there are valency relationships for the pure metal which are very similar to the Hume-Rothery rules and which consider the diameter of the Fermi sphere. Thus a given impurity will behave similarly in Cu, Ag, and Au. One may also calculate that similar impurity interactions occur in aluminum, as is observed.

Seeger found this surprising, saying that it is difficult to believe that the calculations for copper and for aluminum in terms of the Fermi sphere are both equally valid, and asserted that the free-electron method has to be applied with caution. The usual treatment for Al gives in general too small a magnitude for the vacancy-solute interaction, and practically no interactions for next-nearest neighbors, contrary to what the experiments appear to require.

Friedel stressed that these are crude models, and that for aluminum the observed large binding energies between vacancies and impurities may be due to an elastic term not considered in Déplanté's electrostatic analysis.

REFERENCES

1. Powers, R. W., and M. V. Doyle: *Trans. Met. Soc. AIME* **215**:655 (1959).
2. Fisher, J. C.: *Acta Met.* **6**:13 (1958).
3. Smith, R. P.: *J. Am. Chem. Soc.* **70**:2724 (1948).
4. Darken, L. S.: Private communication.
5. Wriedt, H. A., and L. S. Darken: *Trans. Met. Soc. AIME* **233**:111, 122 (1965).
6. Damask, A. C.: In "Diffusion in Body-centered Cubic Metals," chap. 24, American Society for Metals Monograph on International Conference on Diffusion in BCC Materials, Gatlinburg, 1964.
7. Smith, R. P.: *Trans. Met. Soc. AIME* **224**:105 (1962).
5. Homan, C. G.: In "Diffusion in Body-centered Cubic Materials," chap. 5.
9. LeClaire, A. D.: *Ibid.*, chap. 1.
10. Li, J. C. M., R. A. Oriani, and L. S. Darken: *Z. Physik. Chem.* (N.F.) **49**:271 (1966).
11. Bilz, H.: *Z. Physik* **153**:338 (1958).
12. Ern, V., and A. C. Switendick: *Phys. Rev.* **137**:A1927 (1965).
13. Van Arkel, A. E.: Solid State Chemistry Meeting, Philips Laboratory, Aachen, West Germany, May 20–21, 1963.

14. Oriani, R. A., E. McCliment, and J. F. Youngblood: *J. Chem. Phys.* **27**:330 (1957).
15. Rossier, C.: Thèse de 3ème Cycle, Orsay, to be published.
16. Seeger, A.: In R. R. Hasiguti (ed.), "Lattice Defects and Their Interactions," Gordon and Breach, Science Publishers, Inc., New York, to be published.
17. Blandin, A., and J. L. Déplanté: *J. Phys. Radium* **23**:609 (1962); *J. Phys. Soc. Japan* **18**:SII, 85 (1963); J. L. Déplanté: Thèse de 3ème Cycle, Orsay, 1963; A. Blandin and J. L. Déplanté: *J. Phys. Chem. Solids* **26**:381 (1965).

Part Five

LIQUIDS AND DEFECT SOLIDS

BINDING ENERGIES AND ENTROPIES OF MULTIPLE VACANCIES IN FCC METALS

K. P. Chik, Dieter Schumacher, and Alfred Seeger
Max-Planck-Institut für Metallforschung, Stuttgart, West Germany

1 INTRODUCTION

Vacancy-type defects are the predominant crystal imperfections in fcc metals under thermal equilibrium conditions. At temperatures near the melting point the total defect concentrations are of the order of 10^{-4}. In the noble metals up to 12% of the equilibrium vacancies are associated into divacancies or larger multiples.[1]

Larger fractions of associated vacancies may occur under nonequilibrium conditions such as in quenched or in irradiated metals. As an example, divacancy concentrations may become large enough so that they govern the diffusion processes occurring during the annealing of a quenched metal.[2] Hence the knowledge of the properties of multiple vacancies is rather essential for the understanding of transport phenomena in solids.

The concentrations and stabilities of multiple vacancies are determined by the binding energies E_i^B and association entropies ΔS_i of the defects. These quantities are defined by

$$E_i^B = iE_1^F - E_i^F \qquad (1)$$

and

$$\Delta S_i = S_i^F - iS_1^F = -S_i^B \qquad (2)$$

where E^F and S^F are the energies and entropies of formation, respectively; and $i = 1$ for single vacancies, 2 for divacancies, etc. Note that ΔS_i denotes the *increase* in vibrational entropy upon association of the single vacancies, which is the negative of the definition for the binding energy.

This paper presents a comprehensive survey of methods yielding information on these quantities, and gives a list of data obtained so far for a number of fcc metals. Furthermore currently controversial problems in this field are discussed.

2 METHODS OF ANALYSIS

The determination of the binding parameters of multiple vacancies is to some extent made difficult by the fact that the concentrations of these defects are usually exceeded by those of other defects, such as single vacancies or interstitials. As a consequence, the interpretation of effects due to multiple vacancies, in particular the determination of their concentrations, involves a treatment of interactions between the various types of lattice defects present in the crystal.[3] In fcc metals the following approaches have been successful in obtaining binding parameters:

1. Interpretation of annealing experiments on quenched or irradiated materials in terms of an *effective diffusion coefficient* for vacant sites,

$$D(c) = \frac{c_1 D_1 + 2\beta c_2 D_2 + \cdots}{c_1 + 2\beta c_2 + \cdots} \qquad (3)$$

where c is the total concentration of vacant sites; c_i that of single vacancies, divacancies, etc.; and D_i is the diffusion coefficient of the relevant defect. β equals 1 for Brownian motion of the defects, and 2 for diffusion in a concentration gradient. Familiar special cases of Eq. (3) are those for which single vacancy kinetics ($c_1D_1 \gg 2c_2D_2$, $c_1 \gg 2c_2$) or divacancy kinetics ($c_1D_1 \ll 2c_2D_2$, $c_1 \ll 2c_2$) are observed. Of considerable experimental importance in the present context are defect distributions for which $c_1D_1 \ll 2c_2D_2$ and $c_1 \gg 2c_2$. Neglecting trivacancies and larger clusters, the diffusion coefficient then reads[1]

$$D(c) = 2\beta a^2 \nu_2^0 \exp \frac{S_2^M}{k} \exp \frac{\Delta S_2}{k} \exp \left(-\frac{E_2^M - E_2^B}{kT} \right) c \tag{4}$$

Here a is the lattice parameter, $\nu_2^0 \exp (S_2^M/k)$ the frequency factor for divacancy migration, and E_2^M the divacancy migration energy. Hence, if E_2^M and the entropy terms are known, one can determine E_2^B from the concentration dependence of D according to Eq. (4). See Sec. 3 for examples.

2. A second useful approach has been the investigation of the *agglomeration of vacancies* in the course of the annealing treatment. In quenched Au, for example, the annealing conditions can be chosen so that the majority of the vacancies associate into large agglomerates (stacking-fault tetrahedra) instead of annihilating at dislocations or grain boundaries. The initial stages of agglomeration involve reactions between single and small multiple vacancies leading to stable nuclei for the agglomerates. The nucleation kinetics yield information on the lifetimes and stabilities of the reacting multiple vacancies.

Transmission electron microscopy and X-ray small-angle scattering have been employed in studying the agglomerates and deriving nucleation models for Au[4,5] and Cu,[6] respectively. In other metals, larger varieties of observed agglomerate types or the failure to observe appreciable agglomerate concentrations have so far prevented a successful analysis for multiple vacancy properties. It is felt, however, that in the near future a satisfactory model of agglomerate formation, including data for multiple vacancy stabilities, should also become feasible, at least for Ag and Al, since in these metals our knowledge of vacancy properties is comparable with that in Au and Cu.

3. A useful tool is also *high-temperature thermal equilibrium investigations*. In favorable cases, measurements of self-diffusion or of the temperature dependence of physical properties reflect the presence of multiple vacancies.

4. Several attempts have been made by a number of authors to arrive at *theoretical estimates* for the binding energies of multiple vacancies.[7–11] The wide spread of the results obtained is not surprising in view of the difficulties encountered in finding a reliable model for the calculations. For the noble metals the divacancy binding energy is expected to be in the range of 0.1 to 0.2 eV.[11]

Although the experimental approaches mentioned above yield values or estimates for the binding energies of multiple vacancies, the experimental information in general does not suffice to determine also the *binding entropies*. This gap is filled by theoretical estimates by Schottky et al.[9,12] of the entropy changes associated with defect formation in the noble metals.

These values have been utilized in the interpretation of a large number of experimental data for these metals and have been found to yield correction terms that are both appreciable and reliable.

3 DISCUSSION OF INDIVIDUAL METALS

3.1 Gold

The binding energy of divacancies in Au has been a controversial issue for several years. Two different schools of thought favor values of 0.3 to 0.4 ev and $\approx$0.1 ev, respectively. In the following sections, an attempt is made to rationalize the available experimental data. It will be shown that the low-binding-energy scheme provides a satisfactory interpretation for the majority of the data at present available. On the other hand, models based on values of about 0.3 to 0.4 ev encounter serious difficulties in explaining in detail the experimental observations.

Relevant experimental data originate primarily from electron-microscope studies of the agglomeration of vacancy-type defects in quenched Au, and from investigations of the recovery behavior of quenched Au as detected by electrical resistivity measurements. To some extent thermal equilibrium studies may be used to check the models developed to account for the low-temperature experiments.

3.1.1 Electron-microscope Observations. Vacancy precipitates in quenched Au, namely stacking-fault tetrahedra, were first observed by Silcox and Hirsch[13] in transmission electron-microscope studies. It was soon realized that the occurrence or nonoccurrence of the stacking-fault tetrahedra depends sensitively on the quenching and annealing histories of the quenched specimen, and that the tetrahedra growth period is preceded by a period of nuclei formation at the beginning of the annealing treatment. Mori and Meshii[4,14] pointed out the importance of the nucleation processes for the final structure of the specimen. Westdorp et al.[15] found that the nucleation of stacking-fault tetrahedra is completed after 2 min annealing at room temperature for quenches from 890°C. These schools feel that the occurrence of vacancy clustering after quenching implies the formation of divacancies and/or small multiple vacancy embryos *during* the quench and consequently a large divacancy binding energy of 0.3 to 0.4 ev. Similar arguments are used for the interpretation of quench-hardening phenomena.[16]

Although the preceding conclusion was arrived at by the above-mentioned authors in a somewhat qualitative way, a quantitative model for the nucleation based on a detailed analysis of experimental observations has recently been developed.[5] The diffusion process of the mobile defects in the as-quenched state is described in terms of mobile mono-, di-, and trivacancies that are in local thermal equilibrium. Hexavacancies are chosen as the stable nuclei, so as to reproduce correctly the magnitude of the observed tetrahedra density and to give the right dependence of the tetrahedra density on the vacancy concentration. In this model the hexavacancies are considered to be formed by a simultaneous collision of mono-, di-, and trivacancies. Recent investigations[17] showed that trivacancy-trivacancy collisions must be considered as well. This latter nucleation mechanism can be easily allowed for in the nucleation model of Chik[5] and leads to further information on the mobility of trivacancies. Consistent agreement between theory and experiments is found with the following vacancy properties: $E_2^B = 0.09$ ev, $\Delta S_2 = 2.2k$, $E_3^B = 0.3$ ev, $\Delta S_3 = 3.5k$, and $E_3^M - E_2^M \approx 0.1$ ev. Quadri- and pentavacancies are probably relatively unstable. Although the above analysis does not determine E_2^M and E_2^B separately, the self-consistency of the model gives additional support for a low divacancy binding energy (cf. Sec. 3.1.2). The electron-microscope observations in Refs. 4, 14, and 15 are in general agreement with those of Chik.[5] (The somewhat faster nucleation rates observed by Westdorp et al.[15] may be due to their quenching technique. The specimens were sandwiched between asbestos paper disks during quenching. Hence the cooling rate between 100°C and the quenching bath may be rather small, giving rise to an enhancement of the nucleation.) The nucleation model proposed by de Jong and Koehler[18] is inadequate to account quantitatively for the observed nucleation data. Here the stable nuclei are considered to be tetravacancies formed by the migration of single vacancies to trivacancies. This model leads to a gross overestimate of the cluster concentration and an underestimate of the precipitation rate.

The assumption of homogenous nucleation has been questioned by several authors who have reported impurity effects on the precipitate structure.[19–21] It is at present difficult to judge the significance of the addition of impurities to *pure* samples. Siegel[21] recently found that the density of visible tetrahedra decreased with decreasing residual resistivity at 4.2°K, with no saturation effect. He concluded that the nucleation process should be considered as heterogeneous, i.e., precipitation of vacancies at impurities. It is questionable, however, whether the residual resistivity gives the right measure for the purity of the samples, since the impurity atoms may rearrange themselves during high-temperature treatment to give different contributions to the residual resistivity. Furthermore, as is pointed out by Siegel, only certain types of impurity species are likely to cause the heterogeneous nucleation. The residual resistivity is obviously not directly proportional to the impurity concentration responsible for the heterogeneous nucleation. In fact, the tetrahedra densities reported by Siegel[21] for samples of 99.9999% purity show a similar correlation to those of Chik[5] with the vacancy concentration, if the latter is calculated from the mean size and the density of the tetrahedra observed, irrespective of the quenching temperature. The existence of such a correlation supports the idea that the nucleation in pure Au occurs by encounters of vacant sites.

3.1.2 Electrical Resistivity Measurements. The first detailed electrical resistivity recovery measurements on quenched Au were performed by Koehler et al.[22,23] These authors observed a decrease of the apparent effective activation energy E_{eff} from 0.82 to 0.6 ev as the quenching temperature T_Q was raised from 700 to 1000°C. They attributed this decrease to the increasing concentration and participation of divacancies in the recovery process, and deduced an upper limit of 0.3 ev for $E_2{}^B$. Kimura et al.[24] argued that the observed variation of E_{eff} and the occurrence of S-shaped isothermal annealing curves[22] implied a large divacancy binding energy ($\sim$0.4 ev) and a short-range repulsive potential between two vacancies. It was shown by Schottky[25] that the latter assumptions are not necessary for explaining the Bauerle data. In fact, Schottky estimated from the Bauerle data that $E_1{}^M - E_2{}^M$ should be smaller than 0.2 ev for $E_2{}^B = 0.1$ ev; i.e., $E_1{}^M - E_2{}^M + E_2{}^B < 0.3$ ev. The latter estimate was several years later verified by two different experimental approaches, namely by the above-mentioned vacancy precipitation studies of Chik[5] and the electrical resistivity recovery measurements of de Jong and Koehler.[26] A discussion of the results of de Jong and Koehler[26] is given in the appendix.

Meshii et al.[27] measured the electrical resistivity change of gold wires quenched from below 500°C. Isothermal annealing curves at 62.4°C showed first a maximum and then decayed exponentially with approximately the same decay time for different quenching temperatures. The resistivity maximum was explained qualitatively as arising from the association of vacancies into tightly bound di- and trivacancies. The analysis was based on the controversial model that the electrical resistivity of di- and trivacancies is larger than that of two or three isolated vacancies.[28] Recent experimental evidence suggests that in fact the divacancy resistivity is somewhat smaller than the twofold single vacancy resistivity,[26] in agreement with theoretical estimates.[29] Hence the analysis of Meshii et al. and its implications appear unlikely. An alternative explanation for the initial resistivity increase might be possible in terms of the association of vacancies and impurity atoms, since the vacancy concentrations are comparable with or even less than that of the impurities. Meshii et al.[27] have argued that the observed resistivity maxima are independent of the purity of the samples. It should be noted, however, that not all impurity atoms have a strong interaction with vacancies. Cattaneo and Germagnoli[30] reported a binding energy of 0.3 ev for a vacancy-Ag complex in Au, and a negligible binding energy for a vacancy-Ni complex in Au. Thus, if the impurity level of Ag, for example, remains approximately constant for specimens with different total impurity concentrations, a resistivity increase due to the formation of vacancy-Ag complexes would be observed irrespective of the total impurity content. The theory of Damask and Dienes[31] for the annealing of vacancies at impurities is qualitatively consistent with the above results; i.e., a quick resistivity increase due to the formation of vacancy-impurity complexes at the beginning of the annealing treatment is found to be followed by an exponential decay, the reaction times of which are independent of the initial vacancy concentration, provided that the concentration of impurity atoms is larger than that of vacancy-impurity complexes.

Another group of experiments that have been considered to present evidence for a large divacancy binding energy have been reported by Balluffi et al.[21,32–34] and by Meshii and Kauffman.[35,36] They observed in quenched Au an effective activation energy of ≈ 0.71 ev at annealing temperatures between 40 and 170°C in a wide range of frozen-in vacancy concentrations ranging from 10^{-4} to 10^{-6}, i.e., even at rather low concentrations.

The only positive assignment in their interpretation is the assumption that the observed activation energy must be associated with the energy of migration of one particular species of vacancy-type defects, for example, a tightly bound divacancy (since 0.71 ev is too low to be ascribed to the migration of single vacancies). Beyond that the authors offer detailed interpretations neither for the vacancy precipitation nor for electrical resistivity or high-temperature experiments, in contrast to the low-binding-energy model. In fact, their view runs into considerable interpretation difficulties, as will be seen presently.

An immediate difficulty of this model is the failure to provide a satisfactory explanation for the fact that the effective migration energy decreases from 0.71 ev to values around 0.60 ev at higher vacancy concentrations. Balluffi et al.[32] propose that the decrease of E_{eff} may be due to a higher-order vacancy cluster which is loosely bound and more mobile than the divacancy. Using $E_1{}^M = 0.84$ ev and $E_2{}^M = 0.71$ ev, a reasonable fit of the data was obtained for $E_2{}^B = 0.35$ ev, $E_3{}^M = 0.45$ ev, and $E_3{}^B = 0.40$ ev, provided nonequilibrium conditions were assumed.

It was, however, shown by Schottky[9,25] that a trivacancy which is more mobile than the divacancy is hard to conceive on theoretical grounds and in view of the Bauerle and Koehler data.[22] More recently it was furthermore shown that the recovery of Au wires quenched to very low temperatures does not exhibit any evidence for a low-temperature recovery process below the temperature at which divacancies become mobile.[37] Hence any highly mobile trivacancies (or larger multiple vacancies) would not be present in the as-quenched state, but would rather have to be formed as the result of divacancy and/or single-vacancy migration and encounters.

However, upon formation they would migrate instantaneously to sinks and annihilate. It would then be extremely difficult to find processes which lead to the formation of stable nuclei for stacking-fault tetrahedra during room-temperature annealing.

It is shown in the appendix that the low-binding-energy model provides a reasonable explanation for the observed concentration dependence at vacancy concentrations higher than 2×10^{-5} if $E_2^B = 0.07$ to 0.08 ev and $E_2^M < E_3^M < E_1^M$.

For a large divacancy binding energy (>0.3 ev) the concentration of divacancies in quenched Au might be expected to be large enough to give rise to an observable internal friction peak due to the reorientation of the divacancies in an applied stress field. A damping process of this kind has not been found, although admittedly the failure to observe it may be due to the possibility that processes of larger relaxation strengths overlap the divacancy peak.[38]

Another decisive aspect is the kinetics of vacancy annealing at low vacancy concentrations. For $E_2^B > 0.3$ ev, the diffusion and annihilation of divacancies would govern the annealing kinetics. One would expect essentially first-order kinetics for the divacancy migration to sinks. In electrical resistivity annealing experiments on Au quenched from low temperatures, Ytterhus[34] and Siegel[21] observed rather complex annealing kinetics. They associated the recovery with the migration of divacancies, but did not verify how the observed complex kinetics could quantitatively be accounted for by this simple process.

The annealing kinetics for low vacancy concentrations are presumably best reflected by long isothermal recovery runs. It is shown in the Appendix that isothermal annealing curves at low defect concentrations in well-annealed specimens fit the expected kinetic equations with the parameters of the low-binding-energy model[1,5] remarkably well, for example, the isothermal segments of temperature-change experiments. The analysis yields an activation energy close to the accepted value for E_1^M (0.82 ev) as compared to 0.71 ev from the evaluation of the change of slope on the same isothermal segments.

These results are supported by the fact that values around $E_1^M = 0.82$ ev have been found for low vacancy concentrations by techniques other than the change-of-temperature method.[39–41] Hence we may conclude that at low vacancy concentrations the migration energy is found close to E_1^M (implying a low divacancy binding energy) when the activation energy is determined from an analysis of the actual shape of recovery curves, while the change-of-slope experiments may yield activation energy values which cannot be accounted for by the simple systems of kinetic equations so far considered.

It is likely that these discrepancies are due to some fundamental difficulty in the temperature-change experiments. An immediate explanation is suggested by the fact that along an isothermal annealing curve, for example, we have a continuous steady-state recovery, whereas a temperature change involves an abrupt change of defect equilibria and concentrations. It is felt that at least in the case of quenched Au the change-of-temperature technique at intermediate and small vacancy concentrations involves little-known transient processes and is therefore unable to yield a physically meaningful activation energy. On account of the presently available experimental evidence, the value of 0.71 ev cannot be identified with E_1^M, E_2^M, or E_3^M. Neither is it possible to find conditions and reasonable vacancy parameters to interpret this value in terms of an effective migration energy of an equilibrium distribution of single and multiple vacancies. It is recommended that techniques other than the change-of-temperature method be employed to reach a final unambiguous decision on the 0.71-ev problem.

It is entirely possible that the difficulties encountered in the change-of-temperature experiments are particularly evident in the case of Au, since $E_1^M - E_2^M$ is less than 0.2 ev and $E_2^M < E_3^M < E_1^M$. This close margin has the consequence that the relative contributions of the various defects to the effective migration of vacant sites may be affected sensitively by changes in the annealing temperature as small as $\pm 10°C$.

Kauffman et al.[36,42] investigated the influence of transient processes on the evaluation of slope-change experiments by computer calculations. For intermediate and small vacancy concentrations the calculations yield 0.71 ev for the effective activation energy, which is claimed to support the above-mentioned experimental values. In these calculations the assumption of a divacancy binding energy E_2^B as large as 0.3 ev (in addition to $E_2^M = 0.70$ ev) practically determines the outcome of the calculations, so that the result does not constitute an independent support of the ideas used in interpreting the data.

3.1.3 High-temperature Studies. Further estimates for E_2^B may be obtained from high-temperature thermal equilibrium studies. Neglecting vacancy complexes larger than divacancies, the total vacancy concentration in thermal equilibrium is given by

$$c = \exp \frac{S_1^F}{k} \exp \left(\frac{-E_1^F}{kT} \right) \left[1 + 12 \exp \frac{S_1^F + \Delta S_2}{k} \exp \left(- \frac{E_1^F - E_2^B}{kT} \right) \right] \tag{5}$$

where E_1^F and S_1^F are the formation energy and entropy of a single vacancy, respectively. The total vacancy concentration at the melting point in Au has been determined by Simmons and Balluffi[43] as 7.2×10^{-4}. Using this value and $S_1^F = 1.2k$ and $\Delta S_2 = 2.2k$,[12] the influence of increasing E_2^B on E_1^F as determined from Eq. (5) can be calculated (see Table 1). It is seen from the table that E_1^F rises above 1.00 ev for $E_2^B >$ 0.2 ev. Direct determinations of E_1^F by quenching experiments yield values centered around 0.98 ev.[1] Hence the Simmons and Balluffi experiments indicate a value of $E_2^B < 0.2$ ev.

Similarly, the data on self-diffusion in Au[44,45] do not indicate an appreciable contribution of divacancies to the diffusion current. For values of $E_2^M - E_2^B$ as low as 0.3 to 0.4 ev a marked curvature of the Arrhenius plot of self-diffusion coefficients would be expected.[46] This is not observed in the experiments.[44,45] Hence $E_2^M - E_2^B$ is likely to be >0.4 ev.

3.1.4 Summary of the Present Situation. The *high*-divacancy binding-energy scheme ($E_2^B = 0.3$ to 0.4 ev) for Au is based essentially on (1) the fact that vacancy agglomeration occurs in quenched Au (although a detailed model for the precipitation phenomenon has not been offered as yet), and (2) the 0.71-ev phenomenon (although no successful attempt has been made to analyze actual annealing curves). The low-binding-energy model is claimed to be discarded by the impurity effects on vacancy precipitation and by the 0.71-ev phenomenon.

On the other hand, the *low*-binding-energy model provides—in contrast to the high-binding-energy model—consistent explanations for the following phenomena: (1) the nucleation and growth of stacking-fault tetrahedra, including the fact that the nucleation time is short compared with the total annealing time; (2) the concentration dependence of the effective activation energy; (3) the annealing kinetics at low vacancy concentrations; and (4) the results of high temperature investigations.

These achievements in accounting for the experimental phenomena are considered a satisfactory verification of the low-binding-energy model. It is felt that the impurity effects on the vacancy precipitation and the 0.71-ev phenomenon can be incorporated in the low-binding-energy model as soon as systematic data are available.

TABLE 1 ***Influence of E_2^B on E_1^F as Determined from High-temperature Equilibrium Data for Au****

E_2^B†	0.0	0.1	0.2	0.3	0.4
$c_2(T_M)/c_1(T_M)$	0	0.08.	0.17	0.32	0.58
$2c_2(T_M)/c(T_M)$	0	0.14	0.25	0.39	0.54
E_1^F†	0.97	0.99	1.00	1.03	1.06

* See Sec. 3.1.3.
† In electron volts.
T_M = melting point.
$c = c_1 + 2c_2$.

TABLE 2 *List of Best Values for Multiple-vacancy Parameters*

Defect	Property	Au	Ag	Cu	Ni
Divacancy	E_2^M*	0.63	0.57	0.67	0.80
	E_2^B*	0.08	0.19	0.12	0.28
	$\Delta S_2/k$	2.2	1.8	1.8	1.2–1.4
Trivacancy	E_3^B*	0.3	....	....	0.6–0.8
(60° configuration)	$\Delta S_3/k$	4	....	....	4–7

* In electron volts.

The set of multiple-vacancy parameters consistent with the low-binding-energy model is included in Table 2.

3.2 Silver

In noble metals other than Au the knowledge on multiple-vacancy binding parameters has not advanced appreciably since a recent review,[1] in which the relevant investigations are discussed in detail. A brief comment will be given on the data listed in Table 2.

In Ag, the detection of divacancies in quenched specimens is somewhat easier than in Au, since by quenching from high temperatures a relatively large divacancy concentration can be frozen in, so that the migration energy of divacancies ($E_2^M = 0.57$ ev) may be determined directly.[2,47,48] The stability of the divacancies has been investigated in pulsing experiments; i.e., divacancies were broken up by short high-temperature pulses.[2] Although Doyama and Koehler deduced from these experiments a rather large value for E_2^B, namely 0.38 ev, it was shown that an alternative interpretation of the data yields[49] $E_2^B = 0.19$ ev, which appears to be more reasonable in view of the results for the other noble metals and of theoretical estimates.[11] The fact that the binding energy of divacancies in Ag is somewhat larger than that for Au is also reflected by the plots of the logarithm of the quench-induced electrical resistivity $\Delta\rho$ vs. the reciprocal quenching temperature. The straight line (indicating that all vacancies have been retained) extends for Ag only up to quenching temperatures of 750°C,[50] whereas the straight line in the Au plots[22] extends to above 900°C. This suggests that in quenching Ag from temperatures near the melting point appreciable losses of vacancies occur during the quench owing to the formation and the fast migration rate of divacancies, in contrast to the phenomena in quenched Au.

Using the values of Doyama and Koehler,[2] $E_2^M - E_2^B$ would be 0.19 ev. For such a low value one would expect a rather strong contribution of divacancies to self-diffusion in Ag. This is not observed in the experiments.[51] Neither do the high-temperature equilibrium measurements of Simmons and Balluffi[52] indicate a divacancy binding energy in the range 0.3 to 0.4 ev.

Trivacancies or larger multiple vacancies have not been identified yet or taken into account in Ag, since so far there has been no need for this extension. Agglomeration of vacancies does occur during the annealing of quenched Ag, but the resulting defect structure is more complex than in Au.[4] Nucleation models to check multiple-vacancy parameters have not been established yet.

3.3 Copper

The establishment of local equilibria of single and divacancies and the defect migration in quenched Cu during annealing are associated with the precipitation of small vacancy clusters. These phenomena have permitted an extensive combined study using electrical resistivity,[53,54] X-ray small-angle scattering,[6] and transmission electron-microscope techniques.[55] The experimental observations are consistent with a divacancy binding energy[54] of 0.12 ev. In the analysis the binding entropy $\Delta S_2 = 1.8k$ was taken from theory.[12] Experimental information on other multiple-vacancy binding parameters has neither been possible nor yet necessary to account for the data. Theoretical estimates are included in the table.

The reported binding parameters are consistent with the results of self-diffusion and high-temperature equilibrium investigations. Using the equilibrium vacancy concentration at the melting point[56] (2×10^{-4}), the fraction of vacancies associated into divacancies is 4.5% at the melting point.[1] The divacancy contribution to self-diffusion is rather small.[57]

3.4 Nickel

In Ni that was neutron-irradiated at 80°C, a recovery process was observed for which the vacancy concentrations were found to fulfill the conditions of Eq. (4), and $E_2^M - E_2^B$ was determined[58,59] as 0.5 ev. Since E_2^M could be measured directly from magnetic relaxation studies[58,60,61] as 0.80 ev, E_2^B was found to be 0.28 ev. ΔS_2 was taken[59] as $1.4k$. These results are supported also by similar observations on quenched Ni.[62] Vacancy agglomeration phenomena have not yet been studied successfully in Ni. Though a few agglomerates are visible in the electron microscope,[63,64] X-ray small-angle scattering did not show appreciable effect.[62]

The relatively large divacancy binding energy is presumably one of the reasons why it has not been possible to retain the total equilibrium concentration of vacancies by quenching. The experiments indicate that considerable losses of vacancies occur during the quench, as in Ag at high quenching temperatures.[62] The failure to observe appreciable vacancy agglomeration by electron-microscope and X-ray small-angle scattering techniques[62–64] may also be connected with this inefficiency of the quenches. For these reasons it is not justified to consider the low retained vacancy concentrations as implying a large formation energy for single vacancies in Ni. This is seen for example from the value estimated by Wuttig and Birnbaum:[65] For $S_1^F = 1k$ and $E_1^F = 1.9$ ev the concentration of vacancies at the melting point would be 8.6×10^{-6}, which is far below the range observed for other fcc metals.[1]

Both the relatively large divacancy binding energy and the curvature observed in the Arrhenius plots for published self-diffusion coefficients in Ni suggest an appreciable contribution of multiple vacancies to self-diffusion in this metal. A successful analysis of the curvature in terms of single, di-, and trivacancies was indeed possible.[46] This analysis, in connection with the results obtained by other techniques, yielded an internally consistent set of vacancy properties. The data are included in Table 2. Recent critics of this analysis argued that the deviations from a straight line in the $\ln D$ vs. $1/T$ plot might be due to dislocation pipe diffusion in the low-temperature range.[66] Although this effect certainly should be considered in self-diffusion experiments, it does not interfere with the multiple-vacancy analysis, since deviations due to multiple vacancies occur entirely on the *high*-temperature side of the Arrhenius plot. The small diffusion coefficients at low temperature do not influence the analysis of Seeger et al.[46] appreciably. This was checked by running the computer program with and without the low-temperature data.

APPENDIX

Discussion of E_2^B Determination[26]

De Jong and Koehler[26] determined the binding energy of divacancies in Au (1) from the characteristic time constant τ for the establishment of thermal equilibrium between mono- and divacancies, and (2) from the change in divacancy concentration at two different annealing temperatures. They obtained the average value $E_2^B = 0.10 \pm 0.03$ ev. In the discussion of these results by Seeger and Schumacher,[1] the divacancy association entropy ΔS_2 was taken into account, which had been neglected by de Jong and Koehler.[26] The results of the reanalysis[1] are summarized as follows: (1) Allowing for $\Delta S_2 = 2.2k$,[12] the relative values of τ for two annealing temperatures yield $E_2^B = 0.09 \pm 0.06$ ev. (2) The absolute values of τ yield $E_2^B = 0.14$ ev. (3) No reasonable results are obtained for de Jong and Koehler's method 3. (4) From the change in the divacancy concentration at two different annealing temperatures, we obtained $E_2^B = 0.15$ ev.

In the above analysis, the existence of trivacancies was neglected. As is shown by Chik,[5] the trivacancies are probably in equilibrium with mono- and divacancies at high vacancy concentrations and therefore should be considered in the above evaluation. Preliminary estimates of the influence of trivacancies on the establishment of equilibrium show that if the parameters for Au given in this paper are used in the analysis, the value of τ given by de Jong and Koehler[26] is only slightly affected after introducing trivacancies. Therefore the experiments of de Jong and Koehler[26] are in agreement with the model of a low binding energy of divacancies.

Concentration Dependence of the Effective Activation Energy

If single, di-, and trivacancies are in equilibrium before and after a temperature change, the effective migration energy as a function of vacancy concentration is given by the relation

$$\ln \frac{D_{\text{eff}}(T_2, c)}{D_{\text{eff}}(T_1, c)} = \frac{E_{\text{eff}}}{k}\left(\frac{1}{T_1} - \frac{1}{T_2}\right) \tag{A.1}$$

where

$$D_{\text{eff}} = \frac{c_1 D_1 + 2c_2 D_2 + 3c_3 D_3}{c_1 + 2c_2 + 3c_3} \tag{A.2}$$

Using $T_1 = 30°\text{C}$ and $T_2 = 40°\text{C}$, we calculated E_{eff} as a function of c for two

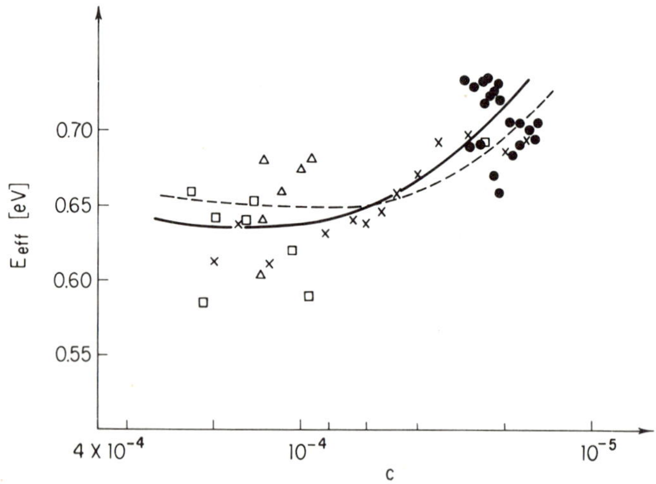

Fig. 1 Concentration dependence of the effective activation energy for two sets of vacancy parameters.

$E_1^M - E_2^M = 0.18$ ev $\quad E_2^M = 0.62$ ev $\quad E_3^M - E_2^M = 0.12$ ev (theoretical estimate[9])

$$\frac{\nu_2{}^0 \exp (S_2{}^M/k)}{\nu_1{}^0 \exp (S_1{}^M/k)} = \frac{1}{6} \qquad \frac{\nu_3{}^0 \exp (S_3{}^M/k)}{\nu_2{}^0 \exp (S_2{}^M/k)} = \frac{1}{2}$$

Full line: $E_2{}^B = 0.07$ ev $\quad E_3{}^B = 0.23$ ev
$\Delta S_2 = 2.2k \quad \Delta S_3 = 3.5k$

Dashed line: $E_2{}^B = 0.07$ ev $\quad E_3{}^B = 0.23$ ev
$\Delta S_2 = 2.2k \quad \Delta S_3 = 3.5k$

Experimental data (□, △, ×, ●) were taken from Ytterhus and Balluffi.[34] (Compare, however, the remarks in Sec. 3.1.2 on the change of slope experiments.)

sets of vacancy parameters with a low divacancy binding energy (see Fig. 1). Also included in Fig. 1 are experimental data from the change-of-slope determinations of Ytterhus and Balluffi.[34] It is seen that the decrease of E_{eff} at high-vacancy concentrations can be accounted for by the low-binding-energy model. The effect of trivacancies is to yield an almost constant effective migration energy at high vacancy concentrations ($c > 6 \times 10^{-5}$). The experimental results at low concentrations ($c < 2 \times 10^{-5}$) are not accounted for by Eqs. (A.1) and (A.2). This is not surprising, since at sufficiently low concentrations the assumption of equilibrium immediately after the change in temperature must break down.

Vacancy Annealing at Low Vacancy Concentrations in Quenched Gold

At low vacancy concentrations ($c < 3 \cdot 10^{-5}$) and at moderate annealing temperatures, we may neglect the formation of clusters larger than divacancies. Under these conditions the dislocations are probably the main sinks for the vacancies. The annealing processes are described by the following rate equations:

$$\frac{dc_1}{dt} = 2K_2c_2 - 2K_1c_1^2 - K_3c_1 \tag{A.3}$$

$$\frac{dc_2}{dt} = K_1c_1^2 - K_2c_2 - K_4c_2 \tag{A.4}$$

with

$$K_1 = 84\nu_1 \exp\left(\frac{-E_1^M}{kT}\right) \tag{A.5}$$

$$K_2 = 14\nu_1' \exp\left(\frac{-(E_1^M + E_2^B)}{kT}\right) \tag{A.6}$$

$$K_3 = \alpha a^2\nu_1 \exp\left(\frac{-E_1^M}{kT}\right) \tag{A.7}$$

$$K_4 = \tfrac{1}{6}\alpha a^2\nu_2 \exp\left(\frac{-E_2^M}{kT}\right) \tag{A.8}$$

α is proportional to the sink concentration, and may be taken for the purpose of the following discussion as equal to the dislocation density. a is the lattice constant. ν_1, ν_1', and ν_2 are the frequency factors for migration.

These equations have been solved by Damask and Dienes[31] for the "steady state" $dc_2/dt = 0$. The steady-state condition gives the relation

$$c_2 = \frac{K_1c_1^2}{K_2 + K_4} \tag{A.9}$$

$$= 6 \exp \frac{\Delta S_2}{k} \exp \frac{E_2^B}{kT} c_1^2 \qquad \text{for } K_4 \ll K_2 \tag{A.10}$$

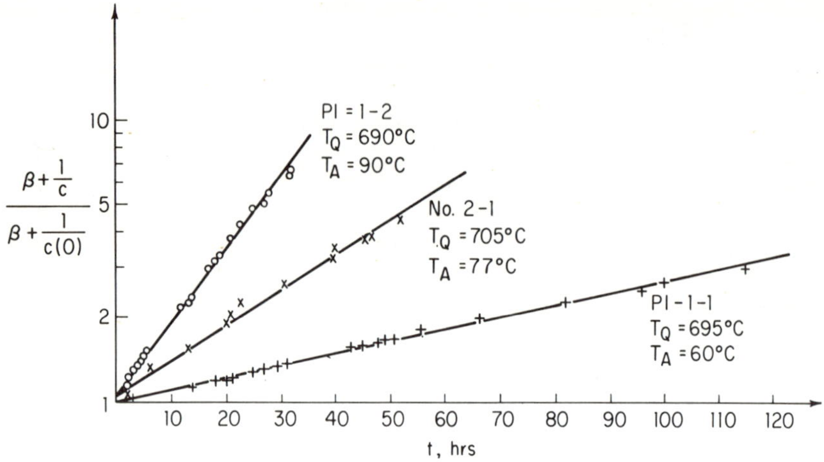

Fig. 2 Analysis of data given in fig. 9 of Ref. 67 according to Eq. (A.11). For details see text.

The solution of the above system of equations can be written as

$$\frac{\beta + 1/c_1(t)}{\beta + 1/c_1(0)} = \exp K_3 t \tag{A.11}$$

where

$$\beta = \frac{2K_1K_4}{K_2K_3} = 3 \exp \frac{E_1^M + E_2^B - E_2^M}{kT} \tag{A.12}$$

If c_2 is small, c_1 can be replaced by the total vacancy concentration c. It can be seen from (A.11) that the annealing is in general faster than first order except for $K_3t \gg 1$. The validity of this model can be checked by comparing Eq. (A.11) with isothermal annealing curves on well-annealed specimens. The first comparison is made with the data obtained by Emrick (fig. 9 of Ref. 67). Specimens P1-1-1 (preannealed for more than 12 hr at 1000°C), 2-1, and P1-1-2 were quenched from around 700°C. In Fig. 2 $\ln[(\beta + 1/c(t))/(\beta + 1/c(0))]$ is plotted vs. the annealing time (annealing temperatures 60, 77, and 90°C, respectively). In the evaluation $c(0)$ is taken as 2×10^{-5}* and $E_1^M + E_2^B - E_2^M = 0.26$ ev.[5] It is seen that the data yield straight lines, as expected from Eq. (A.11), down to vacancy concentrations as low as 2×10^{-6}. The agreement with the theory indicates that the concentration of divacancies should be small throughout the annealing. A very rough estimate of the upper limit of E_2^B can be obtained from the condition $2c_2 \ll c_1$ for the annealing curve of P1-1-1. This condition can be expressed explicitly as follows:

$$12 \exp \frac{\Delta S_2}{k} \exp \frac{E_2^B}{kT} c(0) \ll 1 \tag{A.13}$$

* In the original work, the quenched-in vacancy concentration is not given. Here $c(0)$ is taken approximately equal to the equilibrium vacancy concentration at the quenching temperature $T_q \approx 700$°C.

Taking $T = 60°C$, $c(0) = 2 \cdot 10^{-5}$, $\Delta S_2 = 2.2k$,[12] E_2^B should then be smaller than 0.13 ev. From the slope of the plots in Fig. 2, α is estimated to be $\approx 10^8$, which appears to be of the right order of magnitudes for the dislocation densities in well-annealed specimens.

The temperature dependence of K_3, i.e., of the slope of the plots such as Fig. 2, should give E_1^M, provided α remains constant. In Fig. 3 the isothermal recovery segments for a specimen quenched from 700°C, as taken from fig. 7 of Ref. 34, are plotted according to Eq. (A.11). All points calculated from the isothermal segments of the same annealing temperature should fall on one straight line. Except for the beginning (first isothermal segment) the data are seen to be represented well by Eq. (A.11). The sink density α is of the order of magnitude of 10^9. (The higher value is due to the postquench strain.) From the slope ratios in Fig. 3 an average value of 0.78 ev is found for the activation energy (which is close to the accepted value of 0.82 ev[1]), as compared to 0.71 ev obtained from the original slope-change evaluations. From this analysis it is seen that the vacancy annealing at low vacancy concentrations, as observed in isothermal runs, is consistent with the parameters of the low-binding-energy model and cannot be explained in terms of divacancy migration.

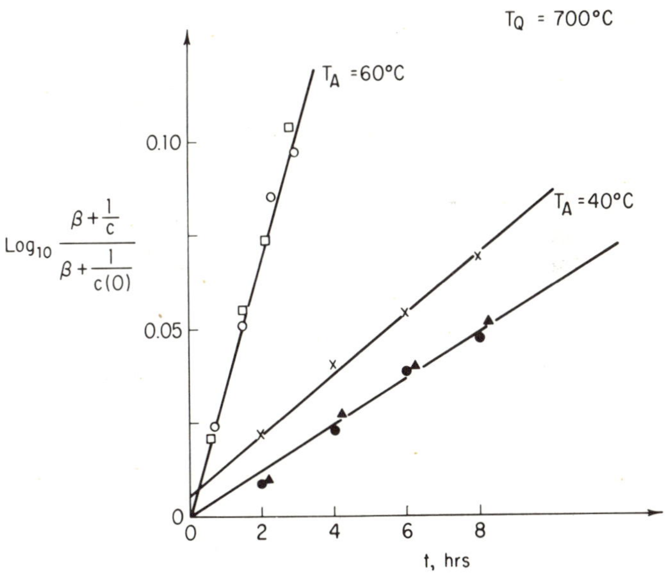

Fig. 3 Analysis of data given in fig. 7 of Ref. 34 according to Eq. (A.11).

× First isothermal segment at 40°C
● Second isothermal segment at 40°C
▲ Third isothermal segment at 40°C
○ First isothermal segment at 60°C
□ Second isothermal segment at 60°C

REFERENCES

1. Seeger, A., and D. Schumacher: In "Lattice Defects in Quenched Metals," p. 15, International Conference, Argonne, 1964, Academic Press Inc., New York and London, 1965.
2. Doyama, M., and J. S. Koehler: *Phys. Rev.* **127**:21 (1962).
3. Doyama, M.: In "Lattice Defects in Quenched Metals," *op. cit.*, p. 167.
4. Meshii, M.: *Ibid.*, p. 387.
5. Chik, K. P.: *Phys. Stat. Solids* **10**:659, 675 (1965).
6. Seeger, A., V. Gerold, and M. Rühle: Z. *Metallk.* **54**:493 (1963).
7. Seeger, A., and H. Bross: *Z. Physik* **145**:161 (1956).
8. Damask, A. C., G. J. Dienes, and V. G. Weizer: *Phys. Rev.* **113**:78 (1959).
9. Schottky, G.: *Z. Physik* **159**:584 (1960).
10. Doyama, M., and R. M. J. Cotterill: *Phys. Rev.* **137**:A994 (1965).
11. Seeger, A.: *J. Phys. Radium* **23**:616 (1962).
12. Schottky, G., A. Seeger, and G. Schmid: *Phys. Stat. Solids* **4**:419 (1964).
13. Silcox, J., and P. B. Hirsch: *Phil. Mag.* **4**:72 (1959).
14. Mori, T., and M. Meshii: *Acta Met.* **12**:104 (1964).
15. Westdorp, W., H. Kimura, and R. Maddin: *Acta Met.* **12**:495 (1964).
16. Mori, T., M. Meshii, and J. W. Kauffman: *Acta Met.* **9**:71 (1961).
17. Chik, K. P.: Unpublished.
18. de Jong, M., and J. S. Koehler: *Phys. Rev.* **129**:49 (1963).
19. Cotterill, R. M. J., and R. L. Segall: *Phil. Mag.* **8**:1105 (1963).
20. Segall, R. L., and L. M. Clarebrough: *Phil. Mag.* **9**:865 (1964).
21. Siegel, R. W.: *Phil. Mag.* **13**:337 (1966).
22. Bauerle, J. E., and J. S. Koehler: *Phys. Rev.* **107**:1493 (1957).
23. Koehler, J. S., F. Seitz, and J. E. Bauerle: *Phys. Rev.* **107**:1499 (1957).
24. Kimura, H., R. Maddin, and D. Kuhlmann-Wilsdorf: *Acta Met.* **7**:145 (1959).
25. Schottky, G.: *Z. Physik* **160**:16 (1960).
26. de Jong, M., and J. S. Koehler: *Phys. Rev.* **129**:40 (1963).
27. Meshii, M., T. Mori, and J. W. Kauffman: *Phys. Rev.* **125**:1239 (1962).
28. Asdente, M., and J. Friedel: *J. Phys. Chem. Solids* **11**:115 (1959).
29. Bross, H., and A. Seeger: *J. Phys. Chem. Solids* **6**:324 (1958).
30. Cattaneo, F., and E. Germagnoli: *Nuovo Cimento* **28**:923 (1963).
31. Damask, A. C., and G. J. Dienes: "Point Defects in Metals," Gordon and Breach, Science Publishers, Inc., New York and London, 1963.
32. Ytterhus, J. A., R. W. Siegel, and R. W. Balluffi: In "Lattice Defects in Quenched Metals," *op. cit.*, p. 679.
33. Balluffi, R. W., and R. W. Siegel: *Ibid.*, p. 694.
34. Ytterhus, J. A., and R. W. Balluffi: *Phil. Mag.* **11**:707 (1965).
35. Meshii, M., and J. W. Kauffman: *Acta Met.* **8**:815 (1960).
36. Kauffman, J. W., and M. Meshii: In "Lattice Defects in Quenched Metals," *op. cit.*, p. 77.
37. Schumacher, D.: *Phys. Letters* **19**:367 (1965).
38. Neuman, C. H.: *J. Phys. Chem. Solids* **27**:427 (1966).
39. Schüle, W., A. Seeger, D. Schumacher, and K. King: *Phys. Stat. Solids* **2**:1199 (1962).
40. Cattaneo, F., and E. Germagnoli: *Phys. Rev.* **124**:414 (1961).
41. Cattaneo, F., E. Germagnoli, and G. Guarini: *Nuovo Cimento*, suppl.1, p. 282, 1963.
42. Kauffman, J. W., and G. J. Dienes: *Acta Met.* **13**:1049 (1965).
43. Simmons, R. O., and R. W. Balluffi: *Phys. Rev.* **125**:862 (1962).
44. Makin, S. M., A. H. Rowe, and A. D. LeClaire: *Proc. Phys. Soc.* **70**:545 (1957).

45. Gilder, H. M., and D. Lazarus: *J. Phys. Chem. Solids* **26**:2081 (1965).
46. Seeger, A., G. Schottky, and D. Schumacher: *Phys. Stat. Solids* **11**:363 (1965).
47. Cuddy, L. J., and E. S. Machlin: *Phil. Mag.* **7**:745 (1962).
48. Quéré, Y.: *Compt. Rend.* **252**:2399 (1961).
49. Seeger, A.: *Phys. Letters* **9**:311 (1964).
50. Doyama, M., and J. S. Koehler: *Phys. Rev.* **119**:939 (1960).
51. Tomizuka, C. T., and E. Sonder: *Phys. Rev.* **103**:1182 (1965).
52. Simmons, R. O., and R. W. Balluffi: *Phys. Rev.* **119**:600 (1960).
53. Seeger, A., V. Gerold, K. P. Chik, and M. Rühle: *Phys. Letters* **5**:107 (1963).
54. Ramsteiner, F., G. Lampert, A. Seeger, and W. Schüle: *Phys. Stat. Solids* **8**:863 (1965).
55. Chik, K. P., A. Seeger, and M. Rühle: *Proc. 5th Intern. Congr. Electron Microscopy* (Academic Press Inc., New York and London) **1**:J-11 (1962).
56. Simmons, R. O., and R. W. Balluffi: *Phys. Rev.* **129**:1533 (1963).
57. Seeger, A.: *Phys. Letters* **12**:176 (1964).
58. Seeger, A., H. Kronmüller, and H. Rieger: *Z. Angew. Phys.* **18**:377 (1965).
59. Mehrer, H., H. Kronmüller, and A. Seeger: *Phys. Stat. Solids* **10**:725 (1965).
60. Beller, M.: *Phys. Verh. DPG* **5**:123 (1965).
61. Mehrer, H.: Unpublished.
62. Mughrabi, H.: *Phys. Verh. DPG* **5**:124 (1965).
63. Mader, S., A. Seeger, and E. Simsch: *Z. Metallk.* **52**:785 (1961).
64. Greenfield, I. G., and H. G. F. Wilsdorf: *J. Phys. Soc. Japan* **18** (suppl. III):20 (1963).
65. Wuttig, M., and H. K. Birnbaum: *J. Phys. Chem. Solids* **27**:225 (1966).
66. Wuttig, M., and H. K. Birnbaum: *Phys. Stat. Solids* **13**:K15 (1966).
67. Emrick, R. M.: *Phys. Rev.* **122**:1720 (1961).

DISCUSSION *on Paper Presented by Alfred Seeger*

ORIANI: If the binding energy of a complex (e.g., a solute vacancy complex or a divacancy) is numerically smaller than the energy for motion of the complex, one might naively think that the complex would dissociate in its attempt to move. Can you formulate the necessary conditions so that the complex may in fact contribute to the overall diffusivity?

SEEGER: Whereas the thermodynamic stability of a multiple vacancy is determined by the binding energy, the energy barrier to be overcome in an actual decay depends both on the binding energy and the mobility of the more mobile of the products of the decay reaction. Consider as an example a divacancy with binding energy E_2^B and migration energy E_2^M. If the attraction between two vacancies is confined to nearest-neighbor sites only, a major part of the binding energy must be supplied in addition to the migration energy E_1^M of a single vacancy in order to make the divacancy disintegrate. The energy barrier for disintegration E_2^D is therefore approximately

$$E_2^D = E_1^M + E_1^B$$

Provided $E_2^D > E_2^M$, the divacancy will be able to make many jumps before it disintegrates. In many cases in the fcc metals this decay process is so difficult that divacancies are transformed into two single vacancies by a complex process involving multiple vacancies.

SOLUTE-VACANCY INTERACTION IN AGE-HARDENING ALLOYS

A. J. Perry

Battelle Memorial Institute, Geneva-Carouge, Switzerland

INTRODUCTION

There have been a number of attempts to measure the vacancy binding energy of solute atoms in fcc metals using various experimental approaches. The results of these experiments are often widely different, and a good example is Mg in Al. Values given include 0.54 ev by Panseri and Federighi,[1] 0.18 ev by Takamura et al.,[2] 0.09 ev by Bartsch,[3] and about zero by Beaman et al.[4] However, these differences are not simply the outcome of using different experimental approaches, as witness the data of Federighi and Bartsch, both of whom studied the effect of Mg on the aging rate of Al-Zn alloys.

We can make some attempt to read order into the apparent chaos shown by the above data by reconsidering the experimental techniques and interpretation of results. In this discussion, we propose to attempt to do this for Mg in Al. As both Beaman's work and also that of Takamura appear, at first sight at least, to be free from serious objection, we shall be even more specific in our aim and see whether the Al-Zn-Mg work can be shown to be consistent with an acceptable interpretation. This is in fact necessary now that Gould and Gerold[5] have shown Federighi's original explanation of his results (Mg-vacancy pairs transporting Zn to zones) to be untenable.

EXPERIMENTAL BASIS

The basis of this type of experiment for determining binding-energy values is to measure the well-known slowing-down effect of impurities on the aging rate of a precipitation-hardening alloy,[6] e.g., the effect of In on Al-Cu. The reason for the effect is that diffusion to *GP* zones is a vacancy-aided process, the vacancies being retained by quenching following solution heat treatment. The impurities trap some of these vacancies so that the zone-forming atoms are starved of them and the aging rate is slower. The reaction involved is simply

$$\text{Vacancy} + \text{impurity} = \text{trapped vacancy} \tag{1}$$

and the reaction constant is

$$K = Z \exp \frac{B^{v-x}}{kT} \tag{2}$$

where B^{v-x} is the required binding energy, Z the coordination number, k Boltzmann's constant, and T the absolute temperature.

Hence, the fraction of trapped vacancies increases as the temperature falls, for example during a quench. By measuring the relative aging rates immediately after quenching an alloy with and without added impurity, the effect of the latter can be measured and the binding energy evaluated.

The approach has been used to study several elements added to Al-Cu (Fig. 1), and the results show that Zn and Mg have a comparable effect but that Mg is somewhat weaker than Zn. The computed curves given in Fig. 1 are derived from Lomer's equation:

$$n = A\left(1 - Zx + Zx \exp \frac{B^{v-x}}{kT}\right) \exp \frac{-E_f{}^v}{kT} \tag{3}$$

where x = atom fraction and $E_f{}^v$ = formation energy of a vacancy in pure Al for the *equilibrium* distribution at 0°C. The curves are the ratio of concentration of free vacancies in the alloy with and without added impurity. Hence we have some reservation about using them here for the interpretation of a nonequilibrium case.

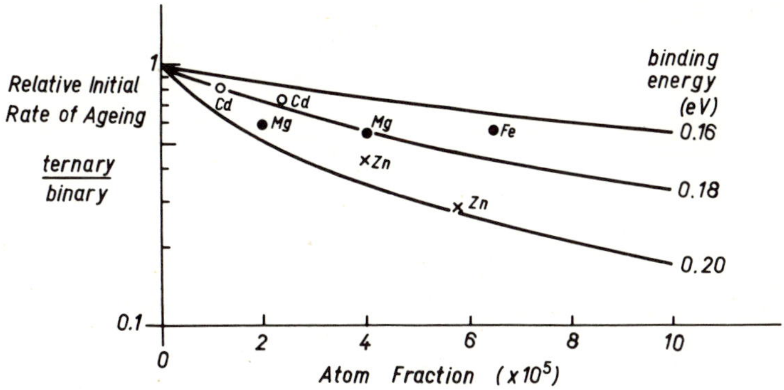

Fig. 1 The effect of added elements on the initial aging rate of a quenched Al-Cu alloy. The curves were computed from the equilibrium vacancy partition given by Eq. (3) at 0°C.

However, if added impurities always act in the same way, then why should Mg slow down the aging of Al-Zn at all? It is clear that we are dealing with two problems here: We have to understand more about the aging alloy before we can attempt to analyze the effect of Mg and hence derive the binding energy.

The first step in the formulation of the present model was the consideration of some data published by Takamura[7] for somewhat more concentrated additions than those in Fig. 1. Figure 2 shows that the binding energy of Mg is still less than that of Zn but only at lower solute fractions. This is no longer true above 0.3 a/o.

We believe that the divergence of Al-Zn from the simple model is caused by solute association, and we can account for the deviation quite well in terms of a Zn-Zn interaction of about 0.1 ev.

The fall in binding energy shown in Fig. 2 cannot be discussed strictly in terms of Eq. (3) because the derivation of it[8] assumes no mutual interference between the species given in Eq. (1). In order to avoid the problem, we have defined an effective vacancy binding energy B^* as the binding energy that each Zn atom would need to exhibit, were they all singly distributed in the lattice, in order to account for the experimental observations. This is an arbitrary and nonphysical definition, but one that is necessary to allow interpretation of our type of experiments.

If we now combine the evidence of association in Al-Zn solid solutions[9] in the literature with the above fall in effective binding energy with concentration, then we have the basis for an explanation of why Mg can slow down the aging rate of Al-Zn. We note also that the degree of association will increase with both concentration and falling temperature. This is not necessarily just Zn pairs, of course, but we assume that any association reduces binding energy.

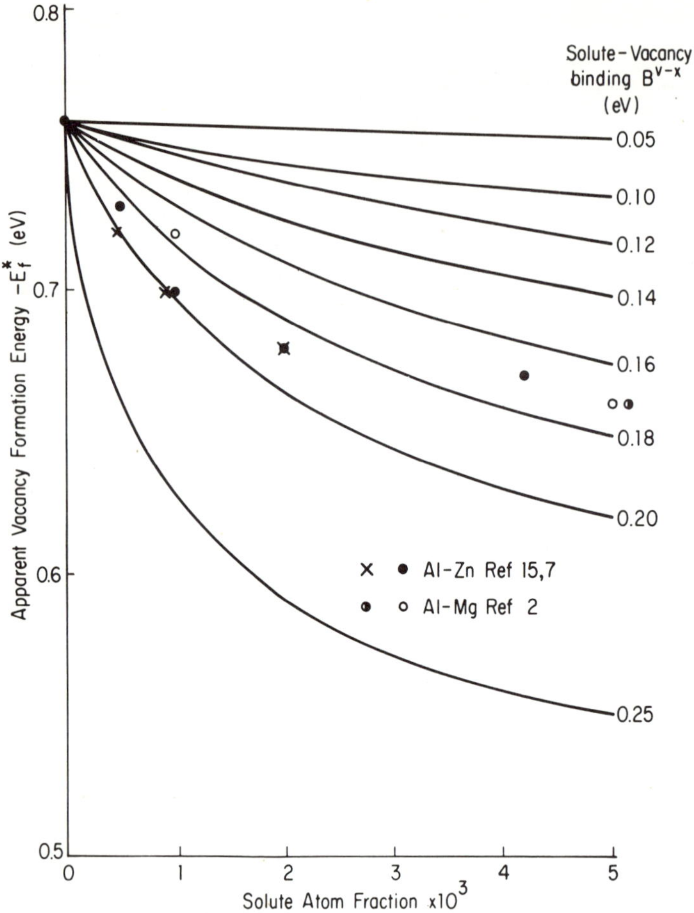

Fig. 2 The apparent vacancy formation energy of dilute Al-Zn and Al-Mg alloys as a function of composition. [*From A. J. Perry, Acta Met.* **14**:719 (1966).]

Our approach is therefore to measure the effective binding energy of Zn in our Al–4 a/o Zn alloy, measure the effect of Mg, and so obtain a binding energy for this latter.

RESULTS AND DISCUSSION

The experimental technique has been given elsewhere,[10] and so this will not be repeated here except to point out that a high-speed reproducible quench was used, and the aging process was followed by resistance measurements during continuous aging at 0°C. As an example of the experimental data, some results for Al-Cu are shown in Fig. 3. Attention is drawn to

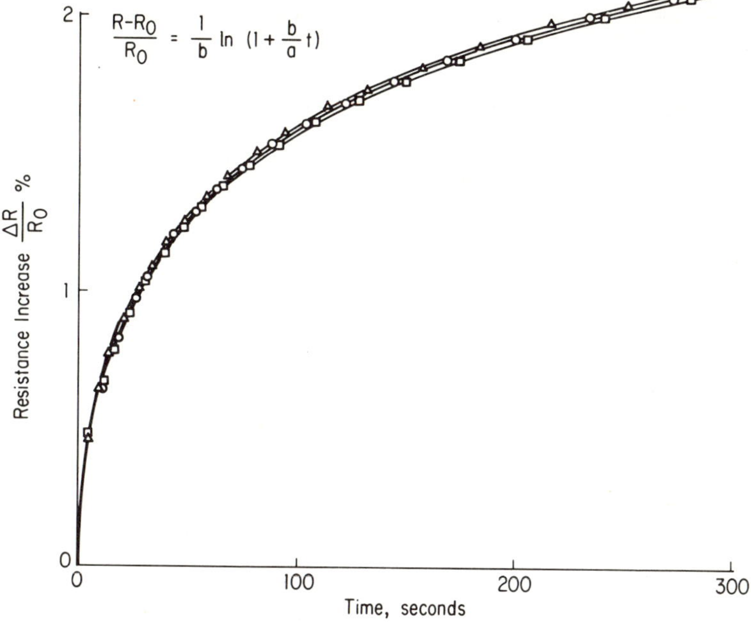

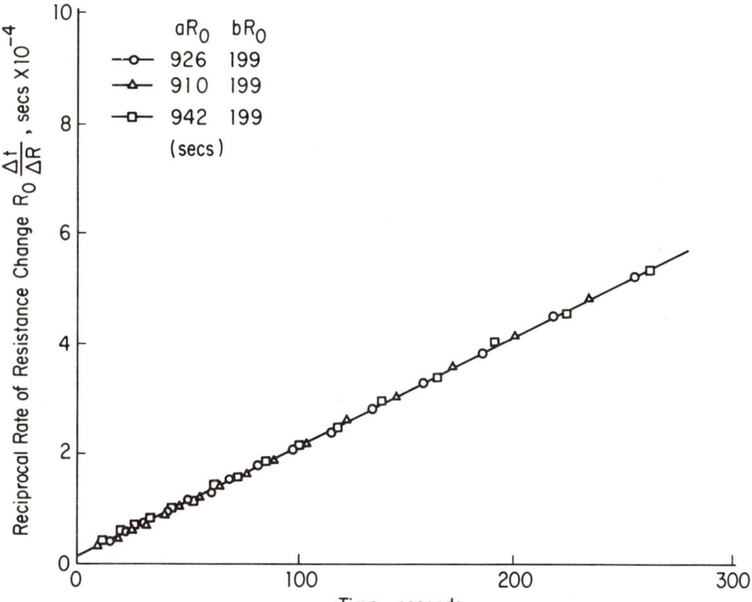

Fig. 3 Some typical data for the resistance increase and its rate of change in a freshly quenched Al-Cu alloy quenched from 520°C and aged at 0°C. [*From A. J. Perry, Acta Met.* **14**:305 (1966).]

the intercept of the lower graph, as this is the measure of the initial aging rate which we have used as an index of the vacancy concentration available for aging. The second parameter used to give the same information was the time taken to reach the resistance peak which corresponds to a critical zone size (examples are given in Fig. 6).

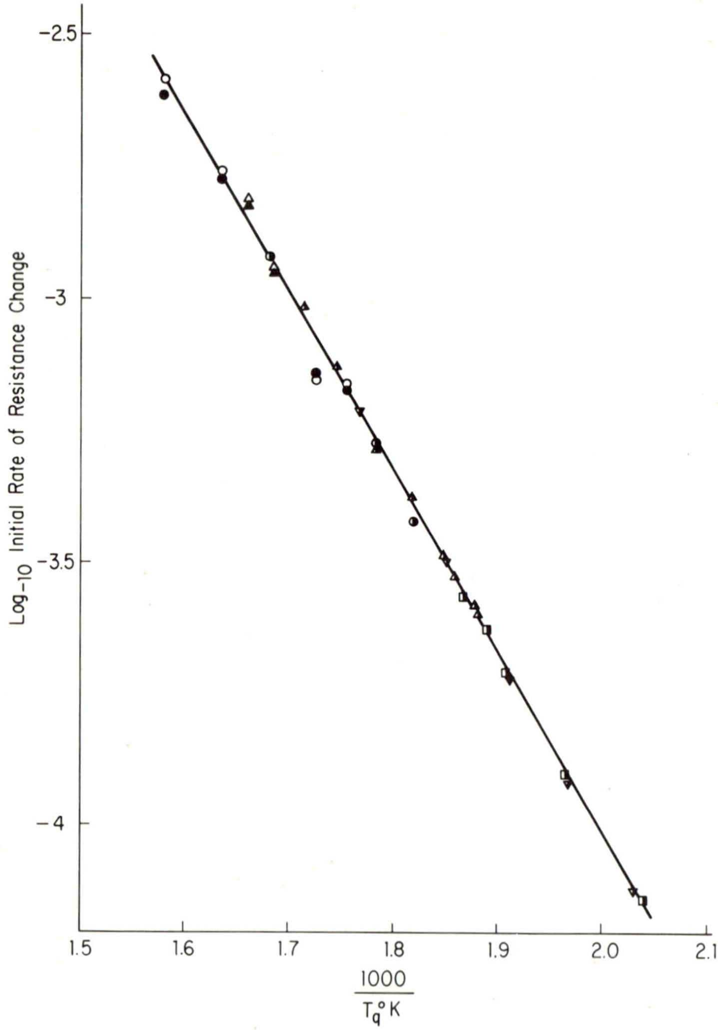

Fig. 4 The effect of quenching temperature on the initial rate of aging of an Al–4 a/o Zn alloy at 0°C. The slope yields an apparent vacancy formation energy of 0.69 ev. [*From A. J. Perry, Acta Met.* (*to be published*).]

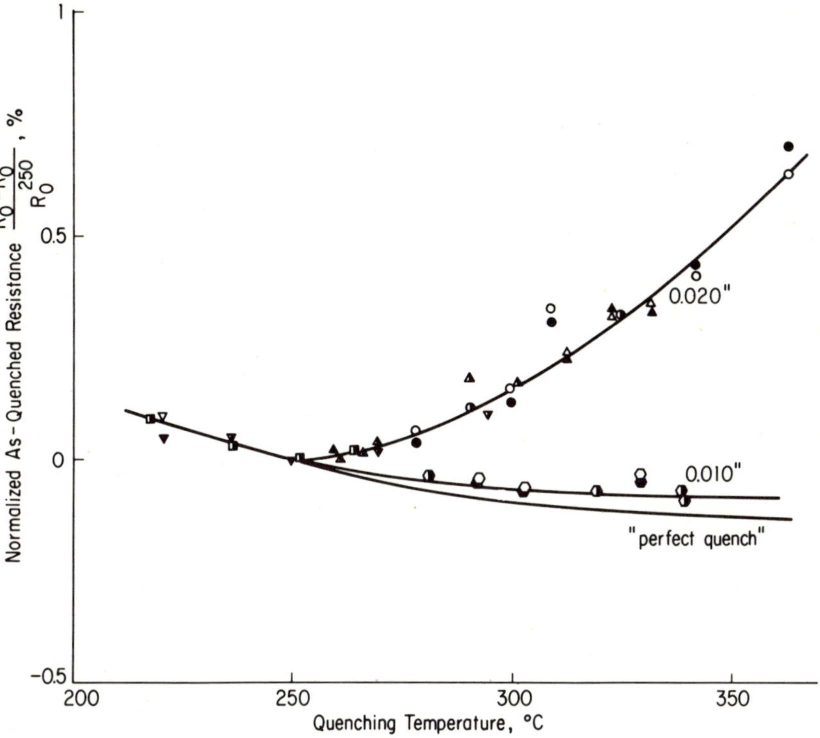

Fig. 5 The variation of the initial-quenched–in resistance with quenching temperature of an Al–4 a/o Zn alloy at 0°C for specimens of different diameter. [*From A. J. Perry, Acta Met. (to be published).*]

The variation of the initial aging rate with temperature for the present Al-Zn alloy is shown in Fig. 4. The slope of the curve gives a formation energy of 0.69 ev. This series of experiments also showed that the resistance of the freshly quenched alloy at 0°C increased with decreasing quenching temperature (Fig. 5). This is taken to mean that solute association increases as the temperature falls in accordance with the ideas already expressed.

We can now consider the effect of Mg. The addition was kept as small as possible (0.009 a/o) in order to avoid the formation of ternary zones as reported by Gerold[11] and encountered by Hashimoto.[12] As expected, the Mg addition slows down the aging process (Fig. 6), but it also lowers the resistance peak. In order to find the true effective binding of the Zn and also the real slowing effect of Mg, we have extrapolated both these parameters to zero aging as a function of peak height (Fig. 7). We also corrected the initial aging-rate data (Fig. 4) for the quench clustering shown in Fig. 5 (this is a small correction). The final results are 0.70 and 0.78 ev, respectively, leading to the values:

Effective binding energy of 4 a/o Zn in Al = 0.06 ev

True binding free energy of Mg in Al at 0°C = 0.17 ev

It must be added that we have neglected Mg-Zn interaction. If this exists, then the value of $B^{v-\mathrm{Mg}}$ must be modified accordingly. The agreement between the present result and that obtained from the Al-Cu studies also aged at 0°C (Fig. 1) suggests that the neglect may not be serious; however, this is by no means certain.

We now appear to understand the effect of Mg on Al-Zn aging. But before we can compare our result with that obtained with other techniques, we must ensure that our model can explain all observations. Firstly, we expect[13] Mg to affect the nucleation of zones by reducing the concentration of vacancies available to Zn atoms. This is borne out in practice because the peak height of the ternary alloy (Fig. 6), which is a function of the number of zones per unit volume of lattice, is lowered.

Secondly, vacancies decay during aging, and hence some must be released by both Zn and Mg atoms in order to maintain the partition described by

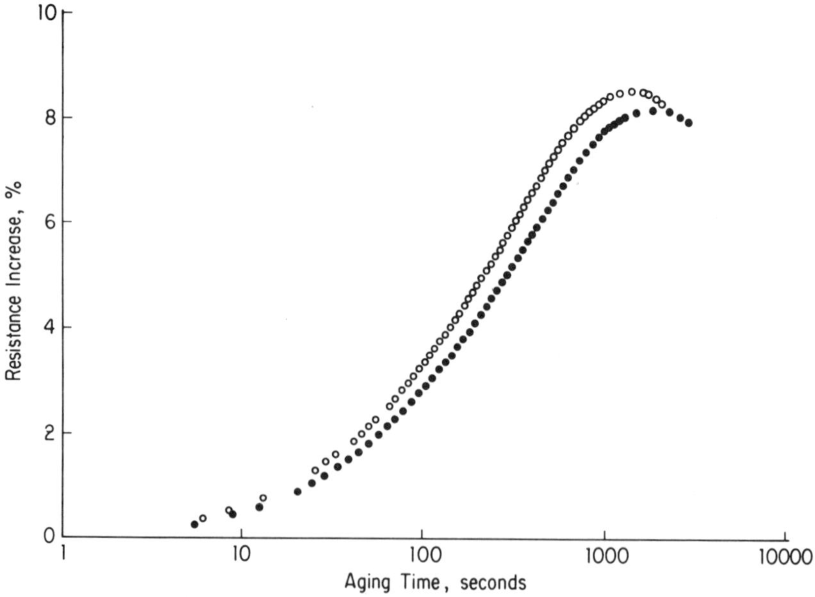

Fig. 6 The effect of 0.009 a/o Mg (filled symbols) on the aging of an Al–4 a/o Zn alloy quenched from 299°C and aged at 0°C. [*From A. J. Perry, J. Phys. Soc. Japan* **21**:549 (1966).]

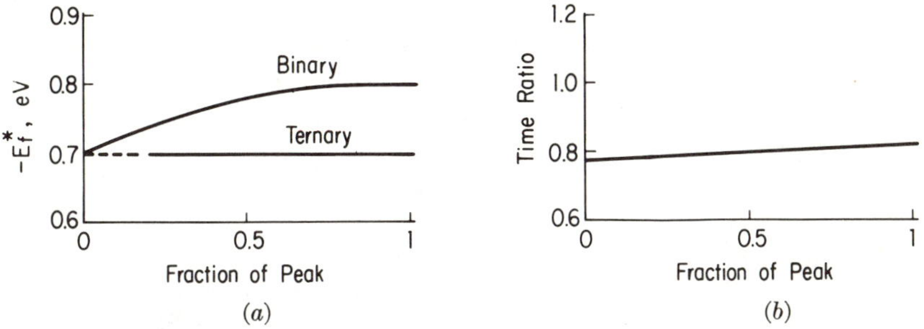

Fig. 7 The effect of aging on (*a*) the apparent vacancy formation energy; (*b*) the relative time to achieve fractions of the resistance peak as a function of peak height for Al–4 a/o Zn with and without 0.009 a/o Mg added. [*From A. J. Perry, Acta Met.* (*to be published*).]

Eq. (1). The release of vacancies by Mg means that fewer are released in parallel by the Zn atoms, and hence the alloy will tend to age slightly faster than it would if no vacancy release by Mg were possible. We expect the effect of Mg to decrease with aging, and this is again found in practice, as shown in Fig. 7*b*. The explanation of the slope of Fig. 7*b* may also be partly due to the following. As the concentration of Zn in the lattice falls, its tendency to associate will also fall, and hence the effective binding may well increase and counteract the slowing effect of Mg.

Thirdly, the lifetime of vacancies in the lattice is affected by Mg as discussed above. We cannot offer a detailed explanation, but we expect the time taken to achieve any degree of aging, such as the resistance peak, to vary in a different manner in the ternary from that observed in the binary Al-Zn alloy. In this way, we are also able to understand the divergence of the two curves shown by Federighi's results.[1]

We are now in a position to compare our binding-energy value for Mg with that reported from other techniques. By comparison with the results of Takamura[2] in the range 350 to 500°C and the about-zero value at 650°C (Beaman[4]), we can conclude tentatively that the binding entropy is positive such that the binding falls as the temperature increases. This conclusion is in accordance with the prediction of Seeger,[14] but in direct contrast with the result for Zn in Al given by Takamura[15] where a negative binding entropy is deduced.

We must note finally, however, that if all three binding-energy values given above are correct, then the temperature dependence of the binding energy is not linear as was assumed by Takamura[15] and Beaman et al.[4] This is interesting because the data of these latter alone would otherwise give an extremely large value of binding enthalpy H_b. It is also quite likely that H_b itself is temperature-dependent.[16]

REFERENCES

1. Panseri, C., and T. Federighi: *Acta Met.* **11**:575 (1963).
2. Takamura, J., K. Okazaki, and I. G. Greenfield: *Trans. Mining Met. Alumni Assoc.* **14**:396 (1962); *J. Phys. Soc. Japan* **18**(suppl. III):78 (1963).
3. Bartsch, G.: *Acta Met.* **12**:270 (1964).
4. Beaman, D. R., R. W. Balluffi, and R. O. Simmons: *Phys. Rev.* **137**:A917 (1965).
5. Gould, R. W., and V. Gerold: *Acta Met.* **12**:954 (1964).
6. Silcock, J. M.: *Phil. Mag.* **4**:1187 (1959).
7. Takamura, J.: In "Lattice Defects in Quenched Metals," p. 521, International Conference, Argonne, 1964, Academic Press Inc., New York and London, 1965.
8. Howard, R. E., and A. B. Lidiard: *Rept. Progr. Phys.* **27**:161 (1964).
9. Rudman, P. S., and B. L. Averbach: *Acta Met.* **2**:576 (1954).
10. Perry, A. J.: *Acta Met.* **14**:305 (1966).
11. Schmalzried, H., and V. Gerold: *Z. Metallk.* **49**:291 (1958).
12. Ohta, M., and F. Hashimoto: *J. Phys. Soc. Japan* **19**:130 (1964).
13. Embury, J. D., and R. B. Nicholson: *Acta Met.* **13**:403 (1965).
14. Seeger, A.: Agenda Discussion of Part IV.
15. Takamura, J., H. Kosuge, and H. H. Shimizu: *Phys. Letters* **16**:223 (1965).
16. McLellan, R. B.: Private communication.

DISCUSSION *on Paper Presented by A. J. Perry*

McLELLAN: You are comparing small energies (0 to 3 kcal/mole) in discussing the experimental binding energies for monovacancy–solute atom complexes. Do you think it is possible that the discrepancies are due to a small variation in binding entropy with temperature? It may be that the entropy could change with temperature owing to a variation in the temperature coefficient of the shear modulus of the solvent lattice, for instance.

PERRY: Yes, I do, but I am not convinced that it is the whole story.

INTERIONIC FORCES AND THE STABILITY OF LIQUID AND SOLID METALS*

J. E. Enderby

Department of Physics, The University, Sheffield, England

N. H. March

Laboratory of Atomic and Solid State Physics, Cornell University, Ithaca, New York

1 INTRODUCTION

Recently Johnson, Hutchinson, and March,[1] referred to below as J-H-M, have demonstrated that by a suitable analysis of the structure factor for liquid metals, valuable information may be obtained on the nature of the interionic force law. In their paper, J-H-M pointed out the importance of the Ornstein-Zernike direct-correlation function $f(r)$ (defined below)

* The authors wish to thank Dr. T. Gaskell for his help in the calculations of $f(r)$. One of us (N.H.M.) wishes to thank Professors Krumhansl, Holcomb, and Parratt for the opportunity of spending the session of 1965–1966 in their laboratory. During this period, partial contractual support has been obtained from the U.S. Atomic Energy Commission.

and its intimate connection with the pair potential $\phi(r)$. This idea has been further developed by the present authors.[2] In this paper we first discuss the determination of f and ϕ from experiment, paying particular attention to the difficulties involved. The characteristic form of ϕ for liquid metals is most readily discussed in terms of electron screening theory. We therefore consider the problem of stability of liquid metals in terms of the screening theory, making particular reference to the melting point and the value of the structure factor at low angles, which is, as demonstrated by Enderby and March,[2] a parameter of basic importance. The possibility of improving the approximate theories of the liquid state on which the derivation of ϕ depends is also briefly discussed.

2 DIRECT AND TOTAL CORRELATION FUNCTIONS

General Remarks

It will be useful to summarize the basic steps in obtaining the correlation functions from the experimental data. The coherent intensity of radiation scattered from a sample containing N atoms is given by

$$I = f^2 \langle \sum_p^N \sum_q \exp i\mathbf{K}.(\mathbf{r}_p - \mathbf{r}_q) \rangle \tag{1}$$

where $\mathbf{r}$ represents the positions of the atoms, $\hbar\mathbf{K}$ is the momentum transfer, f is an appropriate scattering factor, and the brackets indicate a time average. The usual procedure in dealing with liquids is to define a structure factor $S(K)$ by

$$S(K) = \frac{I}{Nf^2} \tag{2}$$

replace the double sum by an integral over the radial distribution function $g(r)$, and obtain finally

$$S(K) = 1 + \frac{4\pi n_0}{K} \int_0^\infty [g(r) - 1] r \sin Kr \, dr$$

$$= 1 + \frac{4\pi n_0}{K} \int_0^\infty rh(r) \sin Kr \, dr \tag{3}$$

Here n_0 is the number of atoms per unit volume, and $h(r) \equiv g(r) - 1$ is known as the total correlation function. Equation (3) is the usual starting point for discussing the structure of liquids, for it follows immediately that

$$h(r) = \frac{1}{2\pi^2 n_0 r} \int_0^\infty [S(K) - 1] K \sin Kr \, dK \tag{4}$$

Another correlation function basic to this paper is the direct-correlation function $f(r)$, first introduced by Ornstein and Zernike. The idea is to split the total correlation function into a direct and an indirect part by means of the equation

$$h(r) = f(r) + n_0 \int f(|\mathbf{r} - \mathbf{r}'|)h(\mathbf{r}')\, d\mathbf{r}' \tag{5}$$

which may be regarded as defining $f(r)$. Combining (4) and (5) yields

$$f(r) = \frac{1}{2\pi^2 n_0 r} \int_0^\infty f(K) K \sin Kr\, dK \tag{6}$$

where
$$f(K) = 1 - \frac{1}{S(K)} \tag{7}$$

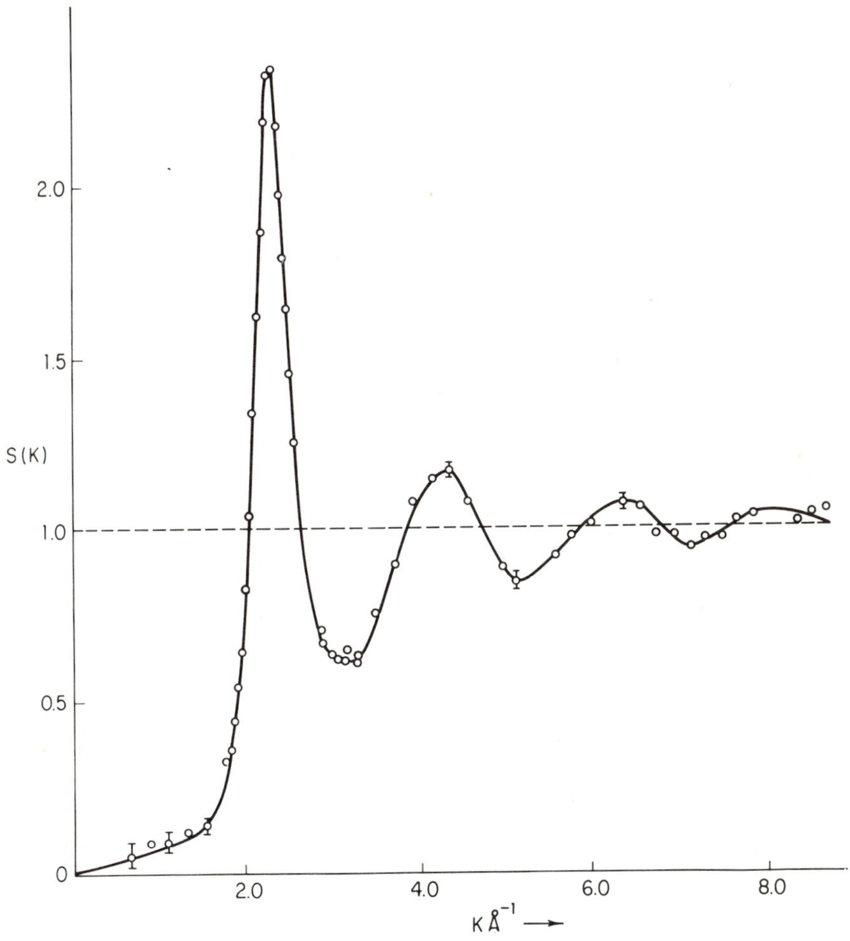

Fig. 1 Structure factor for liquid Tl at 320°C.

In principle then, provided $S(K)$ can be measured with sufficient accuracy over the regions of K space which dominate the integrals in (4) and (6), both $f(r)$ and $h(r)$ can be obtained by experiment.

The Determination of f and h

One of us (J.E.E.) is involved in a new determination of $S(K)$ for liquid metals and alloys using neutron techniques (Egelstaff, Enderby and North[3]). Special attention has been paid to the region $0 < K < 1.5$ A^{-1} which is, as we shall see, basic to the calculation of $f(r)$. Two features

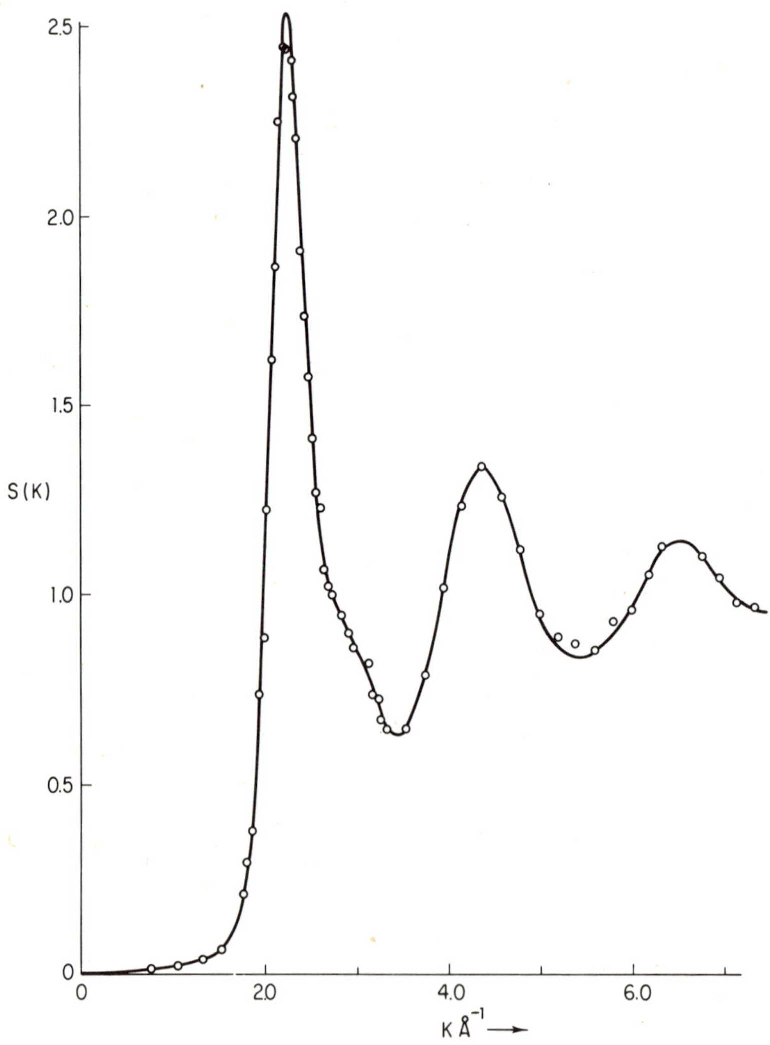

Fig. 2 Structure factor for liquid Sn at 350°C.

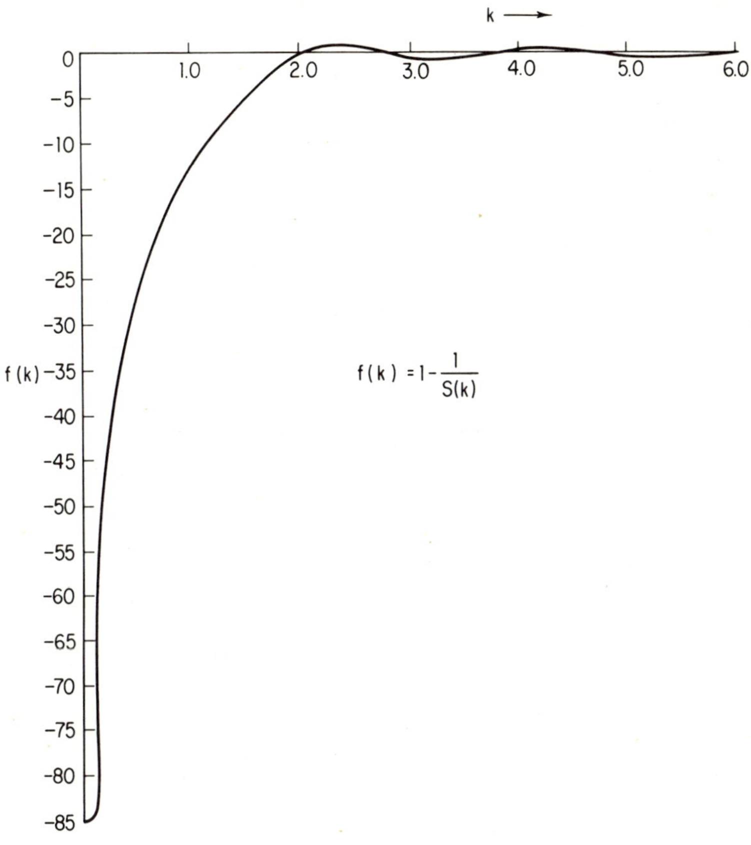

Fig. 3 The Fourier transform of $f(r)$ for liquid Tl.

of our experimental approach need to be stressed. First, in order to minimize the effect of multiple scattering as well as to simplify its discussion, we have used liquid metal samples in the form of thin slabs. Second, we have exploited the fact that the diffraction pattern for V as observed with neutrons is characterized by the almost complete absence of coherent diffraction peaks. The scattered intensity of V may be used, therefore, in deriving the normalization factor converting the observed intensities into absolute intensities for each value of K.

The structure factor for liquid Tl and Sn obtained by this technique is shown in Figs. 1 and 2. We note especially

1. The region $K < 1.5\ \text{A}^{-1}$ in which S rapidly approaches the well-known thermodynamic value

$$S(0) = n_0 kT\chi_T$$

where χ_T is the isothermal compressibility

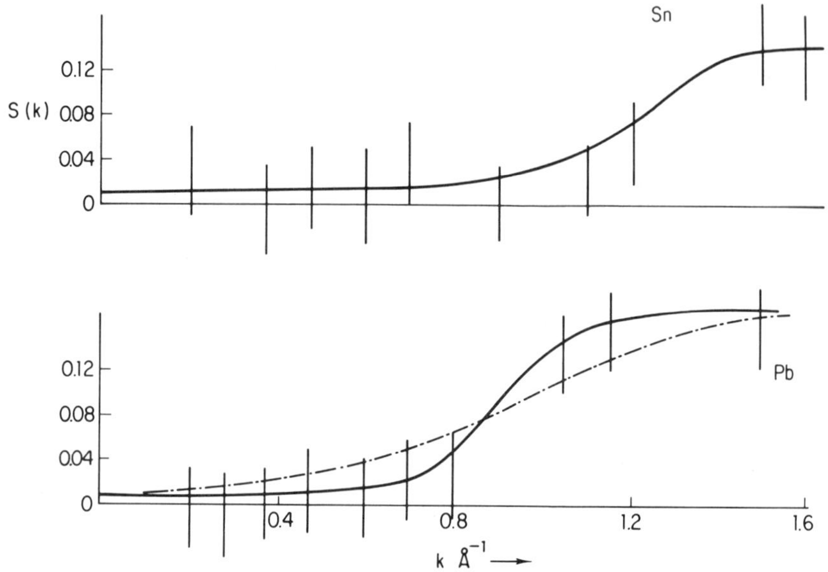

Fig. 4 Low-angle scattering for liquid Pb and Sn.

2. The main peak in $S(K)$ which, to a large extent dominates the form of $h(r)$

The function $f(K)$, by contrast, is dominated by the region $K < 1.5$ A^{-1}, as Fig. 3 shows. To evaluate $f(r)$ with any certainty clearly requires detailed knowledge of $S(K)$ in this region, and careful measurements for Sn and Pb are shown in Fig. 4. The low count rate makes for comparatively poor statistical accuracy although the essential features of $S(K)$ in this region now seem clear. The marked increase in level around $K \sim 0.8$ A^{-1} in liquid Pb is, we believe, a real effect and accords with the observation of Cocking and Egelstaff[4] of cooperative modes of motion in liquid metals. Figure 5 shows $f(K)$ for liquid Pb in the region $0 < K < 1.4$ A^{-1} calculated from the two smooth curves shown in Fig. 4; it is evident that very accurate measurements of $S(K)$ in this region are necessary if precise statements about $f(r)$ are to be made. This is a nontrivial experimental problem.

The calculated values of $f(r)$ for a variety of liquid metals are shown in Fig. 6. For comparison $f(r)$ is shown in Fig. 7 for Ar, using the neutron data of Henshaw.[5] The striking difference is, of course, that beyond $r \sim 5$ A, the f's are of different sign.* This difference is now known to arise from the marked difference in low-angle scattering connected with the big variation of $S(0)$ as between insulators and liquid metals.[2] More

* The position has been further clarified by a communication from Professor Pings in which he informs us that $f(r)$ for A, obtained from 13 sets of data, is always positive for $r > 4$ A (Mikolaj and Pings, to be published).

fundamentally, it reflects the different force law obtaining in the two sorts of systems, and the connection between f and ϕ is the problem to which we now turn.

3 INTERIONIC FORCES AND $S(K)$

It has long been recognized that there exists an exact relationship between g and ϕ. Unfortunately this involves a knowledge of the three-atom correlation function n_3, which is at present experimentally inaccessible. Present methods therefore involve approximating n_3, the best-known theories being those of Born-Green, Percus-Yevick, and the hyperchain diagrammatic analysis.

Even with these approximations, the question arises as to the most satisfactory way to use the limited data available. To illustrate the

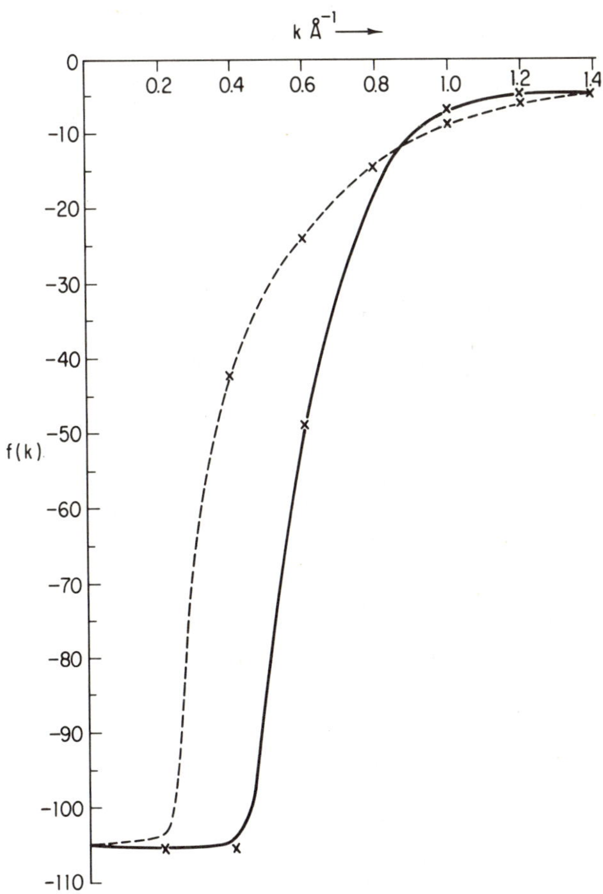

Fig. 5 $f(K)$ for liquid Pb for $0\ \text{A}^{-1} < K < 1.4\ \text{A}^{-1}$.

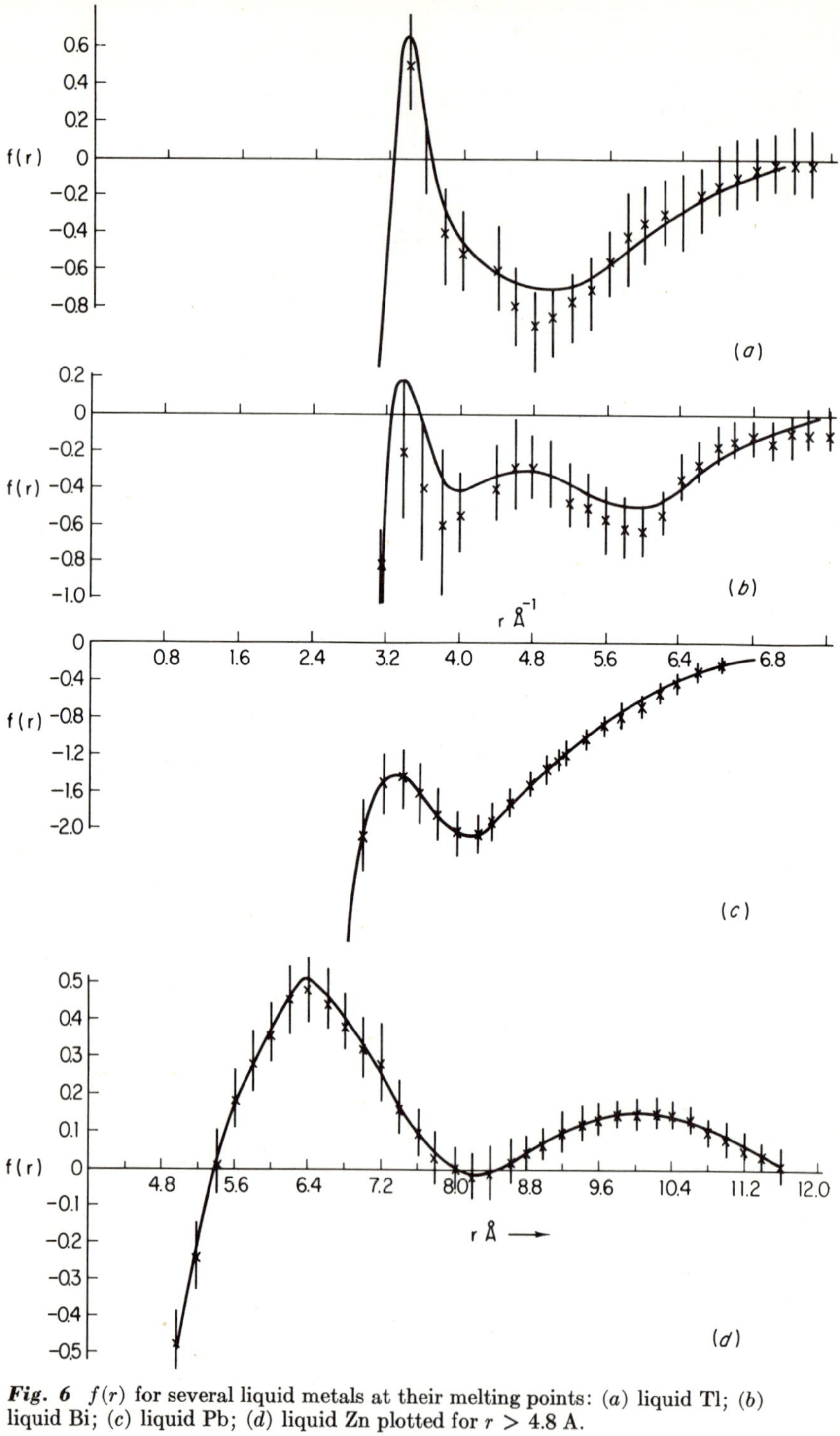

Fig. 6 $f(r)$ for several liquid metals at their melting points: (a) liquid Tl; (b) liquid Bi; (c) liquid Pb; (d) liquid Zn plotted for $r > 4.8$ A.

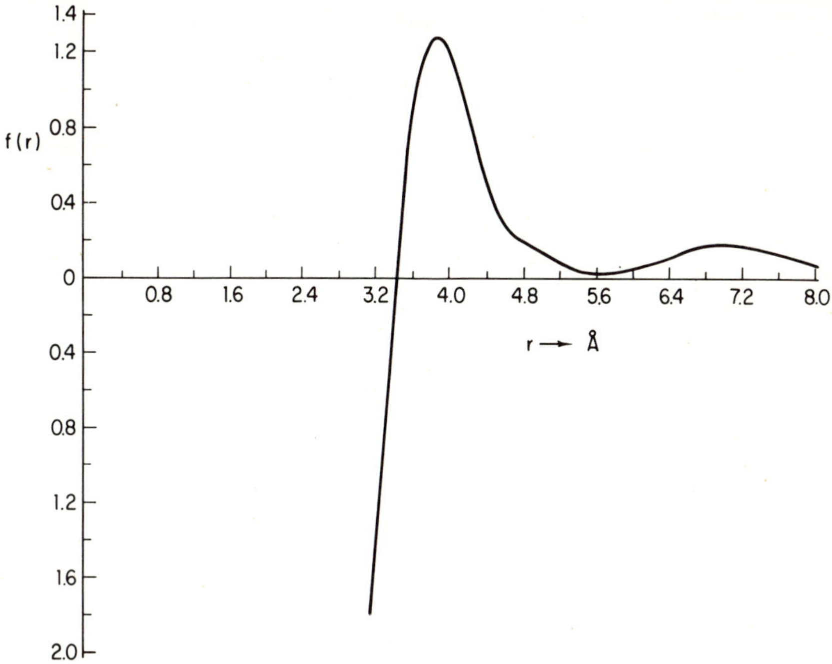

Fig. 7 $f(r)$ for liquid Ar.

problems, let us distinguish two methods, which would be completely equivalent if exact knowledge of $S(K)$ were at hand.

Method Based on $g(r)$

In the original work of J-H-M, the calculation of ϕ was carried out using the approximate theories, from the radial distribution function $g(r)$. In this method, the known physical property that $g(r)$ is essentially zero inside the atomic diameter is invoked, in addition to the raw data for $S(K)$. This information is not contained in the presently available measurements of $S(K)$ at all completely.

That successive transformation between $S(K)$ and $g(r)$ can improve the data has, in fact, been explicitly demonstrated by Kaplow et al.[6]

Method Based Solely on $S(K)$

The alternative procedure (worked out only for the Percus-Yevick and hyperchain theories and not directly applicable to the more appropriate Born-Green equation for liquid metals) is to calculate both $f(r)$ and $g(r)$ directly from $S(K)$ using only the raw data. Then the Percus-Yevick theory, for example, relates ϕ, g, and f by an equation of the form

$$\phi_{\text{P-Y}} = k_B T \ln\left(1 - \frac{f}{g}\right) \tag{8}$$

Using the calculations of $f(r)$ shown in Fig. 6, values of $\phi_{\text{P-Y}}$ can be obtained, and those for liquid Tl and Bi are given in Fig. 8. In general, these do not show the well-defined long-range oscillations characteristic of the J-H-M potentials.

It must be admitted, therefore, that while a major prediction of the

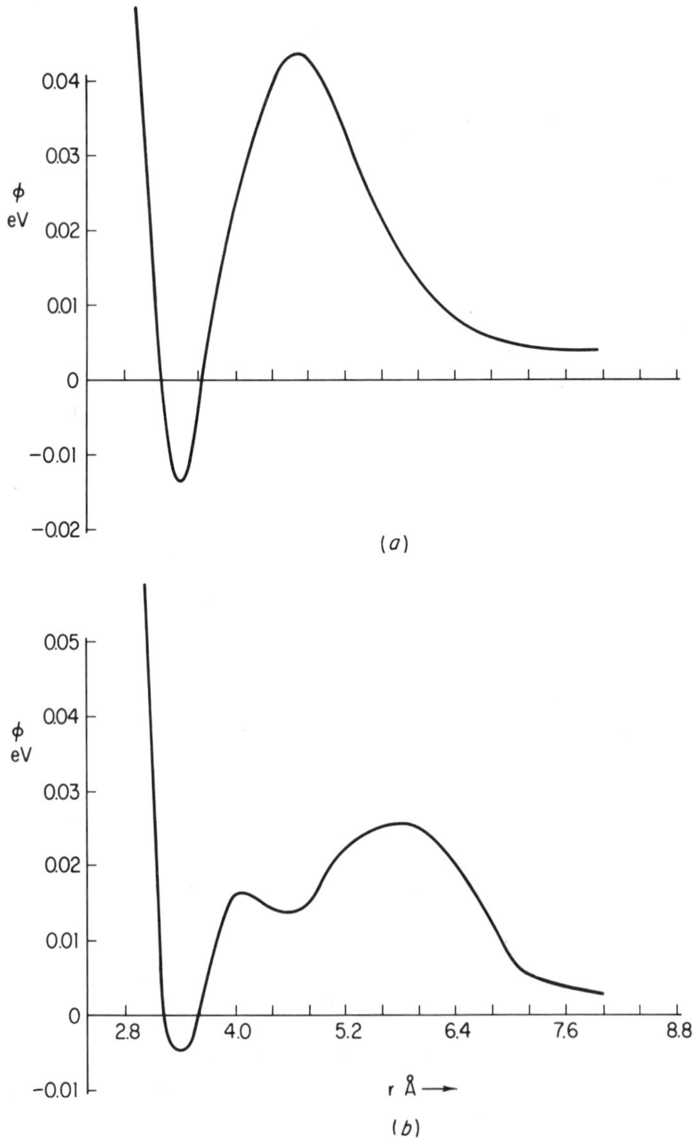

Fig. 8 $\phi(r)$ for (a) liquid Tl; (b) liquid Bi.

J-H-M analysis, namely the presence of a substantial repulsive region in ϕ around 5 to 7 A beyond the first minimum, is completely confirmed, the longer-range behavior of $f(r)$ which emerges depends on the way in which the data are analyzed. But the evidence we have still points to the fact that the original method of J-H-M is the best one available to us, and strongly suggests that no subsequent calculations of ϕ should neglect the region of zero $g(r)$ inside the atomic diameter. Equally, of course, the best available data for small K, together with the thermodynamic value for the long wave limit, must be used in constructing $g(r)$.

In spite of these uncertainties, it now seems firmly established that the forces obtaining in liquid metals, arising from the ion-ion screening by the electron gas, modify $S(K)$ in a way that is experimentally observable. Obviously then it is a matter of some urgency to improve techniques, both experimental and theoretical, so that we can get ϕ from the data.

4 SCREENING THEORY AND THE PROPERTIES OF LIQUID METALS

Introduction

As we have seen, the precise quantitative form of ϕ eludes us not only because of the experimental difficulties involved in determining S but also because of the theoretical problem of relative ϕ and S in the absence of knowledge of n_3. This point is discussed further in Sec. 5. However, elementary screening theory ought to be capable of explaining some of the problems associated with the stability of liquid conductors, and in this section we discuss applications of the basic ideas to:

1. The melting of close-packed metals
2. The value of $S(0)$
3. The liquid vapor transition

The Melting of Close-packed Metals

Recently, Mukherjee[7] has pointed out an empirical relationship between the Debye temperature Θ and the vacancy formation energy E_V for close-packed metals. This relationship has been shown by one of us (March[8]) to follow directly from screening theory. We shall briefly summarize this argument, and then, as again noted by Mukherjee empirically, shall go on to discuss the connection with the Lindemann formula. The Lindemann formula, it will be recalled, relates Θ to the melting temperature T_m by an equation of the form

$$\Theta = D\left(\frac{T_m}{M\Omega^{2/3}}\right)^{1/2} \tag{9}$$

where M is the atomic mass and Ω the atomic volume.

The screening theory treats a vacancy in a simple metal as a point charge $-Z$. Then Z is screened according to the Hartree theory of March and Murray[9] in accordance with the equation for the screened potential $V(r)$

$$\nabla^2 V = \frac{2k_F{}^2}{\pi^2} \int d\mathbf{r}' V(\mathbf{r}') \frac{j_1(2k_F\,|\,\mathbf{r} - \mathbf{r}'\,|\,)}{|\,\mathbf{r} - \mathbf{r}'\,|^{2}} \tag{10}$$

The first-order perturbation energy change is now, if ρ_0 is the electron density,

$$\Delta E = \rho_0 \int V(\mathbf{r})\, d\mathbf{r}$$

and integrating (10) over r, we find

$$\Delta E = \tfrac{2}{3} Z E_F \text{ with } E_F = \frac{k_F{}^2}{2} \tag{11}$$

The energy of formation of a vacancy, in first approximation, is obtained by correcting this with the energy change arising from increase of the crystal volume by one atomic volume, namely $\tfrac{2}{5}ZE_F$ (Fumi[10]). This leaves us with $(E_V = \tfrac{4}{15}ZE_F)$, but since there is some uncertainty in the precise value of the coefficient, we write

$$E_V = \alpha Z E_F \tag{12}$$

Now from the work of Bardeen and Pines[11] we have that the velocity of sound v_s is given by

$$v_s = \sqrt{\frac{Zm}{3M}}\, v_F$$

where v_F is the Fermi velocity.

Using the usual result for Θ in terms of the velocity of sound, namely, for an isotropic solid

$$\Theta = \frac{v_s}{\Omega^{1/3}} \frac{h}{k_B} \left(\frac{3}{4\pi}\right)^{1/3} \tag{13}$$

we find

$$\Theta = \frac{h}{k_B} \left(\frac{3}{4\pi\Omega}\right)^{1/3} \left(\frac{2E_V}{3\alpha M}\right)^{1/2} \tag{14}$$

This is precisely Mukherjee's empirical relation, and if α is taken as $\sim\tfrac{1}{6}$ rather than $\tfrac{4}{15}$ (see Fumi[10] and March[8]), the agreement between theory and experiment becomes almost exact.

Equation (12) calls for some comment, however. This is essentially a "small Z" expression because, as emphasized by March, it is based on first-order perturbation theory. For real metals, especially for $Z > 1$, correction terms are necessary, and these might be quite large. The α of

Eq. (12) is, strictly speaking, $\lim_{Z\to 0}(E_V/ZE_F)$ and will, in this case, tend to the value $\frac{4}{15}$. Figure 9 shows a plot of E_V/ZE_F versus Z, and the extrapolation back to $Z = 0$ confirms this.

The Lindemann melting-point formula can also be obtained by elementary arguments. In this case our basic task is now to connect E_V with T_m. This is possible along the following lines, related to the approach of Lennard-Jones and Devonshire[12] and Frank,[13] though these authors made, of course, no reference to the Fermi energy, being concerned with liquids generally. We argue, as in the Bragg-Williams theory, that we can define an order parameter S, and following Frank we take this as the ratio: number of atoms on lattice sites to total number of atoms. This parameter S is then unity when there is complete order, and by thinking as Frank does, of a skeleton crystal, filled up by vacancies and atoms in disordered positions, we can extend the concept continuously through the melting point.

The second essential feature of the Bragg-Williams theory is that the energy U required to take an atom from an ordered to a disordered site is dependent on the degree of order S. Thus, following Bragg and Williams, we write

$$U = U_0S \tag{15}$$

and notice that U_0 is an energy characteristic of a perfectly ordered crystal, like the vacancy formation energy E_V. Frank has insisted that quadratic

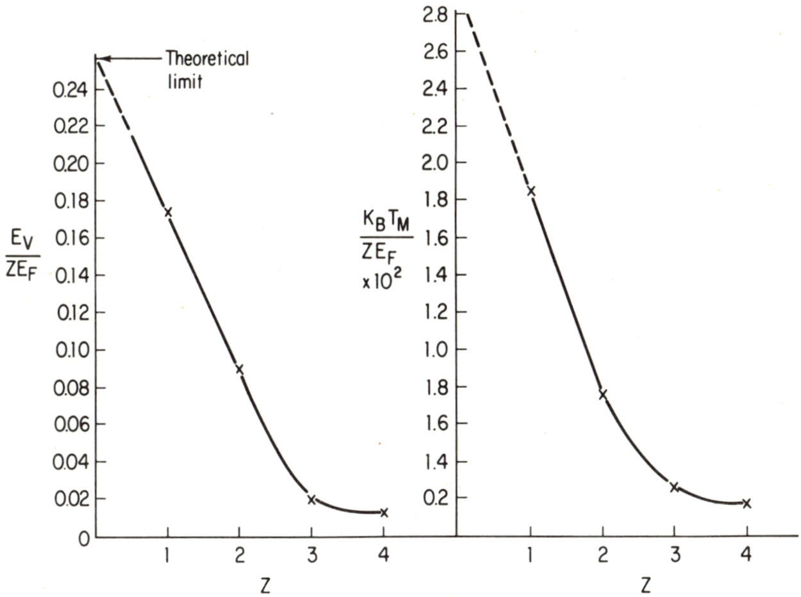

Fig. 9 Variation of E_v and T_m with Z.

and higher terms in S must be included in (15) to explain the fact that we do not observe the phenomenon of superheating of crystals, but this is not essential for the present argument. The Bragg-Williams theory then predicts a critical temperature, the order-disorder transition, which is obviously identified with melting in our model. This critical temperature is directly proportional to U_0, and this is the crucial feature for our argument. We now argue that U_0, like E_V, will be proportional to ZE_F. U_0 is clearly not the vacancy formation energy, but we notice the intimate correlation between E_V and Q, the activation energy for self-diffusion for close-packed metals, namely $E_V = 0.55Q$ (Simmons and Balluffi[14]) and this adds further support to the contention that U_0 is proportional both to valence Z and to the Fermi energy in a simple metal. The actual proportionality constant eludes us, though we expect it of the order of, though substantially less than, unity. For this reason, we need not invoke any precise constant in the Bragg-Williams relation between kT_m and U_0 though their work suggests that in fact

$$k_B T_m \sim \tfrac{1}{4} U_0 \tag{16}$$

Thus

$$k_B T_m = \beta Z E_F \tag{17}$$

where β is less than $\frac{1}{20}(\sim\alpha/4)$. Collecting together these results, we find

$$\Theta = \frac{h}{k_B}\left(\frac{3}{4\pi\Omega}\right)^{1/3}\left(\frac{2k_B T_m}{3M\beta}\right)^{1/2} \tag{18}$$

which is just the Lindemann relation.

This time we have no theoretical value for β, but again we can estimate the $Z = 0$ value by extrapolation (Fig. 9). This turns out to be $\sim\frac{1}{30}$, which yields a theoretical value of the Lindemann constant of 120. This is in excellent agreement with the experimental value, usually quoted for close-packed metals as $\sim$140.

The Value of $S(0)$

This most directly reflects the importance of the screening effect of the electron gas. The structure factor $S(K)$ is, according to thermodynamics, related to the velocity of sound v_s in the long-wavelength limit by

$$S(0) \doteqdot \frac{k_B T}{M v_s^2} \tag{19}$$

assuming the ratio of the specific heats is near unity. Using again the Bardeen-Pines result gives

$$S(0) \doteqdot \frac{3k_B T}{2ZE_F} = \frac{3k_B T\alpha}{2E_V} \tag{20}$$

We now evaluate $S(0)$ just above the melting point; we recall that T_m and E_V are, for close-packed metals, simply related by an equation of the form

$$k_B T_m = \gamma E_V \qquad \text{where} \qquad \gamma \sim 1/9$$

This makes

$$S_{T_m}(0) \doteqdot 3/2\, \alpha\gamma \sim 0.028 \tag{21}$$

Actually, this somewhat overestimates $S_{T_m}(0)$ because the repulsive forces between the ion cores have been neglected. A rough estimate of the correction, made by considering the differences between the velocity of longitudinal and transverse waves in the solid, suggests a factor of ~0.45, which makes $S_{T_m}(0) \sim 0.012$. However, the numerical agreement is less important than the sharp prediction implied in (21), namely that $S_{T_m}(0)$ is constant for all close-packed metals. In Table 1 we collect values of $S_{T_m}(0)$ for both close-packed and bcc metals, and it will be seen that this prediction is roughly confirmed. The bcc values are much closer to (21), which reflects the minor role of core effects for these metals. However, relaxation effects may be large here, and so the correlation with defect energies may be invalid, and it would be unwise to place too much emphasis on this result.

The Boiling Points of Liquid Metals

Unlike the melting temperatures T_m, there is no reasonably simple relation with Z and/or E_F for the boiling points T_b. This is shown in Table 2, where we have tabulated T_m/T_b for 760 mm pressure.

TABLE 1 $S_{T_m}(0)$ ***Values for Liquid Metals***

Metal	Z	T_m	$S_{T_m}(0)$
Cu	1	1356	0.021*
			~0.014†
Ag	1	1233.6	0.018*
			~0.012†
Mg	2	924	~0.019†
Al	3	933	0.016‡
Pb	4	600.5	0.009§
Na	1	371	0.027§
K	1	337	0.024§

* A. Pronin and C. H. Filippov, *Izv. Vysshikh Uchebn. Zavedenii Chernaya Met.* **6**:11 (1963). Also M. B. Gitts and I. G. Mikheilov, *Soviet Physics—Acoustics* **11**:372 (1966).

† Estimated compressibility by increasing value for solid by ~35%.

‡ H. J. Seeman and F. K. Klein, *Z. Angew. Phys.* **19**:368 (1965).

§ O. J. Kleppa, *J. Chem. Phys.* **18**:331 (1950).

TABLE 2 ***Melting-point–Boiling-point Ratios for Liquid Metals***

Valence	Metal	Ratio	Valence	Metal	Ratio
$Z = 1$	Cu	0.52	$Z = 3$	Al	0.37
	Ag	0.52		Ga	0.13
	Au	0.45		In	0.25
$Z = 2$	Zn	0.58		Tl	0.25
	Cd	0.58	$Z = 4$	Sn	0.19
	Hg	0.37		Pb	0.106
	Mg	0.67	$Z = 5$	Sb	0.47
	Ca	0.67		Bi	0.30

The irregular pattern is the strongest indication that we have for supposing that boiling occurs via a marked change in the interatomic force law, caused by the formation of bound states. The qualitative argument would then be that it should not be much more difficult for two electrons to go into an s-like bound state than for just one to go in. This would suggest that the boiling points of monovalent and divalent metals should be grouped together, and Table 2 very roughly bears this out.

If this argument is correct, it would suggest that the stability range for $Z = 3$ and 4 would be greater, as we should need to form two bound states to accommodate all the conduction electrons. One might speculate then that the trivalent metals would become monovalent before boiling.

We have no explanation to offer for Bi and Sb, except that they are complex metals and probably ought not to be discussed on the same basis as the other "simple" metals considered here.

The bound-state formation that we are proposing has probably little to do with changes in screening, but is most likely attributable to the tendency to localized-state formation with increasing disorder. But we have not, so far, been successful in making any useful numerical estimates to confirm, or refute, these tentative suggestions.

It is clearly of interest to inquire about the variation of $f(r)$ with temperature as the normal boiling point is approached. This requires a knowledge of the compressibility and of $S(K)$ at low K over a wide range of temperatures. In the absence of a complete set of data, we focus attention on liquid Hg for which the compressibility is known just below T_b.[15] Combining this with what is known about the structure,[16] we are able to sketch the approximate form of $f(K)$ at two temperatures. Most of the difference between $f(r)$ implied by Fig. 10 at the two temperatures considered appears to arise from the fact that asymptotically $k_B T f(r) \sim -\phi(r)$. The need for careful work over a wide range of temperature for $f(r)$ is

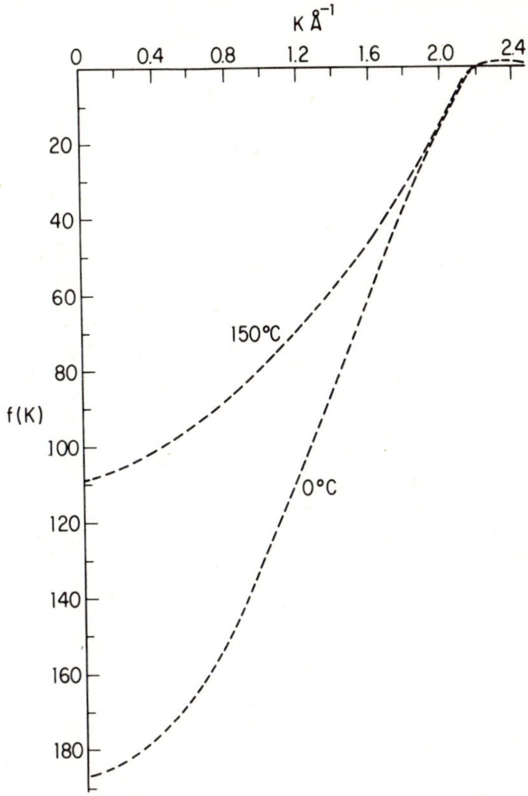

Fig. 10 Approximate forms of $f(K)$ for liquid Hg at two temperatures.

obvious, if changes in $\phi(r)$ of the kind we are proposing are to be tested.

Conclusions

The main conclusion of this section is now quite clear. Elementary screening theory can, in principle, account for both the melting-point formula of Lindemann and the observed values of $S_{T_m}(0)$. It will be recalled that $S_{Tm}(0)$ is of basic importance in deciding the form of $f(r)$ and hence of ϕ, and a consistent viewpoint of the forces in liquid and solid metals based on screening theory is now beginning to emerge. In this connection it is worth mentioning the work of Worster and March.[17] These workers considered the problem of the stability of certain metals if oscillatory pair potentials were used. They concluded that the difficulties associated with the stability of the bcc structure discussed by, for example,

Born[18] disappears when oscillatory as opposed to monotonic pair potentials are used.

However, while the work of Worster and March met with some success, it is clear that the details of the potentials, particularly the phasing of the oscillations, is important in deciding which is the stable configuration. Furthermore, the phonon dispersion curves are also known to be very sensitive to the detailed force curve, especially through the work of Harrison.[19] The current situation has been surveyed by the present authors (Enderby and March[2]). The conclusion appears to emerge that in spite of the present lack of fully quantitative force curves, there is every hope that, for simple metals, a unified treatment of solid and liquid metals may be within our grasp, based on pair forces having features quite characteristic of the conducting state.

5 THREE-ATOM CORRELATION FUNCTION

We shall finally make some brief comments on the important problem of the three-atom correlation function in liquid metals. Although none of the available theories are likely to give fully quantitative accuracy, the work of J-H-M strongly suggests that the Born-Green theory is much the best of them for metallic interaction potentials.

Actually, one difficulty with all the approximate theories, as has been emphasized by Ashcroft and March[20] on the basis of density expansions for hard spheres, is that the approximations give $h(r) - f(r)$ correctly to one higher order in the density n_0 than h and f separately. And by mixing in just a small fraction of $f(K)$ into $h(K)$ at small K for liquid metals, very large errors could be introduced.

In the absence of a tractable first-principles theory of the three-atom correlation function, we propose that a new correlation function $G(\mathbf{r} - \mathbf{r}')$ be studied. This is closely related to the direct-correlation function $f(\mathbf{r} - \mathbf{r}')$ of Ornstein-Zernike at large separations, but is closely related to the three-atom correlation function n_3 (involving integrals over n_3) at intermediate separations. For present purposes, it will suffice to define G through

$$\ln g(r) + \frac{\phi(r)}{k_B T} = n_0 \int G(|\mathbf{r} - \mathbf{r}'|)h(\mathbf{r}')\, d\mathbf{r}' \qquad (22)$$

Since G tells us how to go from the potential of mean force $U(r) = -k_B T \ln g(r)$ to the pair potential $\phi(r)$, we shall refer to it as the force-correlation function. In the Born-Green theory, then, G is approximated by

$$G(r) \doteqdot \int_r^\infty \frac{g(t)\phi'(t)}{k_B T}\, dt$$

while in the hyperchain theory

$$G(r) \doteqdot f(r)$$

Both theories agree that at large r, $G(r) \sim -\phi(r)/k_BT$.

Series expansions for G can be developed, but the most directly useful information that can be obtained at present is from molecular dynamics calculations, such as those of Paskin and Rahman.[21] For Na, for chosen pair potentials $\phi(r)$, these workers have obtained the radial distribution function $g(r)$, essentially exactly. It would be of considerable interest to calculate the force-correlation function G from their work.

It is our belief that the Born-Green theory will have advantages over the diagrammatic methods for the long-range forces in liquid metals, and the procedure discussed above may well indicate then the corrections required. No such advantages over the diagrammatic methods were found by J-H-M for short-range interactions, as in fluid Ar for example.

REFERENCES

1. Johnson, M. D., P. Hutchinson, and N. H. March: *Proc. Roy. Soc.* **A282**:283 (1964).
2. Enderby, J. E., and N. H. March: *Advan. Phys.* **14**:453 (1965).
3. Egelstaff, P. A., et al.: *Phys. Letters* **21**:286 (1966), and to be published.
4. Cocking, S. J., and P. A. Egelstaff: *Phys. Letters* **16**:130 (1965).
5. Henshaw, D. G.: *Phys. Rev.* **105**:976 (1957).
6. Kaplow, R., S. L. Strong, and B. L. Averbach: *Phys. Rev.* **138**:A1336 (1965).
7. Mukherjee, K.: *Phil. Mag.* **12**:915 (1965).
8. March, N. H.: *Phys. Letters* **20**:231 (1966).
9. March, N. H., and A. M. Murray: *Proc. Roy. Soc.* **A266**:559 (1962).
10. Fumi, F. G.: *Phil. Mag.* **46**:1007 (1955).
11. Bardeen, J., and D. Pines: *Phys. Rev.* **99**:1140 (1955).
12. Lennard-Jones, J. E., and A. F. Devonshire: *Proc. Roy. Soc.* **A170**:464 (1939).
13. Frank, F. C.: *Proc. Roy. Soc.* **A170**:182 (1939).
14. Simmons, R. D., and R. W. Balluffi: *Phys. Rev.* **117**:52 (1960).
15. Bett, K. E., K. E. Weale, and D. M. Newitt: *Brit. J. Appl. Phys.* **5**:243 (1954).
16. Kruh, R. F., G. T. Clayton, C. Head, and G. Sardin: *Phys. Rev.* **129**:1479 (1963).
17. Worster, J., and N. H. March: *Solid State Commun.* **2**:245 (1964).
18. Born, M.: *Proc. Cambridge Phil. Soc.* **36**:160 (1940).
19. Harrison, W. A.: *Phys. Rev.* **139**:A179 (1965).
20. Ashcroft, N. W., and N. H. March: *Proc. Roy. Soc.* (in press).
21. Paskin, A., and A. Rahman: *Phys. Rev. Letters* **16**:300 (1966).

DISCUSSION *on Paper Presented by J. E. Enderby*

PICK: As far as I have done computations on the relative stability of different phases on metals, the difference of energy of the phases is some thousandths of the total energy. It does not seem to me quite obvious that a very well defined $\phi(r)$ in a liquid metal will give a good result for any solid phase, even if the actual $\phi(r)$ in the solid is close to the liquid one.

ENDERBY: I absolutely agree that great care is necessary in going from the liquid to the solid; obvious features such as the changes in volume and topology of the Fermi surface all have to be taken in account. Certainly the theory so far is too rough to distinguish small differences of the sort mentioned above. Nevertheless, the importance of a unified approach to both the solid and liquid states is, we believe, worth emphasizing.

DEHLINGER: One of the most important things we know about liquids is that the structure itself is inhomogeneous. One has some sort of short-range ordering in the coordination itself, not only in the concentration and the distribution of vacancies. This inhomogeneity comes out more strongly in kinetic experiments than in equilibrium data. Therefore one expects differences between constants derived from the two sorts of experiments.

ENDERBY: The exact nature of the short-range order in liquids certainly ought to be studied carefully. One useful tool is the inelastic scattering of neutrons in which the time-dependent pair function is studied. The evidence so far points to distinct cooperative modes of motion in liquid metals. The situation for liquid insulators is not so certain, but at the moment it appears that the inelastic spectrum is less clearly defined. This is in accord with the type of potentials for the two systems suggested in our paper.

STATISTICAL GEOMETRY AND THERMODYNAMICS OF LIQUIDS*

Robert Collins

Department of Mathematics, Royal College of Advanced Technology, Salford, England

1 INTRODUCTION

The object of this paper is to give an account of the thermodynamic properties of simple monatomic liquids from the point of view of the statistical geometry of disordered lattices. The particular approach singled out for detailed discussion is that developed by Bernal and his coworkers[2–6] and in a modified form by the author.[8–10]

The type of liquids discussed and the essential assumptions made are similar to those in the recent review article by Rowlinson.[19] It is assumed that classical statistics are adequate, and that the interaction energy between atoms separated by a distance l is given by a scalar pair potential $\phi(l)$. Only monatomic liquids are considered, and the details of the electronic motions are ignored except for their effect on the pair potential $\phi(l)$.

* The author would like to thank Dr. D. C. S. Allison of this department for stimulating discussions. Thanks are also due to Prof. S. F. Edwards, Prof. M. E. Fisher, and Dr. P. Lloyd for several helpful suggestions.

Apart from the geometric approach described here, the main lines of attack on the problem of the equation of state can be roughly divided into three main groups. The first is direct evaluation of the partition function by means of virial expansions. The difficulty of calculating the higher coefficients and the fact that the expansions formally diverge at high densities make it seem unlikely that this approach will be able to account for the properties of real liquids, although Fisher (Ref. 12, p. 944 et seq.) has shown that for lattice gases accurate information on the nature of the critical point can be obtained. For the particular simplified case of hard spheres, Ree and Hoover (Ref. 12, p. 939 et seq.) have obtained results from virial expansion in good agreement with Monte Carlo calculations.

The second type of well-established approach is to calculate the pair-correlation function $g(r)$ via an integral or integrodifferential equation. To this group belong the Kirkwood, Percus-Yevick (P-Y), Born-Green-Yvon (B-G-Y), and hypernetted chain (HNC) approximations. These depend for their validity on the neglect of some or all of the higher-order interatomic correlations, in the form of either the superposition approximation or the selective omission of high-order diagrams. For hard spheres the quantitative results are good, particularly for the P-Y equation, although no phase transition is predicted. For real gases, the validity of the approximations at high densities seems more obscure, and it appears unlikely that in their present form they can account for a liquid-solid transition, particularly since negative values for the pair-correlation function $g(r)$ can result at high densities.

The third form of approximation consists of the setting up of some form of "model" in which an attempt is made to assess, either intuitively or formally, the key properties of the system dominating the thermodynamics, to build these qualitatively into the model, and then to calculate in detail the thermodynamics of the system within the framework of the model. In this class fall the lattice, cell, and hole theories of liquids, discussed in considerable detail by Levelt and Cohen,[18] and the tunnel theory of Barker.[1] As originally introduced, these theories depend on the validity of the ad hoc hypotheses on which they are based, which are difficult to justify formally. Also, in the cell models, the long-range order characteristic of the solid state is incorporated into the coordinate system in the liquid state also, although the liquid is known to have no long-range order.

The statistical geometrical approach differs essentially from the foregoing in that (1) no attempt is made to proceed from the general partition function, and (2) no attempt is made to set up a long-range ordered lattice to describe the liquid state. As a result of extensive experiments with models of hard spheres in two and three dimensions Bernal et al.[2–6] showed that many of the qualitative features of liquid structure [including the general form of the pair-correlation function $g(r)$] could be accounted for by the purely geometrical constraints on the ways in which hard, or nearly hard, spheres can be irregularly packed at high densities. The general conclusions were that the structure of a simple liquid could be represented by an assembly of close-packed irregular polyhedra of various types, and that the relative numbers of these changed in accordance with the transitions from solid to liquid and liquid to gas.

The main advantage of this point of view is that it provides, in principle at least, a common framework within which solid, liquid, and gaseous states can all be described. As originally formulated, however, although hard-sphere systems could be adequately simulated, it was difficult to see how the quantitative effects of temperature variation and a more general $\phi(l)$ could be included. An initial study[8] considered the very simplified two-dimensional model in which the interatomic bonds were all of exactly equal length, and the only polyhedra considered were squares and equilateral triangles. Even with this highly artificial system, phase changes could not be accounted for except by introducing very arbitrary variations of bond energies with local atomic coordination numbers.

In subsequent versions of the theory,[9–10] a number of modifications were made which enabled the calculation to be related to real liquids. The polyhedra were all reduced to tetrahedra by introducing extra bonds where necessary, and the entropy contribution of the bond-length variation was estimated directly, using the information-theory arguments developed by Brillouin and Jaynes.[7,13–15] Using the simplest possible approximations, an equation of state was developed involving $\phi(l)$ and two ad hoc parameters. Using the Lennard-Jones $\phi(l)$ appropriate to Ar by adjustment of the parameters, it was found possible to fit the experimental PV isotherms for liquid Ar over a fair range of densities, and also to obtain the specific heats to within 20% or so. Subsequent investigation has shown that the shape of the C_pP and C_vP isotherms is qualitatively wrong for the equations quoted. Also the wrong form for the virial expansion at low densities was obtained.

In this paper, some additional results are presented which give an improved version of the previous equation of state. These are discussed in Sec. 2, where some aspects of the general coding problem are also reviewed, and also the possibility of accounting for phase changes. In Sec. 3, a more detailed discussion is given of the distribution of local coordination numbers. For the two-dimensional case a system of equations is derived, and the extension to three dimensions is considered. It is found that qualitatively wrong results are obtained for the coordination numbers in hard disks, and that this approximation is inadequate from this point of view, so that local correlation terms will have to be introduced explicitly.

2 CODING THEORY AND THERMODYNAMICS

2.1 Basic Method (B-W Approximation)

The main idea on which the method is based is as follows:

1. Find *any* system of parameters $\xi_1, \xi_2, \xi_3, \ldots, \xi_M$ which completely describes the configuration of some physical system.
2. Let $W(\boldsymbol{\xi})\, d^M\xi$ be the probability of finding the system in a small volume $d^M\xi$ of ξ space subject to any constraints, internal and external, known to be operating on the system.
3. The coding entropy H defined by

$$H = -\int W(\boldsymbol{\xi}) \ln \frac{W(\boldsymbol{\xi})}{m(\boldsymbol{\xi})}\, d^M\xi \tag{1}$$

is then maximized subject to all known constraints, and the result is identified with the thermodynamic entropy S of the system. Here $m(\boldsymbol{\xi})$ is the appropriate measure function for the set $\boldsymbol{\xi}$ which need not necessarily be of the same dimensions.

To carry out this program exactly is no easier than evaluating the usual partition function. However, if the ξ_r are so chosen as to be approximately uncorrelated, then

$$W(\boldsymbol{\xi}) \simeq \prod_{r=1}^{M} w_r(\xi_r) \tag{2}$$

where $w_r(\xi_r)$ is the probability function for ξ_r, and (1) reduces to

$$H = -\sum_{r=1}^{M} \int w_r(\xi_r) \ln \frac{w_r(\xi_r)}{m_r(\xi_r)}\, d\xi_r \tag{3}$$

For a three-dimensional classical monatomic liquid with N atoms, then $M = 6N$, and the first $3N$ of the ξ will be the momentum components p. If $w(p)$ denotes the probability function for any one component, then the total contribution to H is given by

$$H_{\text{mom}} = -3N \int_{-\infty}^{\infty} w(p) \ln w(p)\, dp \tag{4}$$

the appropriate measure function being unity in this case. The remainder of H, arising from the configurational coordinates, is denoted by H_{config}. If we break the structure up into close-packed tetrahedra, then a suitable set of ξ is given by the lengths l of the edges of the tetrahedra (called "links" for short) and the coordination numbers of the atoms.

The first approximation is to consider these links and coordination numbers to be essentially uncorrelated. In its neglect of local particle correlation this is equivalent to the Bragg-Williams order-disorder approximation and hence will be called the B-W approximation. Let w_q be the probability of an atom, selected at random, having local coordination q, and let $\psi(l)$ be the probability function for a link length l. Then in the B-W approximation

$$H_{\text{config}} = -N \sum_q w_q \ln w_q - N_{\text{ind}} \int_0^\infty \psi(l) \ln \frac{\psi(l)}{m(l)}\, dl \tag{5}$$

where N_{ind} is the effective number of link lengths that can be assigned *independently*, and $m(l)$ is the appropriate measure function. Previously[9,10] it was assumed that in three dimensions it was correct to put $N_{\text{ind}} = 3N$, since there are $3N$ nonredundant links in a three-dimensional structure of N vertices. Similarly, in two dimensions this would give $N_{\text{ind}} = 2N$. Also the function $m(l)$ was assumed to be essentially unchanged over relatively small-density ranges, and accordingly dropped, since only changes in entropy affect the thermodynamics. Subsequent consideration has shown that both these assumptions are incorrect.

Since the variable l is one-dimensional, and the atoms are distributed in three-dimensions, then $m(l)$ is effectively a Jacobian transforming from three dimensions to one. The simplest possible approximation is then

$$m(l) = \text{const} \times l^2 \qquad \text{(three dimensions)} \tag{6}$$

with, similarly,

$$m(l) = \text{const} \times l \qquad \text{(two dimensions)} \tag{7}$$

$$m(l) = \text{const} \qquad \text{(one dimension)} \tag{8}$$

Also, it appears that although there are three nonredundant links per atom in three dimensions, they are still strongly correlated, and this reduces the number of nonredundant links that are also effectively uncorrelated to only one per atom. The same result appears to hold for two dimensions and of course is strictly true in one, so that we get

$$N_{\text{ind}} = N \qquad \text{(one, two, and three dimensions)} \tag{9}$$

A brief discussion of possible simple procedures for actually coding the link lengths for transmission over a communication channel is given in the next section.

2.2 Simple Coding Procedure in Two and Three Dimensions

For any irregular two-dimensional array of atoms, we proceed by joining nearest-neighbor pairs, any links so formed being subsequently regarded as an impassable barrier, so that the assembly is eventually converted into close-packed irregular triangles with the atoms at the vertices as in Fig. 1. Consider now the problem of transmitting the various bond lengths over, say, a telegraph wire in such a way as to enable the array to be built up at the receiver. This could be done by the following procedure, it being assumed as a convention that the sides of triangles are listed clockwise. First the lengths l_{12}, l_{23}, l_{31} are transmitted, enabling the atoms 1, 2, and 3 to be located (since we are not here concerned with translation or rotation of the whole structure). We can then proceed by fixing the positions of all the neighbors of 1; then those surrounding atoms 2, 3, 4; and so on. In completing the shell around 1, lengths l_{34}, l_{41}, l_{45}, l_{51}, l_{56}, l_{61} are sent, enabling the positions of atoms 4, 5, and 6 to be fixed. Length l_{62} is now fixed anyway and does not need to be sent. Instead, a single digit can be sent, indicating that the shell around atom 1 is now complete, and the next signals refer to the shell around atom 2. This procedure is then repeated, the diagram indicating the order of the signals up to atom 29, and the dotted lines representing links that are not specified numerically but only as the closure of nearest-neighbor shells. The succession of closure digits is equivalent to specifying the successive coordination numbers q_1, q_2, q_3, q_4, q_5 of atoms 1, 2, 3, 4, 5, . . . , in this case by the numbers 5, 6, 7, 5, 7 . . . . For N atoms, the whole configuration can then be specified by transmitting the numbers $q_1, q_2, \ldots, q_N$ and then all the nonredundant l_{rs} ($r, s = 1, 2, \ldots, N$).

An alternative system would be to specify just one length, say l_{12}, and then only angles. For instance, using an obvious notation, the angles (123) and (312) then fix atom 3; (413) and (134) fix atom 4; and so on. The redundant link l_{62} in the previous coding then is replaced by the redundant angle (216), as shown dotted in Fig. 2 where the angles required

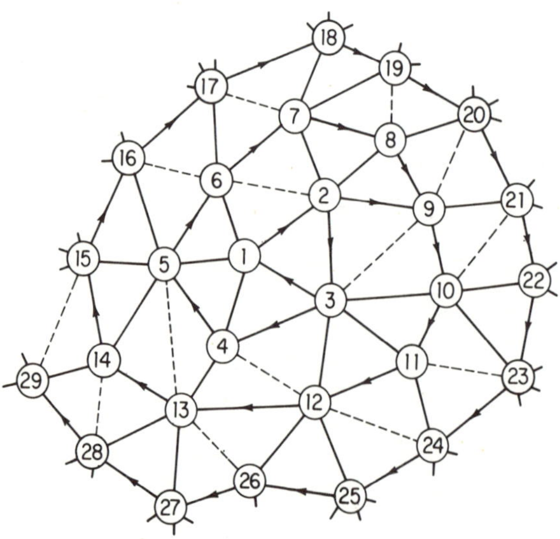

Fig. 1 Coding of a two-dimensional lattice by lengths.

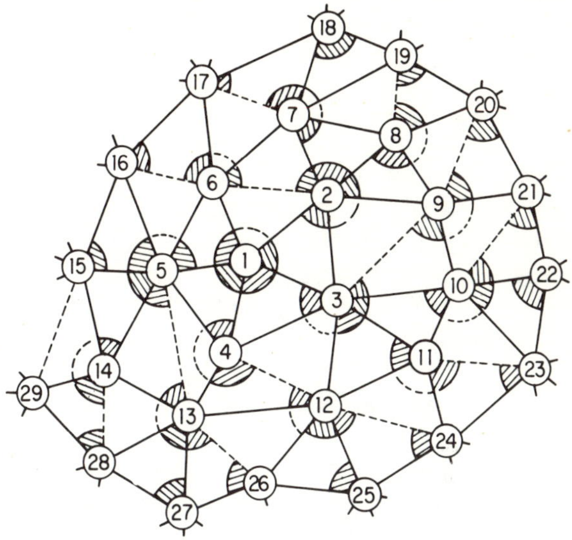

Fig. 2 Coding of a two-dimensional lattice by angles.

to specify the first 29 atoms are shown, the redundant ones indicated by dots.

Although the coding in terms of angles is perfectly feasible, that in terms of lengths has been preferred in the current work as it is easier to express bond energies in terms of lengths rather than as angles.

In three dimensions the triangles become tetrahedra, each completed triangle being regarded as an impassable sheet through which subsequent links may not pass.

Although the resultant structures are too complicated to show clearly on two-dimensional diagrams, the method of coding is similar, requiring a few more conventions on clockwise description of paths on the surface. As in the two-dimensional case, the system configuration can be specified by the nonredundant bond lengths and the coordination numbers of the atoms.

2.3 Equation of State in First B-W Approximation

For ranges of densities to be encountered in liquids, the changes in topological structure may be expected to be small in comparison with other changes, and hence the contribution of the term

$$H_{\text{topol}} = -N \sum_q w_q \ln w_q \tag{10}$$

in (5) will be small in comparison with the second term. To see how far the equation of state can be derived from the entropy of bond-length variation only, we can drop the first term and write [using (9) and (6)]

$$H_{\text{config}} \simeq -N \int_0^\infty \psi(l) \ln \frac{K\psi(l)}{l^2}\, dl \tag{11}$$

where K is some constant. This will be termed a "first" B-W approximation. If $\phi(l)$ is the pair potential, then, as in Ref. 9, the expectation values U, V of the internal energy and volume are given by

$$U = 3N \int_{-\infty}^\infty \frac{p^2}{2m} w(p)\, dp + \tfrac{1}{2} \langle q \rangle N \int_0^\infty \phi(l)\psi(l)\, dl \tag{12}$$

$$V = \mu N \langle l^3 \rangle = \mu N \int_0^\infty l^3 \psi(l)\, dl \tag{13}$$

Here $\langle q \rangle$ is the expectation value for the coordination number of any atom. In Ref. 9, it was supposed that $\langle q \rangle$ would depend on the statistics of the lattice involved and hence was treated as an ad hoc parameter. In two-dimensional lattices (of triangles) it can be shown that

$$\langle q \rangle = 6 \qquad \text{(two dimensions)} \tag{14}$$

and that this is a topological result true for all two-dimensional lattices of triangles. It is also suspected (although the author has so far been unable to prove it rigorously) that the corresponding result for three-dimensional structures of tetrahedra is

$$\langle q \rangle = 14 \qquad \text{(three dimensions)} \tag{15}$$

so that relation (12) becomes

$$U = 3N \int_{-\infty}^\infty \frac{p^2}{2m} w(p)\, dp + 7N \int_0^\infty \phi(l)\psi(l)\, dl \tag{16}$$

and we now have, in this approximation [using (6), (9), and (11)]:

$$H = -3N \int_{-\infty}^\infty w(p) \ln w(p)\, dp - N \int_0^\infty \psi(l) \ln \frac{K\psi(l)}{l^2} dl \tag{17}$$

Maximizing H subject to (13), (16) and the normalizing relations,

$$\int_{-\infty}^\infty w(p)\, dp = 1 \tag{18}$$

$$\int_0^\infty \psi(l)\, dl = 1 \tag{19}$$

gives eventually (cf. Ref. 9)

$$w(p) = (2\pi mT)^{-1/2} \exp\left(\frac{-p^2}{2mT}\right) \tag{20}$$

and

$$\psi(l) = \frac{l^2}{z} \exp\left[-\frac{7\phi(l)}{T} - \frac{\mu P}{T} l^3\right] \tag{21}$$

where

$$z = \int_0^\infty l^2 \exp\left[-\frac{7\phi(l)}{T} - \frac{\mu P}{T} l^3\right] dl \tag{22}$$

so that the equation of state becomes

$$V = z^{-1}\mu N \int_0^\infty l^5 \exp\left[-\frac{7\phi(l)}{T} - \frac{\mu P}{T} l^3\right] dl \tag{23}$$

for a three-dimensional system. Here P denotes the pressure, and T the absolute temperature measured in energy units (Boltzmann's constant equal to unity). In two dimensions the corresponding relations are

$$\psi(l) = z^{-1} l \exp\left[-\frac{3\phi(l)}{T} - \frac{\mu P l^2}{T}\right] \tag{24}$$

with

$$z = \int_0^\infty l \exp\left[-\frac{3\phi(l)}{T} - \frac{\mu P l^2}{T}\right] dl \tag{25}$$

and one dimension

$$\psi(l) = z^{-1} \exp\left[-\frac{\phi(l)}{T} - \frac{Pl}{T}\right] \tag{26}$$

$$z = \int_0^\infty \exp\left[-\frac{\phi(l)}{T} - \frac{Pl}{T}\right] dl \tag{27}$$

In relations (24) to (27) P denotes pressure in the units appropriate to the number of dimensions, i.e., force per unit length in two dimensions, and force in one. Similarly we shall use V to denote area in two dimensions and length in one.

In the case of a one-dimensional system with nearest-neighbor interactions, which is known to be exactly soluble anyway, define a succession of correlation function $\psi_n(r)$ given by

$$\psi_{n+1}(r) = \int_0^r \psi(r - l)\psi_n(l)\, dl \tag{28}$$

with

$$\psi_1(r) = \psi(r) \tag{29}$$

when the one-dimensional pair correlation function $g(r)$ is given by

$$g(r) = \frac{V}{N} \sum_{n=1}^{\infty} \psi_n(r) \tag{30}$$

which provides an easy way of calculating $g(r)$ for any pair potential $\phi(l)$ using (26).

The relation between $g(r)$ and $\psi(l)$ in two and three dimensions is considerably more difficult to find. The author has not succeeded in evaluating this exactly, and various approximations are being studied at the moment.

2.4 The Perfect Gas

Putting $\phi(l) = 0$ in the equation of state gives

$$\frac{PV}{NT} = 1 \tag{31}$$

in one, two, and three dimensions, as expected. The case of the perfect gas is not an entirely trivial check using this approach, however. In three dimensions we have

$$\begin{aligned} \psi(l) &= \frac{3\mu P l^2}{T} \exp\left(\frac{-\mu P l^3}{T}\right) \\ &= \frac{3\mu N l^2}{V} \exp\left(\frac{-\mu N l^3}{V}\right) \end{aligned} \tag{32}$$

For points distributed at random in three dimensions with constant density ρ it is known that the distances $l_1, l_2, \ldots, l_s, \ldots$ of the nearest, next nearest, . . . , sth nearest neighbors to any given point have probability functions given by $f_s(l_s)$ where

$$f_s(l_s) = \frac{3}{(s-1)!}(\tfrac{4}{3}\pi\rho)^s l_s^{3s-1} \exp\left(-\tfrac{4}{3}\pi\rho l_s^3\right) \tag{33}$$

The function $\psi(l)$ will be some weighted average over the $f_s(l)$, although the determination of the exact form of this average depends on the detailed investigation of the statistical geometry, and this has not been carried out yet. However, for very small l, $\psi(l)$ and $f_1(l)$ should become equal. Clearly, from (32) and (33) they have the correct form and do become equal for small l if we take

$$\frac{3\mu N}{V} = 4\pi\rho$$

i.e.,

$$\mu = \frac{4\pi}{3} \tag{34}$$

so that for gases at low density we would expect values of μ of approximately $4\pi/3$, using the above argument.

2.5 Systems of Hard Spheres

If σ denotes the diameter of a sphere, we have

$$\psi(l) = 0 \qquad (l < \sigma) \tag{35}$$

Substituting in the equations of state in one, two, and three dimensions, we get for the equations of state

$$\frac{PV}{NT} = \frac{1}{1 - N\sigma/V} \qquad \text{(one dimension)}$$

$$= \frac{1}{1 - \mu_2 N\sigma^2/V} \qquad \text{(two dimensions)}$$

$$= \frac{1}{1 - \mu_3 N\sigma^3/V} \qquad \text{(three dimensions)} \tag{36}$$

where μ_2 and μ_3 are the appropriate values for μ for the two- and three-dimensional cases. The one-dimensional equation is of course correct. The others differ in form from the cell and tunnel model results. Assume triangular and fcc close packing in two and three dimensions; the corresponding close-packed volume V_0 is given by

$$V_0 = \frac{N\sigma^2\sqrt{3}}{2} \qquad \text{(two dimensions)}$$

$$= \frac{N\sigma^3}{\sqrt{2}} \qquad \text{(three dimensions)} \tag{37}$$

so that the second and third of Eqs. (36) become

$$\frac{PV}{NT} = \frac{1}{1 - 2\mu_2 V_0/V/\sqrt{3}} \qquad \text{(two dimensions)}$$

$$= \frac{1}{1 - \mu_3 V_0 \sqrt{2} V} \qquad \text{(three dimensions)} \tag{38}$$

If we are to obtain V_0 as the minimum possible volume, we must take $\mu_2 = \sqrt{3}/2$ and $\mu_3 = 1/\sqrt{2}$. However, the virial expansion given by (38) in three dimensions is

$$\frac{PV}{NT} = 1 + \frac{\mu_3 V_0 \sqrt{2}}{V} + \left(\frac{\mu_3 V_0 \sqrt{2}}{V}\right)^2 + \cdots \tag{39}$$

The first few terms are known exactly to be (cf. Rowlinson[19]):

$$\frac{PV}{NT} = 1 + \frac{2\pi\sqrt{2}}{3}\frac{V_0}{V} + \frac{5}{8}\left(\frac{2\pi\sqrt{2}}{3}\frac{V_0}{V}\right)^2 + \cdots \tag{40}$$

To obtain the first virial coefficient correctly we have to put $\mu_3 = 2\pi/3$, and the second one is then, in (39), too big by a factor $8/5$. Clearly μ_3 must vary with density, and hence the equation of state derived for any particular value of μ_3 can only be expected to hold over relatively restricted density ranges.

In one dimension the pair-correlation function $g(r)$ can be evaluated exactly, and we get, using (28),

$$\begin{aligned} \psi_n(r) &= 0 \qquad (r < n\sigma) \\ &= \frac{1}{(n-1)!}\left(\frac{P}{T}\right)^n (r - n\sigma)^{n-1} \exp\left[-\frac{P}{T}(r - n\sigma)\right] \qquad (n\sigma \le r) \end{aligned} \tag{41}$$

Substitution from (41) into (30) then gives the standard form for $g(r)$ for a one-dimensional chain of hard spheres.

2.6 Description of Phase Changes in General

In accounting for phase changes within the model, the qualitative differences between boiling and freezing should be reflected in the corresponding changes in the lattices. In the liquid-solid transition, the volume change is relatively small, and the dominant effect is the onset of long-range order. In this case one would expect to find large changes in the w_q, and not much change in the bond lengths. The bulk of the latent heat would then be taken up in the "ordering" entropy change and relatively little in the bond-energy changes. For the liquid-vapor transition, however, both phases have no long-range order, and (below the critical point) the volume change is relatively large. Hence the bulk of the latent heat here will be taken up in the energy changes associated with the increase in bond lengths.

In two dimensions, for a simple monatomic van der Waals liquid, the ordered crystalline state is one in which all the atoms are 6-atoms, i.e.

$$\begin{aligned} w_q &= 1 \qquad (q = 6) \\ &= 0 \qquad (q \neq 6) \end{aligned} \tag{42}$$

and also the structure has long-range order, each atom being displaced a relatively small distance (as a result of thermal agitation) from the corresponding vertex of a lattice of equilateral triangles. In the usual lattice gas and cell models, this long-range order is built into the model in order to account for the crystalline state. However, in this system, it appears as a result of various graphical constructions that a long-range ordered structure is implied automatically by (42) without introducing it formally. A typical lattice satisfying (42) is shown in Fig. 3, in which all the triangle vertices are occupied by 6-atoms. In such a lattice, it is clearly possible to trace the chains of bonds that would be exactly straight in the perfect ordered lattice. For such a lattice which is spatially homogeneous from a statistical point of view, (i.e., one in which a sample of bond lengths from one region containing, say, 1,000 atoms, gives essentially the same histogram of bond lengths as a similar sample taken elsewhere in the lattice), then the departures from chain straightness which are permitted by the topology of the structure are relatively small. The effect of an appreciable bend in a chain is seen in Fig. 4 where the curvature in chain PQ produces large density increases "inside" the bend, thus violating the condition of statistical homogeneity. Hence, if the theory can be made to yield a discontinuous change of the w_q to the values given by (42), then this can be identified as a solid-liquid transition, without the addition of any long-range ordering parameter.

A similar result is believed to hold in three dimensions, although it is more difficult to show. The trouble here is that the usual close-packed structures (hexagonal and fcc close packing) are not well adapted for coding in terms of tetrahedra only, and are better described if octahedra

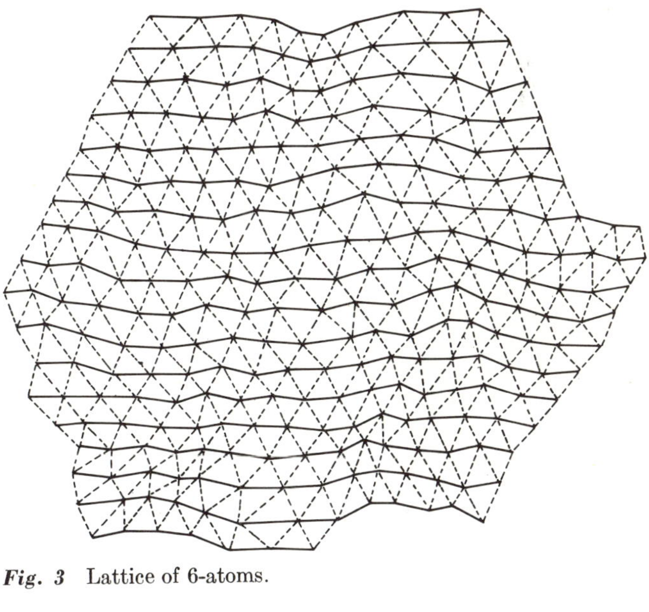

Fig. 3 Lattice of 6-atoms.

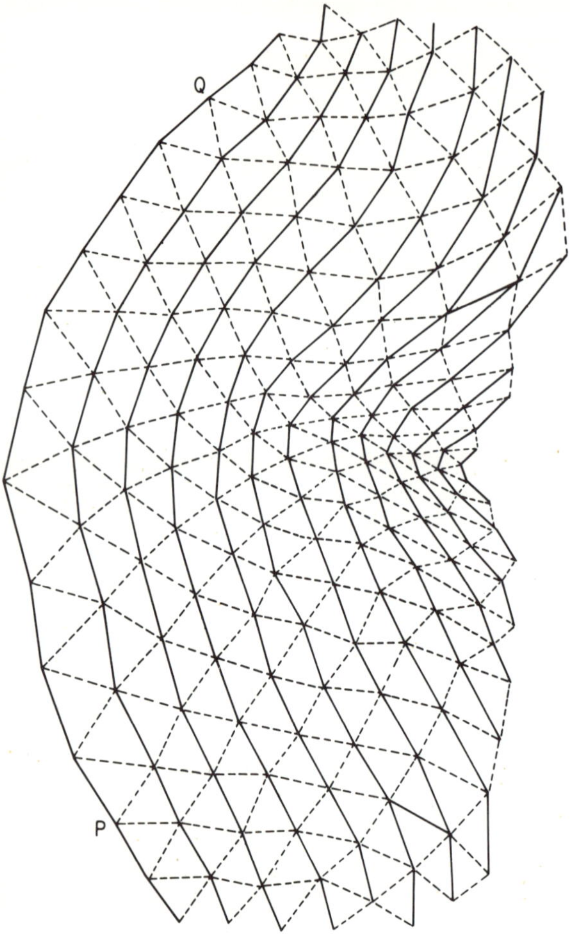

Fig. 4 Density distortion in a 6-atom lattice produced by curvature of a chain of interatomic links.

are also allowed. However, if the fcc or hexagonal close-packed lattices are recoded in terms of tetrahedra by the addition of extra bonds, then they both give coordination numbers of $q = 14$ for the constituent atoms. In fact, this is true of all the other regular lattices as well. This suggests that in three dimensions, with coding in terms of tetrahedra only, (12) should be replaced by

$$w_q = 1 \qquad (q = 14)$$
$$= 0 \qquad (q \neq 14)$$

The behavior of the w_q during the liquid-gas transition is rather more obscure and is being investigated at the moment.

2.7 Results for Real Liquids

A computer program has been developed for evaluating the isotherms given by (16), (21) and (23) for the Lennard-Jones pair potential

$$\phi(l) = \phi_0\left[\left(\frac{a}{l}\right)^{12} - 2\left(\frac{a}{l}\right)^{6}\right]$$

Some results have already been obtained, and the corresponding isotherms are shown in Fig. 5. The isotherms for $T = 100°K$ and 200°K are shown for Ar, the value of μ being taken as $1/\sqrt{2}$, the value indicated by the high-density equation of state for hard spheres. The corresponding experimental isotherms are shown for comparison. It can be seen that the general shape of the experimental curve is moderately well represented by the theory, but of course there can be no sharp phase change in this (B-W) approximation, since no long-range order parameter is available.

To investigate the applicability to metals, a corresponding program is being developed, using the oscillatory pair potential described by March and his collaborators. This assumes that the interaction energy of a metal can be adequately represented by an additive pair potential which has the oscillatory form described by Johnson et al.[17]

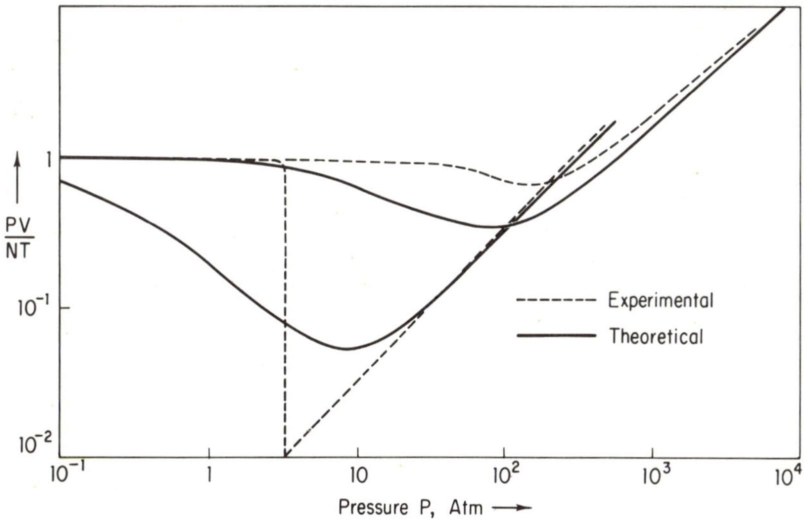

Fig. 5 Theoretical and experimental Ar isotherms with $\mu = 1/\sqrt{2}$.

3 SECOND BRAGG-WILLIAMS APPROXIMATION IN TWO DIMENSIONS

3.1 Assumption and Approximations

The indications are that the probability function $\psi(l)$ obtained by the method of Sec. 2.3 is qualitatively correct, but no information is obtained about the distribution of the w_q. We now consider how they might be calculated. For simplicity we consider first the case of two dimensions, and for a first approximation we use expression (10) for the "topological" or "ordinary" entropy. The approximation can still be considered to be of B-W type in that local correlations are not explicitly introduced, although they are to some extent contained in the relation

$$\sum_q q w_q = 6$$

which is strictly true in two dimensions.

Assume that a link is known to have at least one end at a q-atom. We denote the corresponding probability function for its length by $\psi_q(l)$. Expressions can now be written for the average link energies ψ_q associated with q-atoms and for the link length averages l_q. This enables expressions to be derived for the volume (area) V and energy U of an assembly of N atoms.

3.2 Derivation of the Thermodynamic Relations

If $\psi_q(l)$ is the probability function for the length l of a link with one end at least at a q-atom, then the rms average associated link length l_q and energy ϕ_q are given by

$$l_q^2 = \int_0^\infty l^2 \psi_q(l)\, dl \tag{43}$$

$$\phi_q = \int_0^\infty \phi(l) \psi_q(l)\, dl \tag{44}$$

and the average coding entropy h_q for such a link is given by

$$h_q = -\int_0^\infty \psi_q(l) \ln \frac{K\psi_q(l)}{l}\, dl \tag{45}$$

Taking N links as being independently assignable (one-third of the total), then the average coding entropy h per atom ($h = H/N$ including the momentum contribution) is given by

$$h = -2\int_{-\infty}^\infty W(p) \ln W(p)\, dp + \frac{1}{6} \sum_q w_q h_q - \sum_q w_q \ln w_q \tag{46}$$

To obtain a suitable approximation for $v = V/N$, the area v_q associated with a q-atom is replaced by the area of a regular q-sided polygon with vertices at a distance l_q from the central atom. The total area of the polygon is then $\frac{1}{2}ql_q^2 \sin(2\pi/q)$, and that fraction v_q associated with the central q-atom is one-third of this: i.e.,

$$v_q = \tfrac{1}{6}q\mu_q l_q^2 \tag{47}$$

where

$$\mu_q = \sin\frac{2\pi}{q} \tag{48}$$

The corresponding expression for v is now

$$v = \sum_q w_q v_q = \frac{1}{6}\sum_q q\mu_q w_q l_q^2 \tag{49}$$

The total internal energy u per atom is given by

$$u = 2\int_{-\infty}^{\infty}\frac{p^2}{2m}W(p)\,dp + \frac{1}{2}\sum_q qw_q\phi_q \tag{50}$$

We now maximize h subject to (42), (49), and (50) and also the normalizing relations:

$$1 = \int_0^\infty \psi_q(l)\,dl \tag{51}$$

$$1 = \int_{-\infty}^{\infty} W(p)\,dp \tag{52}$$

$$1 = \sum_q w_q \tag{53}$$

Introducing suitable Lagrange multipliers to form a modified entropy[9] h^* and using the relations

$$\frac{\partial h^*}{\partial w_q} = 0 \tag{54}$$

$$\frac{\delta h^*}{\delta \psi_q(l)} = 0 \tag{55}$$

$$\frac{\delta h^*}{\delta W(p)} = 0 \tag{56}$$

these result eventually in the relations

$$w_q = z^{-1}\lambda^q \exp\left(\tfrac{1}{6}qs_q - \tfrac{1}{2}\beta q\phi_q - \tfrac{1}{6}\gamma q\mu_q l_q^2\right) \tag{57}$$

$$\psi_q(l) = z_q^{-1} l \exp\left[-3\beta\phi(l) - \gamma\mu_q l^2\right] \tag{58}$$

$$W(p) = \left(\frac{2\pi m}{\beta}\right)^{-1/2} \exp\left(-\frac{\beta p^2}{2m}\right) \tag{59}$$

where

$$z = \sum_q \lambda^q \exp\left(\tfrac{1}{6}qs_q - \tfrac{1}{2}\beta q\phi_q - \tfrac{1}{6}\gamma\mu_q l_q^2\right) \tag{60}$$

$$z_q = \int_0^\infty l \exp\left[-3\beta\phi(l) - \gamma\mu_q l^2\right] dl \tag{61}$$

β, γ, and λ are Lagrange multipliers; and h_q has been replaced by s_q since we are now dealing with thermodynamic equilibrium. Replacing h by s in (46), we get eventually

$$s = \ln z - \ln \frac{\beta}{2\pi m} - 6 \ln \lambda + \beta u + \gamma v \tag{62}$$

Using the relations

$$\left(\frac{\partial s}{\partial v}\right)_u = \frac{P}{T} \tag{63}$$

$$\left(\frac{\partial s}{\partial u}\right)_v = \frac{1}{T} \tag{64}$$

these give, after some simplification, the expected results,

$$\beta = \frac{1}{T} \tag{65}$$

$$\gamma = \frac{P}{T} \tag{66}$$

so that we get

$$\psi_q(l) = z_q^{-1} l \exp\left[-\frac{3\phi(l)}{T} - \frac{\mu_q P l^2}{T}\right] \tag{67}$$

with

$$z_q = \int_0^\infty l \exp\left[-\frac{3\phi(l)}{T} - \frac{\mu_q P l^2}{T}\right] dl \tag{68}$$

The parameter λ has to be chosen to satisfy (42), and it can be shown after some rearrangement that this reduces to

$$\sum_q (q - 6) z_q^{q/6} \lambda^q = 0 \tag{69}$$

So far, the analysis is quite general, within the limits of the approximations made.

3.3 Application to Hard Spheres

If σ denotes the sphere diameter, then we have

$$\begin{aligned} \psi_q(l) &= z_q^{-1} l \exp\left(-\frac{\mu_q P l^2}{T}\right) \qquad (\sigma \leq l) \\ &= 0 \qquad (0 \leq l < \sigma) \end{aligned} \tag{70}$$

with

$$z_q = \left(\frac{2\mu_q P}{T}\right)^{-1} \exp\left(-\frac{\mu_q P \sigma^2}{T}\right) \tag{71}$$

Using (70) in (43), we get

$$l_q^2 = \sigma^2 + \frac{T}{\mu_q P} \tag{72}$$

The w_q are obtained by solving for λ the polynomical equation

$$0 = \sum_q (q - 6) z_q^{q/6} \lambda^q \tag{73}$$

and then putting

$$w_q = \frac{z_q^{q/6} \lambda^q}{\sum_q z_q^{q/6} \lambda^q} \tag{74}$$

Using (72) in (49) the equation of state can be written in the form

$$\frac{PV}{NT} = \frac{1}{[1 - (V_0/3V\sqrt{3}) \sum_q q w_q \mu_q]} \tag{75}$$

At high densities, near close-packing, we could expect

$$\begin{aligned} w_6 &\simeq 1 \\ w_q &\simeq 0 \qquad (q \neq 6) \end{aligned} \tag{76}$$

when (75) becomes

$$\frac{PV}{NT} \simeq \frac{1}{1 - 2V_0\mu_6/V\sqrt{3}} = \frac{1}{1 - V_0/V} \tag{77}$$

To actually compute the isotherms, V/V_0 and PV/NT are evaluated in terms of $P\sigma^2/T$ as a parameter.

For the purpose of computation, some upper value Q of q must be assumed in the summations in (73), (74), and (75). Even for the perfect-gas case in two dimensions the occurrence of local coordinations greater than 10 is very rare, because of the particular geometric method of coding adopted, which tends to favor triangles that are not very skew. Hence it should be sufficient to take a value of Q appreciably greater than 10, when it would be expected that the w_q would become vanishingly small as $q \to Q$, and that λ would tend to a finite limiting value as $Q \to \infty$ for any fixed value of $P\sigma^2/T$. The solutions of (74) in fact are found to depend fairly strongly on the particular value of Q chosen, particularly at high densities, where the w_q decrease with q, reach a minimum, and then start to increase again. In particular, the w_q do not "accumulate" at $q = 6$ as $V/V_0 \to 1$.

Hence it is clear that reliable information about the distribution of the w_q (and hence the description of crystallization) cannot be obtained in this approximation, and that a much more detailed account of the local interatomic constraints is required. The foregoing discussion, however, does give an indication of the lines on which the w_q entropy contribution can be included within the framework of the calculations.

4 CONCLUSIONS

It appears that the geometrical approach to the theory of liquids in this B-W approximation gives essentially the correct probability function $\psi(l)$ for the length l of a typical interatomic link. Also, since the equation of state which results from neglecting the "ordering" or "topological" entropy gives fair qualitative agreement with experiment, it would appear that the dominant contribution to the entropy variation of the assembly comes from the variation of link lengths, except near a phase change.

The two-dimensional hard-sphere analysis given in Sec. 3.3, however, shows that this approximation does not yield reliable results for the w_q, and that further geometric constraints implicit in higher-order approximations (analogous to the Bethe-Peierls theory) must be included in the formalism. These might also enable the relative stabilities of various crystalline structures to be assessed.

REFERENCES

1. Barker, J. A.: "Lattice Theories of the Liquid State," Pergamon Press, Oxford, 1953.
2. Bernal, J. D.: *Nature* **183:**141 (1959).
3. Bernal, J. D.: *Nature* **185:**68 (1960).
4. Bernal, J. D.: *Proc. Roy. Soc.* **A280:**299 (1964).

5. Bernal, J. D., and J. Mason: *Nature* **188**:910 (1960).
6. Bernal, J. D., K. R. Knight, and I. Cherry: *Nature* **202**:852 (1964).
7. Brillouin, L.: "Science and Information Theory," Academic Press Inc., New York, 1963.
8. Collins, R.: *Proc. Phys. Soc.* **83**:553 (1964).
9. Collins, R.: *Proc. Phys. Soc.* **86**:199 (1965).
10. Collins, R.: *Proc. Phys. Soc.* **86**:443 (1965).
11. Enderby, J. E., T. Gaskell, and N. H. March: *Proc. Phys. Soc.* **85**:217 (1965).
12. Frisch, H. L., and J. L. Lebowitz (eds.): "The Equilibrium Theory of Classical Fluids," W. A. Benjamin, Inc., New York, 1964.
13. Jaynes, E. T.: *Phys. Rev.* **106**:620 (1957).
14. Jaynes, E. T.: *Phys. Rev.* **108**:171 (1957).
15. Jaynes, E. T.: "Statistical Physics," Brandeis University Lectures, vol. 3, W. A. Benjamin, Inc., New York, 1963.
16. Johnson, M. D., and N. H. March: *Phys. Letters* **3**:313 (1963).
17. Johnson, M. D., P. Hutchinson, and N. H. March: *Proc. Roy. Soc.* **A282**:283 (1964).
18. Levelt, J. M. H., and E. G. D. Cohen: "Studies in Statistical Mechanics, vol. 2, part B, North Holland Publishing Company, Amsterdam, 1964.
19. Rowlinson, J. S.: *Rept. Progr. Phys.* **28**:169 (1965).

DISCUSSION *on Paper Presented by Robert Collins*

DEHLINGER: If one could introduce in the microstate also some information about the impulse (velocity) of the particles, and if then one could average the atomic arrangement over small times, would not the inhomogeneity of the arrangement become greater, in that the nearly regular regions would be more regular, because the movement of the atoms would be directed mostly to the regular sites?

COLLINS: I would not have thought so (although it may be so) because the distribution of moments is uncorrelated with the distribution in l [$w(p)$ and $\psi(l)$ are strictly uncorrelated and this is exact factorization of the usual partition function for example].

MEIJERING: If points are distributed randomly in space (with density n) and Dirichlet cells constructed by drawing bisecting planes, several quantities may be calculated exactly.* The distribution of the number of corner points (where four cells meet) as a function of distance r from the cell nuclei is given by $[(4\pi)^5/105]n^3r^8\exp\,(-\tfrac{4}{3}\pi nr^3)\,dr$. Integration from 0 to ∞ yields $96\pi^2/35$ for the average number of corners of a cell. The number of edges is $1\tfrac{1}{2}$ times this value, and by Euler's theorem one finds the average number of faces to be

$$\frac{48\,\pi^2}{35} + 2 = 15.54$$

thus well above 14, which is sometimes assumed to be the maximum number for Dirichlet-cell aggregates.

By distributing the nuclear points in a more sophisticated way it will be possible to attain at least 16. For instance, the diamond lattice gives 16, but

* *Philips Res. Rept.* **8**:270 (1953).

Professor Laves has pointed out that the cells then have undesirable corners where four edges meet instead of three. However, it is easy to get rid of these by shifting the lattice points just a little bit randomly.

LAVES: Let me first mention again, what a *Wirkungsbereich* (Dirichlet region) is:

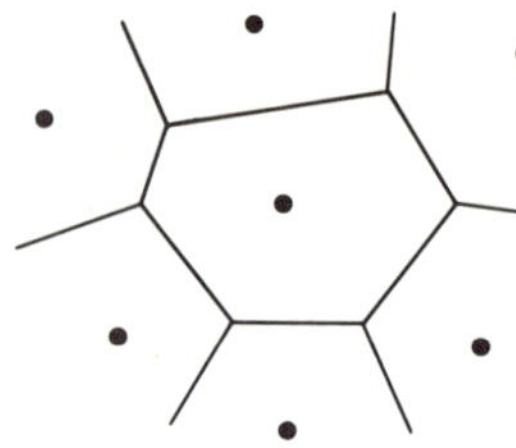

If one has an infinite number of points, randomly distributed on a plane, one has known for decades that the WB's around them are polygons, and that the *average number* of edges is *six*.

Now let's assume that we have randomly distributed points in space.

If we construct the WB's, they will be polyhedra with the following properties:

Two WB meet in *one* face.
Three WB meet in *one* edge.
Four WB meet in *one* corner.

A randomly chosen plane through such an arrangement will be a plane which does not go through corners but only through edges: i.e., it will be a plane divided in polygons with six edges on the average.

Now, we select *randomly* one WB of those cut by the plane. The plane cuts it into two parts.

The two parts meet in a face which is a polygon with *six* edges on the average.

By this procedure we produce an *additional* WB (with an *additional* point, *randomly* inserted into the space).

What is the effect?

1. We produced *one* new WB.
2. We produced *one* new dividing face between the two WB's.
3. In addition we produced (on the average) *six* new dividing faces.

If we call

W = number of WB present originally
$f/2$ = number of dividing faces present originally, i.e.:
f = average number of faces per WB present originally
m = number of partitions, randomly chosen
F = average number of faces per WB after m partitions

then,

$$F = \frac{f + 14m}{W + m}$$

If $m \to \infty$,

$$F = \frac{f + 14m}{W + m} \to \frac{14m}{m} = 14$$

For "hand-waving" reasons I used the concept of Wirkungsbereich. However, from a toplogical point of view the reasoning would hold for any space partitions (I believe) with the conditions 1, 2, and 3 mentioned above (if the average size of the WB's is "uniform" all over the space).

For the sake of priority I would like to add the following: van der Waerden* proved this whole thing in 13 lines. He added:

> "The considerations that led to this proof have been given already very shortly in the Dissertation of A. D. J. Meuse,† but were not noticed before deriving the above equations."

MEIJERING: The model by van der Waerden, although also a random one, is not the same as the one I treated. For instance, one of three dihedral angles between the faces at an edge (where three cells meet) is 180° in his model.

FRANK: We can retain the Dirichlet-cell method for defining coordination numbers, without producing the paradoxical number 16 for diamond, by making the distinction between *direct* and *indirect* neighbors, due, I think, to Meijering. The direct neighbors of an atom X are those for which the corresponding face of its Dirichlet cell is intersected by the straight line joining atom X to the neighbor. In other words, a pair of atoms are neighbors if there is *some* line joining them on which no point is closer to a third atom than to one of these two; they are direct neighbors if no point on the straight line joining them is closer to a third atom than to one of these two. The *indirect* neighbors are those that are not direct neighbors. The number of direct neighbors seems to me to be the most useful of well-defined numbers that can be used to classify the coordination of atoms in various structures. It is only in rather open structures that indirect neighbors arise. Thus, in diamond, a C atom has 4 direct and 12 indirect neighbors.

In metal structures generally, all the neighbors are direct ones. An atom in the bcc structure, or any moderate distortion thereof, has 14 neighbors, all direct.

As a second point, I feel that Laves's "space-filling principle" needs to be divided into two: the packing principle and the covering principle. The difference between these two is clearly brought out in the fcc and bcc lattices. Of all lattices, the former gives the best packing and the latter the best covering. That is, for a given number of lattice points per unit volume, the former allows the maximum radius of equal noninterpenetrating spheres centered on the lattice points; the latter allows the minimum radius of equal interpenetrating spheres to leave no point in the space uncovered. Alternatively stated, it is the latter which for a given number of lattice points per unit volume minimizes the size of the largest sphere empty of lattice points.

LAVES: As to the first paragraph of Frank's comment: I think this new "straight-line" amendment to the earlier Kasper-Frank definitions is in some way very important, as it may sometimes exclude paradoxical coordination numbers.

* *Naturwissenschaften* **53**:8–11 (1966).

† *Rec. Trav. Botan. Neerl.* **38**(41):18–140, especially p. 32.

However, even with the distinction of *direct* and *indirect* bonds, queer results cannot be excluded. Consider the graphite structure: one type of C position has *eleven* direct neighbors, and the other type has at least *nine* direct neighbors (there may be even 15; I do not know at this moment). But who would like to consider the graphite structure to have a CN around 10?—Another example: The fcc structure has 12 neighbors (from the "old-fashioned" *and* from the Kasper-Frank point of view). A slight tetragonal distortion ($c/a = 0.99$) would add *two* direct neighbors, leading to a CN = 14. An opposite distortion ($c/a = 1.01$) would add *four* direct neighbors, leading to a CN = 16. Thus, I am not very happy with Frank's statement that the Kasper-Frank concept (even in the amended forms as outlined above) is the "most useful" way to agree on coordination numbers. With both points of view (the old-fashioned one and the Kasper-Frank one), coordination numbers can be assigned rigorously as far as geometry is concerned. However, both points of view can have their serious shortcomings when it comes to the discussion of special cases. On the other hand, it is interesting to note the following "duality": if the one point of view leads to ridiculous results, the other point of view *may* become reasonable, and vice versa.

As to the second paragraph: I fully agree. I alluded to this problem already in my lecture, and I discussed this point at length in a paper of 1960.* Thus I am fully aware of the fruitfulness of the Frank-Kasper ideas, and I spread them around as far as possible. The names coined by Frank to subdivide the space principle into a "packing" and a "covering" one sounds good to me. However, if such subdivision should be made, I would prefer to chose more specific names to express the geometrical differences of the two aspect: "sphere-packings" and "distance packings."

* "Beiträge zur Physik und Chemie des 20 Jahrhundert," Lise Meistner, Otto Haten, Max von Laue, zum 80 Geburtstag, pp. 152–177, especially p. 167, Friedrich Vieweg & Sohn, Brunswick, Germany.

METASTABLE PHASES*

Pol Duwez

W. M. Keck Laboratory of Engineering Materials, California Institute of Technology, Pasadena, California

INTRODUCTION

All the communications presented at this colloquium on phase stability were obviously focused on the problem of explaining the existence of stable phases in alloy systems. A discussion of the structure of some of the numerous metastable phases obtained by rapid cooling from the liquid state is likely to lead to additional difficulties. Perhaps the only excuse for bringing up the subject is that in a limited number of instances Hume-Rothery phases which are absent in alloys prepared by conventional methods have been synthesized by rapid cooling from the liquid state. Except for these few interesting examples, it is indeed true that most metastable phases (including amorphous ones) will only complicate the problems confronting the metal physicist.

Metastable phases can be obtained through various experimental techniques. In a few cases, cooling from the melt under normal conditions may lead to so-called "metastable" phase diagrams. Alloys of Cd-Sb and Fe-P are classical examples. In fact, one of the most important and also

* Work supported by the U.S. Atomic Energy Commission.

useful alloys, namely Fe-C, may be considered as metastable, but since graphite will not precipitate out within a time scale compatible with human life, the metastable character of steel is of only theoretical interest. The chances of obtaining metastable phases are, of course, increased when the rate of cooling of the melt becomes high enough to prevent the normal process of nucleation and growth. As a result of techniques developed during the last 6 years, a large number of phases have been synthesized by this process. Some of these phases will be discussed in this paper. Vacuum evaporation of alloys, especially on a cold substrate, is another technique for preparing metastable phases. By using this technique, Mader et al.[1] have been able to prepare thin films of amorphous Cu-Ag alloy, which, as the temperature is raised, transformed into a Cu-Ag solid solution, and finally become a two-phase equilibrium alloy. Electrodeposition is a third technique which may lead to nonequilibrium phases. Ni-P alloys may be obtained in an amorphous state by this technique.

PRINCIPLE OF THE RAPID-COOLING TECHNIQUE

It has been shown experimentally that the highest rate of heat removal from a liquid alloy is achieved by conduction to a metal having a high thermal conductivity. It is also obvious that in order to obtain high rates of cooling during solidification the liquid alloy must be spread into a thin layer over the substrate. In addition, it is necessary to ensure a good thermal contact between the liquid and the substrate. Finally the entire process must be accomplished in the shortest possible time. These conditions are satisfied in the quenching apparatus described in Ref. 2. A small liquid globule (weighing about 100 mg) is heated in a suitable container made of either graphite or a ceramic material, and ejected at high speed (several hundred feet per second) by means of a shock wave. The liquid globule is directed almost tangentially onto a concave surface made of Cu and is spread into a thin foil varying in thickness from a fraction of a micron to about 5 to 10 microns. The reason for the curved surface of the target is to provide a centrifugal force contributing to better thermal contact between the liquid and the substrate. Because of the necessity of using a crucible to contain the liquid globule, this method could not be applied to reactive refractory metals. This limitation has been recently removed through an improved design,[3] and it is now possible to quench very high temperature alloys such as W carbide. The actual rate of cooling achieved in this type of apparatus is difficult to measure. An approximate rate, however, was obtained by means of high-speed photography. It was found that the foil was deposited on the target in less than one millisecond, and this would lead to an average rate of the order of 10^6 C°/sec for a globule originally at 1000°C. By locating very small thermocouples on the substrate, Predecki et al.[4] measured rates of cooling ranging from 10^6 to 10^7 C°/sec. These measurements give some information on the rate of cooling at a point of contact with the substrate. Unless the foil is very thin, however, the average rate of cooling is probably obviously less than the measured values.

Another scheme for achieving rapid quenching from the liquid state consists of catching a liquid globule between a Cu-lined anvil and a piston moving at high speed.[5] The foils obtained by this technique are rather thick (30 to 50 microns) and the average rate of cooling is certainly smaller than that obtained by propelling the globule. Many metastable phases, however, have been obtained with the piston-and-anvil technique.

Besides the average rate of cooling, an important difference between the two techniques (gun and piston-and-anvil techniques) is in the shape and thickness of the quenched foils. In the gun technique the foils are about 1×2 cm at the most, are quite uneven in thickness, and contain many holes. Around these holes, the thickness may be less than 0.2 micron, and transmission microscopy can be done in these areas without any mechanical or chemical thinning.[6,7] The foils obtained by the piston-and-anvil technique are quite uniform in thickness. Depending on the type of alloy, the size of the liquid drop, and the velocity of the piston, circular foils up to about 25 mm in diameter and 20 to 50 microns thick can be obtained. The main advantage of this technique is to prepare metastable phases in a form suitable for electrical resistivity measurements and for mechanical properties studies.

METASTABLE HUME-ROTHERY PHASES

The first metastable phase of the Hume-Rothery type was found in Ag-Ge alloys quenched from the liquid state.[8] This phase was discovered by chance, because the purpose of quenching Ag-Ge alloys was to try to obtain an amorphous phase similar to that previously found in Au-Si alloys.[9] The Ag-Ge hexagonal close-packed phase was studied in detail by Klement.[10] Its range of existence is from about 15 to 26 a/o Ge, and within this range the a parameter changes rather abruptly from 2.892 A below about 21 a/o to 2.898 A above this concentration. The c parameter has a rather smooth decrease from 4.724 to 4.721 A. The axial ratio also shows a decrease from 1.634 down to 1.629 with increasing concentrations. It is interesting to note that in the Ag-Ge quenched alloys, the solubility limit of Ge in Ag was increased from 9.6 a/o Ge (equilibrium limit) to about 13.5 a/o Ge, and this metastable limit corresponds to the theoretical electron concentration of 1.40. There is a very small gap between the fcc and the hcp metastable phases. According to Klement, it is probable that a large number of stacking faults exist in alloys in the transition region

TABLE 1 *Metastable Hume-Rothery Type Phases*

Alloy	Structure	Concentration, a/o	Lattice parameter, angstroms		Ratio c/a	Reference
			a	c		
Ag-Ge	hcp	15–26	2.892	4.724	1.634	10
			2.898	4.721	1.629	
			Depending on concentration			
Ag-Si	hcp	10–25	2.87	4.52	1.57	11
Au-Ge	hcp	22.5	Not reported			12
	hcp	21.0	2.87	4.73	1.64	13
	hcp	17.5–22.5	2.885	....	1.642–1.648 Depending on concentration	14
Au-Sb	hcp	13–15	2.898	4.731	1.633	13
Au-Si	γ-Brass	17–20	Not reported			13

between the fcc and hcp phases, but this question has not yet been clarified by electronmicroscopy studies.

A limited number of metastable hcp phases have been reported in rapidly quenched alloys and are summarized in Table 1. Klement reported an hcp phase in Ag-Si alloys in the range of 10 to 25 a/o Si, and approximate lattice parameters are given in Table 1. These alloys, however, were not single phase. Results on Au-Ge alloys were reported in Refs. 12, 13, and 14. An hcp phase seems to exist in this system around about 21 a/o Ge. After quenching from the liquid state, Predecki et al.[13] obtained an hcp metastable phase in Au-Sb alloys between 13 and 15 a/o Sb, corresponding to an electron concentration of 1.4 to 1.5. No hcp phase has been reported in quenched Au-Si alloys, in which an amorphous phase can be obtained around 25 a/o Si.[9] By quenching an alloy containing 20 a/o Si to room temperature, Predecki et al.[13] obtained a crystalline pattern which could be indexed on the basis of a γ-brass structure. If this is so, this would be the first γ-brass type Hume-Rothery phase (at an electron concentration of 21/13) obtained by rapid quenching from the liquid state.

METASTABLE AMORPHOUS PHASES

The question of differentiating between a microcrystalline and an amorphous solid has always been the subject of controversy. In the foregoing discussion, an alloy is called *amorphous* when the following conditions are satisfied: (1) The X-ray diffraction pattern is made of diffused rings which have an intensity vs. sin θ curve very similar to that found for liquid alloys;

(2) the electron-diffraction pattern (taken with 100-kv electrons) is also made of diffused rings; (3) the image obtained in transmission electronmicroscopy under high magnification (60,000 and higher) does not exhibit any detectable grain structure. In addition, there are indirect methods for establishing that a solid is really amorphous. If a specimen is heated at a sufficiently high rate (about above 50°C/min), it will crystallize rapidly within a narrow range of temperature with a large amount of heat release (of the order of 1000 cal/mole). Another method is to observe the transition from the amorphous to the crystalline state in transmission electronmicroscopy. By this technique it is possible to differentiate between crystallization from an amorphous matrix and grain growth from a microcrystalline specimen. The term *amorphous* is used in this paper to describe solid structures for which all the preceding criteria are satisfied.

The amorphous alloys obtained so far by rapid quenching from the liquid state can be classified into two essentially different types: (1) the nonmetallic Te-based alloys, with elements of the IIIB and IVB columns of the periodic table; and (2) the metallic Pd- (or Pt- or Rh-) based alloys with Si, Ge, or Sb.

Te-based Alloys

It is well known that Se can be retained amorphous by relatively slow cooling from the melt. This is not true of Te, which is not amorphous, even when cooled at the highest rates obtainable with presently known techniques. Amorphous structures, however, can be obtained in alloys containing 10 to 30 a/o Ga or In and 10 to 25 a/o Ge.[15,16] The diffraction

TABLE 2 ***Structure of Amorphous* Te*-based Alloys Compared with Crystalline and Liquid* Te**

Nearest-neighbor order	Number of atoms	Interatomic distance, angstroms
Crystalline Te		
1st	2	2.864
2d	4	3.468
3d	6	4.457
4th	2	4.463
Liquid Te (465°C)[12]		
1st	2.3	2.93
2d	....	4.4
Amorphous alloys[12]		
1st	1.3–2.20	2.65–2.80
2d	10.8–12.7	4.00–4.25

patterns of these alloys, obtained with Mo K_α radiation monochromatized by the (200) reflection of an Li fluoride crystal, consist of three or four diffuse rings. Several of these alloys were studied in detail, and the radial distribution functions were determined. Within the limits of uncertainty of the analysis, the average distances for the first and second neighbors were practically the same for all alloys.[16] A comparison is presented in Table 2 of the atomic arrangements in the amorphous alloys, liquid Te,[17] and crystalline Te. Taking into account the uncertainties in the analysis of the diffraction patterns, it is tentatively concluded that the atomic arrangement in the amorphous alloys is very similar to that present in liquid Te, which in turn is very much the same as that in solid hexagonal Te. Why the crystallization of Te is prevented by the addition of elements such as Ga, In, or Ge is still an unsolved problem. The effect of these alloying elements may be an increase in the degree of undercooling of the melt, but experimental studies of this effect have not yet been attempted.

The amorphous nature of Te-based alloys has been confirmed by observing their mode of crystallization in transmission electronmicroscopy.[7] No grain structure is observed in the amorphous alloy, but when it is heated within the microscope, crystallization takes place by dendritic growth, as shown in Fig. 1. Since these alloys are very poor electrical

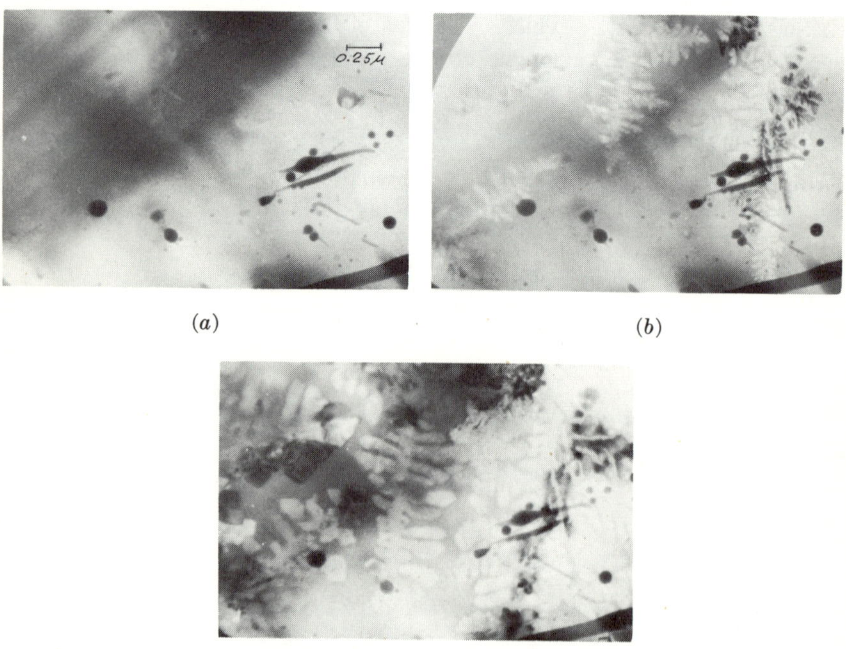

(a) (b)

(c)

Fig. 1 Dendritic crystallization of an amorphous Te_{85}-Ge_{15} alloy during heating in the electronmicroscope.[7]

conductors (and presumably poor heat conductors) Willens[7] concluded that the release of heat during crystallization produced a thermal gradient which is conducive to dendritic growth. This explanation is compatible with the fact that (as explained in the next paper) no dendritic growth has been observed in amorphous metallic alloys which have a low electrical resistivity.

Pd- and Pt-based Alloys

The first amorphous phase having a metallic character was found in Au-Si alloys containing about 25 a/o Si.[9] This phase was not stable at room temperature and transformed into a crystalline metastable form (not yet identified) after about 24 hr. Detailed studies of the electrical properties of this phase were never completed because in the meantime another amorphous phase stable at room temperature was discovered[18] in Pd-Si alloys containing 15 to 23 a/o Si. A typical intensity vs. angle 2θ of a diffraction pattern of a Pd_{80}-Si_{20} alloy is shown in Fig. 2. The analysis of the radial distribution function for this alloy[19] leads to the conclusion that the first shell of neighbors contains approximately 11.6 atoms at an average distance of 2.80 A. The second shell is split into two subshells at distances of 4.78 and 5.38 A. A comparison with the structure of an alloy of the same composition in the liquid state is not possible, because no X-ray diffraction studies of liquid Pd-Si alloys has been reported in the literature. For the same reason, it is not possible to compare this structure with that of pure liquid Pd. It can be pointed out, however, that in solid fcc Pd there are 12 nearest neighbors at 2.75 A. As in Te-based alloys, it appears that the atomic arrangement in the amorphous alloy is very similar to that in the solid, as far as coordination numbers and average interatomic distances are concerned.

Preliminary measurements of the electrical resistivity of a Pd_{83}-Si_{17} amorphous alloy were reported in Ref. 18. The room-temperature resistivity of the amorphous alloy is about 2.3 times that of the crystalline alloy of the same composition. The resistivity decreases with temperature, but the temperature coefficient is very small (about 0.00017/C°) and at 2°K about 96% of the room temperature value is retained. Evidently, the electrical resistivity of the amorphous alloy is mainly governed by scattering resulting from a random structure, and the phonon scattering contribution is of secondary importance. More recently Crewdson[19] has made additional measurement on a Pd_{80}-Si_{20} amorphous alloy covering a range of temperature from liquid N to the melting point and beyond. His results are shown in the graph of Fig. 3. The equilibrium alloy shows a normal behavior up to the eutectic temperature, where, as expected, a sudden resistivity increase occurs. The slope of the curve in the two-phase regions between the eutectic (800°C) and the liquidus (920°C) is much smaller than in the solid state, and is still smaller above the liquidus.

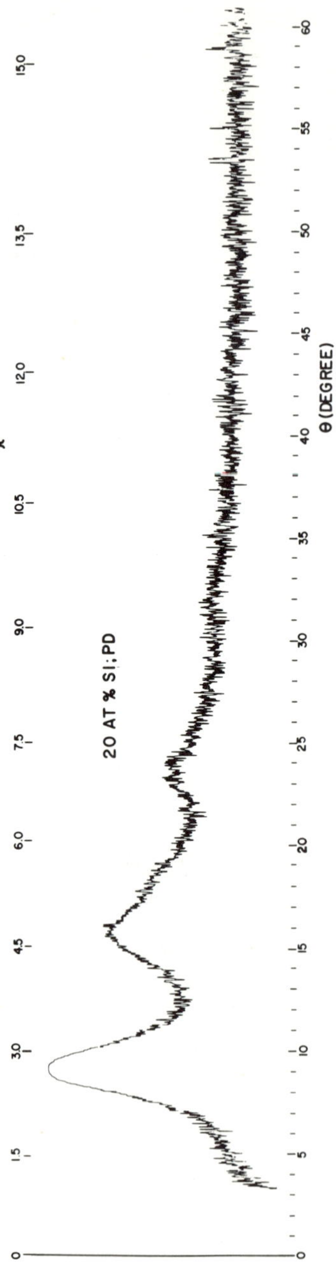

Fig. 2 Microphotometer trace of an X-ray diffraction pattern of a Pd_{80}–Si_{20} amorphous alloy. Monochromatic Mo $K\alpha$ radiation.[19]

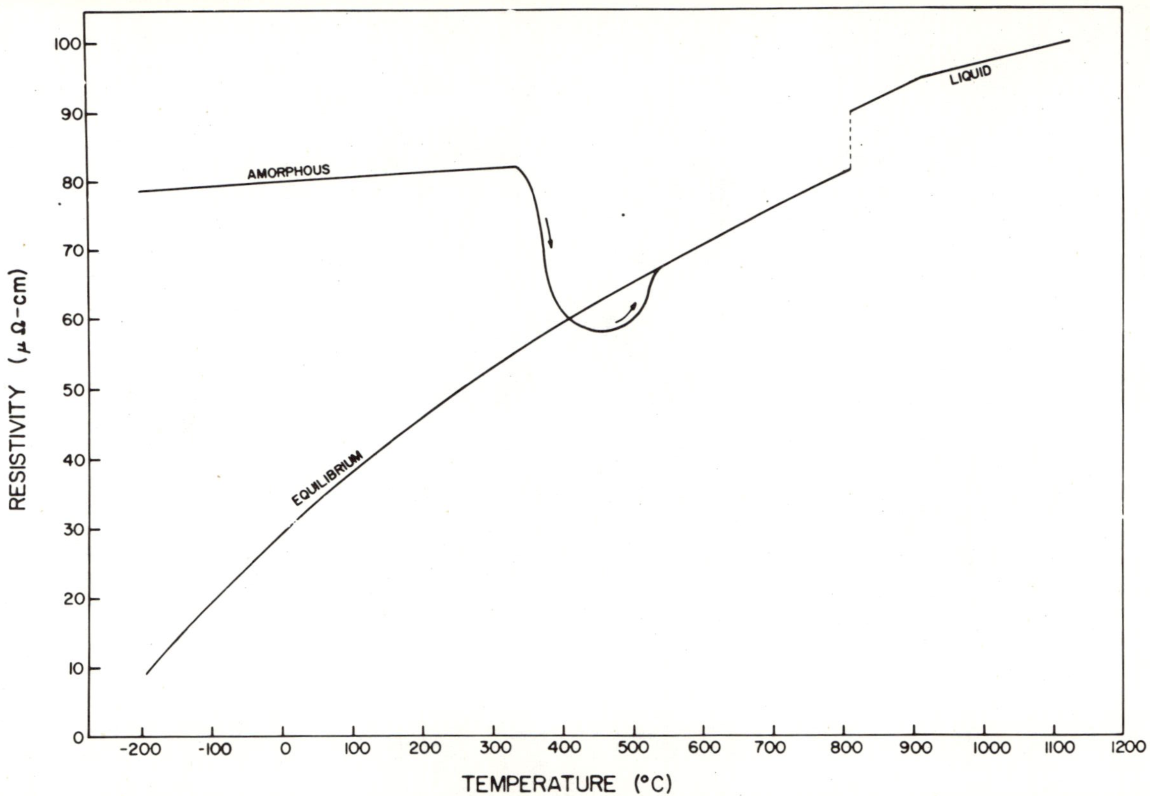

Fig. 3 Electrical resistivity vs. temperature curve of a Pd_{80}-Si_{20} alloy in equilibrium conditions, in the eutectic to liquidus range and in the liquid state. Similar curve shows the resistivity of an amorphous alloy measured at a rate of heating of 1.5°C/min.[19]

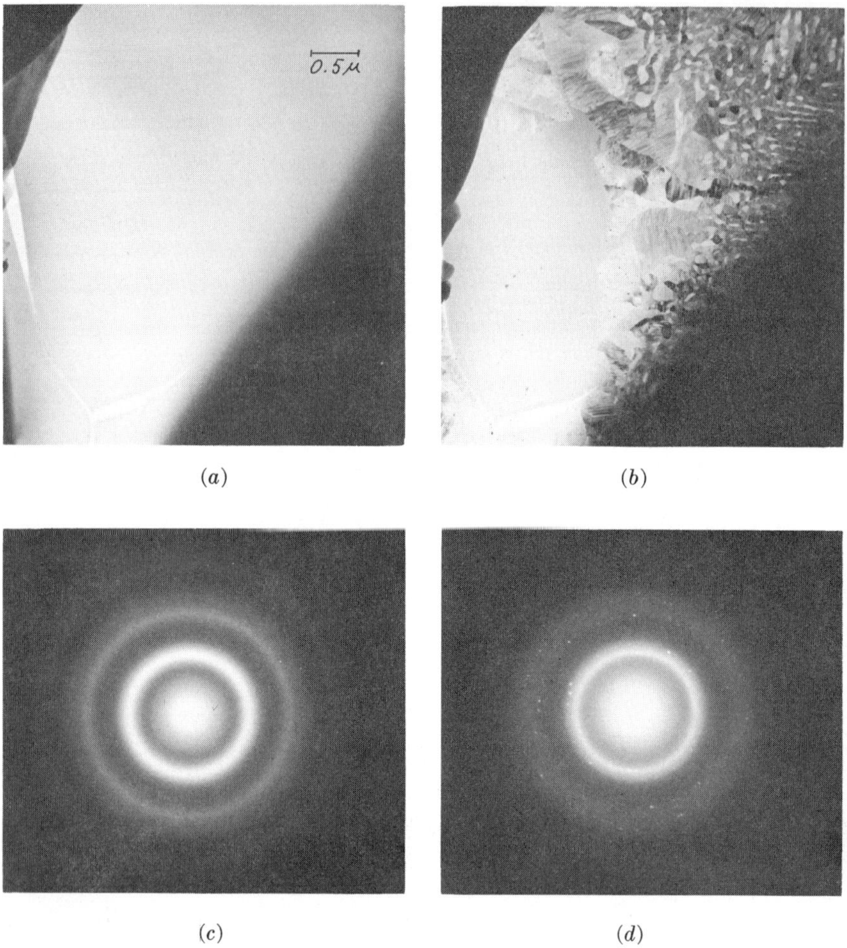

Fig. 4 Electronmicrograph (*a*) and electron diffraction (*c*) of an amorphous Pd_{80}-Si_{20} alloy. After heating in the microscope, partial crystallization took place (*b*). The presence of microcrystals barely visible in the white region of (*b*) is confirmed by the diffraction spots superimposed on the diffused rings in (*d*).[19]

The resistivity of the amorphous phase was measured at a constant rate of heating of 1.5 C°/sec and is a straight line having practically the same very small slope as that reported in Ref. 18 for the Pd_{83}-Si_{17} amorphous alloy. Around 350°C crystallization begins, and obviously a decrease in resistivity occurs. The curve continues to drop, reaches values below those corresponding to the alloy under equilibrium, and finally coincides with the equilibrium curve around 550°C. It is interesting to note that an extrapolation of the curve corresponding to the amorphous alloy will join smoothly with the curve beyond the eutectic temperature provided that the temperature coefficient is allowed to increase steadily with temperature, which is a logical assumption. The fact that the resistivity of the amorphous alloy drops below that of the equilibrium alloy is rather surprising. A study of the crystallization kinetics of the amorphous structure in the range of 300 to 750°C, however, indicates that in the early stages of crystallization, nonequilibrium Pd crystals grow first in the amorphous matrix. At higher temperature, these crystals disappear and are replaced by the equilibrium silicides, whose electrical resistivity is greater than

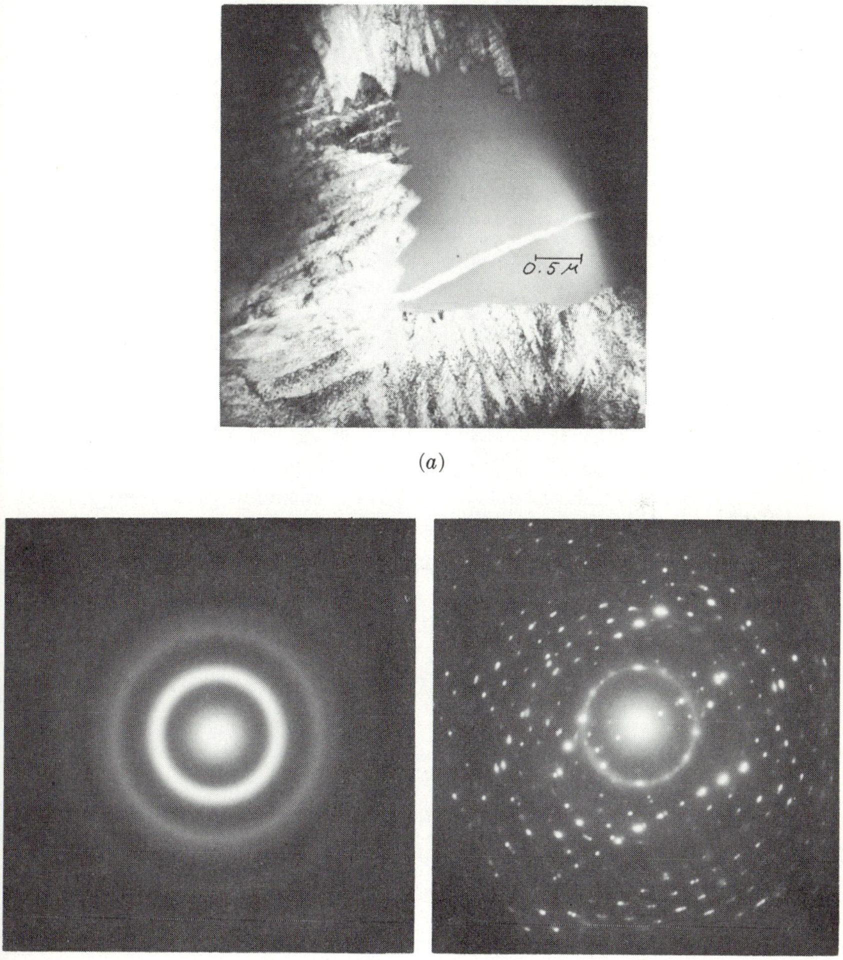

(a)

(b) (c)

Fig. 5 Electronmicrograph of a Pd_{80}-Si_{20} originally amorphous alloy showing the progress of a crystallization front (*a*). The electron-diffraction patterns correspond to the amorphous (*b*) and crystalline (*c*) regions of (*a*).[19]

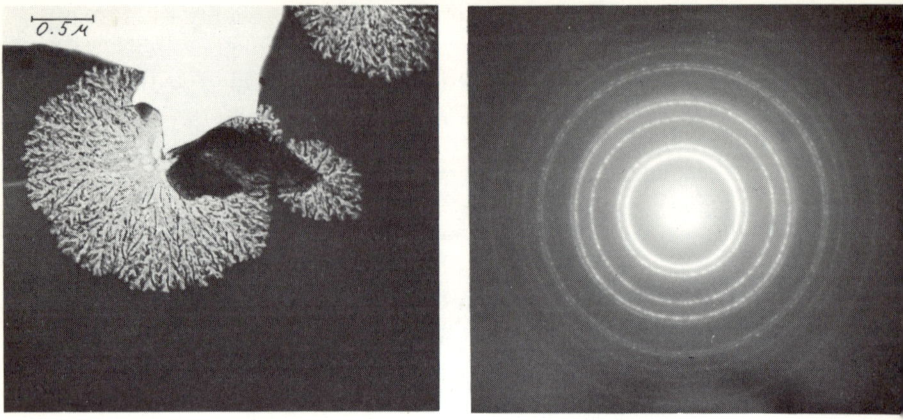

(a) (b)

Fig. 6 Spherulite crystallization in a Pd_{80}-Si_{20} amorphous alloy (*a*). Electron-diffraction pattern (*b*) indicates the presence of very small crystals of Pd or a supersaturated solid solution of Si in Pd.[19]

that of Pd. The details of the transition kinetics of amorphous Pd-Si alloys are not yet known, but it appears that the transformation proceeds through a succession of metastable crystalline phases which are not present under equilibrium conditions. If an alloy is heated at a sufficient rate (above about 20 C°/min), a sudden heat release of the order of 1000 cal/mole is observed[18] at about 400°C. As mentioned previously, this is an indirect evidence of the amorphous nature of the alloy. The structure of the alloy immediately after this "burst" of heat is completely crystalline, but the phase or phases present (not yet fully identified) are not the equilibrium phases. Further heating of the specimens in the range of 600°C and the eutectic temperature (800°C) results in a time-dependent approach to equilibrium.

Several types of crystal growth have been observed by electronmicroscopy in amorphous Pd- and Pt-based alloys.[19] These different morphologies may be related to the temperature at which crystallization takes place and also to the time at temperature. Since the specimens were heated within the electronmicroscope, it was difficult to control the temperature accurately enough to establish a quantitative relationship between temperature and crystal growth. The problem is further complicated by the fact that the rate of crystallization of amorphous alloys increases very rapidly within a rather narrow range of temperature of approximately 100°C. The following description will illustrate some of the crystallization modes most frequently found in amorphous metallic alloys.

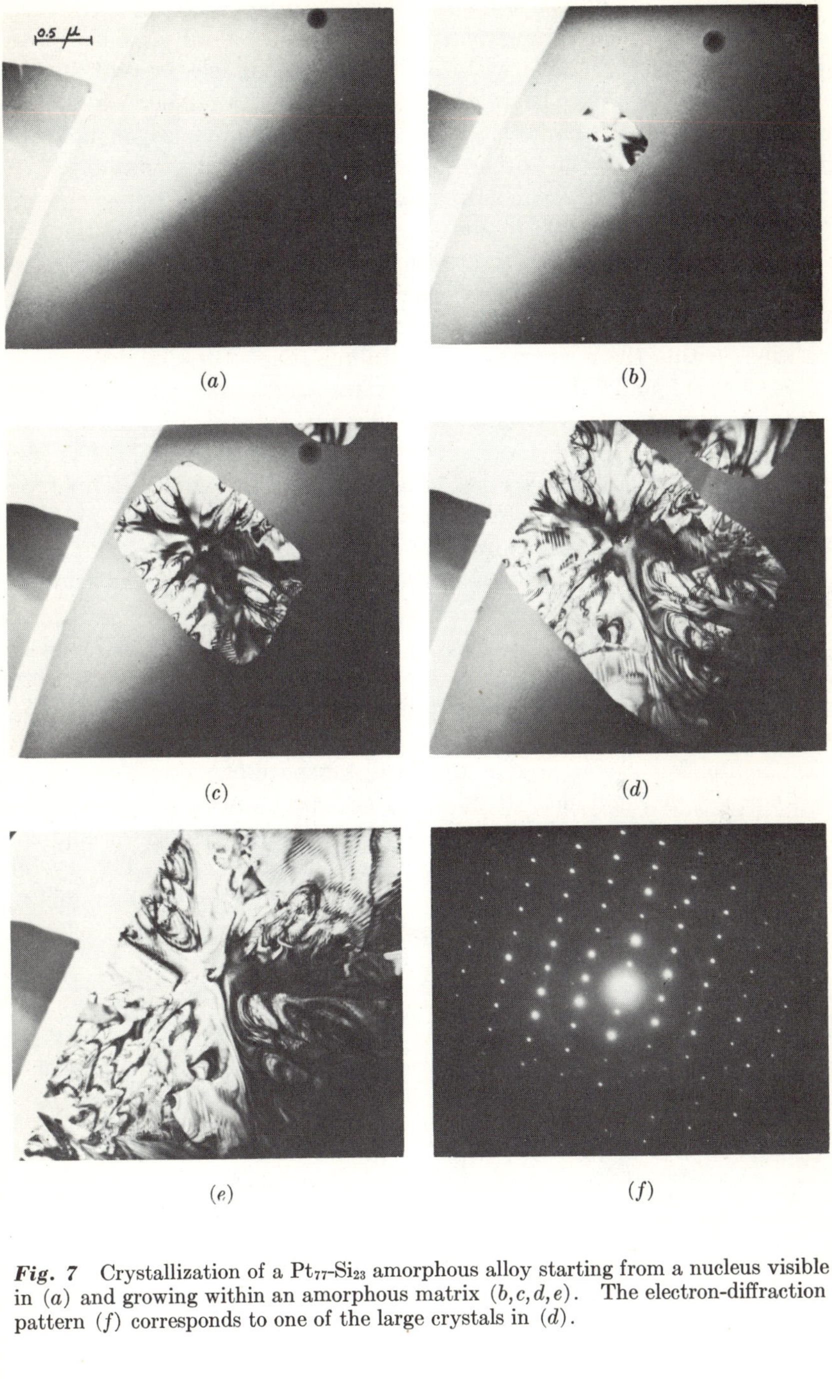

Fig. 7 Crystallization of a Pt_{77}-Si_{23} amorphous alloy starting from a nucleus visible in (*a*) and growing within an amorphous matrix (*b,c,d,e*). The electron-diffraction pattern (*f*) corresponds to one of the large crystals in (*d*).

Crystallization of an amorphous alloy may proceed by an advancing front through the amorphous matrix. Figure 4*a* shows an amorphous region in a Pd_{80}-Si_{20} alloy, with the corresponding electron-diffraction pattern of Fig. 4*c*. After heating for a short time at a temperature in the range of 300 to 350°C, about half of the same region is obviously completely crystalline (Fig. 4*b*). The other half appears to be still amorphous, but close inspection of that region reveals the presence of extremely small crystals of the order of 200 to 300 A. An electron-diffraction pattern in that region (Fig. 4*d*) indeed shows the presence of crystalline diffraction spots superimposed on the diffused rings typical of the amorphous matrix. In an alloy of the same composition, but probably heated under different conditions within the microscope, a crystalline front of the kind shown in Fig. 5 has been found. In this case the crystals have grown with a preferred orientation and are relatively large, as evidenced by the electron-diffraction pattern shown in Fig. 5*c*. Crystallization without a sharp front, has also been found in the same alloy Pd_{80}-Si_{20} and is shown in Fig. 6*a*. This type of crystallization is similar to what has been observed in glass. The electron-diffraction pattern shown in Fig. 6*b* indicates that the crystalline portion of the specimen is made of fcc Pd, or a supersaturated solid solution of Si in Pd.

In Pt-based alloys, similar modes of crystallization have been found in electronmicroscopy. In one case, however, it was possible to observe crystallization starting from what appeared to be a single nucleus. As shown in Fig. 7, crystal growth started around either a microcrystal or a defect in the foil visible in Fig. 7*a*. The sequence of pictures shown in Fig. 7*b*, *c*, *d*, *e* was obtained after heating the specimen for various lengths of time. Three or four crystals are probably present in Fig. 7*e*, and selected area diffraction within one crystal is shown in Fig. 7*f*. The crystal structure of this phase, which is probably a metastable crystalline phase, has not yet been determined. When the Si content of a Pd-Si alloy is

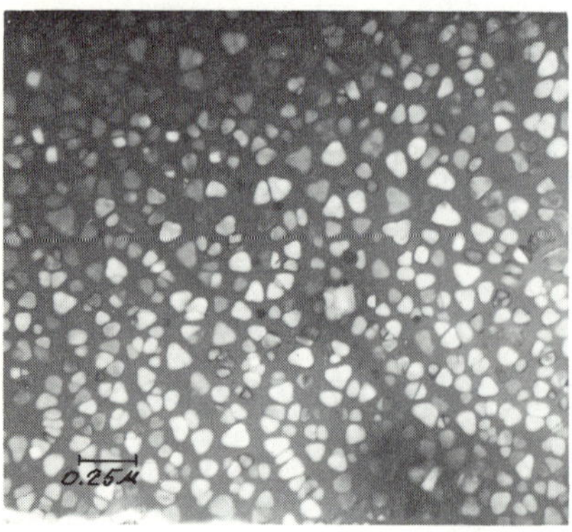

Fig. 8 Electronmicrograph of a Pd_{88}-Si_{12} alloy showing primary Pd (or Pd-Si solid solution) within an amorphous matrix.

below about 15.5 a/o, the quenched alloy is not completely amorphous. In an alloy of composition Pd_{88}-Si_{12} for example, the structure shown in Fig. 8 consists of microcrystals in an amorphous matrix. These crystals have been identified as being either pure Pd or a supersaturated Pd-Si solid solution. In addition, they are oriented with the (111) direction perpendicular to the thickness of the foil.

PREDICTION OF AMORPHOUS PHASES

Immediately after the first metallic amorphous phase in Au-Si alloys was obtained by rapid quenching from the melt, Cohen and Turnbull suggested that the occurrence of an amorphous structure in these alloys was due (in part) to the existence of a low-temperature eutectic located relatively close to the Au end of the diagram. These conditions lead to a sharp drop in temperature from the pure metal to the eutectic concentration. This simple observation served as a guide in the choice of alloys in which amorphous structures could be quenched from the liquid state. Relatively low eutectic temperatures are found in the binary alloys Pd-Si, Pt-Ge, Pt-Sb, and Rh-Ge. In all these systems, amorphous phases were found in alloys whose compositions were located near the eutectic concentration. An amorphous phase can also be obtained in Pd-Ge alloys (around 20 a/o) for which the equilibrium diagram is not known. It is probable, however, that when this diagram is established, a low eutectic will be found near 20 a/o Ge. Unsuccessful attempts to obtain amorphous phases in Ni- and Co-based alloys seem to give support to the low eutectic emperical rule, since in all binary alloys of Ni or Co with the groups IIIB and IVB of the periodic table, eutectic temperatures are relatively high compared with the melting points of Ni and Co.

The existence of a low-temperature eutectic in an alloy system is probably related to the occurrence of undercooling of the melt. If this is true, it could be postulated that as the rate of cooling is increased, the degree of undercooling could eventually be such that crystallization of the melt cannot occur. In at least one of the alloy systems in which amorphous phases can be obtained by rapid quenching from the melt, namely Pd-Si alloys, a large degree of undercooling has been observed, and in order to obtain equilibrium conditions under relatively small rates of cooling (of the order of 10 C°/min) the melt must be inoculated at both the liquidus and the eutectic temperatures.

All these observations do not contribute much to the problem of finding out what are the fundamental reasons for the existence of amorphous phases obtained by rapid cooling from the liquid state, and it is probable that more experimental facts will be required before the problem can even be formulated.

CONCLUSIONS

The recently perfected method of rapidly quenching alloys from the liquid state has lead to the discovery of a large number of metastable phases in alloys, and it is probable that this number will rapidly increase as the technique is more widely used. It is impossible to predict whether or not this relatively new field will either help or hamper the metal physicist in his attempt to predict what phases and what crystal structures should exist in a given alloy system. So far, it appears that the new metastable phases have just added difficulties to an already complex problem. Amorphous metastable phases, however, should be of immediate interest in the study of the liquid state. If it is true that the atomic arrangement in amorphous alloys is very similar to that existing above the melting point, these alloys would greatly simplify the experimental problem connected with the study of liquid alloys at high temperature. It is hoped that in the near future, some of the metallic alloys which can be obtained in the form of an amorphous solid will be carefully studied in both the solid and liquid state, and that a direct correlation will be established between the two states.

REFERENCES

1. Mader, S., H. Widmer, F. M. d'Heurle, and A. S. Nowick: *Appl. Phys. Letters* **3**:201 (1963).
2. Duwez, Pol, and R. H. Willens: *Trans. Met. Soc. AIME* **227**:362 (1963).
3. Willens, R. H., and E. Buehler: *Trans. Met. Soc. AIME* **236**:171 (1966).
4. Predecki, Paul, A. W. Mullendore, and N. J. Grant: *Trans. Met. Soc. AIME* **233**: 1581 (1965).
5. Pietrokowsky, P.: *J. Sci. Instr.* **34**:445 (1962).
6. Willens, R. H.: *Proc. 5th Intern. Congr. Electron Microscopy* (Academic Press Inc., New York and London) **1**:EE-6 (1962).
7. Willens, R. H.: *J. Appl. Phys.* **33**:3269 (1962).
8. Duwez, Pol, R. H. Willens, and W. Klement, Jr.: *J. Appl. Phys.* **31**:1500 (1960).
9. Klement, W., Jr., R. H. Willens, and Pol Duwez: *Nature* **187**:869 (1960).
10. Klement, W., Jr.: *J. Inst. Metals* **90**:27 (1961).
11. Klement, W., Jr.: *Trans. Met. Soc. AIME* **233**:1182 (1965).
12. Luo, H. L., M. F. Merriam, and D. C. Hamilton: *Science* **145**:581 (1964).
13. Predecki, P., B. C. Giessen, and N. J. Grant: *Trans. Met. Soc. AIME* **233**:1438 (1965).
14. Anantharaman, T. R., H. L. Luo, and W. Klement: *Trans. Met. Soc. AIME* **233**:2014 (1965).
15. Luo, H. L., and Pol Duwez: *Appl. Phys. Letter* **2**:21 (1963).
16. Luo, H. L.: Ph.D. thesis, California Institute of Technology, 1964.
17. Buschert, R. C.: Ph.D. thesis, Purdue University, 1957.
18. Duwez, Pol, R. H. Willens, and R. C. Crewdson: *J. Appl. Phys.* **36**:2267 (1965).
19. Crewdson, R. C.: Ph.D. thesis, California Institute of Technology, 1966.
20. Cohen, M. H., and D. Turnbull: *Nature* **189**:131 (1961).
21. Turnbull, D.: *Trans. Met. Soc. AIME* **221**:422 (1961).

A DISLOCATION MODEL OF THE LIQUID STATE

P. S. Rudman

Metal Science Group, Battelle Memorial Institute, Columbus, Ohio

INTRODUCTION

Crystalline matter melts because the liquid state possesses greater entropy. The melting point T_m is defined by

$$\Delta F = \Delta H_m - T_m \, \Delta S_m = 0 \tag{1}$$

where ΔH_m = enthalpy of melting, ΔS_m = entropy of melting, and ΔF = liquid free energy–solid free energy.

X-ray diffraction shows that a considerable degree of lattice order is destroyed on melting, and thus the first tendency is to ascribe the additional liquid-state entropy to configurational origins. However, intuition may be misleading. There is considerable evidence that configurational entropy is, in fact, negligible in liquids in the vicinity of the melting point.

As long ago as 1934, Mott's[1] analysis of the resistivity change in metals on melting provided this clue. Let us briefly reproduce Mott's argument. If we assume that the entropy of melting can be wholly accounted for by a lower effective Debye temperature in the liquid, then Eq. (1) can be written as

$$\Delta H_m - 3RT_m \ln \frac{\theta_S}{\theta_L} = 0 \tag{2}$$

where $\theta_L(\theta_S)$ = Debye temperature of the liquid (solid). This yields

$$\frac{\theta_S}{\theta_L} = \exp \frac{\Delta H_m}{3RT_m} \tag{3}$$

We further assume that it is only the scattering probability that changes on melting, which, being proportional to the mean-square amplitude of vibration, yields

$$\frac{\rho_L(T_m)}{\rho_S(T_m)} = \left(\frac{\theta_S}{\theta_L}\right)^2 = \exp \frac{2\Delta H_m}{3RT_m} \tag{4}$$

Without reproducing Mott's tabulation we can note that it is observed that $\rho_L(T_m)/\rho_S(T_m) \cong 2$ for normal metals and that, using Richard's rule, $\Delta S_m = \Delta H_m/T_m = 2$ cal/(mole)(deg) in Eq. (4) also yields this result.

The extent to which a Debye frequency spectrum is a good approximation

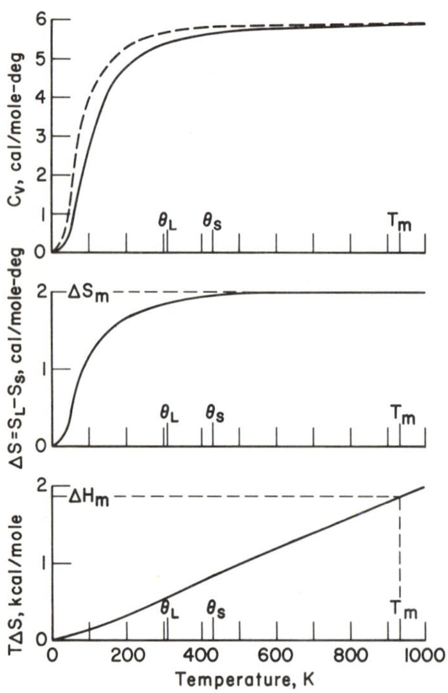

Fig. 1 Thermodynamic relations for Debye vibrational model of liquids using Al as the example.

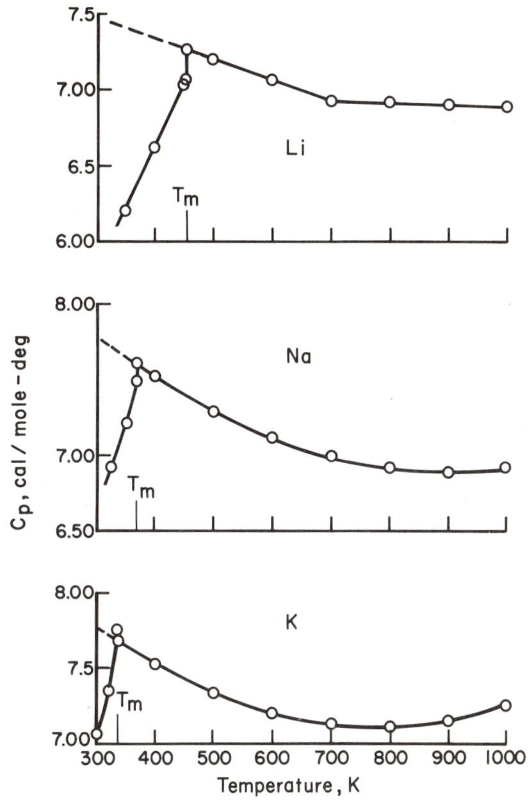

Fig. 2 Specific heat of group IA metals.

can be seen by considering specific heats. In Fig. 1 pertinent thermodynamic functions are plotted, using the values for Al to be specific. Equation (3), again assuming Richard's rule, yields $\theta_S/\theta_L = 1.4$, which was used in these calculations. The essence of this model is that a vibrational specific-heat difference occurring mainly at low temperatures, that is, for $T/\theta < 1$, is the origin of the additional entropy of the liquid state.

In Figs. 2 to 4 are presented some representative specific-heat data for metals. The data have been taken from the compilation of Hultgren et al.[2] It is readily seen that extrapolating c_p(liq) to $T \ll T_m$ always yields $\Delta c_p > 0$. Unfortunately, the extrapolation to the very low temperature region, where in fact the entropy difference is mainly generated, since

$$\Delta S = \int_0^T \frac{\Delta c_p}{T}\,dT$$

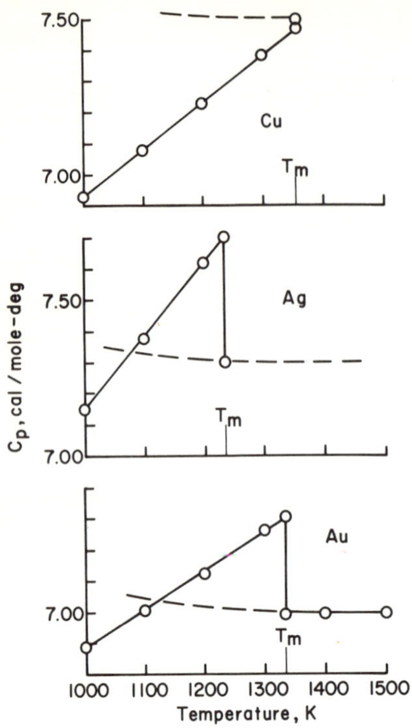

Fig. 3 Specific heat of group IB metals.

cannot be made accurately enough to be meaningful. However, it is easily found that reasonable extrapolation can readily account for the observed entropy of melting. It is also to be noted that, in general, the specific heat of the liquid is not at all well approximated by Debye function behavior. There is the possibility that the differences in solid and liquid specific-heat behavior are only due to the $c_p \rightarrow c_v$ correction:

$$c_p = c_v + \frac{v\beta^2}{\kappa} T \tag{5}$$

where v = molar volume, β = volume expansivity, and κ = compressibility.

However, it does not appear that this correction is adequate to account for the general difference in c_p behavior between liquid and solid, at least near the melting point. Thus our general conclusion is that the entropy of melting is dominantly of thermal rather than configurational origin and that a Debye approximation does not well represent the thermal entropy of the liquid state. In fact, if c_v behaves similarly to the observed c_p, and as idealized in Fig. 5, then many of the thermal modes cannot be described as harmonic vibrations since harmonic vibrations, whatever the frequency spectrum, give rise to a monotonically increasing c_v. Thus our first requirement in a model for the liquid state is the introduction of new thermal modes for which harmonic vibrations are probably a poor approximation.

A second requirement must be geometric: the long-range lattice order must be destroyed. A liquid not far from its melting point is energetically only a relatively small perturbation of the crystalline state when the latter is an elementary structure. In metals for example, the heat of melting is characteristically of the order of only 5% of the heat of sublimation. Thus a common approach to a model for a liquid is to assume a defect crystalline state. The defect first must have the property that it destroys the lattice long-range order. As pointed out by Vineyard,[3] the failure of point defects to accomplish this is a primary failing of models based on vacancies and interstitials. That material of very fine grain size should provide a model for the liquid state was a suggestion by Steward[4] in 1934 derived from his X-ray diffraction studies of organic materials. This model was elaborated on, with particular reference to metals by Mott and Gurney[5] in 1939. With the development of dislocation theory and the equation of grain boundaries to dislocation arrays, models of the liquid state as defected crystalline by dense dislocation arrays naturally emerged. An early

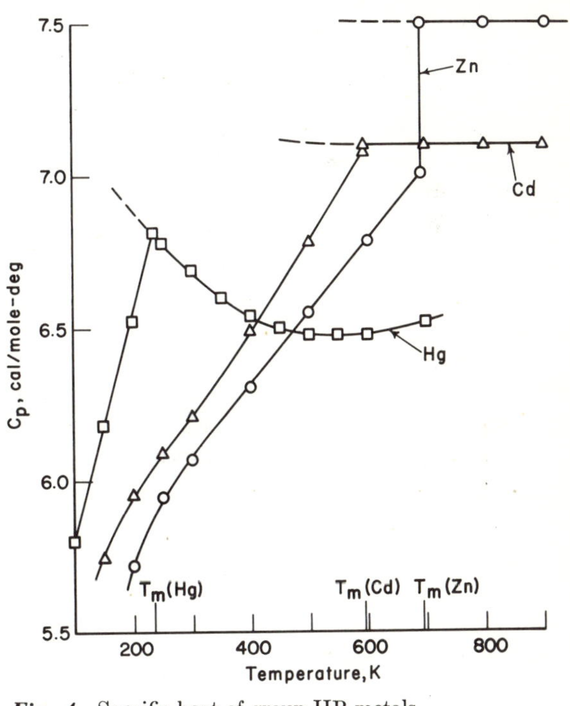

Fig. 4 Specific heat of group IIB metals.

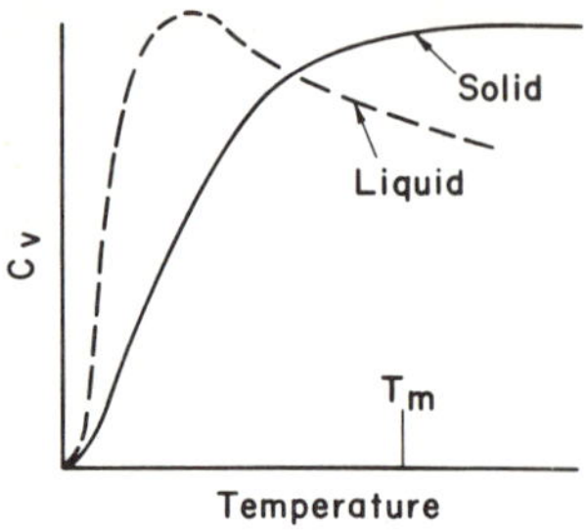

Fig. 5 Idealized specific heat in liquid and solid metals.

detailed treatment of this model was given by Shockley[6] in 1952. However, there is a difference between using dislocation arrays to provide an approximate geometric description of the liquid state and considering dislocations as distinguishible equilibrium defects with an equilibrium concentration that increases precipitously at some temperature that is identified as the melting temperature. It is the latter concept that we shall emphasize here.

DISLOCATION FREE ENERGY

For simplicity let us consider a simple cubic lattice with lattice spacing and Burger's vector b. The dislocation density per mole L' is then

$$L' = Nb^3L \text{ cm/mole} \tag{6}$$

where N = Avogadro's number, and L = dislocation density, cm/cm³.

There will be an associated molar energy

$$\Delta\bar{E}_\perp = \frac{L'}{b}\Delta E_\perp = Nb^2L\Delta E_\perp \tag{7}$$

where $\Delta E_\perp$ = dislocation energy per atom plane.

There is an associated thermal molar entropy

$$\Delta\bar{S}_\perp = Nb^2L\Delta S_\perp \tag{8}$$

where $\Delta S_\perp$ = dislocation thermal entropy per atom plane.

There is an associated configurational molar entropy

$$\Delta\bar{S}_{W\perp} = k \ln W_\perp \tag{9}$$

where k = Boltzmann's constant, and $W_\perp$ = number of configurations per mole associated with a dislocation density L.

The equilibrium dislocation density is then determined by the condition that

$$\frac{\partial\Delta\bar{F}_\perp}{\partial L} = 0 \tag{10}$$

where $$\Delta\bar{F}_\perp = \Delta\bar{E}_\perp - T\,\Delta\bar{S}_\perp - kT\ln W_\perp \tag{11}$$

To estimate $\ln W_\perp$, let us first consider nonflexible dislocations in a cube of N atoms. Each cube face will have $N^{2/3}$ sites, and an area $A = N^{2/3}b^2$. Each face will contain $n = LA/3$ dislocations, so that the number of ways of arranging n dislocations on $N^{2/3}$ sites on each of three faces independently is then

$$W = \left[\frac{(N^{2/3})!}{(N^{2/3} - n)!n!}\right]^3 \tag{12}$$

Were we to assume completely flexible dislocations, that is, able to kink and jog each lattice plane, then the number of configurations would be given by Eq. (12) raised to the power $N^{1/2}$, which is the number of planes threaded by a dislocation. Performing the differentiation of Eq. (10), we obtain $n/N^{2/3} = Lb^2/3 \ll 1$ (typically $b^2 \simeq 6 \times 10^{-16}$, so that $L < 5 \times 10^{14}$),

$$L = \frac{3Z}{b^2} \exp\left\{-N^{1/3} \frac{(\Delta E_\perp^\circ - T\,\Delta S_\perp^\circ) - (L^2/2)(\Delta E_\perp'' - T\,\Delta S_\perp'') - \cdots}{kT}\right\} \tag{13}$$

where $1 < Z < N^{1/3}$; $\Delta E_\perp^\circ$, $\Delta S_\perp^\circ$ are energy and entropy as $L \to 0$; and $\Delta E_\perp'' = \partial^2\,\Delta E_\perp/\partial L^2$, etc.

At low temperatures, and hence low dislocation densities, the first term in the exponent of Eq. (13) suffices. In Fig. 6 the behavior of the free-energy and dislocation-density functions is schematically illustrated by the solid lines for the retaining of these first-order terms only. Because of the presence of the $N^{1/3}$ ($\simeq 10^8$) factor in the exponent, the equilibrium dislocation density will be negligibly small until a critical temperature $T_\perp$ is reached:

$$T_\perp = \frac{\Delta E_\perp^\circ}{\Delta S_\perp^\circ} \tag{14}$$

At this temperature, in first order, the dislocation density would increase precipitously to physically impossible large values. However, for $T > T_\perp$, the higher-order dislocation-dislocation interaction terms will become important, and physically realistic solutions are possible, for example, of the form shown by the dashed curves in Fig. 6.

Whether a dislocation model of melting is reasonable thus resolves to whether it can be shown that $T_\perp = T_m$. This was apparently first considered by Cottrell.[7] For the thermal entropy term Cottrell noted that when a dislocation is present in a crystal, the characteristic vibration frequencies (Einstein model) of the atoms—in the core in particular—are shifted to lower frequencies, owing to the dilation of the lattice, and this leads to a positive entropy term. One method of arriving at an estimate of the magnitude of this entropy term is to note that the core atoms have an environment similar to the atoms surrounding a lattice vacancy. In this latter case, the vibrational entropy has been determined in several metals[8] as about 1 to $2k$ per vacancy. This estimate is comparable to a

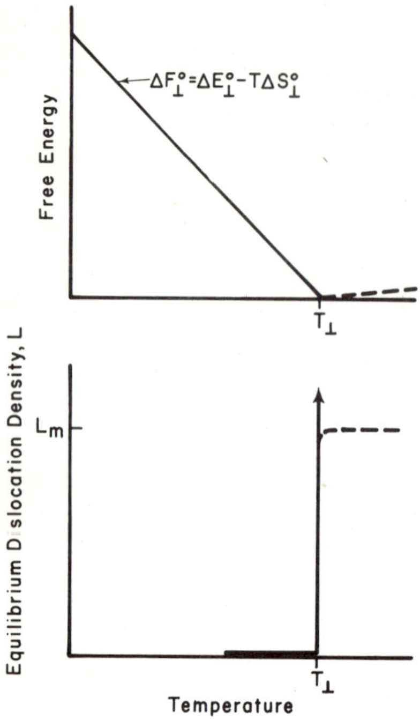

Fig. 6 Dislocation free energy and equilibrium density function.

value arrived at by Cottrell for dislocations by a different procedure. Taking $\Delta E_\perp \simeq Gb^3/2 \simeq 5$ ev, where G = shear modulus, for an isolated dislocation in Cu, and $\Delta S_\perp \simeq 2k$, yields $T_\perp \simeq 30{,}000°K$, from which Cottrell concluded that spontaneous dislocation formation has nothing to do with melting.

In 1956 Friedel[9] reconsidered this calculation with two revisions. First he noted that in a dense array of dislocations, as would be required to describe a liquid, the long-range stress fields would tend to cancel, leaving essentially only the dislocation core energy. Although not easily estimated, Friedel took $\Delta E_\perp \simeq Gb^3/10 \simeq 1$ ev. Of equal numerical importance, but of much greater conceptual significance, Friedel noted that a dislocation is not a static defect but that it will have associated with it thermal modes. Friedel took as his thermal mode model for dislocations, a one-dimensional harmonically vibrating chain of discrete mass points, and did in fact find sufficient vibrational entropy to conclude that $T_\perp = T_m$. This conclusion was, however, criticized by Frank,[10] who noted that Friedel's calculation had not properly conserved degrees of freedom. There are a total of $3N$ degrees of freedom, independent of the state, and hence the dislocation modes must be subtracted from the lattice modes, which Friedel had not done, thereby overestimating the vibrational entropy.

Let us briefly reformulate these considerations. If we consider the vibrating chain problem in the Debye approximation, at high temperatures, the free energy of each harmonic is

$$f'_\nu = kT \ln \frac{h\nu}{kT} \tag{15}$$

and hence the total vibrational free energy is

$$f_\nu = kT \int_0^{\nu_m} g(\nu) \ln \frac{h\nu}{kT} \, d\nu \tag{16}$$

where ν_m = maximum frequency = $v/2b$, v = velocity of sound, $g(\nu)\,d\nu$ = number of harmonics in the internal ν to $\nu + d\nu$.

For the one-dimensional lattice model of a dislocation, $g_\perp(\nu) = \text{constant}$, while for a three-dimensional lattice, $g(\nu) \propto \nu^2$. The maximum frequency is equal for both lattice and dislocation modes since the velocity of sound is equal along a dislocation line and through the crystal, so that for the change in vibrational entropy due to the presence of dislocation modes, we write

$$-T\,\Delta S_\perp \simeq \Delta f_\nu \cong kT \int_0^{\nu_m} [g_\perp(\nu) - g(\nu)] \ln \frac{h\nu}{kT}\, d\nu \tag{17}$$

It is not obvious how lattice modes should be subtracted, and in Eq. (17) we have simply decreased the effective number of atoms of the crystal, otherwise leaving the frequency spectrum unchanged.

$$g_\perp(\nu) = \frac{pl}{\nu_m b} \tag{18}$$

where l = length of dislocation line segment, and p = polarization number = 1 for an edge, = 2 for a screw.

To conserve degrees of freedom

$$\int_0^{\nu_m} g_\perp(\nu)\, d\nu = \int_0^{\nu_m} g(\nu)\, d\nu \tag{19}$$

If now we integrate Eq. (17), taking (18) and (19) into account, we obtain

$$\Delta S_\perp \cong pk \text{ per atom plane} \tag{20}$$

The origin of this entropy change can be described qualitatively as the conversion of high-frequency low-entropy lattice vibrations into low-frequency high entropy dislocation vibrations by virtue of the different weighting functions $g_\perp(\nu)$ and $g(\nu)$.

Let us now reconsider the $T_\perp$ calculation, taking account of the decreased $\Delta E_\perp \simeq 1$ ev, and the sum of the lattice frequency shift and dislocation vibration-mode entropy contributions which can total $\simeq 4k$, yielding $T_\perp = 1 \text{ ev}/4k = 3000°\text{K}$. This is still not $T_m(\text{Cu}) = 1360°\text{K}$, but it is close enough not to be discouraged; perhaps there are other entropy contributions that we have overlooked. Not only can a dislocation vibrate as a flexible chain with its line tension acting as the restoring force, which is what we have considered so far, although the line tension concept was not explicitly invoked, being implicitly included in the wave velocity; it can also vibrate rigidly in the force fields of other dislocations.

For this latter case, the vibrational frequency and hence the entropy will depend on the geometry of the dislocation array. However, to obtain an estimate for this mode, let us consider an edge dipole as schematically illustrated in Fig. 7. The dipole is mechanically stable at $x = y$. If the dipole is now given a perturbing displacement to $x + \Delta x$, there will be a restoring force per atom plane, which after Cottrell[7] is

$$F_x = K\,\Delta x \tag{21}$$

with

$$K = \frac{Gb^3}{4\pi(1 - \sigma)x^2} \tag{22}$$

The dipole characteristic frequency is then given as

$$v = \frac{1}{2\pi}\sqrt{\frac{K}{m}}$$

$$= \frac{1}{4\pi\sqrt{\pi}(1 - \sigma)}\,\frac{b}{x}\sqrt{\frac{Gb}{m}} \tag{23}$$

where σ = Poisson's ratio and m = atomic mass.

If we let $x = 3b$, which corresponds to virtually complete relaxation of long-range elastic strain energy, we obtain $\nu_\perp(\text{Cu}) \simeq 2 \times 10^{11}$. The frequency distribution function for this mode is a delta function at $\nu_\perp$, and thus we obtain from Eq. (17)

$$\Delta S_\perp = k \ln \frac{\nu_m}{\nu_\perp} \qquad \text{per atom plane per dipole} \tag{24}$$

Since $\nu_m(\text{Cu}) = k\theta/h \simeq 7 \times 10^{12}$, where θ = Debye temperature $\simeq 350°\text{K}$, $\Delta S_\perp \simeq 3.5k$ per atom plane per dipole.

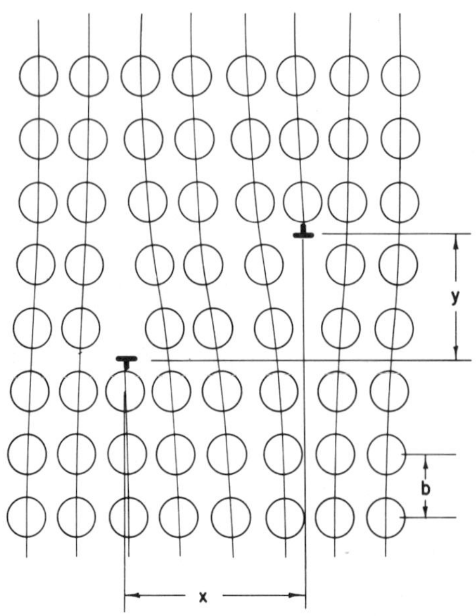

Fig. 7 Edge-dipole dislocation.

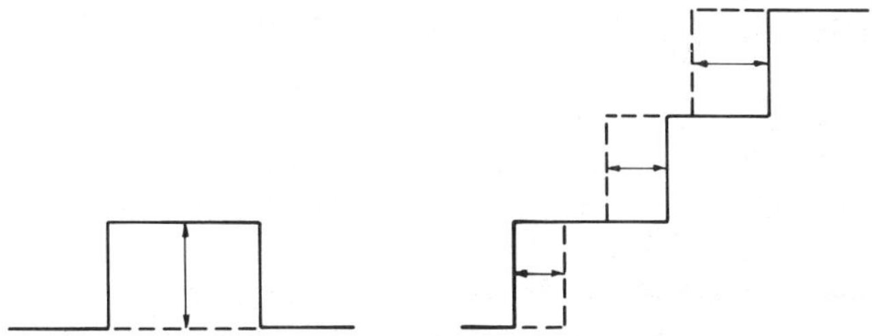

Fig. 8 Kink modes of dislocation motion.

If now we sum each of the entropy contributions: $\simeq 2k$ for lattice frequency shift; $\simeq 2k$ for dislocation flexible-mode vibration; $\simeq 3.5k/2$ for dislocation rigid-mode vibration; and take $\Delta E_\perp \simeq 1$ ev, we obtain $T_\perp \simeq$ 2000°K.

Within the accuracy of our calculations we can consider that we have shown that $T_\perp = T_m$, and hence that melting is just a precipitous increase in the equilibrium dislocation density.

Perhaps there are other dislocation modes that we have not considered, for example kink modes? In Fig. 8 two possible modes are illustrated schematically: double kinking and kink glide. It is questionable whether these should be considered as new modes or simply as alternative descriptions of the behavior that we have already discussed. We shall in fact assume the latter, harmonic vibrations and kink motion simply being limiting descriptions of the same modes. Although the complete specific-heat consequences of kink modes have not been worked out, it is not unreasonable to expect these modes to account for the nonharmonic behavior of thespecific heat of liquids.

A SUGGESTED EXPERIMENTAL TEST

If identifiable dislocations do in fact produce the liquid state, then the liquid state may be studied with the view of learning more about deformed solids and vice versa. In a highly cold-worked metal the dislocation density is of the order of $L \simeq 10^{12}$ cm^{-1}. To obtain an estimate for what would have to be the dislocation density of a liquid, we can use Eq. (7); equating $\Delta H_m = \Delta \bar{E}_\perp$, using values for Cu, and taking $\Delta E_\perp = 1$ ev yields $L \simeq 10^{14}$ cm^{-1}: that is, a dislocation grid of about 10 A or four atom spacings apart. Incidentally, it should be noted that the combination of the T_m and ΔH_m calculations provides a very critical test of self-consistency. Clearly a dislocation concept will not have meaning unless $L \lesssim 10^{15}$, and since $\Delta H_m \propto L \, \Delta E_\perp$, this places a lower limit on $\Delta E_\perp$: for example for Cu, $\Delta E_\perp > 0.1$ ev. Since $T_\perp = \Delta E_\perp / \Delta S,_\perp$ the lower limit for $\Delta E_\perp$ places a lower limit on $\Delta S_\perp$. The values we have used, $L \simeq 10^{14}$, $\Delta E_\perp \simeq 1$ ev, $\Delta S_\perp \simeq 6k$, form a self-consistent set, and no single value can be revised without revising them all.

If as is indicated in Figs. 2 to 5 the specific heat of the liquid is appreciably greater than that for the solid at low temperatures, then even though the dislocation density in a cold-worked metal must be about two orders of magnitude less than in a liquid, there should still be measurable increment to the specific heat due to dislocation modes in cold-worked metals. We are presently setting up to check this experimentally. An alternative and complementary experiment is to measure the specific heat on supercooled or splat-cooled liquids.

REFERENCES

1. Mott, N. F.: *Proc. Roy. Soc.* **A146**:465 (1934).
2. Hultgren, R., R. L. Orr, P. D. Anderson, and K. K. Kelley: "Thermodynamic Properties of Metals and Alloys," John Wiley & Sons, Inc., New York, 1963.
3. Vineyard, G. H.: "Liquid Metals and Solidification," American Society for Metals, Cleveland, 1958.
4. Steward, G. W.: *J. Chem. Phys.* **2**:147 (1934).
5. Mott, N. F., and R. W. Gurney: *Rept. Progr. Phys.* **5**:46 (1939).
6. Shockley, W.: In R. Stoops, "L'Etat solide," Brussels, 1952.
7. Cottrell, A. H.: "Dislocations and Plastic Flow in Crystals," Clarendon Press, Oxford, 1953.
8. Simmons, R. O.: *J. Phys. Soc. Japan* **18** (suppl. II): 172 (1963).
9. Friedel, J.: "Les Dislocations," Gauthier-Villars, Paris, 1956.
10. Frank, F. C.: *Nuovo Cimento* **7**(ser. 10, suppl. 2): 386 (1958).

Agenda Discussion:
LIQUIDS AND DEFECT SOLIDS

chairman **F. C. Frank**
H. H. Wills Physical Laboratory, Bristol University, Bristol, England

secretary **E. I. Salkovitz**
Department of Metallurgy, University of Pittsburgh, Pennsylvania

Frank suggested the following framework for the discussion on liquids:

Direct space		*Fourier space*
Structure		
	Entropy	
	Electrons	

Structure is something we prefer to describe in direct space, though we have to go through Fourier space to discover it experimentally. It is a matter for discussion how much of the entropy of a liquid is configurational, belonging on the left of the diagram, and how much motional, most conveniently dealt with on the right. Most people dealing with electrons work on the right of the diagram, but there are those who still think it useful to consider them on the left.

Frank asked how truly we may regard all liquid metals as alike, with their properties determined by just two parameters, an energy and a size, say? It looked as though this could be a good first approximation, and when there is an anomalously long liquid range, the anomaly is in the solid, having an anomalously low melting point, rather than in the liquid.

Rather surely, one of the dominant principles for liquid structures was what Laves called the principle of space filling: but this was not a simple principle. Maximization of density is not the same thing as minimization of emptiness. Of lattices, in three dimensions, it is the face-centered cubic lattice which gives the greatest density for a given minimal distance between points, but it is the body-centered cubic lattice which for a given density minimizes the size of the largest spheres of emptiness within the structure. Thus, in a sense the latter is the best space-filling lattice. It is one of the many arrangements of spheres (the complex alloy structures exhibiting many more) in which all the interstices are moderately irregular tetrahedra, these in general providing good space filling.

Frank thought that liquids made much use of this multiplicity of good space-filling structures. This brought one back to the question: How much of the entropy of fusion is configurational, how much motional? He suggested perhaps half and half—say about two configurations per atom, and a 20% drop in frequencies of vibration, to account for the entropy of fusion in a typical case.

Collins gave a discussion regarding the calculation of melting entropy.

Neglect, for the time being, the change in density on melting, and consider a chain of atoms of mean spacing a (say) and variance σ^2 (due to thermal agitation). In the liquid, the cumulative variance in the total chain length to the nth atom will be approximately $\sigma^2 n$, while in the solid because of the influence of neighboring chains it will be simply σ^2. Hence, in the solid the standard deviation (S.D.) of the displacement along the chain of the nth atom is σ, while in the liquid it is $\sigma\sqrt{n}$. Consider now the following sequence of operations which rearrange the liquid chains into the solid.

1. Reduce the S.D. σ_2 of the second atom in the chain from $\sigma\sqrt{2}$ to σ. This reduces σ_3 from $\sigma\sqrt{3}$ to $\sigma\sqrt{2}$, σ_4 from $\sigma\sqrt{4}$ to $\sigma\sqrt{3}$, and so on, and requires an entropy change:

$$\Delta S_1 = \ln (\sigma\sqrt{4\pi e}) - \ln (\sigma\sqrt{2\pi e}) = \tfrac{1}{2} \ln 2$$

2. Now reduce σ_3 from $\sigma\sqrt{2}$ to σ, reducing σ_4 from $\sigma\sqrt{3}$ to $\sigma\sqrt{2}$, and so on. This takes an entropy change $\Delta S_2 = \tfrac{1}{2} \ln 2$.

3. Repeat the process until $\sigma_n = \sigma$ ($n = 1, 2, 3M$), where M is the number of atoms in the chain. The total entropy change is then $\tfrac{1}{2}M \ln 2$.

Now make the assumption that the x, y, and z components of a chain may be "sorted" independently. For a cube of N atoms there will be $N^{2/3}$ x chains each of length $N^{1/3}$. The total entropy of rearrangement of all x chains is then

$$\Delta S_x = N^{2/3} \cdot \tfrac{1}{2} N^{1/3} \ln 2 = \tfrac{1}{2} N \ln 2$$

Similar figures for ΔS_y and ΔS_z then give the total entropy change ΔS as $\Delta S/N = \frac{3}{2} \ln 2$ or, in energy units, $\Delta S/N = \frac{3}{2}k \ln 2 \simeq 1.04k$.

Lennard-Jones and Devonshire get $1.70k$ and ter Haar ("Statistical Mechanics") quotes the experimental value as $1.66k$.

In the above we have assumed that probability distributions are essentially gaussian so that we can use the result $\ln \sigma\sqrt{2\pi e}$ for the coding entropy of a gaussian distribution of variance σ^2 (cf. Shannon and Weaver, "Mathematical Theory of Communication").

While very rough, the foregoing does indicate that at least a considerable part of the entropy of melting can be accounted for on a configurational basis, even if the volume change on melting is disregarded.

Rudman said that it should be possible to resolve the controversy of the relative proportions of thermal and configurational entropy in liquids by an experimental test, using the supercooled metals of Duwez. Their structure, he suggested, certainly would not be very different from the structure of a liquid. The possibility exists for doing a third-law test: take this amorphous material, heat it, obtain its entropy change, recrystallize it, obtain the entropy change of transformation and then the entropy change on cooling to 0°K. As illustrated in Fig. 1, this would yield the configurational entropy.

Zener thought it was obvious that at least half the entropy was vibrational and possibly more. His argument was based on the strong corre-

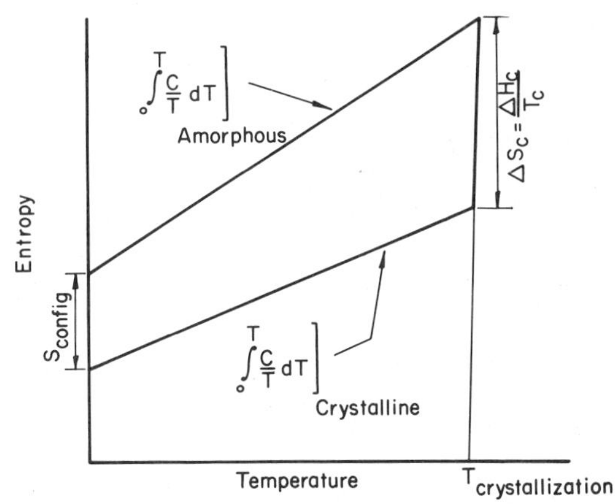

Fig. 1 Third-law determination of the configurational entropy of amorphous materials:

$$S_{\text{config}} = \left(\int_0^{T_c} \frac{c}{T}\, dT\right)_{\text{crystalline}} + \frac{\Delta H_c}{T_c} - \left(\int_0^{T_c} \frac{c}{T}\, dT\right)_{\text{amorphous}}$$

lation between increase in entropy and the work done, i.e., between increase in entropy and increase in energy associated with any shear strain put into the lattice.

(NOTE. Here Zener gave a proof which was challenged by Oriani. See Summary Agenda Discussion for the corrected version.)

Zener showed that $T_m \, \partial S/\partial E \cong \frac{1}{2}$, where T_m is the melting point, and $\partial S/\partial E$ is the change in entropy with change in energy due to shear strain. He then said that if the product were unity, the metal would melt. Therefore he felt it necessary to find a type of deformation which would raise this product from $\frac{1}{2}$ to 1.

Zener mentioned that in talking with Rudman, the point was developed that not much energy was required to put small wiggles into a dislocation line, that is, to form it into a coil. This would make it possible to have a value much greater than $\frac{1}{2}$. This notion reminded Zener of how tungsten wire coils for lamps are coiled upon themselves, making it possible to fill a lot of space.

With regard to the previous comments, Seeger thought a dislocation model for liquids was quite reasonable to explain properties such as viscosity. Several attempts had been made to do so, for example, first by Kochendörfer, and again by Seele. None has been worked out, however, to the theoreticians' satisfaction. One has to study the geometry and motion of partial dislocations which may account for the kinetic properties of a liquid. A dislocation theory of melting is an entirely different matter, as he had said in Melbourne 6 months before. Melting is a first-order phase transition. The free-energy curves for the liquid and solid phases are known to cross. The energy at the point of crossing is related to the latent heat of melting. It is important to note that a phase transition of the first order can occur only if the state stable at low temperature can, at least in principle, be continued to temperatures above the transition temperature. There are a number of examples where this is not so, for example, the order-disorder in β-brass or the ferromagnetic transition. There the order simply goes to zero at the transition point. One gets a transition of the second kind where the two curves just touch. But melting is a transition of the first kind. This means that any theory of vacancy melting is not right. One does not get the right kind of phase transformation. One gets vacancy concentrations up to 10^{-3}.

Seeger also had arguments against the particular dislocation models proposed. It is usual to draw long dislocations and consider their vibrations, he said. On melting, he thought these dislocations would break into small loops of varying sizes, and one would have to consider the vibrational states of these. Normally these loops are unstable, shrinking under their own tension. Whether these loops exist has been investigated by "up-quenching" techniques. This is done by bringing a solid very rapidly to near its melting point. There is then a deficiency of vacancies. These vacancies have to come from some place. The experiment shows that the resistivity of the wire increases in a few milliseconds if the temperature is kept constant. These experiments have been done in Munich and Urbana. The kinetics of vacancy formation can be studied, and then models considered to explain the experiments. The simplest model would be the creation of these vacancies by creation of Frenkel pairs. This is not a good model, because the creation energy is too high. The next guess would be to create dislocation rings. These would not be the dislocation rings talked about this morning which were glissile rings. The rings under discussion now would be rings with a Burgers vector perpendicular to the ring. The energetics would be the same. The rings could climb, and vacancies would be created. This has been proposed by Koehler, and discussed in detail by Baluffi. Seeger went on to say that this picture does not fit the facts, but what does are simply the dislocations inherent in the material which give up vacancies and which migrate long distances. One can go back and estimate how big the energies can be. It turns out that even near the melting point one would not get a sizable concentration of these loops. A third point made by Seeger was that the presence of the small dislocation loops would make the shear modulus go to zero. But, he said, Zener pointed out the shear modulus at the melting point becomes one-half its value at 0°K. This means, on the one hand, that half the entropy of melting arises in the way Zener and Rudman suggest. But, on the other hand, it indicates equally well that it would be well to look for the rest of the entropy in configurational ideas.

Oriani then referred to a remark made by Zener in the latter's introductory talk, that the entropy of a body is related to the uncertainty of location. Oriani said there are uncertainties of two kinds. First there is the uncertainty in the position of the atom about which it oscillates, and secondly the uncertainty of the oscillations themselves. We know two facts. One is that liquids have radial distribution functions which are very fuzzy. The other is that the Duwez frozen liquids have identical core distribution functions. He therefore maintained that the uncertainty of vibrations has essentially gone to zero. Therefore all that could give rise to an entropy contribution is an uncertainty of position.

Frank agreed, but asked how do you make this quantitative? There was no reply.

Enderby said there is another possibility, namely, that one can't separate the entropy of liquids into entropy of vibrations and entropy of disorder. Possibly what matters is how long an atom vibrates before it moves. Further, he has been talking to Duwez, and they have agreed that it was very dangerous to talk about his superquenched material as liquids at absolute zero. The disorder in liquids is of a different sort. The radial distribution function is a very blunt tool. It is an average and conceals many things. Both Enderby and Duwez looked at gold telluride. Duwez's sample was a semiconductor; Enderby's liquid was a metal. Obviously, the disorder cannot be the same in the two cases.

Seeger then proposed that the problem be resolved experimentally. That is, quench pure metal, and study the contents of defects by established techniques (X-ray scattering, resistivity, etc.). He was sure something would be learned about the defect structure, the object being to make the study close to equilibrium.

Duwez commented that Thomas and Willens had observed a sevenfold increase in the vacancy content in aluminum quenched from the liquid. There is room for more work.

Lomer returned to Enderby's concern that the separation of entropy into its two components might not be straightforward. He noted that inelastic neutron scattering can measure both the vibration spectra and the diffusion coefficient. One finds that the diffusion coefficient indicates that the atom vibrates 99% of the time and spends 1% of its time hopping. Presumably this has bearing on the question of balance between configurational and thermal entropy. But he wasn't prepared to say this meant a ratio of 99 to 1 for this balance. The rate at which the vibrational center of an atom diffuses in a liquid is really quite small compared to the vibration.

Frank remarked that Debye himself pointed out that the specific heat of water agrees very well with the law of Dulong and Petit. To this Lomer added that this was in accord with neutron-diffraction studies.

Seeger thought that the question of the relative periods of vibration of the number of jumps had nothing to do with how much the entropy of one kind differs from that of the other. Lomer replied that such information gives an idea of how much energy resides in the vibrational band.

Frank proposed that the discussion could now move to the subject of the electrons in liquid metals. Heine spoke about the elements mercury, gallium, and indium. From the pseudopotential one gets an understanding of three things and sees the relationship between them. First one gets an understanding of the peculiar structure of the metals in the solid state, and the distortions from cubic symmetry. Secondly, the low melting points are explained, and finally some anomalies in the X-ray structure factor. He suggested that the way to look at the problem was first to ask what is the difference in energy between hypothetical structures fcc, bcc, hcp, if the above elements could take up these structures. The formula available involves a term $\Sigma_g W(g) v^2(g) \alpha(g) f(g)$ where we sum over reciprocal lattice vectors a quantity $W(g)$ which is just the structure factor squared times the number of equivalent lattice vectors, times a quantity $v^2(g) \alpha(g) f(g)$ which includes the pseudopotential squared. It is now possible to plot on a linear scale (Fig. 2) the position of the reciprocal lattice vectors (properly normalized for atomic volume, etc.) for the bcc, fcc, and hcp lattices, the lowest sets all coming at about the same distance. For all three metals the pseudopotential passes through zero at about the same place on the diagram (Fig. 3). This means, automatically, that the band gap is very small in all these cases. There would be small energy differences between these structures, and the contributions of the band structure to the cohesive energy is small. Now if the X-ray structure factor squared for the liquid is plotted on the same diagram, its main peak occurs in the same region. The structure in the solid is distorted because the reciprocal lattice vectors can split, making possible bigger band gaps. Also for any of the structures considered, the energy difference between liquid and solid would be small. At low pressures the melting temperature is $T_m = \Delta U/\Delta S$ where ΔS, the change in entropy, is empirically seen to be a roughly constant quantity for all metals. Therefore if the energy difference ΔU is small, then T_m will be small. Finally the measured X-ray structure factors for liquids are about the same for all metals to a first approximation; but small anomalies do exist. Thus at g_0, the wave vector where the pseudopotential is zero, there is a little dip or wiggle in the case

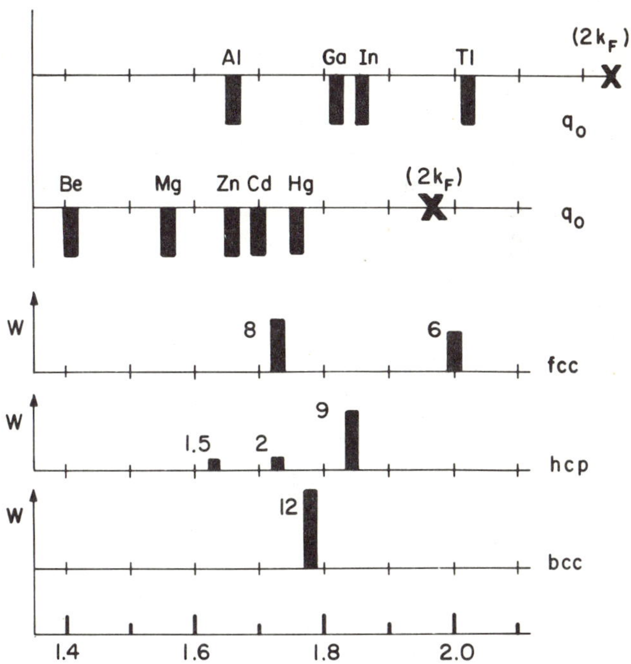

Fig. 2 Position of g_0 for several elements, and of the reciprocal lattice vectors with the weights $W(g)$ for the face-centered cubic, hexagonal close-packed (ideal c/a), and body-centered cubic structures.

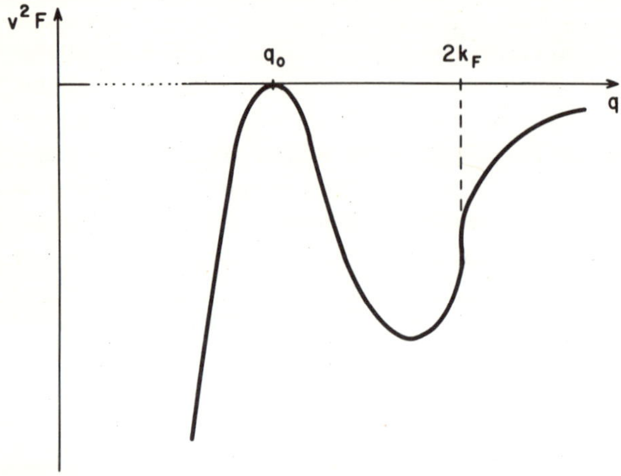

Fig. 3 The function $v^2(g)\alpha(g)f(g)$.

of gallium because the energy is lowered by the structure factor avoiding g_0. This is analogous to the distorted structure in the solid. In indium one gets a little shoulder at g_0. Heine thought the same was true for mercury as well. In tin there is a steepening of the edge. Thus, he believed that all the properties which he had discussed were related to one another and were related to the pseudopotential as well.

Frank asked whether it was possible to relate the wiggles in the structure-factor curve with any anomaly in the structure. Heine thought not, because if the potential made that much difference, the anomalies would be bigger. Frank then asked whether one could say what significance the wiggle or shoulder had in direct space. Heine replied that these merely meant some atoms moved closer together and some farther apart, just as in the solid.

Friedel commented on the nature of interatomic forces (the pair interactions in normal metals), first discussing the form of these interactions in normal metals. He said that in the liquid as in the solid it can be discussed by making use of perturbation theory. He added that we do not know pair interactions too well for two reasons.

1. Disorder exists in the liquid. Electrons therefore have finite mean free paths. As a result one cannot use the sort of interactions in perfect metals which die out as $1/r^3$, but rather a potential which dies out exponentially with the exponent dependent on the mean free path.

2. How to calculate interatomic forces is known if the volume is kept constant. It is only for shear movements that the calculation can be made easily. In liquids, however, there are local variations in density. How to calculate the forces reacting to those variations of density is not as well known. The only justification for using the same type of functions

in liquids as in crystals would be the assumption of fairly close-packed arrangements without large local fluctuations in density. From experimentally derived pair-correlation functions one could argue as Enderby has, that one can work back to the pair interaction. At the moment Friedel does not believe a good enough theory of short-range order in liquids is available to work back to these interaction forces.

Friedel then cited work of Mrs. Cyrot (Orsay) which showed this convincingly: If one works in the Born-Green approximation or the Percus-Yevick approximation, it can be shown that the asymptotic behavior of the pair correlation as measured, say, by neutron diffraction is proportional to an asymptotic form of the pair potential as long as its exponential decrease is not too fast (i.e., as long as the mean free path is not too short). Thus there is a direct proportionality between the two behaviors, but the coefficient of proportionality differs by a factor of about 10 between the two approximations.

Friedel also discussed Mrs. Cyrot's work on liquids of the transition metals, which consisted of a calculation of the form of the d band. Tight binding was used as in the solid state with the usual approximation that the whole wave function in the d band is a linear combination of atomic functions and that overlap between more than two atoms can be neglected. Mrs. Cyrot then obtains a formula for the density of states. $N(E) = \int \delta(E - E_0 - H_k)\, d_3k$, where E_0 is the average energy of the d states in the isolated atom, and H_k is the Fourier transform of H_r. H_r is the product of the pair-correlation function by an overlap integral of two wave functions on different sites: H_r = pair-correlation junction $\langle \Psi_i \mid V_i \mid \Psi_{i+r} \rangle$. The overlap integral is a function of the distance r between the atoms at i and $(i + r)$.

The expression is typical of the tight binding approximation. The correlation function prevents the atoms from getting too close to each other.

One then gets a density-of-states curve with a maximum value at E_0 and having a gaussian or Lorentzian distribution depending on the size of the atomic wave function and on the minimum correlation distance. Friedel did not know of any measurements available to check this but suggested that paramagnetic susceptibility of liquid transitional metals should be useful.

Enderby then made two observations. First, he said, it was necessary to remember that the oscillations in the $rg(r)$ curves for liquids are geared to a sharp Fermi surface. Calculations by Gaskell and March emphasize this. When the mean free path is finite, the Fermi surface is blurred. Thus a 20% blurring of the surface means a mean free path of a few angstroms. This 20% blurring is about what is obtained from positron annihilation experiments on liquid mercury at room temperature. He felt therefore that a proper theory of the stability of liquids would have to take account of the progressive blurring of the Fermi surface as the temperature is increased. The second point which Enderby made was that he was more optimistic than Friedel concerning the Born-Green approximation. Recent calculations of $g(r)$ seemed to indicate this. Friedel replied that he too had the impression that the Born-Green approximation was the better one, and that this was the general impression of the people doing machine calculations at Orsay.

Brewer was now invited by Frank to discuss how a chemist handles electrons in liquid metals. Brewer began his discussion by considering the H_2 molecule. When two hydrogen atoms with $1s$ electrons are brought together, one obtains a bonding state with a bonding energy of 4 ev. If one considers the same problem for excited $2s$ states, one finds that there are limits to the binding energy such that the attractive molecular state resulting from the $2s$ atomic state is highly excited relative to the ground molecular state. For the transition metals, 3 ev would be the maximum and 1 to 2 ev per electron-pair bond would be the normally expected bonding energy. From knowledge of the energies of various electron configurations of the gaseous atom and their respective bonding energies, he pointed out that one can calculate the relative order of stability of the electron configurations in the liquid metal. As an illustration, he listed the experimental energies of the low-lying electronic configurations of gaseous Mo and W as follows:

	Mo		W	
	Configuration	kcal	Configuration	kcal
	d^4sp	80		
	d^5p	74	d^5p	75
	d^6	73	d^4sp	56
	d^4s^2	32	d^5s	9
Gas ground state	d^5s	0	d^4s^2	0
d^5s condensed metal		−157		−203

He posed the question: What is the change in relative energies of the electronic configurations as molybdenum or tungsten atoms are brought together to form a liquid? The d^5s configuration (ground state of Mo or first excited state of W) has six unpaired electrons which can reach out to grab electrons of other atoms as they approach one another. The d^4s^2 or d^6 configurations with only four unpaired electrons available for bonding will be lowered in energy by approximately two-thirds of the bonding energy of the d^5s configuration and will be highly excited states in the liquid metal. The d^5p and d^4sp can bond as well as or even slightly better than the d^5s configuration, but their energies in the gaseous atom are so high that there just isn't enough difference in bonding energy to bring these configurations near the d^5s configuration. The only reasonable configuration then, according to Brewer, is the d^5s forming the metal at -157 kcal or 26 kcal per electron-pair bond (1.13 volts) for Mo and -212 kcal or 35 kcal per electron-pair bond (1.5 ev) for W. As one moves to the left of the sixth group in the periodic table, the known gas data show that the energy of the $d^{n-2}sp$ configuration steadily drops relative to the $d^{n-1}s$ configuration. For fifth-group metals, d^4s is considerably lower in energy than d^3sp, but in the fourth and third groups, both the $d^{n-1}s$ and $d^{n-2}sp$ configurations will be of comparable energy in the liquid metal. Brewer then posed the question of whether there are any other theories which predict contradicting results. (The question was addressed to Friedel.)

Friedel indicated two aspects of Brewer's theory which he did not like. First, he said, Brewer relates the lattice structure of transition metals to the number of *s* electrons, while saying that most of the cohesive energy arises from *d* electrons. This was surprising, for then the *d* electrons should play a role in the choice of lattice structure. His second objection referred to the width of the *d* band. There is direct evidence, he said, that the *d* band is quite wide, say 5 ev in width. X-ray spectra cannot be explained without taking the band to be of such width. Also the occupied part of the conduction band is wide. The energies involved in the kinetic effects under discussion are large compared to the binding energies measured. Therefore he doubted that it was possible to say which electrons would pair. Thus, for molybdenum, Brewer had indicated the difference between d^4s^2 and d^5p configurations was 32 kcal or approximately 1.5 ev, which is small relative to the actual bandwidth. Friedel felt any band formation would automatically mix these two states.

Lomer cautioned that one uses a different language in talking about the atoms and solids. The atomic potentials, however, are very similar in the two cases. In the sixth-group metals, whether we are talking about liquids or solids, we expect one or a few more *s*- or *p*-like electrons and the remainder *d*-like electrons. The configuration energy difference of 32 kcal discussed by Brewer must be compared to binding energies of approximately 150 kcal. The differences are small. Lomer said that this was just another facet of what Friedel had been saying. There is no major conflict in the effective configuration of electrons. That is, there are *about* $(n - 1)$ *d* electrons, and outside of these there is *about* one electron, for any conceivable condensed phase of the group VI transition metals.

Brewer then stated that all the metals under discussion crystallized into bcc which in some cases transformed to hcp, so that one can associate the structure with a particular mix of electrons, e.g., bcc with $d^{n-1}s$. Lomer said there are two ways to look at this. Atomic configurations give us a guide as to the number of d and s electrons to include, roughly. These must then be put into the solid, and the picture examined to see whether opinions must be revised. If observation shows that an sp configuration lies lower, this is an observation. However, theory cannot be used to disprove observations. He did not believe that a configuration obtained by calculation would depend on whether he started with bcc or fcc. He did not believe that structure has much effect on the balance between d and s electrons, or vice versa. Lomer went on to say there was no element in band calculations which leads to the correlation which Brewer wishes to establish.

Brewer again asked whether any theory would give other than bcc structure for the fifth- and sixth-group metals. Friedel replied that in the band theory the d band in the bcc structure is broad, with two well-separated parts separated by a strong minimum. When half filled, as in the fifth- and sixth-group metals, such a band would have a good stability. It is necessary, however, to do calculations for all other structures for all possible phases before concluding conclusively. Band theorists have not done this. Brewer then listed electron configuration and binding energy for rhenium:

Configuration	*kcal*
d^6p	104
d^4p^2	74
d^5sp	44
d^6s	27
d^5s^2	0

Both Lomer and Friedel said that band theory would not include many p electrons. Friedel added that Brewer's model assumed a large amount of energy available from d bonding. It would not be possible to deduce electron structure from atomic energy, because the interaction energies added were large compared to energies already discussed. Thus one could not tell how many s, p electrons there are.

Seeger noted that 10 years ago when band theory was not as popular as today, it was recognized that band theory did not give the full story for cohesion. And today, he said, it still did not suffice for the transition metals. As soon as d electrons enter the picture, s-d distribution and correlations must be considered. It is not enough just to say that band theory says this or that about bonding. Typical one electron-band theory cannot do this. Further he said that it was true that the width of the d band affects the binding energy, but the width of the d band is a measure of overlap of the wave functions when the atoms are brought together. Other methods exist for measuring the effect of d electrons on binding.

The fact that in the middle of the long period the metals are bcc, while at the ends they are fcc, can be explained by "orbital" techniques, and binding energies can be estimated. Band people, however, do not agree with these results either. He then noted that in fact one cannot distinguish between s, p, and d except at high symmetry points or $k = 0$. He concludes that each side is overplaying its hand. It would be more reasonable to conclude that neither picture is complete; each picture should be complemented by the other.

Friedel said that it is clear that cohesion in the middle of the transition series, whether in liquid or solid state, is large. Very few elements have such a large value. Therefore he says this large value which varies gradually through the series is characteristic and is related to unfilled d shells. Since the d electrons are involved in cohesion, they must be employed in discussion of structure. He did not think this entered Brewer's model. Brewer agreed this was a major effect.

Nowotny then discussed carbon which has $2s$ and $2p$ electrons. These must be hybridized to sp^3. The latter is higher in energy but makes more electrons available for bonding. This is quite similar to the situation under discussion, where more bonds are made available. Thus for tungsten maximum bonding is obtained with d^5s, and similarly for rhenium. He concluded it was a very interesting and useful concept.

Lomer agreed with Seeger that significant calculations of cohesion have not been made for transition metals. One must estimate the energy of an assembly of electrons. If five d electrons and one s electron are involved, in a spin-canceled-out state, whether we take the average over observed spectroscopic data as Brewer suggests, or calculate from wave functions, is neither here nor there. The reason for reaction to the Brewer model is that expressed by Friedel.

Brewer replied that nobody complains about the other end of the series. For example, in the case of CoAl and NiAl Hume-Rothery assigns zero valence, which means it doesn't matter how many d electrons are involved. There are several rationalizations: The Altman–Hume-Rothery paper points out that one s electron when hybridized with varying numbers of d electrons always gives the bcc structure. He said there was another chemical point of view which goes as follows. The d orbitals although very strong in bonding do not spread out very far, but the s and p orbitals do. This means that the s and p electrons are related to structure, while the d electrons which are more localized are related to cohesion, and not long-range order.

Friedel's reaction was that if the d electrons are far apart and participate in bonding, this bonding should be affected by distance and direction of bonding. Therefore, obviously the bonding of the d electrons will be affected by lattice structure. Brewer said it depends on how far-ranging they are. They need not be. For example in the transition-metal complexes, *d* bonding is localized between the metal ligands, and one gets very high bonding. Therefore it is not essential that the *d* orbitals be far-ranging. In other words, the rationale is that in all models the *s* and *p* electrons are related to structure, and although the *d* electrons play a role in cohesion, they are not involved in long-range order. If there is a good theory which says this rationale is wrong, then we should throw it away, but if the theory says neither yes or no, then we should continue with it.

Seeger stated that actually there is a theory discussed by Dehlinger in 1954 which attempts to explain the regularity in the structure of the transition metals in terms of *d* bonding and does much more than the usual band theory.

Heine now commented that as earlier stated, the width of the band can be thought about in terms of hopping of *d* electrons. In a certain sense, this is kinetic energy, and although not wrong, nevertheless it is not relevant to the discussion. The width of the *d* band is largely a question of forming bonding or antibonding linear combinations. The shape of the band is a reflection of whether there is overlap. Friedel indicated that this was just another way of looking at the problem. For example, it is well known that the difference in energy between bonding and antibonding states is related to the frequency of the jumping of an electron from one atom to another in a molecule. Heine continued that although, as Brewer said, there is no good theory in the sense of a complete theory of the structure in terms of *d* band, nevertheless there is an outline of a theory which studies the shape of the band as a function of the structure. Since bonding and antibonding states are spread over a range, one can say, let us throw so many in and see which has the lowest energy in the *d* band. If, when the calculations are carefully done, they disagree with experiment, then it is time to think of a new theory. So far the band theory is fully consonant with what is known about the transition metals.

Beck remarked that since the bonding in transition-metal alloys is largely due to the *d* electrons, and since it is present in the liquid state to almost the same extent as in the solid, perhaps it is not unreasonable to ask whether the *d*-band cohesion may be quite insensitive to crystal structure.

Friedel agreed that the differences in energies between various structures for the *d* bands are smaller than the total cohesive energy.

Brewer now raised his third question, which concerned the Hume-Rothery rules. He reminded everyone that up to an electron concentration of $3/2$, the bcc structure prevailed, and around an electron concentration of 2, hcp was found. But trouble arose with the transition metals. Hume-Rothery found it necessary to go to the group numbers. However, if one subtracts from the group number the number of *d* electrons, which will be close to five for most of the electron compounds in the middle of the transition series, one obtains the Hume-Rothery numbers. The same is true for other electron compounds such as the σ, α-Mn, or β-Mn phases. For pure transition-metal alloys, the only difference between saying the bcc structure extends to an average group number of 6.5 and saying the bcc structure extends to the average configuration of 5 *d* and 1.5 *s,p* electrons per atom is that one does not need two sets of rules and can use the same Hume-Rothery numbers for both sets of alloys. If non-transition elements are mixed with transition metals, none of the rule can be made to work unless the *d* electrons are counted independently of the *s,p* electrons. He cited a recent paper by Beck dealing with compounds having β-Mn phases. Beck found it necessary to assign a valence of 9 to tin. Subtracting 5, the group number for tin, from 9 gives 4. This means that in these compounds, the requirement for the transition metal with its half-filled band must be satisfied. Thus what is left of the *s* and *p* electrons combines with the *s* and *p* of the transition elements. As an example Brewer considered Nb and Ru which he said had d^4s and d^6sp configurations. When the two elements are combined, one would have a vacant orbital; the other an excess electron. In the alloy he would argue that the *d* configuration for the two atoms would change to d^5 producing three electrons of the *sp* type for every two atoms, and a resulting bcc structure would be obtained. Thus, he continued, all the Hume-Rothery rules would apply to transition metals if first one uses the concept of electron transfer to obtain the d^5 configuration. Further, here he said was an example where the *d* electron would not determine long-range order.

Heine emphasized that cohesive energy is just one number. He also said he was quite interested in work reported by Beck. The data indicated that in Fe-Al, the top half of the *d* band appeared to be pushed up while the bottom half was pulled down. Sense could be made of these results in terms of interaction with Al. In the case of Cr-Al, the specific heat drops to values comparable to those of alloys of non-transition metals like copper. It appears here as though the two bands are pushed apart.

Massalski granted the *d* electrons in alloys of noble metals are involved in a nonsensitive way. For alloys of low electron concentration, melting at say 1000°C, however, certainly much more bonding is involved than can be attributed to *s,p* electrons. He asked Brewer to explain how it was possible for an alloy of a given electron concentration to exist with as many as three structures.

Brewer replied that this doesn't happen except in cases where electron configurations exist with about the same energy, but the majority of transition metals have only a single structure.

Massalski said he was not referring to transition metals, but rather alloys of the noble metals. For example as mentioned in his talk, CuGa with an electron concentration of 1.4 could be found with bcc, fcc, or hcp

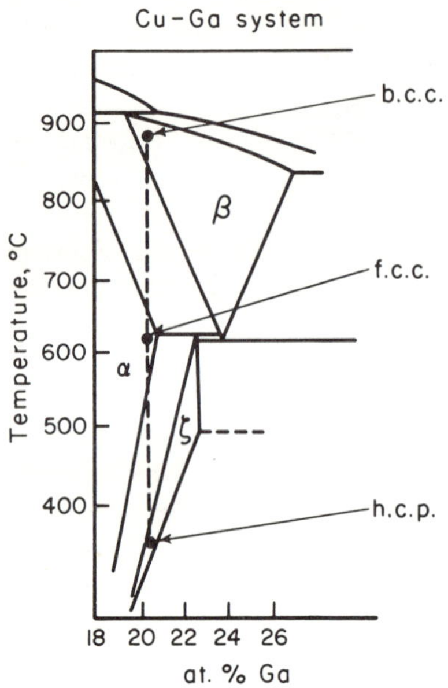

Fig. 4 A part of the copper-gallium system.

structures (Figure 4). The *d* electrons obviously contribute to all three structures in some structure-insensitive way. What do the *s* and *p* electrons do in Brewer's correlation?

Brewer replied that the *d* electrons rarely determine structure unless they decide how many *s* and *p* electrons are going to be left over. Thus a sequence of structures will be obtained, depending on how many *s* and *p* electrons are left over. When platinum is added to molybdenum the bcc becomes unstable owing to the excess electrons, and the next structure is hcp. This occurs because more electrons in the *d* orbitals makes them unavailable for binding, and the additional electrons have to be put into the *s,p* orbitals. This is how he established the compositional ranges and phases for these compounds. That is, he first determined the electron concentration in the pure liquid for each element, then mixed the liquids, making full use of *d* electrons and the remaining *s,p* electrons, as he has already discussed.

Laves observed that in discussions of phase stability one rarely discussed structures with coordination number 10, which does in fact give good sphere packing. For example, this would be obtained with the $MoSi_2$ structure if the two atoms were indistinguishable. Why does one always jump from coordination 8 to 12?

Brewer replied that the $MoSi_2$ structure is observed in some of the compounds of transition metals when the elements are at opposite ends of the periodic table where the size difference is great. In the middle of the table where the sizes of the elements are comparable, the coordination number is 8 or 12. Beck added that Frank had pointed out that the bcc structure represents very close packing, with a coordination number of 14. One may, therefore, conclude that this coordination number is also available at the middle of the transition-metal rows in the periodic table.

Lomer returned to a discussion of the Hume-Rothery rules. Every justification of Hume-Rothery rules rests on Fermi-surface interaction. The Fermi surface for Mo or W is well established. It is not like the Fermi surface for 1 or 1½ electrons of the *s,p* type. It cannot have the same stabilizing and unstabilizing influence as used in the noble metals which are the classical Hume-Rothery alloys. This could only happen if the *s,d* electrons split off to form a completely closed band lying far below the Fermi energy. This would, in turn, require spin pairing which is not observed. For every case the ordinary hybridized *s* and *d* approach used in band theory for Mo and W is fully confirmed by detailed Fermi-surface observation.

Massalski asked Brewer to use his correlations to explain the Hume-Rothery rules for alloys of the noble metals and to explain why it is only in the bcc structure that one finds a very narrow range of electron concentration while the other structures are flexible.

Brewer replied that the Hume-Rothery rules apply at the limits of the phases. They should be related to the Fermi surface at 40 to 50% technetium or rhenium in molybdenum or tungsten. The way the Hume-Rothery rules can be applied at the end of the table of transition metals can be illustrated with nickel-aluminum again. Pure nickel surely involves some *s* and *p* electrons. When nickel is surrounded by aluminum, these electrons are not used, that is, the promotion energy is not being paid for and the nickel must go to a d^{10} configuration. We have therefore d^{10} nickel plus aluminum with three *s,p*. This gives 1½ *s,p* per atom. Similar arguments apply to cobalt-aluminum alloys where again the Hume-Rothery rules apply.

Massalski did not think the arguments applied to alloys of non-transition metals and suggested a return to the discussion of CuGa where at a fixed concentration all three structures are available. Brewer replied that this would have to be looked at in detail. For more weakly bonded systems there are a variety of structures. Strongly bonded structures are needed to establish the close relationship between electron configuration and structure. Brewer indicated that he had been more interested in the cases involving transition metals.

Seeger returned to Heine's discussion to say that bonding and antibonding concepts actually came out of chemistry and, of course, have been used in *d*-band discussion. For the case of germanium, silicon, and diamond, it is agreed that the bonding comes from sp^3, but we still do band calculations. There is no difficulty—we can talk of homopolar bonding in some connections. Finally he repeated that there was a way in which *d* electrons could be seen to determine structure. For example in a cubic or hexagonal environment, the *d* electrons split up into different groups which determine the structure.

Friedel replied to Seeger, saying that in a tight-binding and band calculation, the lower part of the band involves bonding, while the higher is antibonding. In the lower part the electron density is reinforced compared to just adding of the electron density of pure atoms. The reverse is true for antibonding. It can be shown that if the *d* band is treated separately from any other band, it is the same band which gives the bonding and antibonding band. The situation is slightly different in *sp* bonding where because one starts with several orbitals one can build up separate bonding and antibonding bands. Thus one can obtain something looking like molecular bonds using *d* electrons, but if only *d* orbitals are used, it necessarily has metallic character.

Heine wanted to make the analogy between Si, Ge, and diamond more precise. Certainly it is possible to describe these with chemical-type sp^3. But band calculations also describe these properties as shown by the fact that many results of the band-structure calculations can be tied to experiment. Therefore the analogy is not fair; we cannot say we have free s^2p electrons in copper. Instead we must have one *s* electron. Much of our band calculation is tied down to experiment.

Pearson noted that all were agreed that in transition metals the *d* electrons were involved in cohesion. There was disagreement on the role of *s* and *p* electrons. However, Pearson felt there was now no major conflict, since Brewer did not demand correlation for the B group metals if cohesion is not strong.

SUMMARY AGENDA DISCUSSION

chairman **W. M. Lomer**
Solid States Physics Division, Atomic Energy Establishment, Harwell, Didcot, Berks, England

secretary **P. S. Rudman**
Metal Science Group, Battelle Memorial Institute, Columbus, Ohio

The chairman remarked that it seemed to him that the reason for having this conference was the state of depression of physicists with the rate of progress on the major problem of calculating the total energy of any solid system. Ideally one wants to calculate the heat of formation, and the entropy of pure phases and of solid solutions. The idealized calculation that is in mind is complete in the sense that electronegativity factors and size factors would emerge, simply as properties of the atomic potentials involved. If one knew the appropriate potentials, and how to make allowance for electron-electron correlation, there would be no need to introduce other factors. As time goes on, one is learning very much more about both the atomic potentials and the correlation energy estimation, and this formal pathway offers the prospect of a continually improving approach to the desired goal.

Nevertheless, the formal path is difficult, and the chemical/empirical thinkers try to improve things by injecting "commonsense" variables like size, electronegativity, and electronic configuration into the discussion, and most physicists keep trying out little shortcuts. All parties are setting out from a low-lying plain to scale a high mountain; the formal path lies up a tricky shoulder ridge, and is hard going from the start; the chemists' valley goes far into the mountains, giving many perspectives of the peak but never providing a real good climbing base; the physicists have not so much a valley as a set of little terraces where they play with one factor or another in isolation, and they greatly enjoy the game and get to know the boundaries of each terrace rather well. The formal path may still turn out to be the only one which runs from bottom to top; it is not often easy to join it partway up.

On the empirical side one sets out to collect and correlate data, with a view to determining patterns for predictive purposes, or to identify situations of general theoretical interest. Now the question arises: what shall one use as variables to codify the data—atomic radius, atomic volume, electronegativity, free-atom configuration, type of bond, space-filling ability; the list is long. Here the chairman was convinced one must learn from one's biological friends. The technique of multivariate analysis is well known, and it is well known too that no clear measure of the effect of one factor can be obtained if one is at the same time averaging over other significant factors. He did not believe we have yet had a serious enough statistical analysis; for example, the question of whether "size" should be represented by volume or by radius should not be discussed as two rival schemes; they should be tested *together*; if one is irrelevant, it will soon be shown in the results.

On the physics side, the procedure is very different. It is often considered desirable to investigate closely some particular term in the energy expression on the grounds that it is varying rapidly as a function of some structural parameter. The effects discussed usually depend on the outer electrons, not infrequently on the details of the Fermi surface (e.g., Hume-Rothery rules, magnetic structures, etc.). Then one quickly sees that discussion of distortion of structures, or their local stability, can often be managed, but usually comparison between distinct structures is less trustworthy because it rests to an embarrassing extent on assumptions of rigid atomic cores, etc.

The difficult formal path should perhaps be mapped roughly as the following stages:

1. Assume structure.
2. Evaluate self-consistent fields for the component-free atoms.
3. Superpose these potentials to get band structure, and use the resulting wave function to get self-consistent solution for solid.
4. Evaluate correlation energy correction.
5. Evaluate total energy.
6. Repeat for all suspected rival structures.

The tests for local stability of any assumed structure may often be easy to carry through from stage 3 without waiting to do stages 4 and 5.

Empirical checks on the suitability of the atomic potentials, on the elastic constants and phonon frequencies, on the lattice parameters, etc., may be very important for checking intermediate stages, and improving prospects of success. The problems of thermal entropy estimation can be severe, for one really needs all the vibrational spectrum, and also the defect structure of the phase.

In response to Oriani's criticism of Zener's thermodynamic treatment of the previous day, Zener re-presented it, taking volume changes into account:

$$N = \text{torque} \propto \mu\phi v$$

$$\phi = \text{angle}$$

$$\mu = \text{shear modulus}$$

$$d(E - TS) = N\,d\phi - S\,dT$$

$$\left(\frac{\partial S}{\partial \phi}\right)_T = -\left(\frac{\partial N}{\partial T}\right)_\phi$$

$$\frac{T_m \partial S}{\partial W_0} = -\frac{T_m \partial_\mu}{\mu_0 \partial T} - \frac{\Delta v}{v} = 0.5 - \frac{\Delta v}{v}$$

Since from 0°K up to the melting point $\Delta v/v \simeq 1/10$, the volume change effect is significant but relatively small.

After reflecting on the previous day's discussion of dislocations and liquids, he thought of a configuration that he felt was the crux to melting.

The physical basis of melting is the entropy associated with surface vibrations. Rayleigh has discussed the waves on the surface of the earth. These penetrate a distance comparable to the wavelength. The geometric mean of frequency of the surface waves is less than the geometric mean frequency of the interior waves. Thus there is an entropy associated with the surface. This entropy per unit area, S_s, is given by $S_s = -\partial\gamma/\partial T$, where γ = free energy per unit surface area. Surfaces can be internal as well as external. Let us consider a crystal with a surface across which there is no restoring force to shear. Then here also there will be surface waves penetrating a distance comparable to the wavelength. Now let us see how we can form such a surface. Take an fcc metal like copper, and shear it (111) $\langle 11\bar{2} \rangle$; if we shear it enough, we will obtain a bcc configuration. If we plot the energy at 0°K as a function of the shear, we will obtain a periodic potential and, by approximating it by a sine curve and from the elastic modulus, obtain the curvature at the minimum. For copper this amplitude is about 1 kcal/mole.

If the energy of formation per atom at 0°K is ΔE and if the entropy per atom due to the interface vibrations is ΔS, then one will form such a surface at a temperature T, such that $\Delta E - T\Delta S = 0$. The surface is essentially a dislocation whose core has been spread out. If at the melting point we plot the free energy of a dislocation per atomic plane as a function of the width of the core, we should obtain a curve of the following form (Fig. 1).

The advantages of this viewpoint are:

1. It is simple.
2. One can calculate ΔS precisely, simply from the elastic moduli.
3. It predicts that if one would carefully prepare a crystal and heat it parallel to a (111) plane to just above the melting point, a series of internal surfaces spaced several atomic spacings apart would be obtained, while maintaining lattice order between the surfaces.

Seeger remarked that a similar model had been proposed several years ago by Richter and Breitling[1] which says that a liquid of an fcc metal consists of such (111) layers. The interpretation of the experiments is to some extent controversial. In view of Zener's suggestions it would seem of interest to look at the experimental evidence used by Richter and Breitling.

Frank commented that Zener has given another example of a description of a liquid as the breaking down of a solid by introducing defects of various kinds, in this case, by introducing a very large number of stacking faults by splitting dislocations to infinite width at some critical temperature. But the melting point is a first-order transition. Nevertheless the energy of solid must at some temperature go funny for some reason, as at point A in the Fig. 2, since it is hard to imagine the free-energy curve for the solid continuing to infinite temperature. One will reach a temperature when dislocations of high density are a proper component, when the vacancy content will become very large, when the stacking-fault content will increase, and if one has thought of the one which is most important, then one may indeed find a proper critical temperature at which something happens homogeneously to the existence of the solid phase. However, this temperature will be point A in the sketch and thus greater than the melting point.

Friedel noted that the vibrational entropy associated with Zener's surfaces appears to be equivalent to Rudman's mode of dislocations, in this

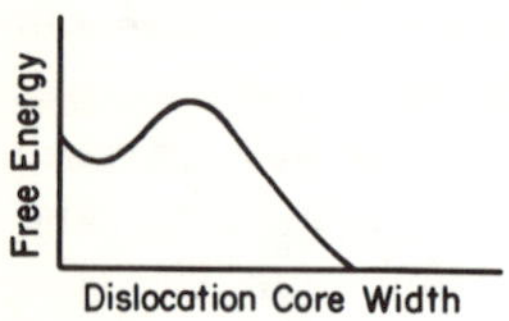

Figure 1

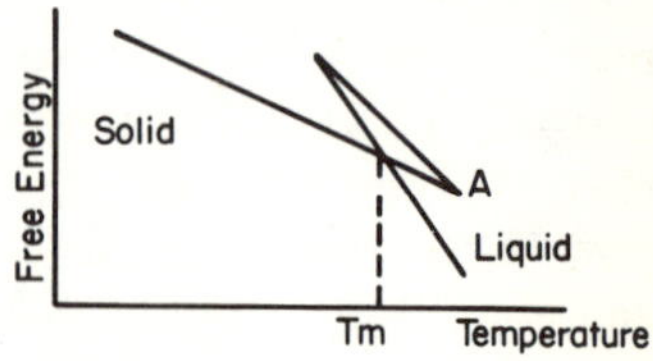

Figure 2

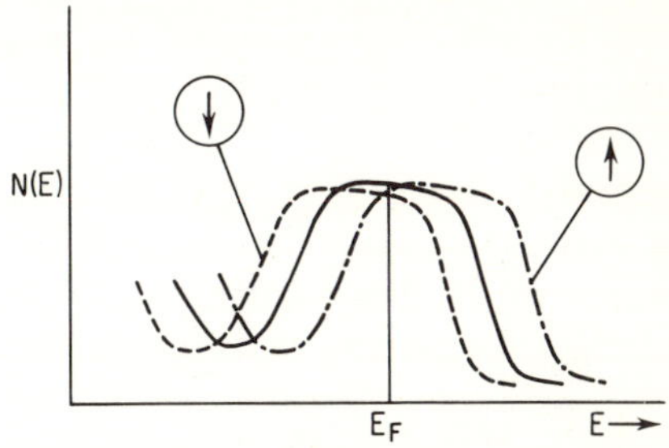

Figure 3

case partials, vibrating not as strings but relative to each other; however, Zener disagreed: the free-surface concept is not equivalent to a description in terms of partials. The low shear surface envisaged is due to a spreading of the core width.

Lomer suggested that the discussion be generalized to include stabilization by other defects, interstitials, and so forth. McQuillan commented that stabilization by hydrogen only occurs in phases of the early transition metals. The increased bonding energy at compositions below 1 a/o is roughly of the order of 10 kcal/mole of hydrogen. Even though it is only 1% of this that must be taken into account, this is still sufficient to give a stabilizing effect. However, it is not a large energy, and only in cases where the relative stabilities of competing phases are close can hydrogen stabilization occur.

Hoch said that titanium, vanadium, and niobium oxides with the NaCl structure have large vacancy concentrations. In these structures vacancies probably play a major role in phase stabilization. The energy gained by promoting conduction and band formation more than compensates the energy required to remove an atom from the lattice. Brewer remarked that it has been found by Rudy that the carbides M_2C ($M = V, Nb, Ta$), which occur at stoichiometry at low temperatures, separate into two nonstoichiometric phases at high temperatures. This is apparently a case of configurational entropy stabilization of the nonstoichiometric phases. The role of vacancies in phase stabilization was also pointed out by Friedel, who noted that nonstoichiometric FeO, which is an insulator rather than metallic, occurs by Fe vacancy formation. At least for some heat treatments, the vacancies assume an ordered arrangement with periodic antiphase domains. This is a sure sign that vacancies have a long-range interaction.

At this point, Lomer discussed the local stability of lattices, to pick up together a few points which have been implicit in some of the discussion. Perhaps the best-known instability is the ferromagnetic one.

Figure 3 shows a density-of-states curve for an unmagnetized metal, with the Fermi energy lying in a peak region. Suppose a small unbalance between spin-up and spin-down states is produced, giving a magnetization

M. Then the spin-up band energy is perturbed upward by some amount αM, and the downspin by $-\alpha M$. It is easy to see the unbalance of spins is then $2\alpha M \times N(E)$, and, further, that if this is greater than the initially assumed number $n = M/\mu$, the paramagnetic situation is unstable. The criterion for local instability is $N(E) > 2\mu\alpha$.

A very similar argument may be brought to bear on the problem of tetragonal distortion of a cubic lattice. Suppose that there is a well-defined peak in the density-of-states curve at the Fermi surface, and that a distortion ϵ of the crystal causes that component of the peak due to one set of crystal directions to increase in energy by $\alpha\epsilon$, and that due to another set to decrease by $\alpha\epsilon$. Electrons are then transferred from the raised peak to the lowered one, with a net saving of energy corresponding to a reduction in the elastic energy of the imposed distortion. It is quite possible for the effect to produce instability, i.e., a negative elastic constant, and it has already been referred to by Friedel apropos the V_3Si phase change.

The change in Fermi energy of the electron gas takes the form

$$dE_F = \frac{1}{N}\frac{dN}{dE}(\alpha\epsilon)^2$$

The normal elastic energy is given by

$$W = \tfrac{1}{2}G\epsilon^2$$

where G is the appropriate modulus.

Instability comes about if

$$\frac{1}{N}\frac{dN}{dE}\alpha^2 > \tfrac{1}{2}G$$

Thus high density of states, or rapidly varying density of states, may well cause structural instability, leading to homogeneous distortion.

Next Lomer discussed spatially inhomogeneous instability—the problem of the frozen-in phonon or spin-density wave. Here one considers the lowering of the energy of the system due to the introduction of a wavelike modulation of the lattice, with a wave vector $\mathbf{q}$. One may illustrate the argument by a new example which follows on from Pearson's talk about ordered β-brass. The central section of the Fermi surface in the repeated reduced zone scheme is shown by calculation and by experiment to be rougly as in Fig. 4*a*. The (E vs. k) plot along a (100) direction is shown in Fig. 4*b*, and it is seen that the introduction of a modulation vector of size $\mathbf{q} = \boldsymbol{\tau}/3$ couples states spanning the Fermi surface, allowing a lowering of energy of the occupied levels. The total lowering of the energy by the introduction of a periodic perturbation is greater if large areas of Fermi surface are parallel, and if the energy gradient normal to the Fermi surface

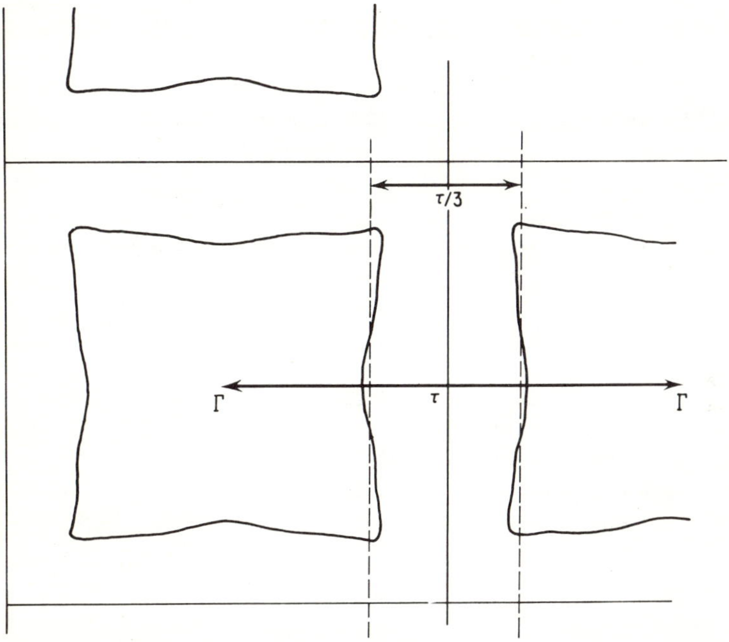

Fig. 4a The central section of the Fermi surface of ordered β-brass in the repeated reduced zone scheme.

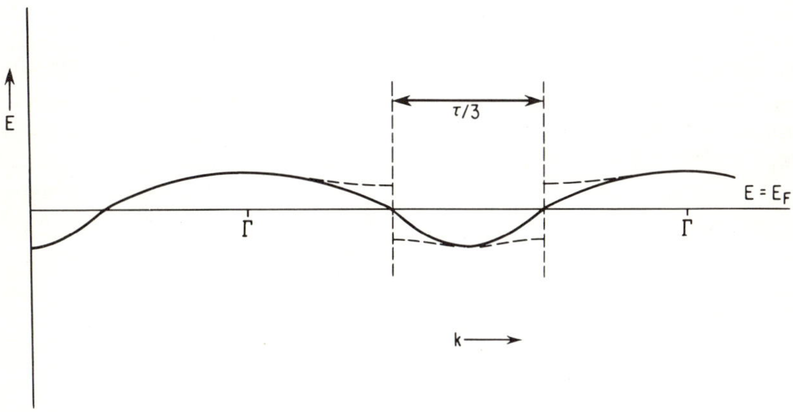

Fig. 4b E versus k along a [100] direction in ordered β-brass.

is small. The β-brass case satisfies this well, and it would seem that the γ-brass structure may be regarded as the one selected to maximize the effective threefold superlattice potential without discussing the complexities of the Jones-zone planes (411) and (330). The expression for the decrease in energy of the electron gas is

$$\Delta E = |V_q|^2 \int |\operatorname{grad} E_q(\mathbf{k})|^{-1} \frac{f(\mathbf{k}) - [1 - f(\mathbf{k} + \mathbf{q})]}{E_q(\mathbf{k})} d^3\mathbf{k}$$

where

$$\begin{aligned} V_q &= \text{modulation matrix element} \\ E(\mathbf{k}) \text{ and } E(\mathbf{k}+\mathbf{q}) &= \text{electron energies at } \mathbf{k} \text{ and } \mathbf{k}+\mathbf{q} \\ E_q(\mathbf{k}) &= E(\mathbf{k}) - E(\mathbf{k}+\mathbf{q}) \\ f(\mathbf{k}) &= \text{Fermi function for } E(\mathbf{k}) \end{aligned}$$

Such an expression may be useful for some circumstances, but correction for correlation and exchange terms is really needed; the sum of the energy parameters is not equal to the energy of the system. This total energy correction is treated for the spin-density wave in Cr.[2]

Nevertheless, the expression shows that an inhomogeneous distortion may be favored if a wave vector $\mathbf{q}$ can be found for which $E_q(\mathbf{k})$ is small for many $\mathbf{k}$ values near the Fermi surface.

Heine recalled that in his paper* he had mentioned that in the α- and β-Hg and the Ga structures there is a reciprocal lattice vector within 2 to 3% of $2k_F$, and this was explained in terms of his formalism. This particular feature of the structures may also be understood from Lomer's view of a frozen-in phonon with wave vector spanning the Fermi surface.

Friedel warned that the way of looking in reciprocal space is very pictorial and helpful in many ways, but it has some dangers.[3] Firstly, if one only looks at what happens at the Fermi surface it is equivalent, if one talks in terms of pair interactions, to say that one is looking at asymptotic behaviors of pair interactions. Therefore, if one wants to look more carefully in real space, this means looking at interactions at short distances; then one must look at the effect of the perturbations on all the density of states, not just at the Fermi surface. This means that Brillouin-zone limits far from the Fermi surface cannot be neglected. Secondly, as long as the magnetic or the shear distortion is small, it is strictly equivalent to talk in terms of Brillouin-zone effects in reciprocal space or pair interactions in real space. This point was made by Pick and Blandin[4] for the nearly free electron, but it is probably also true in all cases. But if one considers pair interactions in real space, as pointed out by Blandin,[5] one has the product of the real, clothed pseudopotential and the bare pseudopotential.

* See also V. Heine and D. Weaire, to be published.

One must not take the clothed pseudopotential squared. This would make an error in counting twice the Coulomb interaction between the screening electrons. So every time one has a Coulomb problem, that is, a distortion, a nonmagnetic problem, one must be careful of this kind of correction which does not appear if one just adds up the Hartree states in reciprocal space. Thus for nonmagnetic, spatial distortions the reasoning in reciprocal space must be corrected, and the correction is as large as the term one gets in reciprocal space for the nearly free electron. It is not so large in tight binding approximation.

Lomer remarked that this emphasizes again that in trying to get from a band model to total energy, one must accurately correct for correlation terms which are averaged out in the elementary and zero-order approximations. In agreement with Friedel, reciprocal space perturbations give only a pictorial method, and it is not adequate for calculating the strength of interaction, the total lowering of energy by integrating over such curves as Lomer has given above, except in very favorable circumstances which may occur in magnetic problems. Nevertheless, the general mode of thought forms a most useful guide to possible instability of simple structures against superlattice formation. Brewer stressed that this possibility of large correlation effects which cannot be accurately calculated is just what persuaded him to approach the problem from the point of view which he has adopted.

Lomer, in order to continue on this subject of Fermi-surface stabilization, which may well be an important part of the electron-compound story, invited Beck to present the results of some of his studies on electron phases.

Beck said that electron phases formed by transition elements, as those formed by the noble metals, are characterized by variable stoichiometry. Figures 5 and 6 show that, with components of different electron concentrations, the composition of α-Mn, and σ phases in various binary systems

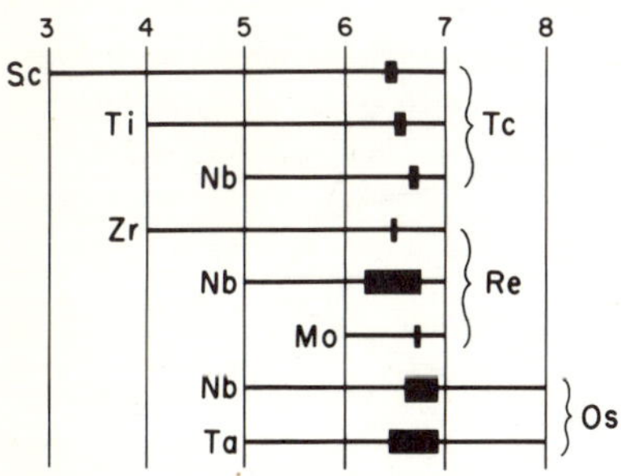

Fig. 5 Binary α-Mn phases versus electron concentration. [*From Bardos et al., Trans. AIME* **236:**40 (1966); *copyright AIME.*]

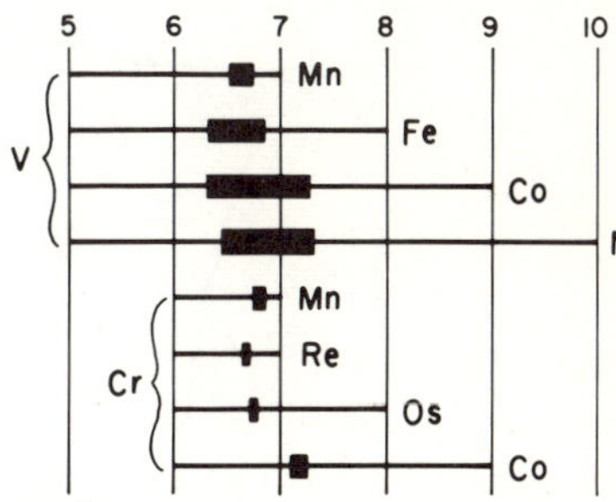

Fig. 6 Binary σ phases versus electron concentration. (*From Bardos et al., loc. cit., copyright AIME.*)

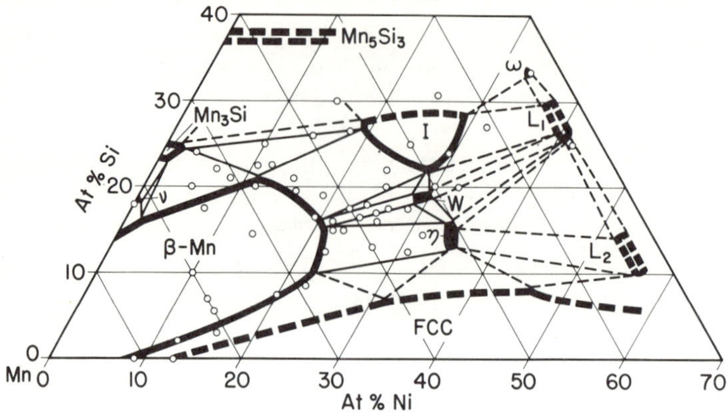

Fig. 7 Manganese-rich portion of the 1000°C isothermal section of the Mn-Ni-Si system. (*From Bardos et al., loc. cit., copyright AIME.*)

shifts in such a way as to correspond approximately to constant electron concentrations. In addition to σ- and α-Mn, a number of other electron phases are now known to occur in alloys of transition metals. Among these are the P-, R-, μ-, and β-Mn types. Common to all these structures is the absence of a high degree of ordering, which would prevent appreciable variations in composition. Another general structural feature of these phases, established through the work of D. P. Shoemaker and of J. Kasper and F. C. Frank, is the close packing in slightly distorted tetrahedral arrangements, which results in high coordination numbers, ranging from 12 to 16. All the characteristics discussed are, of course, consistent with the

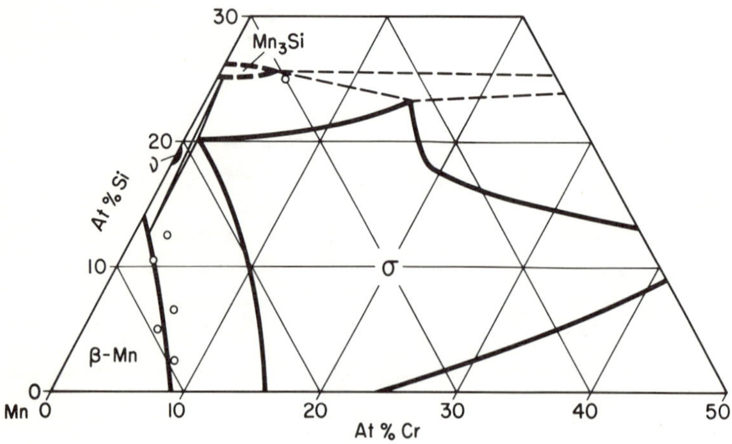

Fig. 8 Manganese-rich portion of the 1000°C isothermal section of the Mn-Cr-Si system. (*From Bardos et al., loc. cit., copyright AIME.*)

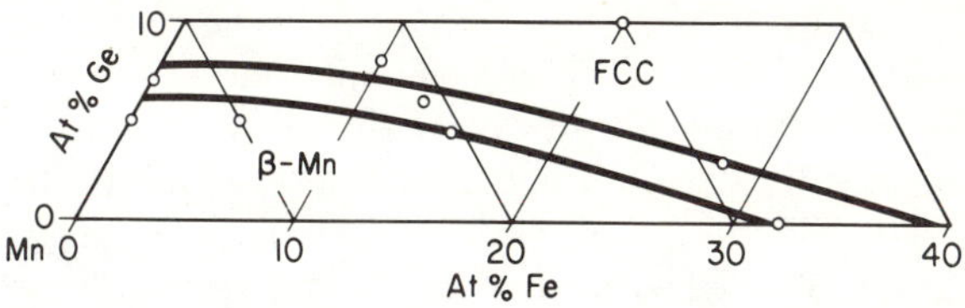

Fig. 9 Manganese-rich portion of the 1000°C isothermal section of the Mn-Fe-Ge system. (*From Bardos et al., loc. cit., copyright AIME.*)

predominantly metallic nature of these phases, the stability of which depends primarily on minimizing the kinetic energy of the electrons.

In a recent investigation[6] the stability of the β-Mn phase was studied in isothermal sections of a number of ternary systems of Mn with another 3*d* metal and Si, Ge, or Sn. The β-Mn phase was found to form elongated phase fields in the systems with Si and Fe, Co or Ni, extending approximately in directions corresponding in each ternary system to $e/a = 7$. An example of such an elongated phase field is shown in Fig. 7. In the systems with Si and Ti, V, or Cr the β-Mn-phase field was found to be cut off and thus restricted to the Mn corner. This is shown in Fig. 8 for the Mn-Cr-Si system. This phenomenon is also consistent with the stability of β-Mn along $e/a = 7$, if one assigns to Si a value of $e/a = 4$. In the corresponding ternary systems with Ge or Sn, the reverse of the above relationships was found. In the Fe-, Co-, or Ni-containing alloy systems the β-Mn phase is cut off as shown in Fig. 9 for the Mn-Fe-Ge alloys. On the other hand, in the Ti-, V-, or Cr-containing alloy systems the β-Mn phase forms elongated fields, as seen in Fig. 10 for Mn-Cr-Sn. That this

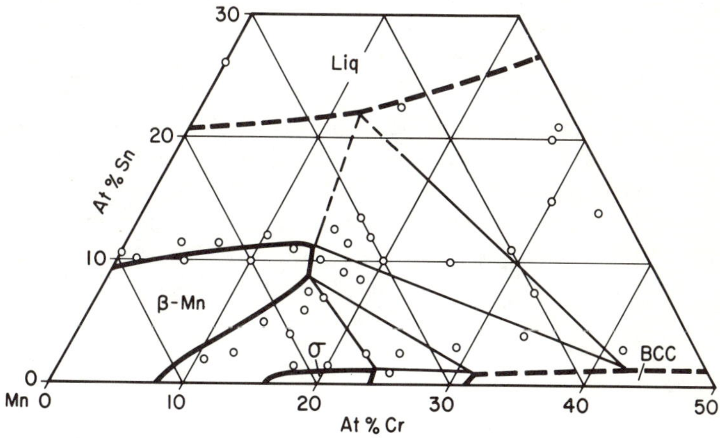

Fig. 10 Manganese-rich portion of the 1000°C isothermal section of the Mn-Cr-Sn system. (*From Bardos et al., loc. cit., copyright AIME.*)

difference in alloying behavior between Si on the one hand and Ge or Sn on the other is not due to the differences in atomic size can be readily seen by comparing, for instance, the alloy systems formed by Fe with those formed by Cr, two metals of nearly equal atomic size. All the observed facts, including the extension directions of the elongated β-Mn phase fields in the systems with Ti, V, or Cr, can be consistently interpreted in terms of electron concentrations, if an effective value of $e/a \simeq 9$ is assigned to Ge and Sn. Although this value may at first glance seem very unlikely, a reasonable interpretation of it may be given.

A value of $e/a = 4$ for Si in these transition-metal-rich alloys, in which the bonding electrons must be predominantly *d*-like, may be interpreted to imply that Si contributes all its valence electrons to the bonding band and that the *d*-like bonding electrons are not excluded from Si sites in the alloy. Alternatively, one may assume that Si contributes only two of its electrons to the bonding band, and that two of the seven electrons per atom in this band are excluded from Si sites. In the first instance one would have to assume that the normally unoccupied *d* states of Si in the alloys considered are lowered sufficiently to be accessible to all the electrons in the band. In the second instance a more moderate degree of accessibility and, hence, a lesser amount of lowering of the *d* states of Si should suffice. For Ge and Sn the separation between the energy of the unoccupied *d* states and the Fermi energy is less than that for Si and, thus, in Ge and Sn one might expect these states to be lowered and become accessible to the *d* electrons of the band more readily than in Si. However, because of the presence of completed *d* shells in Ge and Sn, and the consequent requirement of orthogonality, which in effect represents a repulsive potential to *d* electrons, the 7 electrons per atom of the band are excluded from the sites of these atoms. If, in addition, they contribute two of their valence electrons to the band, then their effect on the electron concentration per transition-metal atom will be equivalent to that of a transition metal with $e/a = 9$.

The use of electron-concentration values based on the average number of electrons outside of closed shells, instead of on ordinary chemical valences, might be justified on the ground that they represent a framework in terms of which not only the condition of stability of electron phases, but also magnetic saturation data in general (Pauling-Slater curve) and electronic specific-heat information can be adequately described.

Nowotny asked Beck what his method of assigning valences was, stressing the need for a uniform system of valences. Beck replied that he uses, as does Hume-Rothery, the group number (except for the Fe, Co, and Ni groups).

At this point, Friedel reported on some work that has been going on for some years at Orsay on rare-earth metals.[7] The rare-earth metals can mostly be described as trivalent ions with a partially filled *f* shell embedded in a conduction band with 3 electrons per atom which in the first approximation behave as free electrons. One knows that they are not really free electrons, but more like the beginning of a transition-metal series with large d character. But, from the study of magnetic coupling between f shells via the conduction electrons, one knows that the free-electron approximation is useful. This f-shell interaction via the conduction electrons, the so-called Ruderman-Kittell coupling, influences the lattice structure. Consider primarily the later rare earths starting at Eu and Gd. For these one knows from the study of magnetic coupling, without going into details, that one can picture the f shells as having an exchange interaction with the conduction electrons. Each f shell polarizes the spins of the conduction electrons, and this polarization is felt by the f shells of neighboring atoms. One can write the coupling energy of one f shell with the conduction electrons, as

$$\text{Coupling energy} = -\Gamma(g-1)Js_e$$

where Γ = exchange constant

g, J = for f shell, to take into account the orbital moment

s_e = spin of the particular conduction electron

From measurements of resistivity and the paramagnetic Curie temperature, in say Gd, one obtains and one can also fix the effective mass required to describe the conduction electrons. This gives a rather large effective mass of the order of 3. Then one can ask if one ferromagnetically aligns all the f shells, what will be the magnetic energy associated with this coupling. The energy is given as

$$E_0 = \frac{9\pi Z^2\Gamma^2(\phi-1)^2}{4V^2E_F} J(J+1)\sum_n F(2k_F R_{0n})$$

where Z = conduction electrons per atom

$$F(\rho) = \text{Ruderman-Kittel coupling} = \frac{\rho\cos\rho - \sin\rho}{\rho^4}$$

Everything is thus known, and one can compute this energy. One can also convince oneself that all the actual magnetic alignments found in the rare earths are, to a good first-order approximation, accounted for by this equation. This very simple formula allows one to understand a number of observations in the rare-earth metals.

The first is the crystal structures. Eu, to fill its half-full 4f shell, is divalent rather than trivalent like most of the rare earths. Thus the equation just given, which involves the Fermi wave number k_F, is slightly different for Eu, and one finds that the bcc is much more stable than either the fcc or hcp structures. And indeed Eu is bcc while most of the other rare earths are hcp. For the trivalent metals one finds that it is the close packed and not the bcc which are stabilized by the magnetic term. Actually, from this purely magnetic term the fcc is slightly more stable than the hcp, but the difference is not large, and can certainly be accounted for by a term due to the conduction electrons, as in (nonmagnetic) lanthanum.

If one goes into more detail about hcp Gd, one can explain by this magnetic term two-well known effects:

1. c/a anomaly
2. Dilatation anomaly

All the rare earths have at high temperature a $c/a \simeq 1.57$, but at low temperatures there is a deviation, becoming larger as Gd is approached. It can be explained numerically by considering that the magnetic term is more stable for a different value of c/a. Of course at high temperatures where all magnetic coupling is suppressed, the magnetic-interaction term disappears, and so does the c/a anomaly.

In the same way one knows from experiment that the strength of the magnetic coupling, in say Gd, increases with increasing volume. Thus the contraction observed in rare earths with increasing temperature, increasing as Gd is approached, is similarly quantitatively explained by the magnetic-interaction term.

Now consider cerium which is anomalous in that it has two fcc phases of different lattice parameters. At low temperatures the transition is first order, but at higher temperatures and pressures the first-order transition disappears. The transition is also a magnetic transition in that only one of the phases has a net magnetic moment. This has been studied at Orsay by Coqblin with the idea that when the f shell finds itself in such a position that the f states lie near the Fermi level, one has resonance effects that give rise to virtual bound levels, a broadening of the f state; and, in principle, the filling of the f state could be progressive, as indeed it probably is at high temperatures and pressures. If one starts filling such an f state, what is the criterion for a magnetic moment to appear? This has been studied by several people for transition-metal d shells, and here usually there is only a magnetic moment from spin, and thus one finds that the onset of the moment is always a second-order transition. But Coqblin finds that the condition for the onset of orbital moment, by a progressive filling of the f shell, leads to a first-order transition: at least at low temperatures, the free-energy function is such that one can go abruptly from a state with an empty f shell to a state with a finite number of f electrons. This is a peculiar property of magnetic moments with an orbital contribution.

Wallace noted that the samarium structure, which is based on the hcp structure but with a repeat distance of $9c$, is formed in rare-earth alloys with certain axial ratios. He asked Friedel whether the magnetic interactions just described explain the origin of this structure and its relationship to c/a. Friedel answered that samarium is one of the early rare earths and thus has a large f shell because the nuclear charge is smaller. This large f shell complicates matters because it sees the lattice potential directly, and this direct effect may well override the indirect effect just considered in determining structure in the early rare earths. The early rare earths have not been studied at Orsay.

Lomer remarked that there are now some excellent band calculations for trivalent hexagonal material, which is the basic structure for most of the rare earths. The work of Loucks* at Iowa State University and the researches of the Slater group at MIT† have given a Fermi surface to work with which is very different from the free-electron one. The Fermi surface shows many $\mathbf{q}$ vectors of the kind which Lomer noted earlier could stabilize modulated structures; the magnetic structures of some of the more complicated rare earths are clearly to be explained in terms of these Fermi surfaces, not free-electron spheres. That simply means that there would be a more complicated function F to put into Friedel's formulation and that most of what he said in qualitative terms will be maintained, but there will be differences requiring further investigation.

At the risk of being repetitive, Lomer felt it necessary to say again that empiricism is a very demanding discipline. Work on many partial correlations of phase-stability data has really got only some limited distance toward identifying an adequate set of factors to correlate against. Much of the difficulty arises from two quite different causes. One is the reluctance to use proper multivariate analysis techniques—it should by now be clear that no one or two factors in isolation is good enough. The second is that there are so few data on how the relative stability of phases varies. One could hope to see far more pattern in the results if, for example, one knew that where σ phase is stable in Fe-Cr, bcc is still much more stable than, say, fcc. The tremendous difficulty of getting enough quantitative data to reveal patterns undoubtedly slows progress.

The question of defining multivariate parameters is quite difficult, but it must always be better initially to include too many rather than too few. For example, one does not have to choose between atomic radius and atomic volume as representations of size; it is perfectly legitimate to use both. Equally, valence, group number, and number of *s-p-d* electrons should all be tried *simultaneously*. Throughout Hume-Rothery's work one will find reference to the importance of the *joint* operation of two factors

* T. L. Loucks, *Phys. Rev. Letters* **14**:693 (1965).

† For example: A. J. Freeman, J. O. Dimmock and R. E. Watson, *Phys. Rev. Letters* **16**:94 (1966).

to produce abnormal stability ranges, etc. One must be careful not to arrive at more than 92 parameters, or one will be correlating just against the chemical element. Nevertheless, it is Lomer's belief that about ten variables might suffice to codify the subject to a very great extent, and might have very high predictive value. There is no guarantee, of course, that it would yield a starting point for theory.

Kubaschewski asked whether the theoretical physicists could choose some particular alloy systems where the experimental data that can be obtained would be of critical importance, and what are the measurements that they require. On the other hand, perhaps the experimentalists and the theoreticians are so far apart that it does not matter to the latter what the former measure.

Jaffee suggested that a sensitive test of the Engel-Brewer theory would be to see how well it predicts stabilization at an allotropic transformation. Titanium transforms from hcp to bcc, that is, by Engel-Brewer theory from $4d^25sp$ to $4d^35s$.

The interstitially dissolving elements would be expected to ionize and contribute *s* and *p* electrons, and thus would be expected to stabilize the hcp phase. This is found for carbon, oxygen, and nitrogen. It is not found for hydrogen, which strongly stabilizes bcc titanium, and this must be considered a failure of the Engel-Brewer correlation.

Substitution alloying elements have different effects. Of the normal metals, aluminum stabilizes hcp in accordance with the Engel-Brewer correlation. However, tin initially slightly decreases the transformation temperature and can be considered to have little effect. Silicon stabilizes bcc, which is an *exception* to the prediction. Zirconium, with the same electronic structure as titanium, would be expected to have little effect, but in fact lowers the transformation range. All transition metals stabilize bcc, in accordance with the Engel-Brewer prediction. Thus although the correlation does a pretty good job of predicting the effect of alloying on the hcp-bcc transformations, there are notable exceptions.

Nowotny developed the point made by Jaffee. The effect of alloying elements on the transformation temperature in titanium and other metals can often be considered in terms of the structure of the next compound. Thus in TiC, C is surrounded by an octahedron, and α-Ti more conveniently accommodates C in an octahedral environment. The same holds for nitrogen and oxygen. Aluminum is a substitutional solute, and the next phase is close packed so that Al should stabilize the α phase. The same holds for tin, and so forth. The first hydride is TiH which is a tetragonally deformed bcc[8] so that the hydrogen stabilization of the β phase is to be expected. Thus in addition to the usual considerations of size and electronegativity there is also a preferred coordination to be taken into account.

McQuillan thought it naïve to expect to find simple numbers for the elements which by simple manipulation will always predict the phase stability on alloying. One always has to consider the environment. A simple and specific example of this is that it looks as though hydrogen is different from carbon, nitrogen, and oxygen until one considers how the metal looks from the point of view of the interstitial atom. In the case of hydrogen in a bcc structure it has six interstices per metal atom; in the hcp structure it has two; so at high temperatures there is a large gain in entropy in the bcc structure, and this tends to stabilize the bcc structure. His measurements suggest that similar considerations apply to carbon, nitrogen, and oxygen, but then size comes in. For example, in the bcc structure, carbon, nitrogen, and oxygen are in an octahedral interstice which is asymmetric. If the oxygen loses an electron to the conduction band, it becomes a flat disk-shaped ion which fits well in the octahedral site of the bcc structure. Thus the geometry of the environment has an important effect on alloying behavior.

Another factor generally neglected is the effect of thermal vibration. When the bonding changes in going from one structure to another, the elastic forces change, and thus so must thermal properties. If these are not taken into account, one cannot understand why, for example, the Ti-Zr system has a transformation temperature minimum in the middle of the diagram. This is entirely a thermal property and has nothing to do with putting a valence number on Ti and Zr.

Lomer felt that unless one has enough variables to cope with a particular system, as for example Ti-Zr-C, one cannot get far by looking at many systems. It clearly requires a multidimensional space before one can do significant correlation: size, valence, environment, electronegativity. Thermal entropy is, of course, a property of each separate phase and can only be brought into consideration after internal energy considerations have determined the zero temperature phases.

Pearson asked whether it is sterile or not to talk about bond type. One knows one is somewhere between ionic and covalent. Further definition is useful only when one relates bond type to conductivity. Not for oneself but for teaching purposes mainly it is useful to classify bond types. The relationship between bond type and conductivity is a sensible way to do this. In fact, had Pauling in 1947 thought about the relationship between bond type and conductivity, he would no doubt have predicted all sorts of semiconductors starting with the group III and V compounds. Briefly, there are the limiting bond models according to spatial distribution: covalent, ionic, and van der Waals. Then one can define two secondary bond types (here one is not discussing configuration in space but the relative filling of the valence subshells). The metallic bond type is such that there are bond paths running continuously through the lattice

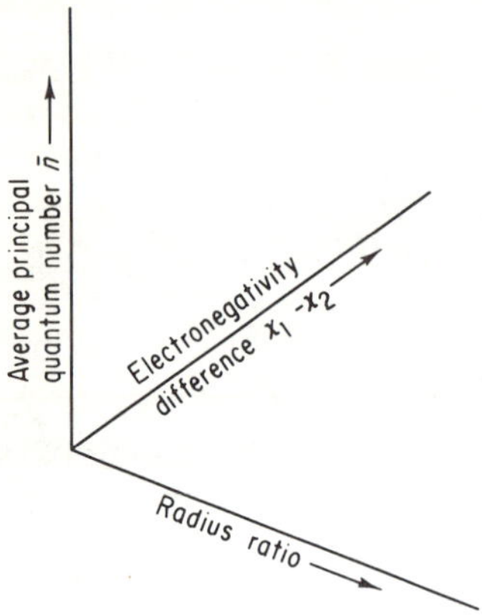

Figure 11

between atoms which do not have filled valence subshells. Some atoms can have filled subshells, but there will still be continuous bond paths through the atoms with nonfilled subshells, so that an electron can move through the lattice under an applied electric field without an activation energy. The other secondary bond type is the semiconducting bond type. In this case there are no direct bond paths through the lattice between atoms which do not have filled valence subshells. In other words, if an electric field is applied, it is necessary to have an activation energy to move an electron through the lattice. This is a chemist's equivalent to Wilson's 1930 definition of a semiconductor. This is the way that Pearson would like to have bond types defined for teaching purposes, with the secondary bond types specifying the filling of the orbitals.

He also gave what he thought is the space most appropriate for correlating properties: electrical, structural, etc. (Fig. 11). Here, $\bar{n}$ is not valence of group number, but the period in the periodic table: i.e., $\bar{n}$(Cu) = 4, $\bar{n}$(Te) = 5, $\bar{n}$(CuTe) = 4.5, and so forth.

Nevitt pointed out that however vague is the concept of the size of an atom, empiricists must for a while continue to use it because it does have predictive capability. For example, one can correlate the occurrence of certain intermediate phases with such simple parameters as radius ratio. One can predict interatomic distances and volumes of unit cells. Furthermore, when phases appear in some instances and disappear in others and this can be correlated with radius ratio, then it is appropriate to seek some parameter that defines a misfit energy. In the absence of another guide one can do this by comparing the volume of an atom in the alloy structure with some intrinsic, unstrained volume for which the pure metal value generally appears to be most appropriate.

Heine previously brought up the point that when an atom is free-electron-like, atomic volume is an appropriate measure of size, but when there is a large and filled d core, as in gold, then the atomic diameter is a more appropriate measure.

Can the theoretical physicists give further guidelines in employing the atomic size concepts?

In answer, Heine had nothing to add to what he said previously: that where d shells are in contact the distance of closest approach appears to be appropriate; otherwise atomic volume is.

Brewer returned to the point made earlier by Jaffee, and remarked that in his brief description of the effect of alloying on the α-β transformation in titanium, he had essentially considered some 80 alloy systems. Without taking into account any other factor than the valence rules that Brewer has given, Jaffee has shown that about 70 of the systems occur as these rules predict. The predicting of 70 out of 80 in a matter of seconds is certainly impressive support for these ideas.

McQuillan said that metallurgists for the past 50 years have been doing the wrong experiments in using primarily the variable temperature rather than pressure in considering phase stability. The advantage in using the pressure variable is that it can be applied at a low temperature where the complicating thermal entropy contributions are minimized. Lomer agreed, and commented that in addition to the pressure variable for getting at the fundamental ΔE (0°K), magnetic field is another variable that can also be exploited at $T \rightarrow 0$°K.

Friedel returned to the point made by Pearson and remarked that the band picture using the tight binding approximation is, for solids and liquids, the equivalent of LCAO molecular orbitals in molecules. In this band picture, there are two types of covalent bonds.[9] There is a covalent bond type with two kinds of orbitals, the *s* and *p* for instance as in diamond, and this can lead to insulators. Then there is a covalent bond type with only one type of orbital, and that in a solid can only give a metal. For example, if a band calculation is made using only *d* electrons, a metal will be obtained although it will still have covalent character. In this molecular-orbital description, covalent character means solutions which reinforce the electron density between the atoms.

Pearson commented that in VO_2 there is one *d* electron, and it is metallic with rutile structure. On cooling, the structure distorts, with V atoms coming together in pairs. The *d* electrons on these two V atoms are paired, and the degeneracy of the *d* electron is lost and the metallic conductivity then disappears. It is then a semiconductor or insulator with a change in conductivity by several orders of magnitude.

Brewer, Nowotny, and Seeger all pointed out that electron pairing is not essential for the formation of a covalent bond.

Brewer gave a few examples of the use of his valency rules to illustrate the need to separate the *s*, *p*, and *d* contribution. If one takes transition metals one can obtain a group-number rule; for example, the σ phase occurs in the interval 6.2 to 6.9. But if one uses non-transition metals, it won't work. On the other hand, if one says 1.2 to 1.9 instead of 6.2 to 6.9, with 5*d* electrons in the transition metal, then non-transition metals can be included in the same electron-concentration rule.

If one considers the β-Mn structure, this normally occurs in alloys of elements from the sixth group and to the left, with elements to the right.

However, iron is an exception. Iron gives β-Mn phases when alloyed with rhenium and technetium for which the only explanation is that iron assumes a valence of one. If one employs the group-number system, there is no way of explaining this occurrence. However, if one recognizes from the bcc structure that iron can have a valence of one, then this is understandable.

Lomer felt that if one is going to try to use valence numbers, one certainly has to distinguish between *d* and *s*. Clearly titanium and silicon are not in any meaningful sense similar in terms of bonding or structure. Although a large measure of correlation has been obtained simply with the number of electrons outside of closed shells, this is certainly not an adequate number to use in all correlations. That means that maybe one should add to it, rather than discard it. The statistical tests for validity of correlations are well-known but not self-evident, and should not give rise to argument.

REFERENCES

1. Richter, H., and G. Breitling: *Z. Naturforsch.* **16***a*:187 (1961) and other papers; a review article will soom appear in *Fortschr Physik.*
2. Mitsudo, R., K. Motizuki, and T. Nagamiya: *J. Phys. Soc. Japan* **20**:710 (1965).
3. Friedel, J.: *Trans. AIME* **230**:616 (1964).
4. Pick, R., and A. Blandin: *J. Cond. Mat.* **3**:1 (1964).
6. Bardos, D. I., R. K. Malik, F. X. Spiegel, and P. A. Beck: *Trans. Met. Soc. AIME* **236**:40 (1966).
7. Rocher, Y.: *Advan. Phys.* **11**:233 (1962); A Blandin, B. Coqblin, and J. Friedel: Tucson Symposium on Physics of Solids at High Pressures, Academic Press, New York, 1965; B. Coqblin: To be published.
8. Jaffe, L. D.: *J. Metals* **8**:861 (1956).
9. Friedel, J.: *Trans AIME* **230**:616 (1964); *J. Phys. Chem. Solids* **1**:175 (1956).

INDEX